Government Printing Office

Seal and Salmon Fisheries and General Resources of Alaska

Volume II

Government Printing Office

Seal and Salmon Fisheries and General Resources of Alaska
Volume II

ISBN/EAN: 9783744719797

Printed in Europe, USA, Canada, Australia, Japan

Cover: Foto ©berggeist007 / pixelio.de

More available books at **www.hansebooks.com**

SEAL AND SALMON FISHERIES

AND

GENERAL RESOURCES

OF

ALASKA.

IN FOUR VOLUMES.
VOLUME II.

WASHINGTON:
GOVERNMENT PRINTING OFFICE.
1898.

REPORTS ON SEAL AND SALMON FISHERIES BY OFFICERS OF THE TREASURY
DEPARTMENT, AND CORRESPONDENCE BETWEEN THE STATE AND
TREASURY DEPARTMENTS ON THE BERING SEA QUESTION
FROM JANUARY 1, 1895, TO JUNE 30, 1896,

WITH ·

COMMENTS ON THAT PORTION THEREOF WHICH RELATES
TO PELAGIC SEALING

BY

DAVID STARR JORDAN.

TABLE OF CONTENTS.

v

[CONTINUATION OF SENATE DOCUMENT NO. 137, PART I, 54TH CONGRESS, 1ST SESSION.]

REPORT

ON

THE SEAL ISLANDS OF ALASKA,

BY

JOSEPH MURRAY,

Special Treasury Agent,

FOR

THE YEAR 1894.

REPORT OF JOSEPH MURRAY, SPECIAL TREASURY AGENT, FOR THE YEAR 1894.

OFFICE OF SPECIAL AGENT,
TREASURY DEPARTMENT,
Washington, D. C., December 30, 1894.

SIR: I have the honor to report that, in compliance with Department instructions dated June 12, 1894, I went to the seal islands of Alaska and inspected the fur-seal rookeries, noting particularly the numbers and present condition of the seals in comparison with what they were every year since I first saw them in 1889.

I afterwards sailed along the American coast from Unalaska to San Francisco, calling at every important settlement on the way; inspecting every salmon stream and cannery on the route; making diligent inquiry into the condition of the native inhabitants of Alaska; the wants and desires of the white settlers who are busy developing the natural resources of the Territory, and noting the views of the people generally on all that appertains to the present and future prosperity of the new country.

On July 10, I left San Francisco on board the U. S. revenue cutter *Rush*, Capt. C. L. Hooper commanding, and arrived on the 15th at Port Townsend, where we were afterwards joined by Hon. C. S. Hamlin, Assistant Secretary of the Treasury, who accompanied us to the seal islands and back as far as Vancouver City, British Columbia.

We sailed on board the *Rush* from Port Townsend July 23 and arrived at the seal islands August 3, first touching at St. George and sailing along the coast, inspecting all the rookeries on that island except Zapadnie, and then sailed over to St. Paul Island, where we landed in a dense fog at 6 o'clock p. m.

The seal islands, commonly called the Pribilof group, consist of four distinct islands in Bering Sea, situated between 55° and 57° north latitude, and about 170° west longitude from Greenwich. They are about 200 miles west from the nearest point on the mainland of Alaska, 200 miles north of the Aleutian chain, and 200 miles south of St. Matthews Island, or, in other words, they are about 200 miles away from any other land.

The seal islands are nearly 2,300 miles from San Francisco, and about 1,600 miles, as the ship sails, directly west from Sitka.

They are known, respectively, as St. Paul, St. George, Otter, and Walrus islands.

Otter and Walrus are small and of no importance, and as the seals do not haul out at present on either of them regularly, and as they are not included in the lease, it will not be necessary to refer to them again.[1]

St. Paul, the larger of the two principal islands, is long, low, and narrow, its extreme length and breadth being 12 and 6 miles, respectively, and its total area being about 36 square miles. Around the greater part of the island runs a long, low, sandy beach, easy of access, where the seals haul out without difficulty, and where they were to be

---

[1] In 1894 about 1,000 seals hauled out on Otter Island.

found for a century in greater numbers than on any other spot on the earth.

St. George Island has an area of about 27 square miles, and its sides rise out of the water so abruptly and so steep that there are only a few places around the whole coast upon which anything coming out of the sea can find a footing, and consequently the number of seals landing must of necessity be limited, which accounts, I think, for the great difference in the numbers to be found on the two islands.

CLIMATE.

The islands are situated in the path of the Japan current, which, on meeting the icy waters of the north at this point, brings forth the dense summer fogs for which Bering Sea is so justly famous, and in which the islands are enveloped from May to September. Owing to difference of altitude, St. George Island being much higher than St. Paul, there is a very marked difference in the amount of rainfall on each—fully five times the volume falling on St. George, although the islands are only 40 miles apart.

The average temperature for the year is about 35°, ranging from 35° to 60° in the summer, and from zero to 15° below in winter.

Both of the islands are of volcanic origin, and there is not a sign of tree, shrub, or vine on either of them. They are covered in season with moss, grass, and wild flowers, but it is impossible to raise anything by cultivation, for, no matter how rich the soil may be, there is not enough sunshine to ripen the crop. Thick fog, leaden sky, drizzly rain, mist, and moisture are the general conditions ruling there, and during a continuous residence of thirty months—fifteen on each island—I saw only six wholly clear, sunshiny days.

The surface of the highlands on St. George is covered with loose and broken rock—rock broken into all shapes and sizes, from that of a pebble to boulders weighing many tons, and thrown together into every imaginable position except a level one.

On St. Paul the winds of centuries have heaped the sands of the seashore into dunes of considerable height and magnitude, and filled up many cavities and rough spots, but, excepting a slight covering of most nutritious reindeer moss, the greater part of the surface of St. George remains to-day as it came from the hands of the Creator.

And yet nature finds a use for those rugged and unshapely rocks, for under and between them, where the prowling, crafty fox can not penetrate, millions of sea birds build their nests, and lay their eggs, and rear their young. I use the word millions advisedly, and I believe I might say billions, and yet be within the bounds of truth.

One of the most beautiful sights to be seen in this otherwise desolate region is the return of the birds from the sea to their nests during the hatching season, when toward evening they fill the air and darken the sun for hours in their flight with their countless numbers.

Here, too, on St. George Island the famous blue fox finds a permanent home, and grows to perfection, for here he has abundance of choice and dainty food, and no one to molest him out of season.

SEALS.

To these islands, notwithstanding their cheerless aspect, their dreary barren shores, their damp and foggy climate, come the fur seals every year with the unerring regularity of the seasons; here they haul out of the water and make their home on land for six months at least, during which time they bring forth and rear their young, after which they return to the sea, and disappear in the depths of the great ocean until the days lengthen out again and nature tells them to return.

Given a few warm, sunny days any time about April 20, and the "first bull" may be seen carefully reconnoitering a rookery and eventually hauling out and taking possession of the identical rock or spot of earth upon which he dwelt with his family last year, and upon which he himself, in all probability, was born.

Early in May the breeding males or bulls begin to arrive in large numbers and select their stations, upon which they lie down and sleep for several weeks, or until about the time the breeding females or cows are expected, when they assume an upright sitting posture and send forth at intervals a cry peculiar to the fur seal, which is supposed to be an invitation or signal to the approaching cows.

About the middle of May, and long before the arrival of the cows, the large young males, or bachelors, begin to arrive at the islands; and they, too, would haul out upon the breeding grounds were it not that the bulls are there to prevent it by driving them off. No male seal can stay on the breeding grounds that is not old enough and strong enough to maintain his position against all comers. The young males are thus naturally forced to herd by themselves at a safe distance from the breeding grounds during the breeding season, and this regulation in turn serves a very good purpose, for, as the breeding and killing seasons run together through the months of June and July, the young males can be easily surrounded and driven to the killing grounds without having to disturb the breeding seals.

None but young male seals are ever killed for food or for skins or for any other purpose on the islands.

About June 10 the cows begin to arrive and haul out and select their stations for the season.

It has been claimed that the bulls meet the cows at the water's edge and fight bloody battles for them, but my observation has convinced me that the cow herself selects her station, and having once made a choice she is certainly compelled to remain there.

Shortly after the arrival of the cows the young seals or pups are to be seen upon the rookeries; and it is safe to say that, with few exceptions, they are all brought forth by July 25.

I have for six years paid particular attention to the formation of the harems or families, and I find that from July 10 to 20 the rookeries are fullest and at their best, and I have counted from 1 to 72 cows in one harem.

After bringing forth their young the cows go into the sea to feed, returning to and nursing their offspring every few hours at first, but gradually lengthening their stay into days and weeks before they return.

When about four or five weeks old the pups begin to stir around and get acquainted with one another, forming pods or crowds, and running in company, at first inclining toward the interior of the rookery, and afterwards, as they advance in age and strength, they direct their steps toward the beach, where they paddle around in the shallows until, step by step, they learn to swim.

About the beginning of August the harems are broken up, the compact formation of the herd is dissolved, and the different sexes mix and mingle together indiscriminately all over the rookeries and hauling grounds.

When the bull hauls out in May he is as "round as a barrel" and as

fat and sleek and glossy as possible; but after a four months' residence
on land, where he never tastes food or drink, he becomes so poor and
gaunt and weak that it is with the utmost difficulty he crawls off into
the sea when he leaves, late in August or early in September, to take
his annual journey through Bering Sea and the North Pacific Ocean.
By September 15 the bulls have disappeared, and by the middle of
October the largest of the young males have followed them.

Early in November the cows begin to leave, and if the weather turns
unusually cold or rough they do not delay their departure.

The pups leave about the middle of November, and the yearlings,
male and female, leave early in December.

In exceptionally fine weather it is common to see a few seals in the
waters around the islands all winter, and in rare instances they have
been taken on shore as late as January; but the great herd follows a
well-defined and (at present) well-known path through the Bering Sea
and the North Pacific Ocean south and east from the seal islands to
the coast of California, nearly opposite Cape St. Lucas, and return
along the American coast and the Aleutian Islands to Bering Sea.

The following very accurate description of the fur seal and its peculiarities is taken from the report of the United States Bering Sea
Commissioners:

1. The northern fur seal (*Callorhinus ursinus*) is an inhabitant of Bering Sea and the
Sea of Okhotsk, where it breeds on rocky islands. Only four breeding colonies are
known, namely, (1) on the Pribilof Islands, belonging to the United States; (2) on
the Commander Islands, belonging to Russia; (3) on Robben Reef, belonging to Russia; and (4) on the Kurile Islands, belonging to Japan. The Pribilof and Commander
Islands are in Bering Sea; Robben Reef is in the Sea of Okhotsk, near the island of
Saghalien, and the Kurile Islands are between Yezo and Kamchatka. The species is
not known to breed in any other part of the world. The fur seals of Lobos Island
and the south seas, and also those of the Galapagos Islands and the islands off Lower
California, belong to widely different species, and are placed in different genera from
the northern fur seal.

2. In winter the fur seals migrate into the North Pacific Ocean. The herds from
the Commander Islands, Robben Reef, and the Kurile Islands move south along the
Japan coast, while the herd belonging to the Pribilof Islands leaves Bering Sea by
the eastern passes of the Aleutian chain.

3. The fur seals of the Pribilof Islands do not mix with those of the Commander
and Kurile islands at any time of the year. In summer the two herds remain
entirely distinct, separated by a water interval of several hundred miles; and in
their winter migrations those from the Pribilof Islands follow the American coast
in a southeasterly direction, while those from the Commander and Kurile islands follow the Siberian and Japan coasts in a southwesterly direction, the two herds being
separated in winter by a water interval of several thousand miles.

This regularity in the movements of the different herds is in obedience to the well-
known law that migratory animals follow definite routes in migration, and return
year after year to the same places to breed. Were it not for this law there would
be no such thing as stability of species, for interbreeding and existence under diverse
physiographic conditions would destroy all specific characters. [1]

The pelage of the Pribilof fur seals differs so markedly from that of the Commander
Islands fur seals that the two are readily distinguished by experts, and have very
different values, the former commanding much higher prices than the latter at the
regular London sales.

4. The old breeding males of the Pribilof herd are not known to range much south
of the Aleutian Islands, but the females and young appear along the American coast
as far south as northern California. Returning, the herds of females move north-

---

[1] The home of a species is the area over which it breeds. It is well known to naturalists that migratory animals, whether mammals, birds, fishes, or members of
other groups, leave their homes for a part of the year because the climatic conditions
or the food supply become unsuited to their needs; and that wherever the home of a
species is so situated as to provide a suitable climate and food supply throughout the
year such species do not migrate. This is the explanation of the fact that the northern fur seals are migrants, while the fur seals of tropical and warm temperate lati-
tudes do not migrate.

ward along the coasts of Oregon, Washington, and British Columbia in January, February, and March, occuring at varying distances from shore. Following the Alaskan coast northward and westward, they leave the North Pacific Ocean in June, traverse the eastern passes in the Aleutian chain, and proceed at once to the Pribilof Islands.

5. The old (breeding) males reach the islands much earlier, the first coming the last week in April or early in May. They at once land and take stands on the rookeries, where they await the arrival of the females. Each male (called a bull) selects a large rock, on or near which he remains until August, unless driven off by stronger bulls, never leaving for a single instant, night or day, and taking neither food nor water. Both before and for some time after the arrival of the females (called cows) the bulls fight savagely among themselves for positions on the rookeries and for possession of the cows, and many are severely wounded. All the bulls are located by June 20.

6. The bachelor seals (holluschickie) begin to arrive early in May, and large numbers are on the hauling grounds by the end of May or first week of June. They begin to leave the islands in November, but many remain into December or January, and sometimes into February.

7. The cows begin arriving early in June, and soon appear in large schools or droves, immense numbers taking their places on the rookeries each day between the middle and the end of the month, the precise dates varying with the weather. They assemble about the old bulls in compact groups, called harems. The harems are complete early in July, at which time the breeding rookeries attain their maximum size and compactness.

8. The cows give birth to their young soon after taking their places on the harems, in the latter part of June and in July, but a few are delayed until August. The period of gestation is between eleven and twelve months.

9. A single young is born in each instance. The young at birth are about equally divided as to sex.

10. The act of nursing is performed on land, never in the water. It is necessary, therefore, for the cows to remain at the islands until the young are weaned, which is not until they are four or five months old. Each mother knows her own pup, and will not permit any other to nurse. This is the reason so many thousand pups starve to death on the rookeries when their mothers are killed at sea. We have repeatedly seen nursing cows come out of the water and search for their young, often traveling considerable distances and visiting group after group of pups before finding their own. On reaching an assemblage of pups, some of which are awake and others asleep, she rapidly moves about among them, sniffing at each, and then gallops off to the next. Those that are awake advance toward her, with the evident purpose of nursing, but she repels them with a snarl and passes on. When she finds her own she fondles it a moment, turns partly over on her side so as to present her nipples, and it promptly begins to suck. In one instance we saw a mother carry her pup back a distance of 15 meters (50 feet) before allowing it to nurse. It is said that the cows sometimes recognize their young by their cry, a sort of bleat.

11. Soon after birth the pups move away from the harems and huddle together in small groups, called "pods," along the borders of the breeding rookeries and at some distance from the water. The small groups gradually unite to form larger groups, which move slowly down to the water's edge. When six or eight weeks old the pups begin to learn to swim. Not only are the young not born at sea, but if soon after birth they are washed into the sea they are drowned.

12. The fur seal is polygamous, and the male is at least five times as large as the female. As a rule each male serves about fifteen or twenty females, but in some cases as many as fifty or more.

13. The act of copulation takes place on land, and lasts from five to ten minutes. Most of the cows are served by the middle of July, or soon after the birth of their pups. They then take to the water, and come and go for food while nursing.

14. Many young bulls succeed in securing a few cows behind or away from the breeding harems, particularly late in the season (after the middle of July, at which time the regular harems begin to break up). It is almost certain that many, if not most, of the cows are served for the first time by these young bulls, either on the hauling grounds or along the water front.

These young bulls may be distinguished at a glance from those on the regular harems by the circumstance that they are fat and in excellent condition, while those that have fasted for three months on the breeding rookeries are much emaciated and exhausted. The young bulls, even when they have succeeded in capturing a number of cows, can be driven from their stands with little difficulty, while (as is well known) the old bulls on the harems will die in their tracks rather than leave.

15. The cows are believed to take the bull first when 2 years old, and deliver their first pup when 3 years old.

16. Bulls first take stands on the breeding rookeries when 6 or 7 years old. Before this they are not powerful enough to fight the older bulls for positions on the harems.

17. Cows, when nursing, regularly travel long distances to feed. They are frequently found 100 or 150 miles from the islands, and sometimes at greater distances

18. The food of the fur seal consists of fish, squids, crustaceans, and probably other forms of marine life.

19. The great majority of cows, pups, and such of the breeding bulls as have not already gone, leave the islands about the middle of November, the date varying considerably with the season.

20. Part of the nonbreeding male seals (holluschickie), together with a few old bulls, remain until January, and in rare instances until February, or even later.

21. The fur seal as a species is present at the Pribilof Islands eight or nine months of the year, or from two-thirds to three-fourths of the time, and in mild winters sometimes during the entire year. The breeding bulls arrive earliest and remain continuously on the islands about four months. The breeding cows remain about six months, and part of the nonbreeding male seals about eight or nine months, and sometimes throughout the entire year.

22. During the northward migration, as has been stated, the last of the body or herd of fur seals leave the North Pacific and enter Bering Sea in the latter part of June. A few scattered individuals, however, are seen during the summer at various points along the northwest coast. These are probably seals that were so badly wounded by pelagic sealers that they could not travel with the rest of the herd to the Pribilof Islands. It has been alleged that young fur seals have been found in early summer on several occasions along the coast of British Columbia and southeastern Alaska. While no authentic case of the kind has come to our notice, it would be expected from the large number of cows that are wounded each winter and spring along these coasts and are thereby rendered unable to reach the breeding rookeries, and must perforce give birth to their young (perhaps prematurely) wherever they may be at the time.

23. The reason the northern fur seal inhabits the Pribilof Islands to the exclusion of all other islands and coasts is that it here finds the climatic and physical conditions necessary to its life wants. This species requires a uniformly low temperature and overcast sky and a foggy atmosphere to prevent the sun's rays from injuring it during the long summer season when it remains upon the rookeries. It requires also rocky beaches on which to bring forth its young. No islands to the northward or southward of the Pribilof Islands, with the possible exception of limited areas on the Aleutian chain, are known to possess the requisite combination of climate and physical conditions.

All statements to the effect that fur seals of this species formerly bred on the coasts and islands of California and Mexico are erroneous, the seals remaining there belonging to widely different species.

DRIVING AND KILLING.

When the first young males, or bachelors, arrive at the islands in May, a drive is made for food for the natives, who are hungry for fresh meat, not having tasted any since the preceding November.

All of the driving is done under the immediate and exclusive directions of the native chief, who is the most experienced and most trustworthy man on the island.

Should the seals happen to lie near the water, it will be necessary to wait till the tide runs out before disturbing them. At the proper time a dozen men are on the ground, and silently and swiftly running in single file along the beach they form a line between the seals and the sea; and then the startled animals will immediately start inland, where they are slowly followed by the men, until they are too far from the beach to escape to the water, when they are put in charge of three or four of the men, who bring them along slowly to the killing grounds, which is never less than half a mile away from the nearest breeding seals. No other part of the work done in taking seal skins is more carefully performed than the driving of seals; they are never driven at a pace greater than about one mile in three hours, and most of the driving is done during the night, so as to take advantage of the dew and moisture, and to avoid the sudden appearance of the sun, which is always more or less injurious to seal life on a drive. The stories told by interested men about careless and reckless driving are not true, and, for obvious reasons, can not be true, because overdriving means overheating, and an over-heated fur seal is one from which the fur has fallen and left the skin valueless, and that means a loss to natives, lessees, and Government alike. As there is no one to benefit by overdriving, it is never indulged in; and during an experience of six years on the islands I never saw a skin injured by overheating or overdriving.

As most of the drives are made in the night, the seals are allowed to lie in the damp grass around the killing grounds for several hours before killing takes place; and it is customary to allow them to rest for a few hours, no matter when they are driven, because it is best for the skin and for the flesh that the animal be killed while it is cool and quiet rather than while it may be warm and excited.

There are four different and well-defined killing grounds on St. Paul Island, from some one of which the most distant hauling ground or rookery is not to exceed 2½ miles.

On St. George there are two killing grounds, from some one of which the most distant rookery or hauling ground is not to exceed 3 miles, and during the past fifteen years there has not been a longer drive made on either island than 3 miles, interested parties to the contrary notwithstanding.

Generally the killing is done just after breakfast, and the whole population turns out and takes part in the work.

The men and boys are divided into grades or classes: Clubbers, stickers, flipperers, and skinners; the women and girls following the skinners and taking care of the blubber and meat.

Two men at opposite sides of the herd will, by advancing till they meet, cut out twenty or thirty seals from the main body and drive them up to the killing ground, where six experienced men stand armed with clubs of ash or hickory about 5½ feet long and about 3 inches thick at the heavier end, which end is generally bound in sheet iron to prevent its destruction by the continuous biting of the seals.

The clubbers are under the immediate orders of the lessee's local agent who is a man of large experience in seal work, one who can tell at a glance how much the skin of any particular seal will weigh, and he points out the seals to be clubbed.

A smart blow on the head knocks the seal down and stuns him, and if the blow has been properly dealt he never recovers; but quite often it requires two to three blows from a bungler to finish him. The clubbed seals are dragged into line and counted, and then "stuck" and "flippered," or, in other words, they are stabbed to the heart and allowed to bleed freely; and then a knife is drawn around the head and flippers, severing the skin and leaving it ready for the skinner, who strips it off in short order and spreads it evenly on the damp grass, flesh side down, to cool.

These several operations are repeated till the desired number are killed, when the remaining seals are allowed to go into the water and return to the hauling grounds.

After the skin has been removed, the women take the carcass and, after stripping off the blubber or fat, cut off the choice meat in strips to dry; and, when dried, they pack it into the dried stomach of the sea lion, where it is kept air-tight and preserved for an indefinite period. The remainder of the seal is boiled and eaten as wanted.

When all the seals killed are skinned, the skins are taken by wagon to the salt house, where they are assorted and carefully inspected by the lessee's agent, who throws out as rejected all skins that do not come up to a certain standard. There are three classes of rejected skins, namely: cut, small, and stagy.

A cut skin is one that has been bitten through by one seal biting another during one of their many battles, or it may have been accidentally cut during the operation of skinning; a small skin is one that weighs a little less than the minimum standard set up by the lessee's agent, generally less than 6 pounds. After July the fur seal sheds his hair, and it is during the shedding season, when the old hair is falling out and before the new hair has attained its full growth, that the skin is said to be stagy.

The fur of a stagy skin is just as good as any other; but the half-grown new hair, being shorter than the fur, can not be plucked out by hand or by machinery, and is therefore considered a blemish on the skin, in consequence of which its price and value are naturally lowered in the market.

Heretofore, and until the adoption of the modus vivendi in 1891, it was customary to allow the natives to kill seals for food at any and all times when they were to be found on the islands. And it was in this way, and in this way only, that stagy skins were ever taken and wasted, because all skins that are rejected by the lessee's agent are wasted so far as Government interests and revenue are concerned.

No killing should be permitted for any purpose whatsoever during the stagy season, say from July 31 to November 15.

After a thorough inspection, the skins are counted one by one in presence of the Treasury agent, who makes a record of the same in a book kept for that purpose, and in which he also enters the date of the drive, the rookery driven from, the hour of driving, the state of the weather, the number of seals killed, the number of skins accepted, the number rejected, and the cause of such rejection.

The accepted skins are then salted by the natives in presence and under the direction of the native chief and the lessee's agent. The skins are spread on the floor, hair side down, and covered with a layer of coarse salt; again a layer of skins is laid on and covered with salt as before, and the operation is repeated until all are salted.

After lying for at least five days in the first salt they are shaken out and examined, and resalted as before, excepting that the top layer is now put down first and the original position of all layers reversed.

When sufficiently cured they are bundled by the natives, who, spreading a thin layer of salt between two skins, lay them flesh side to flesh side, and fold the two into a neat, compact bundle, which they tie securely with strong twine, and throw into the pile for shipping. From the shipping pile they are again counted out, bundle by bundle, by the Treasury agent, in whose presence they are always taken from the salt house to the boat, from which they are again counted by the mate into the steamer that takes them to San Francisco, where they are counted once more by the customs officers, and finally packed into barrels by the lessees and shipped direct to London via New York.

Early in the morning of August 4, 1894, a drive was made from the Reef rookery in presence of Mr. Hamlin, who accompanied the native men who did the work, and who was present throughout the whole operation of driving, killing, and skinning the seals, inspecting, assorting, counting, and salting the skins, just as the same operations have been performed every killing season for the past quarter of a century.[1]

---

[1] The only exception to this is in the method of killing. The olden rule was to allow each man to first knock down his share and then turn in and skin them, but experience taught us that this was bad policy, for the carcasses that were allowed to cool and stiffen before skinning were very apt to have their skins injured in the operation, hence the adoption of the present improved system.

During our five days' stay on St. Paul Island we inspected all the rookeries, walking over many of them, and I carefully noted their condition, the sparsely settled breeding grounds, the deserted hauling grounds, and the desolate appearance of the place in comparison with what I saw there only five years ago, when hundreds of thousands of seals swarmed over the greater portion of the ground that is now bare and abandoned.

Next to the shriveled condition of the seal herd as a whole, the most noted change I observed on the breeding grounds since 1889 was the great number of idle bulls, young and vigorous, lying around in all directions, watching an opportunity to secure cows.

They can not succeed, however, for during the past ten years the cows have been the quarry of the pelagic sealer, whose improved methods of hunting in the open waters, and whose unceasing, unerring, and merciless hunting of them at all seasons, have at length succeeded in destroying at least a million nursing mothers, who, with their starved offspring and unborn young, represent a loss of many millions, which in turn accounts for the acres of bare and unoccupied rookery ground over which we walked without finding a seal. When in 1891 I inspected the same rookeries I counted 1,250 idle bulls at the very height of the rutting season, and I have since observed a steady increase of breeding bulls as the herd continued as steadily to decrease as a whole.

So plain and palpable has this increase of bulls been for the past five years, it has become a topic of general conversation among those who have had opportunities to observe the rookeries from year to year during the breeding season; and in his annual report for 1894 the agent in charge of the islands says:

The only class of seals that showed an increase over last year were the young bulls, who were unable to find a single cow with which to start a harem on the rookeries. There were more idle bulls of breeding age than there were bulls with harems on the breeding grounds. (See Report of Joseph B. Crowley, 1894.)

Another very important feature observed in our inspection of the rookeries in 1894 was the absence of dead pups in the early part of August, for up to our leaving on the 8th I had not seen a dead pup on the island, and the agent in charge, who was on St. Paul Island from June to the latter part of August and who kept a close watch for dead pups, tells me now that it was not till about August 20 there was a dead pup to be seen, but from that date to the close of the season, according to official communications received from the islands, the carcasses of dead pups, starved and emaciated, increased with appalling rapidity until 12,000 were counted by the assistant agents.

The agents report that they actually counted 12,000 dead pups on the accessible portions of the rookeries to which they could go without disturbing the seals, and after making due allowance for the portions not visited at all, they believe that a fair estimate of the total number of dead pups on the two islands of St. Paul and St. George in 1894 would aggregate 20,000. (See report of Joseph B. Crowley, 1894.)

And Mr. Joseph B. Crowley tells us that—

Every precaution was taken to count only such as appeared to have died late in the season. None of the small young pups which showed decay and bore the appearance of having died early in the breeding season were counted. * * * I do not make recklessly the statement of the death of pups from starvation. There is positive proof of it. I witnessed the beginning of its disastrous results the last of August before leaving the islands. Visiting the rookeries in person, I found hundreds of pups which had lately died. They bore every appearance of having died of starvation. Hundreds that were yet alive were so wasted and weak they could scarcely drag themselves over the rocks and would not attempt to get out of the way when approached. (Report of J. B. Crowley, 1894.)

"What is the cause of the death of so many fur-seal pups?" has been asked many times during the past five years' discussion of the seal question, and many conflicting answers have been given. I think the following, under the circumstances, is an answer that can not be contradicted. The pelagic-sealing season opened in Bering Sea on August 1, 1894, in accordance with the international regulations made possible by the Bering Sea Tribunal, under which pelagic sealers are licensed to kill seals, with spears, outside of the 60-mile zone around the seal islands, and immediately we see the result of their work in the thousands of pups starved to death after their mothers had been killed at sea by the men whose right to kill them, at certain seasons, has been established and acknowledged by the very tribunal that was created for the purpose of preventing the destruction of the fur-seal herd.

One of the most horrible and harrowing sights imaginable is that of being surrounded on the bleak and inhospitable shores of the Pribilof Islands by thousands of dead and dying pup seals whose death has been the result of slow starvation, and whose hungry cries and almost human appeals for food and life must be made in vain, for, no matter how willing and anxious one may be to render assistance, one feels it is beyond human power to arrest the gnawing of hunger in an animal who is totally dependent for sustenance on a mother who was killed a month ago by pelagic sealers!

Those who once witnessed such a sight never can forget it, and occasionally I receive letters from some of them which run somewhat like the following:

Do tell me what is to be done with the few remaining seals. * * * If these steps had been taken last year, even, there might have been enough left to tell the tale, but as it is I can not but feel what a pitiable sight the rookeries will present next year. It was discouraging enough last spring when I compared the rookeries with what I had seen just the year before. My heart bled for the poor starving pups so much, the last stroll I took on the rookeries, that I could never go back. I don't see how the judge could stand to see 10,000 dead ones. It would have broken my heart I know. The morning we came into Dutch Harbor on our voyage down we saw three sealing vessels sailing out toward the 60-miles limit. Oh, what a farce, what a snare and delusion that 60-mile limit was! How could anyone who had ever been to the seal islands and noted the habits of the feeding cows ever recommend such a murderous proposition? Even I knew better than that. * * * But 13,000 cows taken staggered me. I had expected about 5,000 or 6,000, and even calculated the terrible consequence upon the rookeries, but 13,000! that was terrible, terrible!

The writer of that letter is the wife of the Treasury agent, an American lady of Christian education, culture, and refinement, who naturally felt horrified at the sight she saw on the rookeries, and, like the tender and merciful woman she is, she denounces the system, regulation, custom, or whatever else it may be called, which makes such suffering possible.

One instance in this connection worth recording is that of a pelagic sealer whose heart was touched by the pitiful cries of an orphan pup, and the story is told by an eyewitness under oath:

Of the seals that were caught off the coast fully 90 out of every 100 had young pups in them. The boats would bring the seals on board the vessel, and we would take the young pups out and skin them. If the pup is good and a nice one, we would skin it and keep it for ourselves. I had eight such skins myself. Four out of five, if caught in May or June, would be alive when we cut them out of their mothers. One of them we kept for pretty near three weeks alive on deck by feeding it on condensed milk. One of the men finally killed it because it cried so pitifully. (Affidavit of Alfred Dardean.)

The reverse side of the question is that held by the average pelagic sealer, who kills the mother seal and cuts out her unborn young or leaves the born young to slowly starve to death on the rookeries.

The British Bering Sea commissioners in this connection stated:

The fur of the female is equally good with that of the male, and under the conditions under which the hunting is carried on, there is room for no sentimental considerations in favor of either sex.

I was informed by the Treasury agent and others who had wintered on the seal islands that the winter of 1893-94 had been one of unusual severity, rigor, and length, and that the seals had been much later in hauling out than for many years past.

This happens occasionally, for whenever it is unusually cold during the spring and early summer months, and the ice hangs around the islands till the latter end of May or early June, the seals will not or can not haul out until passages are made and the rocks and beach cleared of ice; all of which had to be done last season.

From the same source I also learned that never before, since the United States owned the islands, were seals so few upon the rookeries during the killing season of June and July, and that the 20,000 killables allowed to be taken this year were not to be found unless the standard weight and size should be lowered by the lessees and smaller seals taken. As the lessees have not taken any skins weighing less than 7 pounds, and have killed some 16,000 first-class seals, I have no doubt of their being able to get 20,000 had they chosen to take 4,000 skins weighing from 5 to 6 pounds each.

This opens up a question of the utmost importance to our Government, for if we can not find 20,000 young male seals on the seal islands, whose skins will weigh from 7 to 12 pounds each, after a modus vivendi, and a general rest of nearly four years, it is most assuredly time for us to search for the cause of the steady decrease of the fur-seal herd.

To all those whose long and practical experience on the islands and among the seals gives them a right to be heard, the explanation is not hard, but unfortunately, because of many clashing interests, there has been a glamor of secrecy and sacredness thrown around the fur-seal question, by and through which plain, practical, business men have been debarred from expressing an opinion, or, having dared to express one, have been tabooed by interested parties. For years the cause of the decrease in the seal herd has been discussed with unabated vigor; so-called improved methods of all sorts have been suggested, and a few of them tried; and, finally, when the question assumed international proportions, arbitration was resorted to in hopes of forever settling a vexed question and of saving from total extinction the remnants of our seal herd that had, only a few years ago, been numbered by the millions and valued at nearly $100,000,000.

In spite of all that has been done thus far, however, the seal herd is rapidly decreasing, and in the very nature of things must continue to decrease so long as scores of ships and thousands of men are permitted to hunt them in the open sea and kill them without regard to age, sex, or condition.

There is no more mystery about the cause of the decrease and destruction of the fur-seal herd than there would be about the decrease of a herd of cattle on the plains of Colorado if the owner should continue to sell or kill, or allow someone else to sell or kill, his breeding cows for a series of years, or until they were all gone.

Twice since the discovery of the seal islands and during Russian occupation have the seals been almost exterminated because of the indiscriminate slaughter of the female, or mother seal, for it is well known that the Russians continued to slaughter everything on the islands without regard to age, sex, or condition until 1834, when the question of total extermination stared them in the face.

Veniaminov tells us:

> From the time of the discovery of the Pribilof Islands up to 1805 the taking of fur seals progressed without count or lists, and without responsible heads or chiefs, because then (1787 to 1805, inclusive) there were a number of companies, represented by as many agents or leaders, and all of them vied with each other in taking as many as they could before the killing was stopped. After this, in 1806 and 1807, there were no seals taken, and nearly all the people were removed to Unalaska.
>
> In 1808 the killing was again commenced, but the people in this year were allowed to kill only on St. George. On St. Paul hunters were not permitted this year or the next.
>
> It was not until the fourth year after this that as many as half the number previously taken were annually killed.
>
> From this time (St. George 1808 and St. Paul 1810) up to 1822, taking fur seals progressed on both islands without economy and with slight circumspection as if there were a race in killing for the most skins. Cows were taken in drives and killed, and were also driven from the rookeries to places where they were slaughtered. (Elliott's translation.)

And Mr. Elliott, commenting on Veniaminov's zapieska, tells us that—

> A study of this killing throughout the zapooska of 1834 on St. Paul Island shows that for a period of seven years, from 1835 down to the close of the season of 1841, no seals practically were killed save those that were needed for food and clothing by the natives, and that in 1835 for the first time in the history of this industry on these islands, was the vital principle of not killing female seals recognized. It will be noticed that the entry for each and every year distinctly specifies so many bachelor seals or holluschickovkotovie.
>
> The sealing in those days was carried on all through the summer until the seals left in October or November, on account of the tedious method then in vogue of air drying the skins. This caused them in driving after the breaking up of the breeding season by the end of July, to take up at first hundreds, and thousands later on, of the females, but they never spared those cows then when they arrived in the droves on the killing grounds, prior to this date above quoted, of 1835. (Elliott's report, 1890.)

Ignoring for the moment all that has been said about the thoughtlessness and brutality of the Russian methods of driving and killing seals, and of the incalculable waste arising therefrom, which resulted in the almost total destruction of the species on two occasions, it is nevertheless true that after many years of bitter experience they did learn to do better; and when they turned the property over to the United States in 1868 there were nearly 5,000,000[1] of seals on the Pribilof Islands, and that for a period of sixteen years afterwards there was neither decrease nor diminution perceptible in those immense and valuable herds.

Dr. H. H. McIntyre, who was the general superintendent for the Alaska Commercial Company during the whole time of their twenty-year lease of the seal islands, writing, confidentially, to his employers in 1889, says:

> The breeding rookeries from the beginning of the lease till 1882 or 1883 were, I believe, constantly increasing in area and population, and my observations in this direction are in accordance with those of Mr. Morgan, Mr. Webster, and others, who have been for many years with me in your service, and of the late special Treasury agent, J. M. Morton, who was on the islands from 1870 to 1880. (See H. H. McIntyre to Alaska Commercial Company, July 16, 1889, Appendix.)

And Mr. Henry W. Elliott, writing in 1881, fully corroborates the foregoing when he tells us—

> There were no more seals seen here by human eyes in 1786 and 1787 than there are now in 1881, as far as all evidence goes. (Elliott's Seal Islands of Alaska, p. 66.)

---

[1] Grand sum total for the Pribilof Islands (season of 1873), breeding seals and young, 3,193,420. The nonbreeding seals seem nearly equal in number to that of the adult breeding seals; but, without putting them down at a figure quite so high, I may safely say that the sum total of 1,500,000, in round numbers, is a fair enumeration, and quite within bounds of fact. This makes the grand sum total of the fur-seal life on the Pribilof Islands over 4,700,000. (Elliott, The Fur-Seal Islands of Alaska, pp. 61, 62.)

It was in 1873 that Mr. Elliott estimated the number of seals on the Pribilof Islands at 4,700,000, and he again tells us in 1881 that the seals never had been more numerous than they were then; but in 1890 he found them reduced to 959,393 seals, including everything on the islands, or about one-fifth of what the herd had been in 1873.

In 1891 the Treasury agents on the seal islands were instructed to make daily visits to the rookeries during the breeding season for the purpose of noting the peculiar habits of the seals and carefully estimating their numbers at various dates on each rookery, and the highest estimate made, not including the pups, was somewhat less than half a million.

I was one of the agents who did this work in 1891, and I have spent hours and days and weeks, in turn, watching the cows from their first landing. They would often stay away from their offspring for a week at a time.

I have selected a favorable location on the Reef rookery, where I was some 30 feet above the harem and out of danger of being discovered by the seals below, and I have watched one particular pup from its birth until it was a month old; and I found that the cow left it for an hour or two only at first, then for a day, and by the end of the month for four to six days at a time.

This fact, coupled with another that I observed in 1890, convinced me that the fur seals do not digest their food as rapidly as some other animals, and consequently they can live longer without eating or drinking.

The other fact referred to is this: In 1890 we killed for the natives on St. Paul Island some 2,364 pups, after all the cows had been gone from the island for more than two weeks, and we found the stomachs of all those pups full of pure, undigested milk.

I walked over all the rookeries on St. Paul Island twice during the season of 1891, beginning at Halfway Point on July 7, and completing the second journey at Northeast Point on July 22, and the highest estimate I made of the number of seals on each was as follows:

Rookery.	Seals.	Rookery.	Seals.
Northeast Point	149,975	Middle Hill	5,150
Reef	93,150	Ketavie	5,075
Halfway Point	10,500	Lukannon	16,600
Tolstoi and Lagoon	82,650		
Zapadnie and East Zapadnie	86,200	Total, not including pups	481,350
English Bay	32,050		

This estimate was made on the basis of an average of 40 cows to each bull, and it was assumed that only one-half the bulls were in sight at any one time, or, in other words, we could not get close enough to see them without disturbing the seals, so we multiplied the number found by 2, and the product by 40, in order to obtain, approximately, the number of seals on a rookery.

It is possible, of course, that the method of computation adopted was not the best and that we probably missed the real number by many thousands, plus or minus, but for all practical purposes of comparison between the condition of the rookeries in 1891 and 1894 it is as good as perfection, for it is enough to show that no matter how many seals were there in 1891, not to exceed one-half of the number were to be found in 1894.

The same is equally true of St. George, where the rookeries, because of their relatively smaller area, show the decrease at a glance to anyone who was on the island a few years ago, and who ever paid any attention to the seals when the rookeries were filled out to their fullest, and thousands were to be seen sporting in the waters around them.

Indeed, I do not hesitate to say that there was not to exceed 300,000 seals on St. Paul and St. George islands in August, 1894.

It is here the questions naturally arise, "What is the cause of the decrease of the seal herd? Is there a remedy; and if so, how can it be applied?"

I shall attempt to answer the questions in the order in which I have stated them, and I aim to show that all of my own views are in strict accord with those whose disinterestedness, practical knowledge, or scientific attainments warrant them in expressing views on the question at issue.

And it will be found, I think, that while we may differ in our estimates of the number of seals on the islands at any particular time or period, or that our notions about methods and management may never be exactly alike, we are all agreed that the cause of the decrease of the fur-seal herd is pelagic sealing.

Speaking for myself, after an experience of six years on the seal islands, I have no doubt that were it not for pelagic sealing the seal herd would be as numerous and as flourishing to-day as it was in 1868 or 1881, or at any other period since the discovery of the islands; nor is it at random or without long study that I say this, for I have given the subject a great deal of serious thought during the world-wide discussion of the question since 1890.

When the question of the decrease of the seal herd was first mentioned publicly as a reality, theories as numerous as the men who entertained them were offered in explanation of the cause of such decrease, and for awhile it was argued with consummate ability and persistent energy by Mr. H. W. Elliott, who was considered an authority on all that relates to fur seals, that the driving from the hauling to the killing grounds injured the young males to the extent of impotency, and thus unfitted them at maturity for service on the breeding grounds.

A mere idle guess at first, this theory was pushed to the front with energy, although, could angry personal feelings and prejudice have been eliminated from the controversy, the gentleman might have discovered what every scientist, naturalist, and impartial observer saw from the first, that so long as all the cows on the rookeries had pups beside them in season, and every mature cow killed at sea was either a nursing mother or about to become one, the theory of a scarcity of bulls could not be maintained. And after the passions and prejudices existing on the seal islands in 1890 cooled down or had ceased to exist, Mr. Elliott made an affidavit in which he says:

After carefully examining the situation, actual records, and trustworthy testimony of men engaged in sealing with whom I have conversed, and also from knowledge of the migratory habit and peculiar circumstances of seal life, I am of the opinion that unchecked pelagic sealing is sure, speedy destruction of the Pribilof herd of fur seals; that if allowed to continue and the fleet increases in number of vessels and increased skill of hunters, even though the present modus vivendi should remain in force, it would result in the utter commercial ruin of the herd; that in order to preserve the seals from complete destruction, as a commercial factor, it is necessary that pelagic sealing should not only be prohibited in Bering Sea, but also in the North Pacific from the 1st of May until the end of October, annually. The pelagic hunters to-day kill at least 90 per cent cows (the great majority being with young, nearly ready for delivery) in the Pacific Ocean.

As the physical conditions are such that it is utterly impossible to discriminate in matters of sex or age when shooting or spearing in the water, it is evident that pelagic sealing can not be regulated in the slightest degree beyond its complete prohibition within certain limits. (Elliott's affidavit, 1892; see Appendix.)

Of all the testimony collected during the preparation of the United States case for the Tribunal of Arbitration, I know of nothing clearer or more explicit than this of Mr. Elliott, and to me it seems pitiful, indeed, that one who has such a grasp of the subject, and the ability to express it so well, should have been allured for a moment from the plain path of fact to follow the ignis fatuus of theory through so many lanes and byways to the sorrow of so many of his friends and admirers.

Reading his different papers, in the light of subsequent events, their perusal makes one feel sorry, indeed, that he did not adopt Webster's views and follow his advice when the old veteran sealer conversed with him on St. George Island that 26th day of July, 1890, of which Mr. Elliott writes:

Daniel Webster is the veteran white sealer on these islands. He came to St. Paul Island in 1868, and, save the season of 1876 (then on a trip to the Russian seal islands), he has been sealing here ever since, being in charge of the work at Northeast Point annually until this summer of 1890, when he has conducted the killing on St. George.

He spoke very freely to me this afternoon while calling on me, and said there is no use trying to build these rookeries up again so as to seal here, as has been done since 1868, unless these animals are protected in the North Pacific Ocean as well as in Bering Sea; on this point the old man was very emphatic. (Elliott's report for 1890, p. 250.)

What wonder is it that Webster should have been emphatic in his remarks on pelagic sealing? For more than fifty years he has been in Bering Sea, thirty years of which have been spent among the fur seals of which he has had the practical management, and handled and killed more of them than any other living man.

A plain, blunt, rough, practical seaman, honest and patriotic to the core, he could not be wheedled into new-fangled notions or airy theories which are repugnant to good, common sense, and so he makes oath that:

My observation has been that there was an expansion of the rookeries from 1870 to 1879, which fact I attribute to the careful management of the islands by the United States Government. * * * There was never, while I have been upon the islands, any scarcity of vigorous bulls, there always being a sufficient number to fertilize all the cows coming to the islands. * * *

The season of 1891 showed that male seals had certainly been in sufficient number the year before, because the pups on the rookeries were as many as should be for the number of cows landing, the ratio being the same as in former years.

Then, too, there was a surplus of vigorous bulls in 1891 who could obtain no cows. * * * At Zapadnie, on St. George, the drive to the killing grounds is less than a mile. The seals are now being killed there instead of being driven across the island, as they were prior to 1878, when it took three days to make the journey. * * *

At Northeast Point rookery, on St. Paul Island, the longest drive is 2 miles. In former times the Russians used to drive from this rookery to St. Paul village, a distance of 12½ miles. (See Webster's affidavit, Appendix.)

Yes, let it not be forgotten for a moment that from the first taking of fur seals for their skins on the Pribilof Islands to 1868 they were driven a distance of 12½ miles—or from end to end of St. Paul Island—and that no distinction of sex was made, male and female being driven and slaughtered indiscriminately, until the almost total extinction of the species in 1834 compelled the Russian-American Company to investigate the cause of the decrease, which resulted in prohibiting the killing of females forever afterwards.

It seems that in spite of their ignorant and barbarous methods and their possible lack of scientific acumen, these Russians were practical fellows after all, for the sequel certainly shows that the plan adopted by them of saving and protecting the female was the true one. Mr. Elliott's own estimates show that from 1835 to 1881 the herds had steadily increased up to 5,000,000 seals, or up to a point beyond which it was impossible to go. Speaking of the increase of seal life, he tells us:

I am free to say that it is not within the power of human management to promote this end to the slightest appreciable degree over its present extent and condition as it stands in the state of nature heretofore described. It can not fail to be evident, from my detailed narration of the habits and life of the fur seal on these islands during so large a part of every year, that could man have the same supervision and control over this animal during the whole season which he has at his command while they visit the land he might cause them to multiply and increase, as he would so many cattle, to an indefinite number, only limited by time and the means of feeding them. But the case in question, unfortunately, is one where the fur seal is taken, by demands for food, at least six months out of every year, far beyond the reach or even cognizance of any man, where it is all this time exposed to many known powerful and destructive natural enemies, and probably many others, equally so, unknown, which prey upon it, and, in accordance with that well-recognized law of nature, keeps this seal life at a certain number—at a figure which has been reached for ages past, and continue to be in the future, as far as they now are—their present maximum limit of increase, namely, between 4,000,000 and 5,000,000 seals, in round numbers. This law holds good everywhere throughout the animal kingdom, regulating and preserving the equilibrium of life in the state of nature. Did it not hold good these seal islands and all Bering Sea would have been literally covered, and have swarmed like the medusæ of the waters, long before the Russians discovered them. But, according to the silent testimony of the rookeries, which have been abandoned by the seals, and the noisy, emphatic assurance of those now occupied, there were no more seals when first seen here by human eyes in 1786 and 1787 than there are now in 1881, as far as all evidence goes. (Elliott's Seal Islands of Alaska, p. 66.)

What a pity it is that Mr. Elliott should have forgotten in 1890 the fact that the long drives of from 6 to 12 miles were continued by the Russians as long as they were in possession of the islands, and that from 1868 to 1881 the Americans killed, annually, 100,000 young male seals without causing diminution or decrease, and that during the entire forty-seven years, from 1834 to 1881, the herd increased to marvelous proportions in spite of the long drives and the killing of so many young males, until, as he himself says, "there were no more seals when first seen here by human eyes in 1786 and 1787 than there are now in 1881, as far as all evidence goes."

DECREASE OF SEALS—LACK OF MALE LIFE NOT THE CAUSE.

In this connection it may be well to notice some of the testimony bearing on this very question of an excess or a dearth of bulls on the breeding grounds, collected by the United States when preparing their case for submission to the Tribunal of Arbitration, where the British counsel laid such stress upon Mr. Elliott's report of 1890, with his theory of overdriving, impotency, dearth of bulls, innumerable barren females, and a consequent decrease of the seal herd as a whole.

In their report the British Bering Sea Commissioners say:

Upon the Pribilof Islands in 1891 we did not ourselves note any great abundance of barren females, but the facts in this matter would be scarcely apparent to those not intimately connected with the rookeries for more than a single year. In his official report on the condition of the islands in 1890, Mr. Elliott states that there were then 250,000 females "not bearing or not served last year and this," but he does not explain in what way this numerical estimate was arrived at. (Report of British Commissioners, sec. 433, p. 77.)

Not only did they not note "any great abundance of barren females," but it is an open question whether they noted any, for the fact is there were not any such animals there to be seen, but they gladly quote Mr. Elliott's story of 1890 about the 250,000 barren females which he observed on the islands.

There was not a single day of the breeding season of 1891 when some of the four Treasury agents were not out on the rookeries making careful examination of the condition of seal life thereon, and, although I was one of the four, I have yet to hear the first word from any of them, or from any one else who has ever been on a rookery (excepting Mr. Elliott) about barren females.

It has been amply demonstrated by different individuals, and in many ways, that there was not a shadow of truth in Mr. Elliott's theory, and many of his own most intimate friends and fellow-workers, who are well qualified to speak as scientists on the seal question, are among the foremost of those who flatly contradict him on that point.

Prof. B. W. Evermann, of the United States Fish Commission, visited St. Paul Island while I was there in 1892, and he very carefully inspected the seals on many of the rookeries, beginning July 18 and ending on the 31st, and here is what he found:

LUKANNON ROOKERY, *July 19, from 1.30 to 4 p. m.*

Harems.	Bulls.	Cows.	Pups.	Harems.	Bulls.	Cows.	Pups.
1	1	7	26	9	1	5	3
2	1	6	60	10	1	12	20
3	1	4	20	11	1	4	5
4	1	2	5	12	1	5	15
5	1	27	12	13	1	6	30
6	1	10	15				
7	1	2	0	Total	13	90	211
8	1	0	0				

REEF ROOKERY, *July 20, p. m.*

* * * Many quite large bulls were seen among the bachelors, and there is no doubt in my mind but that the number of available bulls is considerably in excess of the number necessary to serve the cows.

NORTHEAST POINT ROOKERY, *July 22.*

Several hours in the middle of the day were spent in examining this rookery. * * *

Just west of this is a bunch of about 10 good-sized bulls that had no cows about them at all. These were not old, superannuated bulls, but young, vigorous ones, and undoubtedly well able to maintain harems were there a greater number of cows. This and numerous other similar sights convince me that there are even now a good many more bulls than are necessary to serve the cows. (Notes on the fur seal, by B. W. Evermann, Counter Case, United States, p. 264.)

And C. H. Townsend, of the United States Fish Commission, who has had many years practical experience among fur seals, afloat and ashore, and who was on duty in Bering Sea during the summer of 1892, makes affidavit as follows:

As already stated above, I was attached to the steamer *Corwin* during the past summer, and I made all the examinations of the stomachs of the seals referred to in Captain Hooper's report, covering in all 33 seals. * * * These seals were taken on the 2d day of August, 1892, at a distance of about 175 miles from the islands. * * * From the fact that among the females thus taken and examined there were found mostly nursing cows, with a small number of virgin cows, it is reasonable to conclude that there are practically no barren females swimming about in the sea unattached to the islands, or that at any rate, if such seals exist, they are rarely, if ever, taken. In all my experience I never saw anything to lead me to the conclusion that there is such a thing as a barren female. In the case of the virgin cows, a careful examination of the uterus proved them to be too immature for conception. (C. H. Townsend; see affidavit in Appendix.)

The testimony of Professor Evermann and Mr. Townsend is a fair sample of that given by naturalists generally, and it is doubly valuable in this instance, because it comes from personal friends of Mr. Elliott, and from friends who rather inclined to his theory until they had opportunity to investigate for themselves, and to demonstrate to their own and the world's satisfaction that there never was an impotent bull or a barren cow seen on the breeding grounds or rookeries of the Pribilof Islands or in the waters adjacent thereto.

Additional testimony of those who have had experience with the fur seals, and whose practical knowledge of the whole subject of seal life, its growth, expansion, and decay, and the causes thereof, entitles them to a hearing on the point at issue, is most respectfully submitted to the earnest consideration of all who are interested in the perpetuation of the Alaskan fur seal.

Mr. Joseph Stanley-Brown, who also was on the seal islands in 1891–92, testifies as follows:

No intelligent observer would be so bold as to assert that during the season of 1892 there was not an abundance of males of competent virility, despite the occurrence of occasional large harems. The accompanying photographs[1] show that even at the height of the season, and just previous to the disintegration of the breeding grounds, there were, unsupplied with cows, old males which had taken their stand, and from which I was unable to drive them with stones.

I should have been extremely glad to have been able to note a great many more large harems, but the work of the pelagic hunter among the females has been so effective, that the average size of the harems is growing smaller and smaller, while the number of idle bulls is steadily increasing.

The abundance of male life for service upon the rookeries was evidenced by the number of young bulls which continually sought lodgment upon the breeding grounds.

It is highly improbable that the rookeries have ever sustained any injury from insufficient service on the part of the males, for any male that did not possess sufficient vitality for sustained potency would inevitably be deprived of his harem by either his neighbor or some lusty young aspirant, and this dispossession would be rendered the more certain by the disloyalty of his consorts.

The seal being polygamous in habit, each male being able to provide for a harem averaging twenty or thirty members, and the proportion of male to female born being equal, there must inevitably be left a reserve of young immature males, the death of a certain proportion of which could not in any way affect the annual supply coming from the breeding grounds. These conditions existing, the Government has permitted the taking, with three exceptions up to 1890, of a quota of about 100,000 of these young male seals annually. When the abundance of seal life, as evidenced by the areas formerly occupied by seals, is considered, I do not believe that this could account for or play any appreciable part in the diminution of the herd. * * *

From my knowledge of the vitality of seals, I do not believe any injury ever occurred to the reproductive powers of the male seals from redriving that would retard the increase of the herd, and that the driving of 1890 necessary to secure about 22,000 skins could not have caused nor played any important part in the decrease that was apparent on every hand last year.

Karp Buterin, native chief of St. Paul Island (see Appendix):

Plenty of bulls all the time on the rookeries, and plenty bulls have no cows. I never seen a 3-year-old cow without a pup in July; only 2-year-olds have no pups.

H. N. Clark, local agent for lessees:

I never noticed any disproportion of the sexes that would lead me to suspect that the bull seals were too few, nor more than an occasional barren cow. These latter were so few as to excite no remark, but if any such disproportion did in fact exist in 1888 and in 1889, it was the fault of those who killed them at sea, because it never occurred at all until the marine hunters became numerous and aggressive. I mention this matter here, because since I left the island I have heard it asserted that the mismanagement there caused the decrease of seal life. The management there was just such as I would follow if all the seals belonged to me.

O. L. Fowler, local agent for lessees:

I never saw any impotent bulls on the rookeries, and do not believe there ever was any, unless it was the result of age; nor do I believe that young male seals were ever rendered impotent by driving. There has always been a plenty of bulls on the rookeries for breeding purposes ever since I have been on the islands.

John Fratis, native sealer, St. Paul Island:

I never knew of a time when there were not plenty of bulls for all the cows, and I never saw a cow seal, except a 2-year-old, without a pup by her side in the proper season. I never heard tell of an impotent bull seal, nor do I believe there is such a thing, excepting the very old and feeble or badly wounded ones. I have seen hundreds of idle vigorous bulls upon the rookeries, and there were no cows for them. I saw many such bulls last year.

---

[1] Not given here.

H. N. Glidden, Treasury agent:

During these years there was always a sufficiency of vigorous male life to serve all the female seals which came to the islands, and certainly during this period seal life was not affected by any deficiency of males.

Alex. Hansson:

The orders of the boss of the gang in which I worked in 1888 and 1889, under the management of the Alaska Commercial Company, were not to kill the 5-year-old bulls, because they were, he said, needed on the rookeries.

Aggei Kushen, native sealer, St. Paul Island:

We noticed idle, vigorous bulls on the breeding rookeries, because of the scarcity of cows, and I have noticed that the cows have decreased steadily every year since 1886, but more particularly so in 1888, 1889, 1890, and 1891.

H. H. McIntyre, general superintendent Alaska Commercial Company:

And I am satisfied a sufficient number of males was always reserved for future breeding purposes.

That during the twenty years I was upon said Pribilof Islands as general agent of said Alaska Commercial Company there were reserved upon the breeding rookeries upon said islands sufficient vigorous bulls to serve the number of females upon said rookeries; that while I was located upon said islands there was at all times a greater number of adult male seals than was necessary to fertilize the females who hauled upon said rookeries, and that there was no time when there were not vigorous bulls on the rookeries who were unable to obtain female consorts.

So well was this necessity for reserving sufficient mature male life recognized that when in 1887, 1888, and 1889 the depleted rookeries (depleted from causes that will be explained further on) would not furnish the quota of 100,000 large skins, 2 and 3 year-old male seals were taken to make up the quota in preference to trenching upon this reserve of maturer male life.

The policy of the Alaska Commercial Company during the whole period of its lease was, as might be naturally expected, to obtain the best possible skins for market and at the same time preserve the rookeries against injury, for it was not only in their interests to be able to secure every year, until the expiration of the lease, the full quota allowed by law, but they confidently expected, by reason of their good management of the business and faithful fulfillment of every obligation to the Government, to obtain the franchise for a second term. I was, therefore, always alert to see that the due proportion of breeding males of serviceable age was allowed to return to the rookeries. This was a comparatively easy task prior to 1882, but became from year to year more difficult as the seals decreased. No very explicit orders were given to the bosses upon this point until 1888, because the bulls seemed to be plentiful enough, and because it was easier to kill and skin a small seal than a large one, and the natives were inclined, for this reason, to allow the large ones to escape; but in 1888 and 1889 there was such a marked scarcity of breeding males upon the rookeries that I gave strict orders to spare all 5-year-old bulls and confine the killing to smaller animals.

Anton Melovedoff, native chief of St. Paul Island:

I have never known or heard tell of a time when there were not bulls enough and to spare on the breeding rookeries. I never saw a cow 3 years old or over in August without a pup by her side. The only cows on a breeding rookery without pups are the virgin cows who have come there for the first time. I never went onto a rookery in the breeding season when I could not have counted plenty of the idle, vigorous bulls who had no cows.

Talk of epidemics among seals and of impotent bulls on the rookeries, but those who have spent a lifetime on the seal islands and whose business and duty it has been to guard and observe them have no knowledge of the existence of either. An impotent bull dare not attempt to go on a rookery even had he a desire to do so. Excepting the extremely old and feeble, I have never seen a bull that was impotent.

Simeon Melovidov, native school teacher, St. Paul Island:

Nor is there any shadow of fact for the idle statement made from time to time about a dearth of bulls on the rookeries or of impotent bulls.

I have talked to the old men of our people, men who can remember back over fifty years, and not one of them knows of a time when there was not plenty of bulls, and more than enough on the breeding rookeries, and no one here ever heard of an impotent bull. * * * It has been said that cows are barren sometimes because of the dearth of bulls, but such is not the case at all, for the only cows on the breeding rookeries in July or August without pups are the 2-year-olds (virgins), which have come on the rookeries for the first time.

T. F. Morgan, foreman on Pribilof Islands for Alaska Commercial Company:

Despite the lowering on the standard weight of skins, care was taken annually on St. George that the residue of available male breeders was sufficient for the needs of the rookeries, and instructions to that effect were given to the assistants by the superintendent of the Alaska Commercial Company. In this we were aided by the inaccessible character of some of the hauling grounds.

I. H. Moulton, Treasury agent:

During these years there were always a sufficiency of male seals for breeding purposes, and in every year I saw great numbers of idle, vigorous bulls about and back of the breeding grounds which were unable to obtain females.

S. R. Nettleton, Treasury agent:

During my stay on the islands I have never seen a time during the breeding season when there has not been a number of large, vigorous young bulls hanging about the borders of the rookeries watching for an opportunity to get a position of their own.

L. A. Noyes, M. D. (see affidavit in Appendix):

The "dearth of bulls theory" has been thoroughly and impartially investigated without discovering a cow of 3 years old or over on the rookeries without a pup by her side at the proper time, and I am convinced that the virgin females coming onto the rookeries for the first time are the only ones to be found there without pups.

The investigation established the additional fact that hundreds of vigorous bulls were lying idle on the rookeries without cows, and many others had to content themselves with only one or two.

The theory of "impotency of the bull through overdriving" while young was also found to be untrue, and it was shown that after 1878 all long drives on both islands had been abolished, and instead of driving seals from 6 to 12 miles, as was done in Russian times, none were driven to exceed 2½ miles.

It is also a well-known fact that none but the physically strong and aggressive bulls can hold a position on the rookeries, and that a weak or an impotent animal has no desire to go there.

J. C. Redpath, lessee's agent at the seal islands (see affidavit in Appendix):

A dearth of bulls on the breeding rookeries was a pet theory of one or two transient visitors, but it only needed a thorough investigation of the condition of the rookeries to convince the most skeptical that there were plenty of bulls, and to spare, and that hardly a cow could be found on the rookeries without a pup at her side.

For five years I have given this particular subject my most earnest attention, and every succeeding year's experience has convinced me that there is not and never was a dearth of bulls. The theory of impotency of the young bulls because of overdriving when young is not worthy of consideration by any sane or honest man who has ever seen a bull seal on a breeding rookery; and as I have already answered the question of overdriving, I will only add here that no young bull ever goes upon the breeding rookery until he is able to fight his way in, and an impotent bull has no desire to fight, nor could he win a position on the rookery were he to attempt it. The man is not alive who over saw a 6 or 7 year old seal impotent.

B. F. Scribner, Treasury agent:

There was always in both seasons a great sufficiency of adult males to serve all the females coming to the island, and I noticed each year a great number of idle, vigorous bulls behind the breeding grounds who could not obtain consorts, and one of these extra bulls always took the place of an old male unable longer to be of use for breeding purposes.

Daniel Webster, lessees' agent at the seal islands (see affidavit in Appendix):

There was never while I have been on the islands any scarcity of vigorous bulls, there always being a sufficient number to fertilize all the cows coming to the islands. It was always borne in mind by those on the islands that a sufficient number of males must be preserved for breeding purposes, and this accounts partly for the lowering of the standard weight of skins in 1888. The season of 1891 showed that male seals had certainly been in sufficient number the year before, because the pups on the rookeries were as many as should be for the number of cows landing, the ratio being the same as in former years. Then, too, there was a surplus of vigorous bulls in 1891 who could obtain no cows.

W. H. Williams, Treasury agent:

During the season of 1891 nearly every mature female coming upon the rookeries gave birth to a young seal, and there was great abundance of males of sufficient age to again go upon the breeding grounds that year, as was shown by the inability of large numbers of them to secure more than one to five cows each, while quite a number could secure none at all. My investigation confirms what has been so often said by others who have reported upon this subject, and that is that the Pribilof Islands are the great breeding grounds of the fur seals, and that they can be reared in great numbers on said islands, and at the same time, under wise and judicious restrictions, a certain number of male seals can be killed from year to year without injury to the breeding herds, and their skins disposed of for commercial purposes, thereby building up and perpetuating this great industry indefinitely, and thus adding to the wealth, happiness, and comfort of the civilized world, while, on the other hand, if pelagic hunting of this animal is to continue, and the barbarous practice of killing the mother seal with her unborn young, or when she is rearing it, is to go on, it will be but a very short time before the fur seal will practically become extinct and this valuable industry will pass out of existence.

There is the testimony of twenty men who have been on the seal islands for years, some of them being born and raised there, and several of them having had from fifteen to twenty-five years experience, while every one of them have been directly interested in the business either for the Government or for the lessees, and two of them, at least, are naturalists of repute, who could not be induced under any circumstances to vary from the truth and facts as they found them.

I have made the quotations from the "Case of the United States," as it was prepared for the United States counsel before the Tribunal of Arbitration, and I could quote many others to the same purpose were it necessary.

Being personally acquainted with most of the gentlemen named, and knowing the truth of their several statements, I deem it quite unnecessary to add another name from the scores at hand.

DEAD PUPS.

Assuming then that the "dearth of bulls" theory has been disproved and disposed of, we will now take up the subject of dead pups on the islands, and show that until the work of the pelagic sealer in Bering Sea became an industry of some importance, dead pups by the thousands, or by the acre, were unheard of and unknown; but as the pelagic sealing industry flourished and grew, and the fleet of schooners multiplied and doubled in numbers from year to year, the number of dead pups was found to increase on the rookeries in the same proportion.

That this proposition has been, and may still be, denied by the interested ones; that men may be found who will swear to the contrary is already conceded by me, for I have met them who did it; but, in every instance, they were men whose whole interest, capital, and labor were engaged in the business of seal hunting, and who would follow a seal wherever it went, on land or water, unless the strong hand of a power superior to their own intervened to prevent them.

Another class, in which are to be found men of the highest intelligence and personal honor, argue that possibly a stampede or an epidemic, or something else of which we may not be aware, is at the bottom of the trouble.

Of the latter class are the British Bering Sea Commissioners, and I quote them in full:

(L)—MORTALITY OF YOUNG SEALS IN 1891.

344. In the season of 1891 considerable numbers of dead pups were found in certain places upon the rookery grounds or in their vicinity and various hypotheses were advanced to account for this unusual mortality. As some of these have special bearings on the general question of seal preservation, it may be well to devote a few words to this particular subject.

345. In order to exhibit the circumstances surrounding this fact and to arrive at a probable explanation of its true meaning, it will be necessary in the first instance to give in summarized form the observations and notes bearing upon it made on the ground by ourselves.

346. When visiting Tolstoi rookery, St. Paul Island, on the 29th of July, we observed and called attention to several hundred dead pups which lay scattered about in a limited area on a smooth slope near the northern or inland end of the rookery ground and at some little distance from the shore. The bodies were partly decomposed and appeared to have lain where found for a week or more, which would place the actual date of the death of the pups, say, between the 15th and 20th of July. Neither the Government agent who was with us, nor the natives forming our boat's crew at the time, would at first believe that the objects seen on the rookery were dead pups, affirming that they were stones; but when it became clearly apparent that this was not the case they could suggest as causes of death only overrunning by bulls or surf along the shore, neither one of which appeared to us at the time to be satisfactory. Mr. D. Webster, interrogated on the subject some days later on St. George Island, offered merely the same suggestions, but a few days still later, both whites and natives on the islands were found to have developed quite other opinions and to be ready to attribute the deaths to the operations of pelagic sealers killing mothers while off at sea and leading to the death of pups from starvation consequent on such killing.

347. Believing the matter to be one of considerable importance, however, it might be explained, particular attention was paid to it on subsequent visits to rookeries. On the 31st of July and the 1st of August the rookeries of St. George were inspected, but no similar appearances were found, nor was anything of the same kind again seen till the 4th of August, on Polavina rookery, St. Paul Island, where, near the southern extremity of the rookery, several hundred dead pups were again found by us, here also covering an area of limited size, which we were able to examine carefully without disturbing the breeding seals. It was estimated that the pups here found had died between ten days and two weeks before, which would place the actual date of death at about the same time with that of those first referred to.

348. On the following day the extensive rookeries of Northeast Point were visited and examined, but very few dead pups were anywhere seen. Mr. Fowler, in charge of these rookeries for the company, was specially questioned on this point, and fully confirmed the negative observations made by ourselves at the time. It may here be mentioned that the vicinity of Northeast Point had been the principal and only notable locality from which, up to this date, sealing vessels had been sighted in the offing or had been reported as shooting seals within hearing of the shore.

349. On the 19th of August, after a cruise to the northward of about a fortnight's duration, we returned to St. Paul and on the same day revisited Tolstoi rookery. On this occasion the dead pups previously noted were still to be seen, but the bodies were flattened out and more or less covered with sand by the continuous movement of the living seals. There were, however, on and near the same place, and particularly near the angle between Tolstoi rookery and the sands of English Bay, many more dead pups, larger in size than those first noted and scarcely distinguishable in this respect from the living pups, which were then "podded out" in great numbers in the immediate neighborhood. Messrs. Fowler and Murray, who accompanied us on this occasion, admitted the mortality to be local, and the first-named gentleman stated that in his long experience he had never seen anything of the kind before, and suggested that the mothers from this special locality might have gone to some particular "feeding bank" and have there been killed together by sea sealers. On the same day we visited the Reef rookery again, and a search was made there for dead pups, which resulted in the discovery of approximately the same size with those last mentioned, but probably not more than an eighth, and certainly not more than one-fourth in number as compared with the inner end of the Tolstoi rookery ground, and proportionately in both cases to the number of living pups.

350. While making a third inspection of the St. Paul rookeries in September, on the 15th of that month, the Reef and Northeast Point rookeries were again specially examined. The rookery ground of the southeastern side of the Reef Point was carefully inspected, area by area, with field glasses, from the various rocky points which overlook it, and from which the whole field is visible in detail save certain narrow, stony slopes close to the sea edge, where dead pups might have been hidden from view among the bowlders. Subsequently, the northeastern sloping ground, named Garbotch on the plans, being at that date merely occupied by scattered groups of seals, was walked over. The result of the inspection was to show that there were on the southeast side a few dozen dead pups at the most in sight, while on the opposite side perhaps a hundred in all were found in the area gone over, being, probably, the same with those seen here the previous month, and in number or contiguity not in any way comparable with those seen at the inner end of Tolstoi.

351. On the same day a final visit was made to the Northeast Point rookeries, then in charge of three natives only. Two of these men went over the ground with us

and were questioned on various subjects, including that of dead pups, through our Aleut interpreter. They would not admit that they had seen any great number of dead pups on the northeast part this season, and did not seem to be in any way impressed with the idea that there had been any unusual mortality there. The ground to the north of Hutchinson Hill was, however, carefully examined by us from the slopes of the hill, and a few dead pups were made out there. Again, at a place to the north of Sea Lion Neck of the plaus, and beyond the sand beach upon which hollusckickie generally haul out, a slow advance was made among a large herd of females and pups, though part of these were necessarily driven off the ground in so doing. An occupied area of rookery was thus walked over, and the dead pups which appeared at this spot to be unusually abundant were counted with approximate accuracy. A very few were found scattered over the general surface, but on approaching the shore edge an area of about 20,000 square feet was noted, in which about 100 dead pups were assembled. Some of these lay within reach of the surf at high tide. Most appeared to have been dead for at least ten days, and several were broken up and mangled by the movement of the living seals on and about them. This particular locality showed a greater number of dead pups to area than any other seen at this time either on the Northeast or Reef rookeries, but in number in no respect comparable to that previously noted at Tolstoi, or even to that on the south part of Polavina.

352. We were informed on this our last visit to the Pribilof Islands that subsequent to our discovery of and comments upon the dead pups at the two last-mentioned places, the attention of Mr. J. Stanley Brown (who was engaged during the summer in making a special examination of the rookeries for the United States Government) was called to the circumstance, and that he undertook some further examination of it, of which the result will no doubt eventually be rendered available. Dr. Acland, who had just been installed as medical officer on St. Paul, also told us that he had, within a few days, examined the bodies of six of the pups from Tolstoi, and that though rather too much decomposed for correct autopsy, he had been unable to find any signs of disease, but that all these examined were very thin and without food in the stomachs.

353. It may be noted here that the carcasses thus examined must have been those of pups which had died in the month of September, or when no sealing schooners remained in Bering Sea.

354. The body of a pup found by us on the Northeast rookery on the 5th of August, which was still undecomposed, was preserved in alcohol, and has since been submitted to Dr. A. Gunther, F. R. S., of the British Museum, who kindly offered to make an examination of it. This is quoted at length in Appendix (D). The stomach was found to contain no food. The body was well nourished, with a fair amount of fat in the subcutaneous tissue, but no fat about the abdominal organs. The lungs and windpipe were found in an inflammatory condition. Respecting the actual cause of death, Dr. Gunther says: "Both the absence of food as well as the condition of the respiratory organs are sufficient to account for the death of the animal; but which of the two was the primary cause, preceding the other, it is impossible to say."

355. It would be inappropriate here to enter into any lengthened discussion of the bearings of the above facts on the methods of sealing at sea; but as, after the tentative adoption of various hypotheses, the mortality of the young seals was with a remarkable unanimity attributed to pelagic sealing by the gentlemen in any way connected with the breeding islands, and as it has since been widely and consistently advertised in the press as a further and striking proof of the destructiveness of pelagic sealing, it may be permissible to allude to a few cogent reasons, because of which the subject seems at least to require consideration of a much more careful and searching kind:

(1) The death of so many young seals on the islands in 1891 was wholly exceptional and unprecedented, and it occurred in the very season during which, in accordance with the modus vivendi, every effort was being made to drive all pelagic sealers from Bering Sea. Those familiar with the islands were evidently puzzled and surprised when their attention was first drawn to it, and were for some time in doubt as to what cause it might be attributed.

(2) The explanation at length very unanimously concurred in by them, viz, that the young had died because their mothers had been killed at sea, rests wholly upon the assumption that each female will suckle only its own young one, an assumption which appears to be at least very doubtful, and which has already been discussed.

(3) The mortality was at first local, and though later a certain number of dead pups were found on various rookeries examined, nothing of a character comparable with that on Tolstoi rookery was discovered.

(4) The mortality first observed on Tolstoi and Polavina was at too early a date to enable it to be reasonably explained by the killing of mothers at sea. It occurred, as already explained, about the 15th or 20th of July, at a time at which, according to the generally accepted dates as well as our own observations in 1891, the females

had not begun to leave the rookeries in large numbers, or, when leaving them, to do no more than swim or play about close to the shore. It has already been stated that Bryant gives the 25th of July as the opening of the period in which the females begin to leave the rookeries. Maynard states that the bulls, cows, and pups remain within the rookery limits to the same date, while Elliott places this change in the rookeries between the end of July and the 5th and 8th of August. It is, moreover, acknowledged by the best authorities that the dates upon seal life upon the islands have become later rather than earlier in recent years, as compared with those in which the dates above cited were ascertained. In the case of the death of pups after the middle of August, it might be an admissible hypothesis that the mothers had been killed at sea and that subsequently to such killing the young had had time to starve to death, but not at dates earlier than this. In the present case the mortality began long before that date, and it seems probable that the deaths which occurred later must be explained by the same cause, whatever it may have been, extending from the original localities and becoming more general.

356. The causes to which the mortality noted may be attributed with greatest probability are the following, but the evidence at present at disposal scarcely admits of a final attribution to one or other of them. If, however, the examination made by Dr. Acland of several of the carcasses be considered as indicative of the state of the whole, one of the two first is likely to afford the correct explanation:

(a) It is well known that in consequence of the decreased number of killables found on the hauling grounds in late years it has been found necessary to collect these close to and even on the edges of the breeding rookeries, and that it has thus been impossible to avoid the collection and driving to the killing grounds, with the killables, of all sorts of seals not required, including seecatchie and females. It is also known that the driving and killing in the early part of the season of 1891 was pushed with unwonted energy, taking into consideration the reduced number of seals, and it appears to be quite possible that the females thus driven from their young, though afterwards turned away from the killing grounds in an exhausted and thoroughly terrified state, never afterwards found their way back to their original breeding places, but either went off to sea or landed elsewhere. The places where the greatest number of dead pups were first seen on Tolstoi and Polavina were just those from the immediate vicinity of which drives were most frequently made.

(b) The appearances, indicating a local beginning and greatest intensity of mortality, with its subsequent extension to greater areas, might reasonably be explained by the origination and transmission of some disease of an epidemic character.

(c) The circumstances where the mortality was observed to be greatest appeared to be such as to be explicable by a panic and stampede, with consequent overrunning of the young; but, if so, such stampedes must have occurred more than once. They might not improbably have resulted from attempts to collect drives too near the breeding rookeries.

(d) It is entirely within the bounds of probability that raiders may have landed on at least Tolstoi and Polavina rookeries without anyone upon the islands becoming cognizant of the fact. Females would in such a case be killed in greatest numbers, for these occupy the stations most easily got at from the seaside, and the killing upon the rookery ground would also unavoidably have resulted in stampeding large numbers of seals of all classes. (Report of British Bering Sea Commissioners, pp. 61–64.)

A brief review of the salient points of the foregoing will not be out of place at the present time, even though the Tribunal of Arbitration, before which they were considered and upon which they exerted an influence perhaps, is now a thing of the past.

In section 346 they tell us:

We observed and called attention to several hundred dead pups. * * * The bodies were partly decomposed and appeared to have lain where found for a week or more. * * * Neither the Government agent who was with us nor the natives forming our boat's crew at the time would at first believe that the objects seen on the rookery were dead pups, affirming that they were stones.

Now, all that seems plain enough, but does it not sound rather ludicrous, to say the least, when it is alleged by any man that a boat's crew of native sealers, whose life work is the handling of seals, could not tell the difference between the decomposed carcass of a pup seal and a stone, when those who had never been to the seal islands before saw the difference at a glance? The commissioners continue:

The bodies were partly decomposed and appeared to have lain where found for a week or more.

To anyone not knowing the real conditions existing at Tolstoi rookery on that particular 29th of July, the words quoted would imply that the men who "found" the bodies of the "decomposed pups" were walking around on the rookery, but the truth is we did not land on Tolstoi rookery at all during the 29th of July, nor did we find any dead pups that had been lying there for a week or more, nor did we find any.

As I was the Government agent who accompanied the commissioners and was in charge of the boat's crew of natives, I affirm that we sailed from the village landing to Zapadnie or Southwest Bay, where we landed and walked on the rookery without seeing any dead pups; and afterwards we sailed from Zapadnie and followed the trend of the shore all around English Bay and over to Tolstoi, without making a landing till we arrived home at the village. It was while we were passing Tolstoi someone asked the question, "What is that up there on the side-hill?" Field glasses were used by several of the men, and some said the objects pointed at were dead seals, some said "dead pups," and some claimed they were not certain whether they were bones or rocks.

Let it be borne in mind that we were looking at a very steep hill, broken and rocky; that we were from 200 to 300 yards out from land, and in a boat that was on a choppy sea, and therefore in constant motion, and it will be readily understood why the native sealers were so dull about dead pups on Tolstoi rookery.

In section 349 they tell us that—

On the 19th of August * * * we returned to St. Paul, and on the same day revisited Tolstoi rookery. * * * Messrs. Fowler and Murray, who accompanied us on this occasion, admitted the mortality to be local, and the first-named gentleman stated that in his long experience he had never seen anything of the kind before, and suggested that the mothers from this special locality might have gone to some particular "feeding bank" and have been killed together by sea sealers.

Without attempting to enter into an argument of what we actually saw and said that day on Tolstoi rookery, I will say that it is true we, Fowler, Murray, and Barnes, were astonished at the number of dead pups we beheld, a number far exceeding anything we had ever seen before, and it was in that spirit of astonishment that Mr. Fowler said he never saw the like, meaning that he never saw so many at one time, which is very easily accounted for now by the well-known fact that in no year previous to 1891 were so many seals killed and taken by pelagic sealers, as may be seen by a reference to the following table:

Table of pelagic catch from 1868 to 1894, both inclusive, from the best authorities and sources of information, revised and corrected to date.

Year.	Number.	Year.	Number.	Year.	Number.	Year.	Number.
1868	4,867	1875	5,033	1882	15,551	1889	43,158
1869	4,430	1876	5,515	1883	16,585	1890	51,814
1870	8,686	1877	5,210	1884	17,183	1891	69,788
1871	16,911	1878	5,544	1885	24,960	1892	73,394
1872	5,336	1879	8,867	1886	38,094	1893	109,000
1873	5,229	1880	8,910	1887	46,628	1894	142,000
1874	5,873	1881	10,382	1888	26,915		

The real number taken in 1891 was 78,000, but only those actually sold in London are counted here, and, as there is no doubt that from 80 to 90 per cent of the total catch were female seals, it is not to be wondered at that from 20,000 to 30,000 pups were found dead on the rookeries in the fall of that year.

What Mr. Murray did say on that memorable 19th of August, 1891, is a matter of record, as follows:

Accompanied by Agent Barnes, Mr. Fowler, of the North American Commercial Company, and by the British commissioners, I visited Tolstoi rookery on August 19, and we found thousands of dead pups, covering a space of about 5 acres, and their mothers had disappeared. Dr. Dawson, one of the commissioners, took kodak views of the place, and when he asked me what I thought was the cause of their death, I answered, "Their mothers have been killed at sea."

Since I left St. Paul Island I have received a letter from Agent Barnes, in which he says: "You remember the appearance of Tolstoi? I visited Halfway Point along with Mr. Fowler and found the same state of affairs, or worse; and those who have been to Northeast Point say it is still worse there."

Bearing in mind that Northeast Point is the largest rookery in the world, it is no exaggeration to say that between 20,000 and 30,000 pups are lying dead at St. Paul Island whose mothers were slaughtered by sealing schooners in the open sea and the pups left to starve upon the rookeries.

The theory of an occasional epidemic among the seals has been broached, and plausible arguments advanced to prove that the decrease in seal life can be accounted for without blaming the sealing schooners, but as the "oldest inhabitant" on the islands has no recollection of anything of the sort, and as no one ever saw a dozen dead cows on any rookery, it is safe to say there is no foundation for or truth in the epidemic theory. (Murray's Report, 1891, Senate Ex. Doc. No. 107, Fifty-second Congress, second session.)

It does not seem possible that the person who wrote in 1891 the report from which the foregoing has been copied could have "admitted the mortality to be local," and, as a matter of fact, he never did. On the contrary, because of a thirty months' continuous residence on the islands and a personal acquaintance and very intimate and friendly relations with every person on both, I was well aware of the annual increase of dead pups on the rookeries from the time of the first-confirmed shrinkage of the seal herd in 1886. That the terrible sight which met our gaze on Tolstoi rookery should have caused exclamations of surprise from all of us, who knew its real meaning, is not to be wondered at, I think, for the starved carcasses emphasized the fact that in spite of the efforts of the fleets of the United States and of Great Britain, the pelagic sealers' deadly work was being done with an energy and success beyond all preceding seasons, and that unless some other mode of protection could be devised by the nations directly interested the seal herd would soon be annihilated.

In section 362 the commissioners say:

Dr. Acland, who had just been installed as medical officer on St. Paul, also told us that he had within a few days examined the bodies of six of the pups from Tolstoi; * * * he had been unable to find any signs of disease, but that all those examined were very thin and without food in the stomachs.

Dr. Akerly it was who visited the rookeries and examined the dead pups, and whose affidavit will be found in the Appendix.

Commenting on section 255, they say:

(1) The death of so many young seals on the islands in 1891 was wholly exceptional and unprecedented, and it occurred in the very season which, in accordance with the modus vivendi, every effort was being made to drive all pelagic sealers from Bering Sea. Those familiar with the islands were evidently puzzled and surprised when their attention was first drawn to it, and were for some time in doubt as to what cause it might be attributed.

It is true we were rather astonished at the number of dead pups on the rookeries, and being aware "that every effort was being made to prevent pelagic sealing," we were puzzled to account for it at the time, for we knew of no cause other than the killing of the females at sea by which it could be accounted for.

Subsequently, however, we learned of the unprecedented catch made that season by the sealing fleet, and, naturally, we concluded that our conjectures were confirmed. Nor have I had any information since sufficiently reliable to cause a change of opinion.

(3) The mortality was at first local, and though later a certain number of dead pups were found at various rookeries examined, nothing of a character comparable with that on Tolstoi rookery was discovered.

Treasury Agent Barnes, who was on St. Paul Island long after the commissioners left in the fall, is my authority for saying that the " same state of affairs or worse was found later on Polavina and on Northeast Point," the two rookeries visited by him.

(4) The mortality first observed on Tolstoi and Polavina was at too early a date to enable it to be reasonably explained by the killing of mothers at sea. It occurred, as already explained, about the 15th or 20th of July, at a time at which, according to the generally accepted dates, as well as our own observations in 1891, the females had not begun to leave the rookeries in large numbers, or when leaving them, to do more than swim or play about close to the shore.

As already shown, there were no dead pups seen—most certainly nothing worth noting—until August 19.

As the commissioners did not land on either of the seal islands till the latter end of July—about the 28th, if I remember rightly—I can not see how they could have personally observed the movements of the females or the condition of the breeding grounds about the 15th or 20th of July.

Section 356. (a) It is well known that in consequence of the decreased number of killables found on the hauling grounds in late years, it has been found necessary to collect these close to and even on the edges of the breeding rookeries. * * *
It is also known that the driving and killing in the early part of the season of 1891 was pushed with unwonted energy, * * * and it appears to be quite possible that the females thus driven from their young, though afterwards turned away from the killing grounds in an exhausted and thoroughly terrified state, never afterwards found their way back to their original breeding places, but either went off to sea or landed elsewhere. The places where the greatest number of dead pups were first seen on Tolstoi and Polavina were just those from the immediate vicinity of which drives were most frequently made.

The reading of the above quotation staggers one who ever had experience on a rookery or a killing ground.

The "it is well known," is surely unworthy of the commissioners. Which of the men on the islands ever said such a thing? Whoever said that seals were collected close up to the breeding grounds? No man who ever made a drive or saw one made. Who is responsible for the story of the driving of females in the early part of the season of 1891? No sealer, of course, for he would know that there are no females on the rookeries in the early part of the season.

The commissioners found more dead pups on Tolstoi than on any other rookery, and they endeavor to show that collecting and driving of seals from near the breeding rookeries and the consequent taking of some females or the disturbance of the herds caused the death of the pups.

The island records of all the drives made on St. Paul Island in 1891 are at hand, and I will produce a copy for the purpose of showing that no drives whatever were made from either Tolstoi or Polavina during the year 1891!

And yet the commissioners, who were supposed to make an impartial report, say:

The places where the greatest number of dead pups were first seen on Tolstoi and Polavina were just those from the immediate vicinity of which drives were most frequently made.

The fact is, as may be seen by consulting the records for 1890-91, in the Appendix, that no drives were made from Polavina since July 13, nor from Tolstoi since July 20, 1890; so that, instead of being the places most driven from, they are the places not driven from at all in 1891.

The following table, from the official records of the Treasury Department, gives all of the killings for food and for skins (quota and modu ; vivendi) from the close of the season July 20, 1890, to the close of the season August 10, 1891:

Date.	Rookery.	Seals killed.	Remarks.
1890.			
July 28	Lukannon	120	For food.
Aug. 5	Reef	123	Do.
14	Lukannon	124	Do.
23	Reef	155	Do.
30	.....do	110	Do.
Sept. 6	Lukannon	83	Do.
13	.....do	93	Do.
22	.....do	110	Do.
29	Middle Hill	109	Do.
Oct. 4	Lukannon	109	Do.
14	Middle Hill	114	Do.
22	.....do	95	Do.
29	.....do	134	Do.
Nov. 14	.....do	255	Do.
Dec. 4	Reef	283	Do.
1891.			
May 15	Reef	233	For food.
29	.....do	114	Do.
June 4	Zapadnie	463	Do.
11	Zapadnie and Reef	718	Quota.
11	Northeast Point	1,112	Do.
12	Zapadnie	428	Do.
13	Northeast Point	430	Do.
13	Middle Hill	232	Do.
15	Northeast Point	866	Modus vivendi.
16	Reef	842	Do.
17	Southwest Bay	186	Do
18	Reef	1,027	Do.
20	Middle Hill	119	Do.
25	Reef	215	Do.
29	.....do	400	Do.
July 8	.....do	100	Do.
13	.....do	121	Do.
15	Lukannon	122	Do.
21	Middle Hill	178	Do.
27	.....do	248	Do.
Aug. 3	Reef	118	Do.
5	Northeast Point	407	Do.
10	Lukannon	100	Do.
	Total	10,805	

Surely this is sufficient to convince every reasonable man that instead of impartially inquiring into the causes of the decrease of the seal herd and the best possible method of protection, as was originally intended, the commissioners have endeavored to screen the result of the work of the pelagic sealer by making statements about drives, stampedes, and epidemics on the islands which the facts do not warrant.

In another chapter will be found quotations, bearing on this phase of the case, from the argument of United States counsel before the Tribunal of Arbitration, to which I respectfully call the most earnest attention.

In order to show that the "dead-pup" problem was not a new thing on the islands before the British commissioners "discovered" it in 1891, I quote from the testimony of intelligent native chiefs and sealers and of many agents of the Government and of the lessees, who have had many years' experience on the seal islands, and they are unanimous in saying that previous to 1884 there were practically no dead pups to be seen on the rookeries; agents who were on the islands previous to 1884 saw but few or none. That, although there were some few drowned in the surf during heavy wind storms, or trampled to death occasionally by the fighting bulls, it was not until the pelagic sealer appeared in Bering Sea that dead pups were found by hundreds and by thousands and sometimes by the acre.

DEAD PUPS ON THE ROOKERIES.

Dead pups, which seemed to have starved to death, grew very numerous on the rookeries these latter years, and I noticed when driving the bachelor seal for killing, as we started them up from the beach, that many small pups, half starved, apparently motherless, had wandered away from the breeding grounds and become mixed with the killable seals. The natives called my attention to these waifs, saying that it did not use to be so, and that the mothers were dead, otherwise they would be upon the breeding grounds. (H. N. Clark, lessees' agent.)

There were a good many dead pups on the rookeries every year I was on the island, and they seemed to grow more numerous from year to year, because the rookeries were all the time growing smaller, and the dead pups in the latter years were more numerous in proportion to the live ones. (Alex. Hansson, sealer.)

The seals were apparently subject to no diseases; the pups were always fat and healthy, the dead ones very rarely seen on or about the rookeries prior to 1884. Upon my return to the islands in 1886 I was told by my assistants and the natives that a very large number of pups had perished the preceding season, a part of them dying upon the islands and others being washed ashore, all seeming to have starved to death. The same thing occurred in 1886 and in each of the following years to and including 1889. Even before I left the islands, in August, 1886, 1887, and 1888, I saw hundreds of half-starved, bleating, emaciated pups wandering aimlessly about in search of their dams, and presenting a most pitiable appearance. (H. H. McIntyre, general manager.)

But facts came under my observation that soon led me to what I believe to be the true cause of destruction. For instance, during the period of my residence on St. George Island, down to the year 1884, there was always a number of dead pups, the number of which I can not give exactly, as it varied from year to year and was dependent upon accidents or the destructiveness of storms. Young seals do not know how to swim at birth, nor do they learn how for six weeks or two months after birth, and therefore are at the mercy of the waves during stormy weather. But from the year 1884 down to the period when I left St. George Island there was a marked increase in the number of dead pups, amounting, perhaps, to a trebling of the numbers observed in former years, so that I would estimate the number of dead pups in the year 1887 at about 5,000 or 7,000 as a maximum.

During my last two or three years I also noticed among the number of dead pups an increase of at least 70 per cent of those which were emaciated and poor, and in my judgment they died from want of nourishment, their mothers having been killed while away from the island feeding, because it is a fact that pups drowned or killed by accidents were most invariably fat. Learning further, through the London sales, of the increase in the pelagic sealing, it became my firm conviction that the constant increase in the number of dead pups and the decrease in the number of marketable seals and breeding females found on the islands during the years 1885, 1886, and 1887 were caused by the destruction of female seals in the open sea, either before or after giving birth to the pups. The mother seals go to feeding grounds distant from the islands, and I can only account for the number of starved pups by supposing that their mothers are killed while feeding. (T. F. Morgan, lessees' agent.)

TIME OF APPEARANCE OF DEAD PUPS.

The loss of life of pup seals on the rookeries up to about 1884 or 1885 was comparatively slight, and was generally attributed to the death of the mother seal from natural causes or from their natural enemies in the water, or, as sometimes happened, sudden storms with heavy surfs rolling in from certain directions onto the breeding rookeries; but never at any time would a sufficient number of pups be killed to make it the subject of special comment either among the natives or the employees of the company. (W. S. Hereford, M. D., resident physician.)

Between 1874 and 1883 predatory vessels occasionally appeared in Bering Sea, among them the *Cygnet* in 1874 and the *San Diego* in 1876, but the whole number of seals destroyed by such vessels was small, and had no appreciable effect upon the rookeries. In 1884 about 4,000 skins were taken in Bering Sea by three vessels, and starved pups were noticed upon the islands that year for the first time. In 1885 about 10,000 skins were taken in this sea, and the dead pups upon the rookeries became so numerous as to evoke comment from the natives and others upon the islands. (H. H. McIntyre.)

NO DEAD PUPS PRIOR TO 1884.

Poaching in Bering Sea had not begun in those years (from 1868 to 1876), and it was a rare thing to find a dead pup about the shores or on the rookeries. I had frequent occasion, after the close of the breeding season, to visit all parts of the island, and there was no appearance of gaunt or starved seals. Occasionally a dead pup was found that had been crushed to death by the bulls in their encounters with each other. (George R. Adams, lessees' agent.)

A dead pup was rarely seen, the dead being a small fraction of 1 per cent to the whole number of pups. I do not think while I was there I saw in any one season 50 dead pups on the rookeries, and the majority of dead pups were along the shore, having been killed by the surf. (Charles Bryant, Treasury agent.)

There were not, in 1880, sufficient dead pups scattered over the rookeries to attract attention or to form a feature on the rookery. (W. H. Dall, naturalist.)

During the time I was on the islands I only saw a very few dead pups on the rookeries, but the number in 1884 was slightly more than in former years. I never noticed or examined dead pups on the rookeries before 1884, the number being so small. (H. A. Glidden, Treasury agent.)

In performing my official duty I frequently visited the breeding rookeries, and during my entire stay on the island I never saw more than 400 dead pups on all the rookeries. (Louis Kimmel, Treasury agent, 1882–83.)

I never saw but a few dead pups on the rookeries until the schooners came into the sea and shot the cows when they went out to feed, and then the dead pups began to increase on the rookeries. (Nicoli Krukoff, native chief, St. Paul Island.)

I am informed that of late years thousands of young pups have died on the islands while the season was in progress. Certainly such condition did not exist during my residence on the Pribilof group. The pups were sometimes trampled upon by the larger animals, and dead ones might be seen here and there on the rookeries; but the loss in this particular was never enough or important enough to excite any special comment. (J. M. Morton, Treasury agent, 1877–78.)

Never while I was on St. George Island did I see a dead pup on the rookeries, and I certainly should have noticed if there had been any number on the island. (B. F. Scribner, Treasury agent, 1879–80.)

While I was on the island I never saw more than 25 dead pups on the rookeries during any one season. I have seen occasionally a dead one among the bowlders along the shore, which had probably been killed by the surf; but these dead pups were in no instance emaciated. (George Wardman, Treasury agent, 1881–1885.)

While on St. George Island there were practically no dead pups on the rookeries. I do not think I saw during any one season more than a dozen. On St. Paul Island I never saw any dead pups to amount to anything until 1884, and then the number was quite noticeable. (J. H. Moulton, Treasury agent.)

NUMBER OF DEAD PUPS IN 1891.

One thing which attracted my attention was the immense number of dead young seals; another was the presence of quite a number of young seals on all the rookeries in an emaciated and apparently very weak condition. I was requested by the Government agent to examine some of the carcasses for the purpose of determining the cause or causes of their death. I visited and walked over all the rookeries. On all, dead seals were to be found in great numbers. Their number was more apparent on those rookeries, such as Tolstoi and Halfway Point, the water sides of which were on smooth ground, and the eye could glance over patches of ground hundreds of feet in extent which were thickly strewn with carcasses.

Where the water side of the rookeries, as at Northwest Point and the reef (south of the village), were on rocky ground the immense number of dead was not so apparent, but a closer examination showed that the dead were there in equally great numbers scattered among the rocks. In some localities the ground was so thickly strewn with the dead that one had to pick his way carefully in order to avoid stepping on the carcasses. The great mass of dead in all cases was within a short distance of the water's edge. The patches of dead would commence at the water's edge and stretch in a wide swath up into the rookery. Among the immense masses of dead were seldom to be found the carcasses of full-grown seals, but the carcasses were those of pups, or young seals born that year. I can give no idea of the exact number of dead, but I believe that they could only be numbered by the thousands on each rookery. Along the water's edge and scattered among the dead were quite a number of live pups, which were in an emaciated condition. Many had hardly the strength to drag themselves out of one's way; thus contrasting strongly, both in appearance and actions, with the plump condition and active, aggressive conduct of the healthy appearing pups. (J. C. S. Akerly, M. D., resident physician.)

In the latter part of July, 1891, my attention was called to a source of waste, the efficiency of which was most startlingly illustrated. In my conversations with the natives I had learned that dead pups had been seen upon the rookeries in the past few years in such numbers as to cause much concern. In the middle of July they pointed out to me here and there dead pups and others so weak and emaciated that their death was but a matter of a few days. By the time the British commissioners arrived the dead pups were in sufficient abundance to attract their attention, and they are, I believe, under the impression that they first discovered them.

By the latter part of August deaths were rare, the mortality having practically ceased. An examination of the warning lists of the combined fleets of British and American cruisers will show that before the middle of August the last sealing schooner was sent out of Bering Sea. These vessels had entered the sea about July 1 and had done much effective work by July 15. The mortality among the pups and its cessation is synchronous with the sealing fleet's arrival and departure from Bering Sea.

There are several of the rookeries upon which level areas are so disposed as to be seen by the eye at a glance. In September Dr. Akerly and I walked directly across the rookery of Tolstoi, St. Paul, and in addition to the dead pups in sight, they lay in groups of from three to a dozen among the obscuring rocks on the hillside. From a careful examination of every rookery upon the two islands made by me in August and September, I place the minimum estimate of the dead pups to be 15,000, and that some number between that and 30,000 would represent more nearly a true statement of the facts. (J. Stanley-Brown, Treasury agent, 1891-92.)

No mention was ever made of any unusual number of dead pups upon the rookeries having been noticed at any time prior to my visit in 1870, but when I again visited the islands in 1890, I found it a subject of much solicitude by those interested in the perpetuation, and in 1891 it had assumed such proportions as to cause serious alarm. The natives making the drives first discovered this trouble, then special agents took note, and later on I think almost everyone who was allowed to visit the rookeries could not close their eyes or nostrils to the great numbers of dead pups to be seen on all sides. In company with Special Agent Murray, Captain Hooper, and Engineer Brerton, of the *Corwin*, I visited the Reef and Garbotch rookeries, St. Paul Island, in August, 1891, and saw one of the most pitiable sights that I have ever witnessed. Thousands of dead and dying pups were scattered over the rookeries, while the shores were lined with emaciated, hungry little fellows, with their eyes turned toward the sea uttering plaintive cries for their mothers, which were destined never to return. Numbers of them were opened, their stomachs examined, and the fact revealed that starvation was the cause of death, no organic disease being apparent. (W. C. Coulson, captain, revenue marine.

The schooners increased every year from the time I first noticed them, until in 1884 there was a fleet of 20 or 30, and then I began to see more and more dead pups on the rookeries, until in 1891 the fleet of sealing schooners numbered more than 100 and the rookeries were covered with dead pups. (John Fratis.)

It was during these years that dead, emaciated pups were first noticed on the rookeries, and they increased in numbers until 1891, in which year, in August and September, the rookeries were covered with dead pups. (Edward Hughes, employee of lessees, 1888-1894.)

On the 19th of August, 1891, I saw the young pups lying dead upon the rookeries of St. Paul, and I estimated their number to be not less than 30,000; and they had died from starvation, their mothers having been killed at the feeding grounds by pelagic hunters. (Joseph Murray, Treasury agent, 1889-1894.)

Q. Have you noticed any dead pups on the rookeries this past season, and in what proportion to former years?—A. I have seen an unusual number of dead pups this year on the breeding grounds; I may say twice as many as formerly. (J. C. Redpath, lessees' agent, 1875-1894.)

H. Doc. 92, pt. 2——3

CAUSE OF DEATH OF PUPS.

Q. Did you see any dead pups on the rookeries this season?—A. Yes; my attention was called to the matter by J. Stanley-Brown, who requested me to examine them with a view to determining the cause of their death. I examined a number which had apparently recently died. Their bodies were entirely destitute of fat, and no food to be found in their stomachs. After a careful examination I found no evidence of disease.

Q. What do you assign as the cause of their death?—A. I believe them to have died of starvation.

Q. Why do you think they died of starvation?—A. From the fact that nearly all the dead on the rookery were pups, and from absence of all signs of disease, emaciated condition of their bodies, and absence of food from their stomachs. (J. C. S. Akerly, M. D.)

There were a great many dead pups on the rookeries during my last three years on St. Paul Island. Many of them wandered helplessly about, away from the groups or pods where they were accustomed to lie, and finally starved to death. We knew at the time what killed them, for the vessels and boats were several times plainly in sight from the island shooting seals in water, and the revenue cutters and company's vessels arriving at the island frequently reported their presence in Bering Sea, and sometimes the capture of these marauding crews. If all had been captured and the business broken up the seal rookeries would be healthy and prosperous to-day, instead of being depleted and broken up. I speak positively about it, because no other cause can be assigned for their depletion upon any reasonable hypothesis. (W. C. Allis, lessees' agent.

Dr. Akerly, the lessees' physician at the time, made an autopsy of some of the carcasses and reported that he could find no traces of any diseased condition whatever, but there was an entire absence of food or any signs of nourishment in the stomach. Before Dr. Dawson left I called his attention to what Dr. Akerly had done, but whether he saw him on the subject I can not tell. (Milton Barnes, Treasury agent.)

I procured a number of these pups, and Dr. Akerly, at my request, made autopsies, not only at the village, but later on upon the rookeries themselves. The lungs of these dead pups floated in water. There was no organic disease of heart, liver, lungs, stomach, or alimentary canal. In the latter there was but little and often no fecal matter, and the stomach was entirely empty. Pups in the last stage of emaciation were seen by me upon the rookeries, and their condition, as well as that of the dead ones, left no room to doubt that their death was caused by starvation. (J. Stanley-Brown.)

The pups on the rookeries were fat and healthy, and while I was on the islands no epidemic disease ever appeared among them, nor did the natives have stories of an epidemic ever destroying them. (Charles Bryant, Treasury agent, 1869–1877.)

I was informed at the time (November, 1891) that the stomachs of dead pups had been examined by the medical officers at the island and no traces of food found therein. From personal observation I am of the opinion that fully 90 per cent of them died of starvation, great emaciation being apparent. (John C. Cantwell, revenue marine.)

I have never known of any sickness or epidemic among the seals, and I am of the opinion that the thousands of dead pups on the rookeries last year died of starvation on account of their mothers being shot and killed while feeding at the fishing banks in the sea. I was present last year and saw some of the dead pups examined. Their stomachs were empty, and they presented all the appearances of starvation. I also noticed on the rookeries a great many emaciated pups, which on a later visit would be dead. It has always been the practice prior to 1891 for the natives to kill 3,000 to 4,000 pups in November for food, and we always find their stomachs filled with milk. (C. L. Fowler, lessees' agent.)

It is my opinion that the cows are killed by the hunters when they go out in the sea to feed, and the pups are left to die and do die on the island. (John Fratis, native sealer.)

They were thin, poor, and appeared to have starved to death. (Alex. Hanssen, sealer.)

It is a well-known fact that the female seals leave the islands and go great distances for food, and it is clearly proven that many of them do not return, as the number of pups starved to death on the rookeries demonstrates. (W. S. Hereford, M. D.)

For if the mother seals are destroyed their young can not but perish; no other dam will suckle them; nor can they subsist until at least 3 or 4 months old without the mother's milk. The loss of this vast number of pups, amounting to many thousands, we could attribute to no other cause than the death of the mother at the hands of pelagic seal hunters. (H. H. McIntyre.)

Q. How do you account for this?—A. I think the cows were killed by the poachers while away from the rookeries, and as mother seals nurse none but their own young, consequently the pups whose mothers were killed die from starvation. (Antone Melovedoff, native chief.)

The seals are never visited by physical disorders of any kind, so far as I could ascertain, and I have never seen on their bodies any blemishes, humors, or eruptions which might be attributed to disease. (John M. Morton.)

These latter pups I examined, and they seemed to be very much emaciated. In my opinion, they died of starvation, caused by the mothers having been shot while absent from the islands feeding. Another cause of their starving is because a cow refuses to give suck to any pup but her own, and she recognizes her offspring by its cry, distinguishing its voice from that of hundreds of others which are constantly bleating. (J. H. Moulton.)

The epidemic theory was urged very strongly in 1891, when the rookeries were found covered with dead pups, but a careful and technical examination was made of several of the dead bodies without discovering a trace of organic disease, while starvation was so apparent that those who examined them decided that it was the true cause of their death. Had sickness or disease attacked the seal herd, it is only reasonable to suppose a few grown seals would be found dead where so many young ones had died so suddenly, but the most diligent search has failed to find a grown seal dead upon the islands from unknown causes. From the discovery of the islands until the present time the flesh of the fur seal has been the daily meat ration of the natives and of the white people, and yet it is a fact that a tainted or diseased carcass has never been known. (L. A. Noyes, M. D., resident physician, 1880–1894.)

Some of these losses were due to their perhaps too early attempts to swim. When the pup is a few months old the mother seal conducts it to the water and teaches it to swim near the shore. If a heavy sea is encountered the weak little pup is liable to be thrown by the surf against the rocks and killed, but under natural conditions, and with the protection to the rookeries formerly enforced at the islands, the losses from this cause and all others combined (save alone the authorized killing) amounted to an infinitesimal percentage of the whole numbers in the herds. (H. G. Otis, Treasury agent, 1879–1881.)

Another theory, equally untrue, was that an epidemic had seized the herd; but investigations of the closest kind have never revealed the death on the islands of a full-grown seal from unknown causes. Let it be remembered that the flesh of the seal is the staple diet of the natives, and that it is eaten daily by most of the white employees as well; and yet it is true that a sign of taint or disease has never been found on a seal carcass in the memory of man. It was not until so many thousands of dead pups were found upon the rookeries that the problem was solved. The truth is, that when the cows go out to the feeding grounds to feed, they are shot and killed by the pelagic hunter, and the pups, deprived of sustenance, die upon the rookeries. Excepting a few pups killed by the surf occasionally, it has been demonstrated that all the pups found dead are poor and starved, and when examined their stomachs are found to be without a sign of food of any sort. The resident physician, Dr. Akerly, examined many of them, and found in every instance that starvation was the cause of death. (J. C. Redpath.)

A double waste occurs when the mother seal is killed, as the pup will surely starve to death. A mother seal will give sustenance to no pup but her own. I saw sad evidences of this waste on St. Paul Island last season, where large numbers of pups were lying about the rookeries, where they had died of starvation. (Commander Z. L. Tanner, U. S. N.)

I never heard of any disease among the seal herd, nor of an epidemic of any sort or at any time in the history of the islands. (Daniel Webster, lessees' agent, 1868–1894.)

If the mother of a young seal is killed, the pup is very likely to die. It will be so weak that the storm will dash it ashore and kill it, or it may die of starvation. I have seen pups hardly larger than a rat from lack of nourishment. A starved or neglected orphan pup is nearly sure to die. At one storm the natives found over 300 pups washed ashore in a little cove, and the water around was full of dead pups. It is certain that nearly all the dead pups were orphans. The female seal when suckling her young has to go out into the ocean in search of food, and it is those animals, or females on the way to the breeding grounds to give birth to the young, that we kill in the Bering Sea. (T. T. Williams, quoting Captain Olsen.)

The foregoing quotations are from the affidavits and reports of men who, through years of experience, gained a practical knowledge of fur-seal life in all its details, and who therefore know of what they speak beyond the possibility of successful contradiction.

It may be urged by our opponents that the testimony is that of men who are neither learned nor scientific, and who, being employed by either the Government or the lessees, had private and personal interests to subserve.

For the purpose of meeting such objections, and to show how the practical and scientific are agreed in this matter, I will here introduce a paper written by a well-known naturalist, who has had many years practical experience among the fur seals on the Commander Islands, and who has not now, nor ever had, any interest in the Pribilof Islands or the Alaskan seal herd.

His testimony is therefore the more valuable, and it will be found that it confirms my position in every particular:

DEPOSITION OF NICHOLAS A. GREBNITZKI,[1] RUSSIAN MILITARY CHIEF OF THE COMMANDER ISLANDS.

I, Nicholas A. Grebnitzki, Russian military chief of the Commander Islands district, with the rank of colonel, make the following statement:

I have been residing on the Commander Islands and have directed all sealing operations there for the last fifteen years, and during this whole period have been absent from the islands but very little. I have carefully observed seal life, the condition of the rookeries, and the method of taking seals at all seasons and under all conditions, with the object of keeping the Russian Government thoroughly informed as to its sealing interests and the proper management of the same.

While I have never had the opportunity to examine the Pribilof Islands seals, yet I do not hesitate to express the opinion that that herd and the Commander Islands herd are distinct and do not mingle at all. There are some natives on the islands who are familiar with both, and who state that there is a marked difference in the animals. Besides, my studies as a naturalist enable me to state that it would be contrary to all reason to suppose that they mingle with one another. The Commander herd approaches very closely to the Robben Island herd in winter, and yet it does not mingle with it. Of this I am sure, for I have charge of Robben Island as well as of the Commander Islands, and know the skins of the two herds to be different. The skin of the Commander seal is thicker, has coarser hair, is of a lighter color, and weighs about 20 per cent more than a Robben skin of the same size.

It is wholly improbable that the seals of the Commander herd visit any land other than the Commander Islands. I believe they regard these as their home, these islands being peculiarly adapted to their needs at the period to bring forth their young and of breeding. The fact that the Robben Island herd still frequents Robben Island to the exclusion of any other land, notwithstanding it has been subjected there to the utmost persecution, shows to my mind conclusively that the presence of man will not prevent a seal herd from returning to the same land year after year. Even if isolated cases have occurred (I know of none) in which for various causes a few of the Commander Islands seals reached other shores, such exceptions would not disprove the general rule above stated. I can readily understand that a female which had been wounded in the water might be subject (sic) to seek the nearest land and there give birth to her pup.

Annually, at almost stated periods, they arrive at the islands and immediately proceed to occupy the same grounds which have been occupied during past years in a way which makes it impossible to doubt that they are familiar with the locality. I believe that at some time during the year every seal comes ashore. There is no reason to believe that a certain number of any class remain swimming about in the neighborhood of the islands all summer without landing, although there is considerable difference in the time at which different classes arrive.

Soon after landing at the Commander Islands those cows which were fertilized the year previous give birth to their young. A cow does not, except in very rare instances, give birth to more than one pup in a season. The birth of pups can only take place

---

[1] No written evidence having been produced in the report of the British commissioners in support of the various views attributed to Mr. Grebnitzki, the United States have deemed it desirable to obtain from that official a written expression from his views upon seal life in general.

on shore. Cows never arrive at the islands with new-born pups. But the impossibility of birth in the water is best proved by the fact that the pup when first born is purely a land animal in all its habits. It does not voluntarily approach the water till it is several weeks old, and then it is obliged to learn to swim. A surf will sometimes wash the young pups off the rocks, when they are sure to be drowned. The pups can not swim at birth, but must be taught by their mothers. A pup would drown if thrown into the sea before learning to swim.

Copulation in the water I believe to be impossible, for the act is violent, of long duration, and in general character similar to that performed by land animals.

I believe that the seals leave the vicinity of the islands mainly on account of the severity of the winter. Of course, I do not mean to say that they would remain on the shore all the year round, as many of them do throughout the whole of the summer, for they would be obliged to take to the water to obtain food. What I mean is that they would not go so far away as they now do, but would remain around the islands, and thus give additional proof of the unquestionable fact that they regard them as their home. I base this statement upon the fact that during mild winters I have myself seen them in large numbers off the Commander Islands. They are often reported about 50 miles south of the westernmost of the Aleutian Islands and the Kamchatka Coast. This would be in accord with the habits of the seals of the Southern Hemisphere, which, I am informed, are found in the same locality, more or less, at all seasons. The seals generally leave the Commander Islands by the middle of November, by which time it has become cold and stormy, but in mild winters they have been on the islands as late as December.

I do not think that fur seals should be classed with wild animals any more than sheep or cattle when out on large pasturing grounds. Seals, unless needlessly frightened, become more or less accustomed to the sight of man among them on the rookeries, and while on land are at all times under his complete control. A few men can drive a large number of them without difficulty. They are intelligent to a very high degree, and can be made to become in a short time pets. The breeding males or bulls are alone aggressive.

Seals are polygamous, and the powers of fertilization of the male are very great. Since the births are about equally distributed between males and females, it follows that under natural conditions there would be a great excess of male life over that actually needed for the propagation of the species, and it is, as in the case of so many other animals, for the positive benefit of the herd as a whole that a portion of this excess of male life be killed off before it is of sufficient age to go on the rookeries. If not killed off the competition by the bulls upon the rookeries for females would be destructive of much life. This competition is already fierce enough.

During some of the years prior to the time of my arrival on the islands there had been considerable indiscriminate killing of seals without regard to age or sex; but during the fifteen years of my management of the Commander Islands rookeries all seals which have been killed constituted a portion of the excess of males above referred to, and known as bachelors, or holluschickie. This is why the rookeries are to-day in a much better condition than when I first went to the Commander Islands, notwithstanding that until the year 1891 a gradually increasing number of large skins has been taken. From 1886 to 1890 the average annual catch was about 50,000, the skins all being large. The last two years I have reduced the catches, because I now think 50,000 skins somewhat in excess of what the rookeries can yield, and for other causes which I will mention later. I feel very sure that the great cause of this diminution is pelagic sealing.

This year I have counted over 3,500 skins seized on poaching vessels, and have found 96 per cent to be skins of females. They were skins taken from Commander Island seals.

As to skins taken near Pribilof Islands I counted the skins seized in the *Rosa Olsen* and found two-thirds of them were skins of females. These were taken, as the log book of the *Rosa Olsen* shows, over 80 miles from shore.

I consider it a false argument to say that the killing of a proper portion of the excess of male life is bad, merely because it is an interference with the order of nature. If not interfered with, nature will produce an overpopulation of the rookeries, which would, of course, be a bad thing. By the present mode of killing a certain number of young males, population is regulated. No facts can be brought forward to show that this method is not the right one. Past experience shows that it is right.

The method is not proved to be bad by showing that during some years too many males may have been killed, and that the rookeries have thereby suffered. When such mistakes have been made they can be corrected by reducing the number of males to be killed for a few years; for the most absolute control can be exercised over the herd while it is on land. I claim that the method now pursued, when executed under proper regulations, is in theory and practice the only one by which sealing can be carried on commercially without injuring the vitality of the herd and its ability to maintain its numbers at the proper limit. It does not cause the seals to change their habits in any way, and I do not believe that even an excessive killing of young males on the islands would have the effect of altering the habits of the female seals with regard to landing and cause them to remain about the islands instead of coming on shore.

Cows, except, perhaps, in rare cases of accident or for scientific purposes, are never allowed to be killed on the islands, and the reason for this is that all cows are needed for breeding purposes. To kill, therefore, any cow except a barren one (and there are few barren ones except amongst the very old cows) inflicts a much greater injury on the herd than the loss of a single life. It is not true that because it is proper to kill a certain number of males it is also proper to kill a certain number of females. But assuming that it might at some time become desirable to kill some females, it would still be wholly improper to kill them without regard to size or condition, as is the case when they are killed in the water.

There is at the present time upon the Commander Islands an abundance of male life for breeding purposes, and there is no fear that any female will not be served from lack of virile males. On the other hand, it is undoubtedly true that there were in 1892 relatively fewer females than in former years, and I attribute this to two causes: First, to killing of seals in the water, and, second, raids upon the islands. The first of these causes is by far the more important.

The raids have, owing to the great amount of foggy weather, taken place, to a certain extent, notwithstanding the greatest precautions to guard against them. The raiders kill males, females, and pups without discrimination; but however injuriously the raids have affected the rookeries still they are of much less importance than the killing of Commander Islands seals in the water. During the past two summers, and especially during the last one, this killing in the waters has become so great that if allowed to continue in future years the herd will be in danger of ultimate extinction.

I do not know exactly how wasteful this method may be, from the fact that all the animals wounded or killed are not captured, though I am told that much loss occurs in that way, and I know that under certain conditions a seal shot dead will sink at once. I can state positively, however, from actual experience and personal examination, that a vast proportion, fully 96 per cent, of the skins taken by this method during the present year are those of female animals. In addition, a certain number of the skins so taken are those of very young seals, probably of both sexes, such as are never killed on land.

Very few of the females killed are barren, no matter when or where they are killed. Females taken early in the season are generally heavy with young, in which condition they travel slowly as compared with the other seals. The killing of such a female involves, of course, the immediate loss of two lives. But even when the female is taken after she has been on shore and given birth to her young this same result follows eventually, for a seal will suckle only her own pup, and the pups are for the first three to five months dependent altogether on their mothers for food. Consequently when the mothers, who, after the birth of their pups, leave the rookeries in search of food (traveling sometimes considerable distances, I do not know exactly how far), fail to return, their pups must necessarily die.

There are always a few dead pups to be found on the rookeries whose death is not due to that of their mothers; but during the last year or two a greater number of dead pups have been actually noticed than heretofore, and have attracted the attention of all persons on the islands who are at all familiar with seal life. It can not be successfully contended that they all died of natural causes. There is no disease among the Commander Island seals, and while a certain number of young pups are always exposed to the danger of being crushed to death (but not as a result of the drives which are made to collect seals for killing) or of being drowned by the surf, yet these causes of death will not account for the greater mortality of pups which took place during the past summer. Besides, the bodies of the dead pups I refer to are those of starved animals, being greatly emaciated.

It is chiefly during the next few years that the effects of the recent killing of females will become most noticeable, because many of the pups which in those years would have become bachelors or holluschickie have never been born or died soon after birth.

With regard to the driving of the seals from the beaches to the places of slaughter, while it does not benefit them, yet I believe that there are very few cases in which it does them any harm even if they are redriven. I am sure it does not render them impotent. It should be remembered that, unlike the hair seals, they are fairly adapted to movement on land, as is proved by the fact that they are in some cases actually driven considerable distances over ground that is both rough and steep.

Since the killing of seals in the water is wasteful, and in every sense contrary to the laws of nature (which require that special protection be afforded to the females and young of all animals), I am of the opinion that it should be entirely forbidden. If it is only partly suppressed or prohibited within a certain distance from the islands, the evil would not be cured, although its effects might be less noticeable, for the killing of females, many of them heavy with young, would necessarily continue, since all experience shows that female animals always constitute the chief catch of the open-sea sealer.

NICHOLAS A. GREBNITZKI,
District Chief of the Commander Islands, District St. Petersburg.

(Counter case, United States, p. 362.)

Here we have the testimony of one who is at once a scientist and one of the most practical of men; a man who has been officially interested in the fur-seal industry for many years, and who has devoted a great part of his life to the scientific and practical study of the species.

Every word he utters shows his intimate knowledge of the subject treated, and his practical common sense and scientific acumen, coupled with a breadth of view all his own, gives an extraordinary value to everything he says on the subject of fur seals.

True, he is interested in the fur-seal industry on the Commander Islands, belonging to Russia, and for that reason he may fall under the ban of the hypercritical who seem to suspect the honesty and the motives of all who have, or ever did have, any connection with the fur-seal islands on either side of Bering Sea.

That the class of critics alluded to may be silenced on this point I will introduce the testimony of leading naturalists, which is in full accord with all that I have already quoted. It will be seen that Dr. Merriam briefly stated the question at issue to the naturalists of Europe and asked for their views, which were freely given and which I take the pleasure of quoting in full.

CIRCULAR LETTER OF DR. C. HART MERRIAM.

Dr. C. Hart Merriam, one of the American Bering Sea commissioners, addressed the following circular letter to various leading naturalists in different parts of the world, for the purpose of obtaining their views as to the best method of preserving the fur seals of Alaska:

WASHINGTON, D. C., *April 2, 1892.*

DEAR SIR: The Government of the United States having selected me as a naturalist to investigate and report upon the condition of the fur-seal rookeries on the Pribilof Islands, in Bering Sea, with special reference to the causes of decrease and the measures necessary for the restoration and permanent preservation of the seal herd, I visited the Pribilof Islands and made an extended investigation of the subject, the results of which are here briefly outlined.

FACTS IN THE LIFE HISTORY OF THE NORTHERN FUR SEAL (CALLORHINUS URSINUS).

(1) The fur seal is an inhabitant of Bering Sea and the Sea of Okhotsk, where it breeds on rocky islands. But four breeding colonies are known, namely, (1) the Pribilof Islands, belonging to the United States; (2) the Commander Islands, belonging to Russia; (3) Robben Reef, belonging to Russia; and (4) the Kurile Islands, belonging to Japan. The Pribilof and Commander islands are in Bering Sea; Robben Reef in the Sea of Okhotsk, near the island of Saghalien, and the Kurile Islands between Yezo and Kamtchatka. The species is not known to breed in any other part of the world.

(2) In winter the fur seal migrates into the North Pacific Ocean. The herds from the Commander Islands, Robben Reef, and the Kurile Islands move south along the Japan Coast. The Pribilof Islands herd move south through the passes in the Aleutian chain. The old breeding males are not known to range much south of these islands. The females and young reach the American Coast as far south as California.

(3) Returning, the herds of females move northward along the coast of California, Oregon, Washington, and British Columbia in January, February, and March, occurring at varying distances from shore Following the Alaska coast northward and westward they leave the North Pacific Ocean in June, traversing the passes in the Aleutian chain, and proceed at once to the Pribilof Islands.

(4) The old (breeding) males reach the islands much earlier, the first coming the first week in April or early in May. They at once land and take stands on the rookeries, where they await the arrival of the females. Each male (called a bull) selects

a large rock, on or near which he remains, unless driven off by stronger bulls, until August, never leaving for a single instant, night or day, and taking neither food nor water. Before the arrival of the females (called cows) the bulls fight savagely among themselves for positions on the rookeries, and many are severely wounded. All the bulls are located by June 20.

(5) The pregnant cows begin arriving early in June, and soon appear in large schools or droves, immense numbers taking their places on the rookeries each day between June 12 and the end of the month, varying with the weather. They assemble about the old bulls in compact groups called harems. The harems are complete early in July, at which time the breeding rookeries attain their maximum size and compactness.

(6) The cows give birth to their young soon after taking their places on the harems. The period of gestation is between eleven and twelve months.

(7) A single young is born in each instance. The young at birth are about equally divided as to sex.

(8) The act of nursing is performed on land; never in the water. It is necessary, therefore, for the cows to remain at the islands until the young are weaned, which is when they are 4 or 5 months old.

(9) The fur seal is polygamous, and the male is at least three times as large as the female. Each male serves 15 to 25 females.

(10) Copulation takes place on land. Most of the cows are served by the middle of July, or soon after the birth of their pups. They then take to the water, and come and go for food while nursing.

(11) The pups huddle together in small groups called pods, at some distance from the water. When 6 or 8 weeks old they move down to the water's edge and learn to swim. The pups are not born at sea, and if soon after birth they are washed into the sea they are drowned.

(12) The cows are believed to take the bull first when two years old, and deliver their first pup when 3 years old.

(13) Bulls first take stands on the breeding rookeries when 6 or 7 years old. Before this they are not powerful enough to fight the older bulls for positions on the harems.

(14) Cows when nursing, and the nonbreeding seals, regularly travel long distances to feed. They are commonly found 100 to 150 miles from the islands and sometimes at greater distances.

(15) The food of the fur seal consists of fish, squids, crustaceans, and probably other forms of marine life also.

(16) The great majority of cows, pups, and such of the breeding bulls as have not already gone, leave the islands about the middle of November, the date varying considerably with the season.

(17) The nonbreeding male seals (holluschickie), together with a few old bulls, remain until January, and in rare instances even until February.

(18) The fur seal as a species is present at the Pribilof Islands eight or nine months of the year, or from two-thirds to three-fourths of the time, and in mild winters sometimes during the entire year. The breeding bulls arrive earliest and remain continuously on the islands about four months; the breeding cows remain about six months, and the nonbreeding male seals about eight or nine months, and sometimes during the entire year.

SEALS KILLED ON THE PRIBILOF ISLANDS.

(19) The only seals killed for commercial purposes at the seal islands are nonbreeding males (under 5 or 6 years of age, called holluschickie). They come up on the rookeries apart from the breeding seals, and large numbers are present by the latter part of May. They constantly pass back and forth from the water to the hauling grounds. These animals are driven by the natives (Aleuts) from the hauling grounds to the killing grounds, where they are divided up into little groups. Those selected as of suitable size are killed with a club by a blow on the head; the others go into the water and soon reappear on the hauling grounds. In this way about 100,000 young males have been killed annually on the Pribilof Islands for twenty years.

(20) In addition to the commercial killing above described, a number of male pups were formerly killed each year to furnish food for the natives, but the killing of pups is now prohibited by the Government.

PRESENT NUMBERS COMPARED WITH FORMER ABUNDANCE.

The rookeries on both St. Paul and St. George islands bear unmistakable evidence of having undergone great reduction in size during the past few years. This evidence consists (1) in the universal testimony of all who saw them at an earlier period, and (2) in the presence upon the back part of each rookery of a well-marked strip or zone of grass-covered land, varying from 100 to 500 feet in width, on which the stones and bowlders are flipper-worn and polished by the former movements of the seals, and the grass is yellowish-green in color and of a different genus (*Glyceria angustata*) from the rank, high grass usually growing immediately behind it (*Elymus mollis*). In many places the ground between the tussocks and hummocks of grass is covered with a thin layer of felting, composed of the shed hairs of the seals matted down and mixed with excrement, urine, and surface soil. The exact year when this yellow-grass zone was last occupied by seals is difficult to ascertain, but the bulk of testimony points to 1886 or 1887. The aggregate size of the areas formerly occupied is at least four times as great as that of the present rookeries.

CAUSES WHICH LED TO THE DEPLETION OF THE ROOKERIES.

The seals which move northward along the coast of the northwestern United States, British Columbia, and southeastern Alaska from January until late in June are chiefly pregnant females, and about 90 per cent of the seals killed by pelagic sealers in the North Pacific are females heavy with young. For obvious reasons many more seals are wounded than killed outright, and many more that are killed sink before they can be reached, and consequently are lost. As each of these contains a young, it is evident that several are destroyed to every one secured.

For several years the pelagic sealers were content to pursue their destructive work in the North Pacific, but of late they have entered Bering Sea, where they continue to capture seals in the water throughout the entire summer. The females killed during this period are giving milk and are away from the islands in search of food. Their young starve to death on the rookeries. I saw vast numbers of such dead pups on the island of St. Paul last summer (1891), and the total number of their carcasses remaining on the Pribilof Islands at the end of the season of 1891 has been estimated by the United States Treasury agents at not less than 20,000.

The number of seal skins actually secured and sold as a result of pelagic sealing is shown in the following table:

Year.	Number of skins.	Year.	Number of skins.	Year.	Number of skins.	Year.	Number of skins.
1872	1,029	1877	(?)	1882	17,700	1887	33,800
1873	(?)	1878	264	1883	9,195	1888	36,818
1874	4,949	1879	12,500	1884	14,000	1889	39,563
1875	1,646	1880	13,600	1885	13,000	1890	51,404
1876	2,042	1881	13,541	1886	38,907	1891	62,500

Inasmuch as the number of seals annually secured by pelagic sealing represents but a fraction of the total number killed, a glance at the above figures is enough to show that the destruction of seal life thus produced is alone sufficient to explain the present depleted condition of the rookeries.

Pelagic sealing as now conducted is carried on in the North Pacific Ocean from January until late in June, and in Bering Sea in July, August, and September. Some sealing schooners remain as late as November, but they do so for the purpose of raiding the rookeries.

It has been alleged that overkilling of young males at the islands is a principal cause of the depleted condition of the rookeries.

In reply to this contention, it is only necessary to bear in mind that the number of male and female fur seals is equal at birth, that the species is polygamous, and that each male serves on an average at least 15 to 25 females. It is evident, therefore, that there must be a great superabundance of males, of which a large percentage may be killed annually forever without in the slightest degree endangering the productiveness of the herd. Furthermore, it has been shown that the killing of seals at the Pribilof Islands is completely under the control of man and is restricted to the superfluous males, for selection as to sex and age can be and is exercised, so that neither females nor breeding males are killed. It is evident that this killing of nonbreeding males could in no way affect the size or annual product of the breeding rookeries unless the number killed was so great that enough males were not left to mature for breeding purposes. There is no evidence that this has ever been the case. Moreover, all seals killed or wounded are invariably secured and their skins marketed; in other words, there is neither waste of the seal herd nor impairment of the productiveness of the breeding stock.

Pelagic sealing, on the other hand, is wasteful in the extreme and is directed to the fountain head or source of supply. From the very nature of the case, selection can not be exercised, and a large percentage of seals wounded are lost. Owing to the peculiar movements of the seal herds, it so happens that about 90 per cent of the seals killed in the North Pacific are females heavy with young, entailing a destruction of two seal lives for every adult seal killed. In Bering Sea, also, large numbers of females are taken; these females are in milk, and their young die of starvation on the rookeries.

Pelagic sealing as an industry is of recent origin, and may be said to date from 1879. The number of vessels engaged has steadily increased, as has the number of seals killed, until it appears that unless checked by international legislation the commercial extermination of the seal is only a matter of a few years. It seems a fair inference, therefore, that the only way to restore the depleted rookeries to their former condition is to stop taking seals at sea, and not only in Bering Sea, but in the North Pacific as well.

Having been selected by my Government solely as a naturalist, and having investigated the facts and arrived at the above conclusions and recommendations from the standpoint of a naturalist, I desire to know if you agree or differ with me in considering these conclusions and recommendations justified and necessitated by the facts in the case.

I shall be greatly obliged if you will favor me with a reply.

Very truly, yours,

C. HART MERRIAM.

REPLIES TO C. HART MERRIAM.

REPLY OF DR. ALPHONSE MILNE EDWARDS.

PARIS, *April 20, 1892.*

SIR: I have read with great interest the letter you addressed me with reference to the fur seals of Bering Sea, and I think it would be of real advantage to have concerted international measures so as to insure an effective protection to those valuable animals.

To-day the means of transportation at the disposal of the fishermen are so great, the processes of destruction which they employ are so improved, that the animal species, the object of their desire, can not escape them. We know that our migratory birds are during their travels exposed to a real war of extermination, and an ornithological international commission has already examined, not unprofitably, all the questions relating to their preservation.

Would it not be possible to put fur seals under the protection of the navy of civilized nations?

What has happened in the Southern Ocean may serve as a warning to us.

Less than a century ago these amphibia existed there in countless herds. In 1808, when Fanning visited the islands of South Georgia, one ship left those shores carrying away 14,000 seal skins belonging to the species *Arctocephalus australis.* He himself obtained 57,000 of them, and he estimated at 112,000 the number of these animals killed during the few weeks the sailors spent there that year.

In 1822 Weddell visits these islands, and he estimates at 1,200,000 the number of skins obtained in that locality. The same year 320,000 fur seals were killed in the South Shetlands. The inevitable consequences of this slaughter were a rapid decrease in the number of these animals. So, in spite of the measures of protection taken during the last few years by the governor of the Falkland Islands, these seals are still very rare, and the naturalists of the French expedition of the *Romanche* remained for nearly a year at Tierra del Fuego and the Falkland Islands without being able to capture a single specimen.

It is a source of wealth which is now exhausted.

It will soon be thus with the *Callorhinus ursinus* in the North Pacific Ocean, and it is time to insure to these animals a security which may allow them regular reproduction.

I have followed with much attention the investigation which has been made by the Government of the United States on this subject. The reports of the commissioners sent to the Pribilof Islands have made known to naturalists a very large number of facts of great scientific interest, and have demonstrated that a regulated system of killing may be safely applied in the case of these herds of seals when there is a superfluity of males. What might be called a tax on celibacy was applied in this way in the most satisfactory manner, and the indefinite preservation of the species would have been assured if the emigrants, on their way back to their breeding places, had not been attacked and pursued in every way.

There is, then, every reason to turn to account the very complete information which we possess on the conditions of fur-seal life in order to prevent their annihilation, and an international commission can alone determine the rules, from which the fishermen should not depart.

Accept, etc.,

A. MILNE EDWARDS,
Director of the Museum of Natural History.

REPLY OF DR. CARLOS BERG, OF BUENOS AYRES.

JUNE 4, 1892.

SIR: In answer to your circular dated April 2, and directed to Dr. Hermann Bur-meister, I regret to let you know that same died shortly before the transmission of your circular by D. N. Bertolette, esq.

Having been named director of the national museum in the place of the deceased, I have read with great interest your report and conclusions about the causes of the decrease and the measures necessary for the restoration and permanent preservation of the seal herd on the Pribilof Islands, in Bering Sea, and according to your wish I have the pleasure to let you know that from the standpoint of a naturalist I per-fectly agree with you in considering your conclusions and recommendations justified and necessitated by the facts stated by you as a result of your special investigation on the above-named islands.

Very truly, yours,

CARLOS BERG.

REPLY OF PROF. DR. ALFRED NEHRING, ROYAL AGRICULTURAL COLLEGE OF BERLIN.

BERLIN, April 21, 1892.

Mr. C. HART MERRIAM,
United States Department of Agriculture, Washington, D. C.:

HIGHLY ESTEEMED SIR: I have carefully read and considered your elaborate and very interesting letter of the 2d instant, which I received yesterday through Mr. John Brinkerhoff Jackson, secretary of legation of the North American legation in this city, and, in reply, I send you a statement of my views with regard to its contents.

What you say concerning the mode of life, and especially the annual migrations of the fur seal (*Callorhinus ursinus*), whose breeding places are the Pribilof Islands, is so clear and convincing, and harmonizes so perfectly with what has been observed by other reliable scientists, that I fully agree with your deductions. I am, like your-self, of the opinion that the remarkable decrease of fur seals on the rookeries of the Pribilof Islands, which has, of late years, become more and more evident, is to be attributed mainly, or perhaps exclusively, to the unreasonable destruction caused by the sealers who ply their avocation in the open sea. The only rational method of taking the fur seal, and the only one that is not likely to result in the extermi-nation of this valuable animal, is the one which has hitherto been employed on the Pribilof Islands under the supervision of the Government. Any other method of taking the northern fur seal should, in my opinion, be prohibited by international agreement. I should, at furthest, approve a local pursuit of the fur seal, where it is destructive of the fisheries in its southern winter quarters. I regard pelagic fur sealing as very unwise; it must soon lead to a decrease, bordering on extermination of the fur seal.

With great respect,

Prof. Dr. ALFRED NEHRING,
Professor of Zoology in the Royal Agricultural College of Berlin.

REPLY OF PROF. COUNT TOMMASO SALVADORI.

ZOOLOGICAL MUSEUM, *Turin, April 25, 1892.*

C. HART MERRIAM,
United States Department of Agriculture,
Division of Ornithology, Washington, D. C.

DEAR SIR: I have received your letter concerning the northern fur seal, on the condition of which you have been selected as naturalist to investigate and report by the Government of the United States.

As a whole I agree with you as to the facts and conclusions drawn on your report, although the increasing number of seal skins actually secured and sold, as a result of pelagic sealing shown in your table, does not sufficiently prove, in my mind, that we are already in the period of a decided diminution of the number of living seals. Still, I quite admit that it is absolutely necessary to adopt some measures for the preservation of the seal herds.

No doubt the free pelagic sealing is a cause which will act to the destruction of the seal herds, and to that it must be put a stop as soon as possible. But at the same time I think that the yearly killing of about 100,000 young males on the Pribilof Islands must have some influence on the diminution of the herds, especially pre-venting the natural or sexual selection of the stronger males, which would follow if the young males were not killed in such a great number. So that, with the stop-ping of the pelagic sealing, I think, at least for a few years, also the slaughter of so many young males in the Pribilof Islands should be prohibited.

I remain, very truly, yours,

PROF. T. SALVADORI.

REPLY OF DR. G. HARTLAUB.

BREMEN, *April 23, 1892.*

Mr. C. HART MERRIAM.

DEAR SIR: Your excellent report on the northern fur seal I have read and reread with intense interst.

I am far from attributing to myself a competent judgment regarding this matter, but considering all facts which you have so clearly and convincingly combined and expressed, it seems to me that the measures you propose in order to prohibit the threatening decay of the northern fur seal are the only correct ones promising an effective result.

I sincerely regret that for practical reasons it can not be thought of to prohibit fur-seal hunting for a few years entirely, as this would naturally assist numerically the menaced animal.

There is at any rate danger in view, and it can not be too strongly emphasized that your so well-founded proposals should be executed at the earliest time possible.

With sincere thanks for the confidence you have placed in my judgment, I am, dear sir, your most obedient,

Dr. G. HARTLAUB.

REPLY OF PROF. ROBERT COLLETT, OF THE ZOOLOGICAL MUSEUM OF THE UNIVERSITY OF CHRISTIANIA, NORWAY.

CHRISTIANIA, *April 22, 1892.*

MY DEAR SIR: It would be a very easy reply to your highly interesting treatise of the fur seal, which you have been kind enough to send us, when I only answered you that I agree with you entirely in all points. No doubt it would be the greatest value for the rookeries on the Pribilof Islands, as well as for the preservation of the existence of the seal, if it would be possible to stop the sealing at sea at all. But that will no doubt be very difficult when so many nations partake in the sealing, and how that is to go about I can not know. My own countrymen are killing every year many thousands of seals, *Cysto phorœ,* on the ice barrier between Spitzbergen and Greenland, but never females with young; either are the old ones caught or—and that is the greatest number—the young seals. But there is a close time, accepted by the different nations, just to prohibit the killing of the females with young. Perhaps a similar close time could be accepted in the Bering Sea, but that is a question about which I can not have any opinion.

Many thanks for the paper.

Yours, very truly,

R. COLLETT.

REPLY OF LEOPOLD VON SCHRENCK.

ST. PETERSBURG, *April 13/25, 1892.*

DEAR SIR: Having read with eager and critical attention the memoir you have addressed to me upon the condition of the fur-seal rookeries on the Pribilof Islands in Bering Sea, the causes of decrease and the measures necessary for the restoration and permanent preservation of the seal herd, I can not but completely agree with you in considering the conclusions and recommendations you arrived at quite justified and necessitated by the facts. I am also persuaded that the pelagic sealing, if pursued in the same manner in future, will necessarily end with the extermination of the fur seal.

Very truly, yours,

LEOPOLD VON SCHRENCK,
Member of the Imperial Academy of Sciences, St. Petersburg.

REPLY OF DR. HENRY H. GIGLIOLI.

FIRENZE, 19 VIA ROMANA, *May 2, 1892.*

DEAR SIR: Years ago, in November, 1867, I had the good fortune to be able to visit an extensive rookery of one of the South Pacific eared seals, the well-known *Otaria jubata.* It was during my voyage round the world on the *Magenta.* The rookery in question lies just behind Cape Stokes in the Gulf of Penas, on the southern coast of Chile, and is the one seen by Darwin during his memorable voyage in the *Beagle.* I shall never forget that day, when my astonished gaze rested on hundreds of these eared seals lying about in every attitude of repose on the beach and rocks of the shore, or gracefully, and without showing the slightest fear, performing the most acrobatic evolutions in the water round our boat. That day I had my first experience of these singular creatures, and from that day dates the special interest I have ever since taken in the study of the life history of the *Otariidæ*, which is one of the most marvelous in zoology.

In the spring of 1880, while commissioner for Italy at the grand "Fischerei-Ausstellung" held at Berlin, I first had occasion to admire, in the United States exhibit, the beautiful and spirited drawings of Henry W. Elliott. I have since then taken a keen interest in the wonderful life history of the North Pacific fur seal (*Callorhinus ursinus*), as best exemplified on the Pribilof Islands. Later on I have carefully read and commented on the various accounts which have appeared in print on the subject; thus, in J. A. Allen's North American Pinnipeds, Washington, 1880 (p. 312 et seq.), but more especially the detailed and graphic descriptions which have been published by Henry W. Elliott in his masterly monograph, The Seal Islands of Alaska, in that grand work by G. Brown Goode and associates, The Fisheries and Fishery Industries of the United States (vol. 1, p. 75 et seq.), Washington, 1884, and again in his most interesting volume, An Arctic Province, Alaska and the Seal Islands, London, 1886.

After these precedents you can easily imagine how great an interest I take in that "vexata quæstio," the fur-seal fishery in the Bering Sea; with what pleasure I received through the United States Government and Mr. Long, the United States consul in this city, your communication, and how glad I am of the opportunity thus afforded me of giving my unbiased opinion in the case and aiding you in your noble effort to preserve from utter destruction one of the most interesting of living creatures, and to save at the same time a most valuable source of human industry and profit.

I have read with great attention your condensed but very complete statement of the salient points regarding the life history of the North Pacific fur seal (*Callorhinus ursinus*). I have carefully considered the results of your investigation upon the condition of the fur-seal rookeries on the Pribilof Islands, your conclusions regarding the cause of their decrease, and the measures you suggest as necessary for the restoration and permanent preservation of the seal herd; and I am happy to state that I entirely agree with you on all points.

The first and most important point for consideration is evidently the cause of the unquestionable decrease ascertained in the fur-seal rookeries on the Pribilof Islands during the few past years. The stringently enforced rules which strictly limit the killing for commercial purposes to nonbreeding males or holluschickies, carefully selected, which selection can only be made on land, entirely preclude to my mind the suggestion that the lamented decrease may be attributed in any degree to the killing of too large a number of nonbreeding males. Such a decrease might have been in some slight measure attributed to the former custom of killing each year a certain number of male pups to furnish food for the natives, but that practice has been wisely prohibited. Therefore, I feel positive that the notable decrease in the number of fur seals resorting to the rookeries on the Kurile Islands, on the Robben Reef (Saghalien), and more especially on the Commander Islands, as being in the Bering Sea.

Having conclusively shown that the lamented decrease in the herd of fur seals resorting to the Pribilof Islands can in no way be accounted for by the selective killing of nonbreeding males for commercial purposes, which takes place on those islands under special rules and active surveillance, we must look elsewhere for its cause, and I can see it nowhere but in the indiscriminate slaughter, principally practiced on breeding or pregnant females, as most clearly shown in your condensed report, by pelagic sealers.

In any case, all who are competent in the matter will admit that no method of capture could be more uselessly destructive in the case of pinnipedia than that called "pelagic sealing;" not only any kind of selection of the victims is impossible, but it is admitting much to assert that out of three destroyed one is secured and utilized, and this for obvious and well-known reasons. In the case of the North Pacific fur seal, this mode of capture and destruction falls nearly exclusively on those—the nursing or pregnant females—which ought on no account to be killed. It is greatly to be deplored that any civilized nation possessing fishery laws and regulations should allow such indiscriminate waste and destruction. The statistical data you give are painfully eloquent, and when we come to the conclusion that the 62,500 skins secured by pelagic sealing in 1891 represent at a minimum one-sixth of the fur seals destroyed, viz, 375,000—that is, calculating one in three secured and each of the three suckling a pup or big with young—we most undoubtedly need not look elsewhere to account for the rapid decrease in the rookeries on the Pribilof Islands; and I quite agree with you in maintaining that, unless the malpractice of pelagic sealing be prevented or greatly checked, both in the North Pacific and in the Bering Sea, the economic extermination of *Callorhinus ursinus* is merely the matter of a few years.

International legislation ought to intervene, and without delay, in this case and suggest the means of possibly preventing or at least considerably limiting the pelagic capture and killing of the northern fur seal—a destructive and ultimately fatal industry, which forcibly recalls the well-known fable of the peasant who killed the hen which laid the golden eggs. The industry derived from the rational killing of fur seals, as practiced on the Pribilof Islands, has an economic value which extends far beyond the limits, though vast, of the United States; and it must be remembered that the commercial extermination of the fur seal must also put an end to those industries which are connected with the preparation of the much valued seal-skin fur.

It is both as a naturalist and as an old commissioner of fisheries that I beg to say once more that I most entirely and most emphatically agree with you in the conclusions and recommendations you come to in your report on the present condition of the fur-seal industry in the Bering Sea, with special reference to the causes of decrease and the measures necessary for the restoration and permanent preservation of that industry, which conclusions and recommendations are fully supported and justified by the facts in the case.

With much regard, believe me, dear sir, very truly, yours,

HENRY H. GIGLIOLI.

REPLY OF DR. RAPHAEL BLANCHARD.

Dr. C. HART MERRIAM,
    Bureau of Animal Industry,
        Department of Agriculture, Washington, D. C.

SIR AND HONORED COLLEAGUE: I have read with the deepest interest the learned memoir which you have done me the honor to send me concerning the biological history of the fur seal (Callorhinus ursinus).

The very precise observations which you made at the Pribilof Islands and the no less certain information based on official statistics which you give on the subject of the capture of the females on the high seas at the moment when they are returning to the Pribilof Islands to give birth to their young, have suggested to you conclusions with which I fully agree.

I will go even further than you, for I think it urgent not only to rigidly prohibit the taking of the migratory Callorhinus in the open sea, but also to regulate and limit severely the hunting on land of males still too young to have a harem.

According to your own observation the male does not pair off before the age of 6 or 7 years, and the females give birth to only one pup at a time. It can be said then that the species increases slowly and multiplies with difficulty. These are unfavorable conditions, which do not allow it to repair the hecatombs which for several years past have been and are decimating the species.

By reason of the massacres of which it is the victim this species is advancing rapidly toward its total and final destruction, following the fatal road on which the Rhytina stelleri, the Monachus tropicalis, and the Macrorhinus angustirostris have preceded it, to cite only the great mammifers which but recently abounded in the American seas.

Now, the irremediable destruction of an eminently useful animal species, such as this one, is, to speak plainly, a crime of which we are rendering ourselves guilty toward our descendants. To satisfy our instincts of cupidity we voluntarily exhaust, and that forever, a source of wealth which, properly regulated, ought on the contrary to contribute to the prosperity of our own generation and of those which will succeed it.

When we live on our capital we can undoubtedly lead a gay and extravagant life; but how long does this foolish extravagance last? And what is its to-morrow? Inextricable poverty. On the other hand, in causing our capital to be properly productive we draw from it constantly a splendid income, which does not, perhaps, give the large means dreamed of, but at least assures an honorable competency, to which the wise man knows how to accommodate himself. By prudent ventures or by a well-regulated economy he can even increase progressively his inheritance and leave to his children a greater fortune than he had himself received from his parents. It is evidently the same with the question which occupies us, and it is for our generation an imperious duty to prevent the destruction of the fur seal, to regulate strictly its capture—in a word, to perpetuate this source of wealth and to bequeath it to our descendants.

To these considerations of an economic character I will add another of a nature purely sentimental. It is not without profound sadness that the naturalist sees a large number of animal species disappear, the destruction of which this century will have seen accomplished. When our seas are no longer inhabited by the cetacea and the great pinnipeds, when the air is no longer furrowed in all directions by little insectivorous birds, who knows if the equilibrium of nature will not be broken—an equilibrium to which the creatures on the way to extinction have greatly contributed?

With his harpoons, his firearms, and his machines of every kind, man, with whom the instinct of destruction attains its highest point, is the most cruel enemy of nature and of mankind itself.

Happily, while yet in time, the savants sound the alarm. In this century, when we believe in science, we must hope that their voice will not be lost in the desert.

Above all, I have the conviction that the very wise measures which you propose with the view of preserving the *Callorhinus ursinus* from an impending destruction will be submitted to an international commission, which will ratify them and give them the force of a law.

Will you accept, sir and honored colleague, the expression of my most distinguished sentiments.

Dr. RAPHAEL BLANCHARD.

REPLY OF DR. WILHELM LILLJEBORG.

STOCKHOLM, *May 14, 1892.*

Dr. C. HART MERRIAM.

DEAR SIR: In answer to your letter of 2d of April, asking our opinion as to the causes of the decrease of the stock of northern fur seals (*Callorhinus ursinus*) on the rookeries of the islands in the North Pacific or Bering Sea, and concerning the means proposed by you to arrest this decrease, allow us to state the following:

Your description of the life of the northern fur seal corresponds generally with similar descriptions by former authors, from the celebrated Dr. Steller, who (1741–42) visited the Commander Islands with Vitus Bering, to our days, and also with our own personal experiences of the animal life in the Arctic seas, and with the informations one of us gathered from the inhabitants during a short stay in the Bering Sea.

We do not, therefore, hesitate to declare that the facts about the life and habits of the fur seal stated by you in your said letter under 1–20 should serve as a base for the regulations necessary to preserve this gregarious animal from its threatened extinction in a comparatively short time.

These regulations may be divided into two categories, viz: (1) Regulations for the killing, etc., of the fur seals on the rookeries, in order to prevent the gradual diminution of the stock; (2) regulations for the pelagic sealing, or for the hunting of the seals swimming in the ocean in large herds to and from the rookeries, or around the rookeries during the time when the females are suckling the pups on land.

As to the former question, the killing of the seals on the rookeries, it seems at present regulated in a suitable manner to effectually prevent the gradual diminution of the stock. If a wider experience should require some modifications in these regulations, there is no danger but that such modifications will be adopted. It is evidently in the interest of the owners of the rookeries to take care that this source of wealth should not be lessened by excessive exploitation. Nor will there be any difficulty for studying the conditions for health and thriving of the animals during the rookery season.

As to the pelagic sealing, it is evident that a systematic hunting of the seals in the open sea on the way to and from or around the rookeries will very soon cause the complete extinction of this valuable and, from a scientific point of view, extremely interesting and important animal, especially as a great number of the animals killed in this manner are pregnant cows, or cows temporarily separated from their pups while seeking food in the vicinity of the rookery. Everyone having some experience in seal hunting can also attest that only a relatively small part of the seals killed or seriously wounded in the open sea can in this manner be caught. We are therefore persuaded that a prohibition of pelagic sealing is a necessary condition for the prevention of the total extermination of the fur seal.

Very truly, yours,

W. LILLJEBORG.

REPLY OF DR. A. V. MIDDENDORFF.

Mr. W. WURTS.

SIR: My delay in answering your letter is due to illness. I am very glad that the United States have selected so competent a person as Dr. Merriam for the purpose of ascertaining the causes of the rapid decrease of seals. The facts of the case have now been scientifically explained, so that they may be readily understood even by an unscientific person. The method of treating these animals which was originally adopted by the Russian American Company at their home on the Pribilof Islands is still continued in the same rational manner, and has, for more than half a century been found to be excellent, both on account of the large number of seals taken, and because they are not exterminated. So long as superfluous young males only are killed, not only the existence, but even the increase of the herd is assured.

Seals are, unfortunately, migratory animals, and set out on their journey during the winter months. This is especially true of the pregnant females. They are then hunted with constantly increasing rapacity, and are killed in the open sea by free-booters from all parts of the world. It is evident that the only remedy for such a state of things can be afforded by international protection.

How rapidly extermination progresses is shown by the disappearances of millions of bisons. With these, however, the case is quite different, since their destruction is of no importance in an economical point of view. Its importance is merely of an esthetical character, and from this standpoint only does modern civilization demand the preservation of two specimen herds, numbering a few hundred head each—one in Lithuania and the other in North America. Since the attempts to domesticate the bison, and to produce a cross between it and our domestic cattle have proved a failure, it is plain that the ground where the bison formerly grazed can be more advantageously occupied and yield milk abundantly.

The case is quite otherwise with the seal. This animal is of economical impor-tance, and was created for a domestic animal, as I pointed out many years ago. (See my Siberian Journey, vol. iv, part 1, p. 846.) It is in fact the most useful of all domestic animals, since it requires no care and no expense, and consequently yields the largest net profit. If we suppose the seal to have disappeared, what could take its place as converter of the immense supply of fish in the ocean into choice furs to stock the markets of the world?

Bering Island, which has been deserted for one hundred and fifty years, now stands as a warning. Has modern progress succeeded in any way in supplying the place of the seal cow (*Rhytina stelleri*), that huge monster which, as a consumer of marine plants, was intended to convert useless sea weed into savory meat?

If you will communicate (as you say you propose to do) the contents of this letter to Dr. Merriam, whose address I do not know, you will oblige me greatly.

I have the honor to be, sir, your most obedient servant,

Dr. A. V. MIDDENDORFF.

REPLY OF DR. EMIL HOLUB.

PRAGUE, *May 18, 1892.*

Dr. C. H. MERRIAM, Esq.

DEAR SIR: With sincere attention I have perused the records of your investigation of the habits, the present decrease, and regarding the future of the fur seal (*Callor-hinus ursinus* Gray). Having well considered the matter, I will pass my opinion without any prejudice whatever.

The Government of the United States may be congratulated upon the action taken in having sent out for the investigation of a matter which falls into the department of the board of trade a scientist, and in this special case a man who has taken such great pains with the object of his researches.

Our age makes it a duty for all civilized nations to bring trade and commerce in a close contact with science. This becomes quite a necessity, like in the present case, in which commercial customs, even international agreements, laws, etc., become insufficient to secure a sound decision. Such scientific investigations can supply the desired conclusions; they do advice the measures to be taken, and provide the basis upon which an international understanding can be established.

Regarding the object of your researches, I indorse your opinion that the decrease of the numbers of the fur seal on the Pribilof Islands has been caused by pelagic sealing in the North Pacific and in the Bering Sea, and that this taking of the seals at sea has to be stopped as early as possible.

To restore in time the numbers of former years, I take the liberty to name the fol-lowing measures for the sake of consideration:

A. Concerning certain agreements with other powers.

(1) A mutual understanding upon the question between the United States, Russia, and Japan. These three States are concerned primo loco in this matter as being the proprietors of the breeding places as well, like also of the fishing grounds of the said animals during their yearly wanderings to and fro.

(2) For the sake of brevity in action and a speedy settlement, these three States (after having agreed upon the foregoing) to select but one representative.

(3) The United States having given impulse to the matter to gain the prestige, that a United States man shall be selected to this honor.

(4) A congress to be called together, invitations to be sent to those of the Euro-pean and American powers, whose subjects indulge in pelagic sealing in the North Pacific and the Bering Sea.

(5) In the congress the representative of the three powers to have six voices, resulting in two voices for every one of these powers, which concession to be granted upon the facts of paragraph 1.

(6) The congress to deal with the stoppage of pelagic sealing of the fur seal, and possibly to come to an understanding upon it and to enforce it.

B. Concerning certain laws and precautions in the dominions of the United States.

(1) To prohibit taking seals at sea by home vessels and by small boats along the coast during the wanderings of the animals. I think that a great many fur seals are killed on their way to the south and their return to their breeding places in the north before ever they do reach the neighborhood of the latter. The fact that these wandering animals are chiefly pregnant females, which as game are protected by laws among all civilized nations, may grant them safety also along the coasts of British Columbia.

(2) To see that the existing laws at present in use on both St. Paul and St. George islands regarding the protection of male pups are strictly observed.

(3) To investigate the nutritious necessities of the fur seal. I believe that the animals feed, besides on fish and crustaceans, also on different forms of mollusca, especially on mussels, and also on certain seaweeds.

(4) In ascertaining the foregoing, to try to increase the quantity of food in the sea of the Pribilof Islands, especially for the reason that females, when nursing, may be not compelled to stray as far as 100 to 150 miles from shore, deserting their pups for so long and being also exposed to the weapons of the pelagic sealers.

(5) In ascertaining the nutritious necessities to pay special attention to mussels belonging to the families of the *Mytilidæ* and *Ariculidæ* (to the genera of *Mytilus, Modiola, Lithodomis, Pinna*, and others), who have thin shells, or to other species of the North Pacific, which would promise a good prolification; further, also, to certain seaweeds, for submarine plantation, the species to which I allude containing a great deal of eatable gelatinous matter.

(6) These measures, besides to be taken from economical reasons on behalf of home commerce and home trade, to be recommended also from a scientific point of view, as an act of preservation of a sea mammal and from the common laws of humanity, that species of large and wild living mammal may be guarded against utter annihilation.

Mankind, never to forget that, being the master among the living creatures on earth, it has the power of re-creation.

If the pelagic sealing of the fur seal is carried on still longer, like it has been executed during the last years, the pelagic sealing as a business matter and a living will soon cease by the full extermination of the useful animal.

The objections brought forward by the friends of the pelagic sealing against its stoppage, that the latter will ruin a great many families of seamen and fishers, can not be taken as sound arguments. It is a well-known and a common thing in our age, but a weekly occurrence during the last years, that a new trade springing up ruins two other trades, and hardly in one case out of hundreds can a compensation be given or is asked for.

In concluding my note, I thank you, my dear sir, as my esteemed fellow-worker in another transatlantic sphere, for the excellent work which you have executed during your weary investigations in the Bering Sea. May this noble and important work be crowned with the deserved success that that piratic hunt may be stopped forever. The opportunity of the Columbian Exhibition in Chicago might be used to call the congress to Washington, and then to give to the delegates the treat of a visit to the monstrous exhibition.

I should feel very happy if one day to come I can make your personal acquaintance and can shake hands with you, my dear sir.

With my humble respects, I remain, your most obedient,

Dr. EMIL HOLUB.

LETTERS AND STATEMENTS OF NATURALISTS.

STATEMENT BY PROF. T. H. HUXLEY.

The following statement by Prof. T. H. Huxley, F. R. S., etc., the eminent naturalist, was prepared at the request of the counsel for the United States. As appears from the statement itself, it was given by Professor Huxley as a scientist, not as a retained advocate:

(1) The problem of the fur-seal fishery appears to me to be exactly analogous to that which is presented by salmon fisheries. The Pribilof Islands answer to the upper waters of a salmon river; the Bering Sea south of them and the waters of the

Northwest Pacific from California to the Shumagin Islands to the rest of the course of the river, its estuary, and adjacent seacoast. The animals breed in the former and feed in the latter, migrating at regular periods from the one to the other. (The question whether the fur seals have any breeding places on the Northwest Coast outside of Bering Sea may be left open, as there seems to be no doubt that the main body breeds at the Pribilofs.) *

(2) An important difference is that the females, bachelors, and yearling fur seals feed largely within a radius of, say, 50 miles of the Pribilof Islands, while the adult salmon do not feed (sensibly, at any rate) in the upper waters.

(3) It is clear in the case of fur seals, as in that of the salmon, that man is an agent of destruction of very great potency, probably outweighing all others. It would be possible in the case of a salmon river to fish it in such a fashion that every ascending or descending fish should be caught, and the fishery be in this way surely and completely destroyed. All our salmon-fishery legislation is directed toward the end of preserving the breeding grounds on the one hand, on the other of preventing the lower-water fishermen from capturing too large a proportion of the ascending fish.

(4) Our fishery regulations are strict and minute. Every salmon river has its fishery board, composed of representatives of both the upper and the lower water fisheries, whose business it is to make by-laws under the acts of Parliament and to see that they are carried out. A Government inspector of fisheries looks after them, and holds inquiries under the authority of the home secretary in case of disputes. On the whole, the system works well. The fisheries of rivers which have been pretty nearly depopulated have been restored, and the yield of the best is maintained. But the upper-water and lower-water proprietors are everlastingly at war, each vowing that the other is ruining the fisheries, and the inspector has large opportunities of estimating the value of diametrically opposite assertions about matters of fact.

(5) In the case of the fur-seal fisheries the destructive agency of man is prepotent on the Pribilof Islands. It is obvious that the seals might be destroyed and driven away completely in two or three seasons. Moreover, as the number of bachelors in any given season is easily ascertained, it is possible to keep down the take to such a percentage as shall do no harm to the stock. The conditions for efficient regulation are here quite ideal.

(6) But in Bering Sea and on the Northwest coast the case is totally altered. In order to get rid of all complications, let it be supposed that western North America, from Bering Straits to California, is in the possession of one power, and that we have only to consider the questions of the regulations which that power should make and enforce in order to preserve the fur-seal fisheries. Suppose, further, that the authority of that power extended over Bering Sea and over all the Northwest Pacific east of a line drawn from the Shumagin Islands to California.

Under such conditions I should say, looking at nothing but the preservation of the seals, that the best course would be to prohibit the taking of the fur seals anywhere except on the Pribilof Islands, and to limit the take to such percentage as experience proved to be consistent with the preservation of a good, average stock. The furs would be in the best order, the waste of life would be least, and, if the system were honestly worked, there could be no danger of overfishing.

(7) However, since northwest America does not belong to one power, and since international law does not acknowledge Bering Sea to be a mare clausum, nor recognize the jurisdiction of a Riparian power beyond the 3-mile limit, it is quite clear that this ideal arrangement is impracticable.

The cause of the fur-seal fisheries is, in fact, even more difficult than that of the salmon fisheries in such a river as the Rhine, where the upper waters belong to one power and the lower to another.

(8) The Northwest Pacific, from California to Shumagin at any rate, is open to all the world, and, according to the evidence, the seals keep mainly outside the 3-mile limit. A convention between Great Britain and the United States (backed by a number of active cruisers) might restrain the subjects of both. But what about ships under another flag?

(9) Moreover, I do not see how the Canadians could be reasonably expected to give up their fishery for the sake of preserving the Pribilof fisheries, in which they have no interest.

(10) If, however, it is admitted that the Canadians can not be asked to give up their fisheries, I see no way out of the difficulty except one, and I do not know that it is practicable. It is that the Pribilof, Bering, and Northwest coast fur-seal fisheries shall be considered national property on the part of the United States and Great Britain, to be worked by a joint fishery commission, which shall have power to make by-laws under the terms of a general treaty, to which I suppose other powers (who have hardly any interest in the matter) could be got to agree.

(11) I am free to confess that my experience of the proceedings of fishery boards does not encourage me to hope that the proceedings of such a commission would be

altogether harmonious; but if it were composed of sensible men they would, sooner or later, struggle out into a modus vivendi, for, after all, it is as much the Canadian interest that the Pribilof fisheries should be preserved as it is the United States interest that the seals should not be extirpated in Bering Sea and the Northwest Pacific.

(12) In such a case as this I do not believe that the enforcement of a close time, either in Bering Sea or on the Northwest coast, would be of any practical utility unless the fishing is absolutely prohibited (which I take to be out of the question). It must be permitted while the seals are in the sea; and if it is permitted, there is no limit to the destruction which may be effected.

Numerous as the seals may be, they are a trifle compared with herring schools and cod walls, and human agency is relatively a far more important factor in destruction in their case than in that of herrings and cod. Up to this time fishing has made no sensible impression on the great herring and cod fisheries; but it has been easy to extirpate seal fisheries.

(13) Finally, I venture to remark that there are only two alternative courses worth pursuing.

One is to let the fur seals be extirpated. Mankind will not suffer much if the ladies are obliged to do without seal-skin jackets, and the fraction of the English, Canadian, and American population which lives on the seal-skin industry will be no worse off than the vastly greater multitude who have had to suffer for the vagaries of fashion times out of number. Certainly, if the seals are to be a source of constant bickering between two nations, the sooner they are abolished the better.

The other course is to tread down all merely personal and trade interest in pursuit of an arrangement that will work and be fair all round, and to sink all the stupidities of national vanity and political self-seeking along with them.

There is a great deal too much of all these undeniable elements apparent in the documents which I have been studying.

T. H. Huxley.

April 25, 1892.

AFFIDAVIT BY DR. PHILIP LUTLEY SCLATER.

Philip Lutley Sclater, Ph. D., secretary of the Zoological Society of London, being duly sworn, doth depose and say that in his opinion as a naturalist:

(1) Unless proper measures are taken to restrict the indiscriminate capture of the fur seal in the North Pacific he is of opinion that the extermination of this species will take place in a few years, as it has already done in the case of other species of the same group in other parts of the world.

(2) It seems to him that the proper way of proceeding would be to stop the killing of females and young of the fur seal altogether or as far as possible, and to restrict the killing of the males to a certain number in each year.

(3) The only way he can imagine by which these rules could be carried out is by the killing the seals only in the islands at the breeding time (at which time it appears that the young males keep apart from the females and old males) and by preventing altogether, as far as possible, the destruction of the fur seals at all other times and in other places.

Philip Lutley Sclater, Ph. D., F. R. S.

City of Washington, *District of Columbia, ss:*

C. H. Townsend, being duly sworn, deposes and says:

I am 33 years of age, and my profession is that of a naturalist. I am attached to the United States Fish Commission steamer *Albatross*, with which Commission I have been connected for nine years. Occupying the position of resident naturalist on that vessel, as I did, I have collected constantly during this period and have hunted with all kinds of firearms and under various conditions. I have made seven voyages to Alaska.

I visited the Pribilof Islands for the first time in 1885, spending the months of June and September thereon in making collections of natural-history specimens, including those of the fur seal, of which I brought down twenty. In the year 1891 I again visited the Island of St. Paul, arriving there July 28 and remaining there about ten days. The British commissioners were on the island at that time. I made frequent observations as to the conditions of the rookeries during this period. Early in the summer of 1892 I visited, at the request of the United States Government, Guadalope Island, for the purpose of acquainting myself with seal life there and of obtaining skulls of the fur seals which formerly frequented those regions. Later in 1892 I once more visited the Island of St. Paul, arriving there June 30. I was there on the island and on the United States Revenue steamer *Corwin*, cruising to the west of the islands, continuously until about August 15, and was engaged during all of this time in the study of seal life, either on land or in the waters of Bering Sea, and have shot seals from a small boat.

I carefully noted the fact this year that the young seal is at birth attached to a large placenta, equal parts to one-third of its weight and of a bright red color. It is sometimes not expelled until an hour or so after birth, remaining attached in the meanwhile by the umbilical cord to the pup. It frequently remains attached to the pup a day or more. After parturition the female takes an immediate interest in her young, and if it has fallen into some slight rock crevice she gently draws it toward her, taking its nape in her teeth. She repeatedly turns to it with manifestation of affection.

Prior to July 27, 1892, many of the females had taken to water to feed and could thereafter be seen returning at all times to suckle their young. I quote the following written memorandum made by me on St. Paul at that date: "Bulls on rookeries getting exhausted and quiet, mostly sleeping. Cows largely at sea. Some bulls have hauled out on sand beaches that so far have been bare. Four-fifths of the seals on rookeries to-day are pups."

July 28 I made the following note: "Many females coming from the water bleating for their young."

I have killed sea lions at the following localities, where they breed in considerable numbers, and found their breeding ground impregnated with the same rank, disagreeable smell that is so noticeable a feature of the breeding grounds of the Pribilof fur seal: Light-house Rock, Alaska Peninsula, Farallon Islands, and Monterey Rock, California; San Benito Islands, Lower California, and San Luis Islands, in the Gulf of California. The soil and rocks at these places is as foul with seal excrement as at the Pribilofs, where urine, excrement, decaying placentas, and other filth rubbed and trodden into the soil and rock depressions cause the odors so characteristic of this vicinity. The rocks at Monterey may be used in illustration: They lie near Cypress Point, 400 or 500 yards off the shore, which the carriage drive follows, and are covered with hair seals, which breed there. They are conspicuously stained with excrement, and where the animals lie thickest the ground is smeared and slippery with it. I collected sea lions there in January of the present year, and after my shooting had frightened all the animals off to the sea the rank smell of the place itself drifted across the channel into the nostrils of the tourists of the Hotel del Monte, who witnessed our operations. It would indeed be an extraordinary occurrence if fur seals did not deposit excrement upon their breeding grounds in the same way that all other animals of this class do.

As already stated above, I was attached to the steamer *Corwin* during the past summer, and I made all the examinations of the stomachs of the seals referred to in Captain Hooper's report, covering, in all, 33 seals. I annex hereto photographs of two of the seals which were dissected and examined by me on the deck of the steamer *Corwin*. These seals were taken on the 2d day of August, 1892, at a distance of about 175 miles from the islands. The photographs exhibit the mammary glands and convey a good idea of the considerable size of these glands, which in all cases were filled with milk. The inference is unavoidable that the pup is a voracious feeder, and this inference is in keeping with the observations I have made on the rookeries, where I have repeatedly seen pups suckle for half an hour at a time. The mammary gland is very widely spread over the lower surface of the animal; beginning between the fore flippers, in fact at the anterior of the sternum, it extends well up under the armpits and back to the pubic bones. The milk glands are quite thick and completely charged with milk. The photographs, especially the first one, exhibit the milk streaming from the glands on to the deck.

Annexed to the report of Captain Hooper is a table giving the results of the examination of 41 seals which were killed in Bering Sea in 1892. It appears that of this number 22 were nursing seals. The photographs hereto annexed show exactly the way all of these nursing female seals looked when cut open on the deck of the *Corwin*.

From the fact that among the females thus taken and examined there were found mostly nursing cows, with a small number of virgin cows, it is reasonable to conclude that there are practically no barren females swimming about in the sea unattached to the islands, or that, at any rate, if such seals exist they are rarely, if ever, taken. In all my experience I have never seen anything to lead me to the conclusion that there is such a thing as a barren female. In the case of the virgin cows, a careful examination of the uterus proved them to be too immature for conception.

In the stomachs of many of the seals examined as above stated there were found large quantities of fish, mainly codfish. There is nothing surprising in this fact, that codfish should be found in the stomachs of surface feeders such as seals are. While taken at the bottom, the codfish is not restricted to deep water. It is found from the shallows along the shore out to the banks where fishermen usually take them. They are often taken at intermediate depths, but fish taken at the bottom are, as a rule, larger.

The cod is a voracious feeder upon squid, which abound at the surface. In Alaskan waters I have taken hundreds with the dip net, after attracting them with the electric light of the *Albatross*. In its frequent migrations from bank to bank the cod passes over tracks of ocean where the water is of profound depth. It is a regular feeder upon herring and many other fishes which school at the surface, and in Alaskan waters frequently follows the fisherman's bate from the bottom to the surface.

As a result of my combined observations upon land and water, as hereinbefore detailed, I have no hesitation in stating positively that soon after a female gives birth to her young she leaves the island in quest of food, that she travels great distances in search of it, and that she returns to the islands heavily laden with milk.

While hunting in the *Corwin's* boat many seals were fired upon when asleep. They usually sleep with their head to leeward and keep it moving uneasily from side to side, but with the nose held clear of the water. A sleeping seal has his vital parts pretty well submerged—the nose, lower jaw, and flippers being usually held above the surface, although a little more appears at times according to the condition of the sea and the movements of the animal.

One has to be very close to get a shot at the head that will kill it. Many times the animal is wounded sufficiently to get out of reach of the hunter before it dies. I had very little difficulty in approaching sleeping seals close enough for a fair shot, but much in killing them. Fair shots that scattered the charge all about them, hitting the flippers, I firmly believe, and in some cases drawing plenty of blood, were usually without result, until I learned to fire directly at the head. Then the shots began to prove fatal; but even then, unless hit in a vital part, the animals got away, though bleeding freely. At first I blamed the ineffectual firing on the cartridges, but the cartridges proved all right as soon as I learned to aim at the head and not at the animal as a whole.

I learned after some experiments that seals which dashed away apparently uninjured were usually hurt, and after following them persistently, at great labor to the boat pullers, found that they were bleeding.

I believe that the majority of sleeping seals fired at are struck. The number killed at the islands with buckshot in them bears out this claim to a considerable extent. I do not see how an ordinary marksman can shoot at so large a target as a seal at short range with a double-barrel gun loaded with 21 buckshot without striking some of the exposed portions of the animal.

It is from the instantly killed that seals are secured; the wounded animal uses its death struggle to get out of reach. What proportion of the seals reaching the Pribilofs with shot in them bear to those which are fired at and escape (wounded, as I state above) is not known, but I believe that fully as many perish leaving no trace, as recover sufficiently to reach the islands.

Feeding seals shot when raising their heads about the boats from curiosity are more likely to be killed instantly than sleeping seals, but they sink more quickly. A clear shot at the head is afforded which knocks the life completely out of them, and the rest of the body being under water at the time it would seem that the pressure upon the limp body forces the air from it. As a rule, seals killed instantly, when the head is entirely clear of the water, go down quickly, sinking stern foremost. Sleeping seals killed when the head is low in the water float for a time, the head settling into the water first, the air is retained in the body and it floats. I shot a seal off Quadeloupe Island in May when it raised its head close to the boat, killing it instantly. It sank before we could reach it with the gaff, and continued sinking, stern first, as we could plainly see far below in the clear water.

Another illustration of the wastefulness of the pelagic sealing might be found in the number of cartridges expended. During the work of the *Corwin* no record of this kind was kept. The hunter usually carried two or three dozen cartridges, which were, as a rule, expended before they returned to the ship. The number of seals lost by sinking, number wounded, and number secured were recorded.

Repeated firing from the boats was often heard on board ship, and a large number of empty shells would be returned, when comparatively few seals were definitely reported as secured, lost, or wounded, all other shots being supposed to be misses. I do not think this feature has received proper consideration. The hunters were certainly average marksmen, and it is my belief that the great majority of the sleeping seals fired at were struck. The guns used were 10-bore Parkers, loaded with 21 buckshot. Time after time I have seen the heavy charge strike about the sleeping seal fully expecting to see it killed, when, to my utter surprise, it would dive and come up beyond our reach. It is incredible that the great number of seals thus escaping were uninjured. How can one always find traces of blood or signs of injury when the frightened animal is retreating at a rate so rapid that it is soon out of sight, and especially as its course is mainly under water and it only appears at the surface with a porpoise-like leap to catch its breath and then dives again?

C. H. TOWNSEND.

The foregoing testimony is that of scientists whose knowledge of the subject under discussion can not well be questioned. Speaking for myself, personally, I am pleased to find my own conclusions (based on a practical knowledge solely) so fully indorsed by learned and disinterested men.

In addition to the testimony already quoted, however, and in order to strengthen the position taken, I append to my report the testimony of statesmen, jurists, scientists, naturalists, shipmasters, sealers, seal hunters, pelagic sealers, naval officers (American and British), merchantmen, seamen, Indian hunters, native sealers, Treasury agents, company agents, British and American Bering Sea Commissioners, fur traders, furriers, fur experts, customs officers, and men of all classes, native and foreign, friends and enemies, who have had either the practical experience, the general information, or the scientific knowledge to warrant them in making sworn statements on the subject at issue; and a careful reading of the testimony introduced will show that their views in general are in accord with mine, and sustain my position in every particular.

The quotations above referred to are taken from the American case and counter case.

RETROSPECTIVE AND EXPLANATORY.

So much has already been said in contradiction of the theories advanced by honest but mistaken men about overdriving of the young males and its consequent result of impotency, of stampedes on the rookeries, and epidemics in the herd, by which so many pups were supposed to be destroyed annually during the past decade, it is necessary for a correct understanding of the contention that I go back a few years and give a sketch of the causes which gave rise to such, until then, unheard of theories which have been the direct cause of more than one-half the troubles growing out of the fur-seal question in Bering Sea.

As already shown by the testimony of Messrs. H. H. McIntyre, T. F. Morgan, Daniel Webster, J. C. Redpath, Dr. Noyes, and others who were on the seal islands for many years, it was not until 1886 the first unmistakable decrease of the seal herd was apparent. Had the facts been reported immediately to the Department and the true cause of such a sudden shrinkage shown, steps might have been taken which would have prevented further pelagic sealing, or at least an addition to the sealing fleet; but unfortunately an overzealous Treasury agent reported an increase of nearly 2,000,000 since Elliott's measurements and estimates, some fourteen years earlier; and again, in 1888, he tells the Department:

I am happy to be able to report that although late landing the breeding rookeries are filled out to the lines of measurement heretofore made, and some of them much beyond those lines, showing conclusively that seal life is not being depleted, but is fully up to the estimates given in my report of 1887. (Report of G. R. Tingle, 1888.)

When that report was written, and before it was written, everyone on the seal islands knew there were indications of a decrease of the seal herd, and the employees of the lessees so reported at the time to the superintendent, Dr. H. H. McIntyre, who tells us:

I repeatedly pointed out to our company and to the special Treasury agents during the seasons of 1887, 1888, and 1889 that the seals were rapidly diminishing, and that in order to get the full quota allowed by law we were obliged to kill, in increasing numbers in each of those years, animals that should have been allowed to attain greater size; and, finally, the catch of 1889 was mostly of this class. (See McIntyre to Jeffries, December 15, 1890, Appendix.)

Mr. Daniel Webster, the oldest and one of the most reliable and practical of sealers, tells, under oath:

In 1884 and 1885 I noticed a decrease, and it became so marked in 1886 that everyone on the islands saw it. This marked decrease in 1886 showed itself on all the rookeries on both islands. (See affidavit in Appendix.)

And Mr. J. C. Redpath, the local agent for the lessees, after an experience of twenty years on the islands, says:

As the schooners (pelagic hunters) increased, the seals decreased, and the lines of contraction on the rookeries were noticed to draw nearer and nearer to the beach, and the killable seals became fewer in numbers and harder to find. In 1886 the decrease was so plain that the natives and all the agents were startled. (Ibid.)

In 1889 the usual annual quota of 100,000 could not be found without taking 50,000 young seals whose skins did not average more than 4 pounds each.

It was then that the apparent and appalling suddenness of the decrease aroused in the minds of those who were neither practical sealers nor had definite knowledge of seal life on the rookeries doubts as to the true cause of the decrease, and of the actual conditions existing on the seal islands so soon after an official report had appeared affirming the fact of an increase of over 2,000,000 seals in fourteen years.

Theories, as numerous as the men who broached them, were launched forth to a still doubting world; from press and platform came an array of argument and statistics as erroneous as they were bewildering; and when the Treasury agent's reports reached the Department it was decided to send an extra special agent to the islands to thoroughly investigate the conditions existing there and if possible to find the cause of the sudden decrease of the fur-seal herd; and Mr. Henry W. Elliott was selected for that important work.

When, in 1890, Mr. Elliott reached the seal islands after an absence of fourteen years, and found only a scant one-fifth of the seals that he saw there in 1876, he impulsively and erroneously concluded that the driving of the young males from the hauling grounds was injurious to their healthy growth and full development; that it produced impotency and destroyed their usefulness as breeders on the rookeries, thus producing a dearth of breeding males and a surplus of barren cows, and, without a shadow of proof to sustain him, he made out a most elaborate report in which he labored to show the truth of his new and wonderful theory, and then felt personally hurt and wronged because the Government refused to indorse or approve it.[1]

Every enemy of the United States in both hemispheres, however, hailed it with delight, and quoted from it against us with much approbation until, after years of patient research and scientific investigation on the part of the United States and of Great Britain, it was demonstrated that Elliott was in error, and that pelagic sealing is the cause of the sudden and rapid destruction of the American fur seal.

In his overanxiety to prove his theory he persistently continues to reiterate the story of a time when no drives were made from a number of places on St. Paul Island where a great "reservoir of surplus male life" was held in reserve; but I will let him tell his own story:

In 1872-1874 when no driving was made from Southwest Point, Zapadnie, and all English Bay to the westward of Neahrpahskie Kammen, from Polavina, or anywhere between it and the hauling grounds of Lukannon, then there were reservoirs of

---

[1] See letter of Secretary of the Treasury, Appendix.

young male life which were not drawn upon or disturbed, from which a steady stream of new male blood for the breeding grounds could and did flow. (Elliott's report (Paris print), 1890, p. 237.)

Again, he says:

Nobody, in 1872, ever thought of such a thing as coming over from the village to make a killing at Zapadnie. (Ibid., p. 246.)

He continues:

I had this point in my thought during my studies in 1872-1874, but at that time no holluschickie were driven from Southwest Point, from Zapadnie, from Tonkee Mees or Stony Point, or from Polavinia—no seals were driven from these places where everybody admitted that full half of the entire number belonging to the islands, laid. (Ibid., 271.)

Then that immense spread of hauling ground covered by swarms of young male seals, at Zapadnie, at Southwest Point, at English Bay, beyond Middle Hill, west, at Polavinia, and over all that 8 long miles of beach and upland hauling ground between Lukannon Bay and Webster's house at Novastoshnah, all of this extensive sealing area was not visited by sealing gangs, or spoken of by them as necessary to be driven from. (Elliott's letter to the Secretary of the Treasury, report of 1890, p. iv.)

In 1872-1874 I observed that all the young male seals needed for the annual quota of 75,000 or 90,000, as it was ordered in the latter year, were easily obtained every season, between the 1st of June and the 20th of July following, from the hauling grounds of Tolstoi, Lukannon, and Zoltoi Sands—from these hauling grounds adjacent to the rookeries or breeding grounds of Tolstoi, Lukannon, Reef, and Garbotch. All of these points of supply being not more than 1½ miles distant from the St. Paul village killing grounds, the Zoltoi drive being less than 600 feet away. (Ibid.)

Therefore, when attentively studying in 1872-1874, the subject of what was the effect of killing annually 100,000 young male seals on these islands (90,000 on St. Paul and 10,000 on St. George), in view of the foregoing statement of fact, I was unable to see how any harm was being done to the regular supply of fresh blood for the breeding rookeries, since those large reservoirs of surplus male life, above named, held at least just half of the young male seal life then belonging to the islands—these large sources of supply were never driven from, never even visited by the sealers, and out of their overwhelming abundance I thought that surely enough fresh male seal life must, did annually mature for service on the breeding rookeries. (Ibid.)

That day in 1879, when it became necessary to send a sealing gang from St. Paul village over to Zapadnie to regularly drive from that hitherto untouched reserve, was the day that danger first appeared in tangible form since 1870—since 1857 for that matter. (Ibid.)

For the good of the public service the truth must be told; and that is that the official records of the drives and killings on the islands of St. Paul and St. George are in direct opposition to Mr. Elliott. They show that, beginning in 1871, there are no records of the daily killings for 1870—drives were made regularly from every hauling ground on the islands; and a close inspection will reveal the fact that an aggregate of 102 drives were made, before 1879, from Zapadnie or Southwest Bay, Polavinia or Halfway Point, and from English Bay, during the very period of which Mr. Elliott speaks when he tells us "they were never driven from, never even visited by the sealers."

For convenience of reference I quote from the official island records the daily drives and killings made between 1870 and 1879 from the three principal rookeries of which Mr. Elliott speaks so positively; and I think it will be sufficient to show every fair-minded man in the country that the large reservoir of "surplus male life" so often spoken of by Mr. Elliott was unknown to everyone else on the seal islands, and never had an existence outside his own fertile imagination.

Here are the drives made each year from 1871 to 1878, both inclusive, from the rookeries in question:

Year.	Zapadnie, or Southwest Bay.	Polavina, or Halfway Point.	English Bay.	Total.
1871	2	1	6	9
1872	1	1	11	13
1873	5		7	12
1874	5		10	15
1875	7	1	10	18
1876	6	1	4	11
1877	4	3	5	12
1878	4	3	5	12
Total	34	10	58	102

(See Senate Ex. Doc. No. 107, Fifty-second Congress, second session, Appendix.)

In an attempt to show that it was not until 1879 that drives were made from certain rookeries which he is pleased to call a "large reservoir of male life," which had not been disturbed or touched before 1879, Mr. Elliott quotes the Island Journal as follows:

Page 92, June 9, 1879: Antone Melovedov started with a gang to make a drive at Halfway Point, Polavina. (Elliott's report (Paris print), 158.)

Page 93, June 10, 1879: The drive to-day (at Polavina) resulted in the taking of 1,118 skins. (H. G. Otis.) (Ibid., 159.)

Page 93, June 11, 1879: The drive from Southwest Bay (Zapadnie) to-day, and 1,462 skins taken. (H. G. Otis.) (Ibid., 159.)

There is not a word in the foregoing, nor is there a word in the journal, to show that the drives mentioned were the first that were made from those rookeries, but Mr. Elliott is determined to show that overdriving is the principal cause of the destruction of the seals, and he continues:

From this day (June 11, 1879) on to the close of that sealing season's work, July 20, Zapadnie was driven often, and Polavina also; but in 1880 only one drive was made from this reservoir at Zapadnie, * * * and, again in 1881, it was not driven from at all, and only one drive that year made from the Polavina reserve. (Elliott's report for 1890, p. 159.)

Here the gentleman is again in error, for I find that drives were made from Zapadnie or Southwest Bay on May 19 and June 7, 1879, only a few days before he discovered that the first drive had been made on the 9th of June.

And in 1880 four drives were made from Zapadnie and five from Polavina, and in 1881 six drives were made from Zapadnie and five from Polavina, as the following table, taken from the island records, will show:

Zapadnie or Southwest Bay:

1880.	Drives.
May 14	1
June 8	1
12	1
16	1
Total	4

1881.	
June 7	1
15	1
28	1
July 6	1
14	1
Dec. 7	1
Total	6

Polavina or Halfway Point:

1880.	Drives.
June 14	1
21	1
28	1
July 5	1
30	1
Total	5

1881.	
June 10	1
17	1
24	1
July 2	1
8	1
Total	5

(See Senate Ex. Doc. No. 107. Fifty-second Congress, second session, Appendix.)

Many inaccuracies are to be found in Mr. Elliott's report of 1890, due, perhaps, to the hurried manner in which it was prepared, and the bitterness, excitement, and many disappointments attending it all the way through; nearly all of which were of a private character, and which can not well be made public, even had I a desire to do so, which I have not by any means. One instance more and I am done.

In his "field notes" on the state of the rookeries in 1890, Mr. Elliott writes:

June 19.—Not a single holluschak of any age whatsoever on Zoltoi Sands this day, and there has not been a killable seal thus far there this season. (Elliott's report, 1890 (Paris print), pp. 263–264.)

June 22.—Fine weather for seals to haul in continues, but the seals do not haul; not a single seal on Zoltoi Sands this morning; has not been a holluschak there yet. (Ibid., p. 264.)

June 22.—Now, not a single young male seal has hauled on Zoltoi thus far this season. (June 22, 6 a. m.) (Ibid., p. 265.)

June 22.—Not a seal on Zoltoi Sands this morning, and not one since during the day. (Ibid., p. 266.)

June 22.—Not a holluschak or any other class of fur seal on Zoltoi Sands this morning or noon. (Ibid., p. 274.)

June 30.—Not a holluschak on Zoltoi Sands to-day. (Ibid., p. 276.)

July 8.—Also, not a holluschak has as yet hauled upon Zoltoi Sands. (Ibid., p. 284.)

July 19.—I observe that not a single young male is on Zoltoi Sands this morning— not one has hauled there thus far this season. (Ibid., p. 295.)

The official records of the drives and killings made on the seal islands in 1890 are on file in the Treasury Department, and a copy will be found in the appendix to this report. I quote from the records the following drives from Zoltoi in 1890: "May 24, 1 drive; July 19, 1 drive."

According to Mr. Elliott there was not a seal on Zoltoi on the 19th of July; according to the island records a drive was made from Zoltoi on that very same day.

Another error of like importance are the two passages in the same report which read as follows:

The importance of understanding this fact as to the readiness of the holluschickie to haul promptly out on steadily "swept" ground, provided the weather is inviting, is very great, because when not understood, it was deemed necessary, even as late as the season of 1872, to "rest" the hauling grounds near the village (from which all the driving has been made since), and make trips to far-away Polavina and distant Zapadnie, an unnecessary expenditure of human time and a causeless infliction of physical misery upon phocine backs and flippers. (Elliott's report, 1890, p. 122.)

Nobody in 1872 ever thought of such a thing as coming over from the village to make a killing at Zapadnie. (Ibid., p. 246.)

At page 122 Mr. Elliott remembered and acknowledged that drives were made in 1872 from Zapadnie and Polavina, and the records confirm his story.

He might have included 1871, for the records show drives were made from both places in that year also.

At page 246 he seems to have forgotten some of what he had already written, for he gravely tells us: "Nobody in 1872 ever thought of such a thing as coming over from the village to make a killing at Zapadnie."

Enough has been said, I think, for the purpose of showing the public how it happens, sometimes, that matters of small moment in themselves may beget questions so momentous that it requires international arbitration to settle them; and that the report of one overzealous officer and the official report of another, made in anger and bitterness, have cost the United States a whole fur-seal herd, worth, originally, nearly $100,000,000.

So numerous and so palpable were the inaccuracies all through the report that Mr. Foster, the then Secretary of the Treasury, refused to have it published, and subsequently, in a letter to the State Department, gave good reasons for such action.[1]

That the theory of injury of the young males to the extent of impotency by driving on the islands, so forcibly presented by Mr. Elliott, has been denied by naturalists generally and disproved by facts adduced by both the scientific and the practical world, has already been amply demonstrated; that Mr. Elliott himself, in several very able papers subsequently written, has adopted the views of every scientist of note, from our own American, Dr. Merriam, to Prof. T. II. Huxley, is satisfactory evidence, I think, that the bitter contention is practically ended, and the claim of the United States, that pelagic sealing is the cause of the decrease of the seal herd, is generally acknowledged.

PELAGIC SEALING AND DIPLOMACY.

When the actual condition of the seal herd became known in 1890–91, and the ravages of the pelagic sealer could no longer be hidden, it was suggested that arbitration be tried for a final adjustment of all differences between the United States and the pelagic sealer. The seals being born and reared on United States territory, and never landing anywhere else, it was naturally supposed they were the property of the United States, and until their skins became commercially and exceedingly valuable no one questioned our absolute ownership of the herd. So sure were we of our unquestioned title to the seals that, on taking possession of our newly acquired Territory of Alaska, Congress enacted laws for the protection of Alaskan interests and particularly for the protection of all "fur-bearing animals." A few sections of the statute law, in the light of subsequent events, are interesting:

SEC. 1960. It shall be unlawful to kill any fur seal upon the islands of St. Paul and St. George, or in the waters adjacent thereto, except during the months of June, July, September, and October in each year; and it shall be unlawful to kill such seals at any time by the use of firearms or by other means tending to drive the seals away from those islands; but the natives of the islands shall have the privilege of killing such young seals as may be necessary for their own food and clothing during other months, and also such old seals as may be required for their own clothing and for the manufacture of boats for their own use; and the killing in such cases shall be limited and controlled by such regulations as may be prescribed by the Secretary of the Treasury.

SEC. 1961. It shall be unlawful to kill any female seal, or any seal less than one year old, at any season of the year, except as above provided; and it shall also be unlawful to kill any seal in the waters adjacent to the islands of St. Paul and St. George, or on the beaches, cliffs, or rocks where they haul up from the sea to remain; and every person who violates the provisions of this or the preceding section shall be punished for each offence by a fine of not less than two hundred dollars nor more than one thousand dollars, or by imprisonment not more than six months, or by both such fine and imprisonment; and all vessels, their tackle, apparel, and furniture, whose crews are found engaged in the violation of either this or the preceding section, shall be forfeited to the United States.

SEC. 1962. For the period of twenty years from the first of July, eighteen hundred and seventy, the number of fur seals which may be killed for their skins upon the island of St. Paul is limited to seventy-five thousand per annum; and the number of fur seals which may be killed for their skins upon the island of St. George is limited to twenty-five thousand per annum; but the Secretary of the Treasury may limit the right of killing if it becomes necessary for the preservation of such seals, with such proportionate reduction of the rents reserved to the Government as may be proper; and every person who knowingly violates either of the provisions of this section shall be punished as provided in the preceding section.

SEC. 1967. Every person who kills any fur seal on either of those islands, or in the waters adjacent thereto, without authority of the lessees thereof, and every person

---

[1] See letter in Appendix.

who molests, disturbs, or interferes with the lessees, or either of them, or their agents, or employees, in the lawful prosecution of their business, under the provisions of this chapter, shall for each offence be punished as prescribed in section nineteen hundred and sixty-one, and all vessels, their tackle, apparel, appurtenances, and cargo, whose crews are found engaged in any violation of the provisions of section nineteen hundred and sixty-five to nineteen hundred and sixty-eight, inclusive, shall be forfeited to the United States.

SEC. 1968. If any person or company, under any lease herein authorized, knowingly kills, or permits to be killed, any number of seals exceeding the number for each island in this chapter prescribed, such person or company shall, in addition to the penalties and forfeitures herein provided, forfeit the whole number of the skins of seals killed in that year, or, in case the same have been disposed of, then such person or company shall forfeit the value of the same.

Thus for a quarter of a century did the United States throw every possible safeguard of law around the seals and other fur-bearing animals of Alaska, which, under the fostering care of the Government, and the good management of the lessees on the seal islands, produced the grand results of "growth and expansion" in the herd and on the rookeries, sworn to by so many disinterested witnesses who have had ocular knowledge of every fact to which they testified, while during the same period of time the sea otter, which, owing to its pelagic habits, was necessarily left to the tender mercies of the pelagic hunter, who knows no law higher or holier than avarice and selfishness, has been practically exterminated. Laws were enacted from time to time as occasion required them; regulations in accordance with law were made annually for the proper enforcement of the statutes and for the betterment of the natives of the seal islands and the industry upon which they depended for a livelihood, and on which millions of civilized people depended for one of the most beautiful, valuable, and useful furs known to commerce.

Who else, among the thousands now claiming an interest in the seals, ever offered to protect them as we have done?

Where was the pelagic sealer in the days gone by, when the United States were spending millions of money to protect the seal islands, and when our statutes of protection to the female seal were being enacted? Echo answers, "Where?"

Immediately after the treaty of cession, and before we could bring order out of chaos, the marauder of those days landed on the seal islands and slaughtered seals indiscriminately, killing a quarter of a million in one season, and only stopping the ruinous work when the salt was exhausted.

Afterwards the United States statutes were enforced by Government agents sent to the islands for the purpose, and, until 1884, the seals increased in numbers and in value under the fostering care of the Government.

For a period of thirteen years, from 1871 to 1884, inclusive, we had taken 100,000 male seals annually without a sign of decrease or diminution on the rookeries or the slightest injury to the herd, but, on the contrary, a well-known and generally acknowledged growth and expansion.

Dr. H. H. McIntyre, general superintendent for the Alaska Commercial Company at the seal islands during the entire term of their twenty years lease, when writing confidentially to his company in 1889, says:

The breeding rookeries from the beginning of the lease to 1882 or 1883 were, I believe, constantly increasing in area and population, and my observations in this direction are in accordance with those of Mr. Morgan, Mr. Webster, and others, who have been for many years with me in your service, and of the late special Treasury agent, J. M. Morton, who was on the islands from 1870 to 1880. (See letter in Appendix.)

In 1884 an increased fleet of pelagic sealers appear upon the scene, and with vessels specially designed and fully equipped for the work, they follow the seals from year's end to year's end, shooting, spearing, and ripping up all they overtake, without a thought or care for age, sex, or condition; and immediately the rookeries show signs of diminution to those who, like Morgan and Webster, had the experience and the opportunity to observe it.

Dr. McIntyre, in the letter already quoted, continues:

The contrast between the present condition of seal life and that of the first decade of the lease is so marked that the most inexpert can not fail to notice it. Just when the change commenced I am unable from personal observation to say, for, as you will remember, I was in ill health, and unable to visit the islands in 1883, 1884, and 1885. I left the rookeries in 1882 in their fullest and best condition, and found them in 1886 already showing a slight falling off, and experienced that year for the first time some difficulty in securing just the class of animals in every case that we desired. * * * For the cause of the present diminution of seal life we have not far to look. It is directly traceable to the illicit killing of seals of every age and sex during the last few years in the waters of the North Pacific and Bering Sea. We are in no way responsible for it. During the first thirteen years of the lease comparatively few seals were killed by marauders, and we were then able, * * * under our careful management, to produce a decided expansion of the breeding rookeries.

Dr. McIntyre's letter was written in 1889, when the effects of pelagic sealing first startled the civilized world, and his statements were met with doubt or open denial from all who were ignorant of the situation, and with the charge, from pelagic sealers and their apologists, that the Americans had destroyed the seals by overdriving on land.

The absurdity and the injustice of this idle charge have been shown in many ways during the discussion of the seal question, but it is reiterated again and again by those who have established what they are pleased to call an "industry," the chief corner stone of which is the killing of the female seals at sea—of seals about to become mothers, from whose suddenly ripped bellies the unborn young are cut, or torn out alive and thrown into the ocean—of mothers whose young have been left upon the rookeries during their absence on the feeding grounds, left to die of slow starvation where, as Captain Coulson truly says, "the shores are lined with emaciated, hungry little fellows, with their eyes turned toward the sea, uttering plaintive cries for their mothers, which were destined never to return."

And, hard as it may seem, and difficult to believe though it may be, it was with this same pelagic sealer, or for his sake at least, we were asked to arbitrate the question of our exclusive right of property in the seal herd, and of our right to protect them outside of the ordinary, "3 miles," limit from the land upon which they were born and which they made their home.

Even Mr. Elliott was induced to lend his influence to the scheme for arbitration, and, after his return from the seal islands in 1890, we find him addressing Mr. Blaine, who was then Secretary of State, as follows:

Let me again, just before I leave, earnestly urge that you do not hesitate to invite an English commission to meet us, and jointly visit and view the Pribilof seal rookeries next summer at the height of the breeding season in July. That wreck and ruin thereon, which I saw last summer, will be there, and still more pronounced on the same ground next year (1891); it will not fail to arouse the interest and sympathy of the British agents, and the sight of these dwindling herds will be a most eloquent and satisfactory proof of the correctness of your position taken in your leading letter of January 22, 1890, and upon the truth of which your whole argument in the Bering Sea question rests. It is not quite fair to ask John Bull to believe me now, * * * but I assure you that if he gets up there he will soon see enough to make him respect me, and be our sworn friend in cooperating to save the fur seal from impending extermination. Indeed, he should be allowed to see for himself now; it is only manly and fair in us to allow him to do so under the circumstances. (Elliott to Blaine, December 19, 1890.)

The English commission was invited as suggested by Mr. Elliott; the commissioners arrived at the seal islands in the latter part of July, 1891; they visited the rookeries and saw the "wreck and ruin thereon;" they noted the "dwindling herds," and they saw new grass growing on acres of ground where, a few years earlier, hundreds of thousands of seals swarmed in season and brought forth their young. The commissioners found acres of ground covered with dead pup seals as thick as they could lie—"emaciated little fellows"—whose mothers had gone out to the feeding banks, and were captured by the pelagic sealers.

Whether the visit induced them to believe or respect Mr. Elliott remains to be seen, but it certainly did not "arouse their interest or sympathy" for the seals, or for the nation that claimed the right to protect them. Nor did it make them "our sworn friends in cooperating to save the fur seal from impending extermination."

On the contrary, though, they adopted Mr. Elliott's own exploded theories of overdriving, impotency, dearth of bulls, lack of young male blood, redriving, scraping the rookeries, stampeding, and added two or three more of their own, almost as absurd and. nonsensical; and they wound up their sympathetic and impartial labor in behalf of protection for fur seals by the following regulations suggested by the British Bering Sea commissioners:

(B) SPECIFIC SCHEME OF REGULATIONS RECOMMENDED.

155. In view of the actual condition of seal life as it presents itself to us at the present time, we believe that the requisite degree of protection would be afforded by the application of the following specific limitations at shore and at sea:

(a) The maximum number of seals to be taken on the Pribilof Islands to be fixed at 50,000.

(b) A zone of protected waters to be established, extending to a distance of 20 nautical miles from the islands.

(c) A close season to be provided, extending from the 15th of September to the 1st of May in each year, during which all killing of seals shall be prohibited, with the additional provision that no sealing vessel shall enter Bering Sea before the 1st of July in each year.

156. Respecting the compensatory feature of such specific regulations, it is believed that a just scale of equivalency as between shore and sea sealing would be found, and a complete check established against any undue diminution of seals, by adopting the following as a unit of compensatory regulation:

For each decrease of 10,000 in the number fixed for killing on the islands, an increase of 10 nautical miles to be given to the width of protected waters about the islands. The minimum number to be fixed for killing on the islands to be 10,000, corresponding to a maximum width of protected waters of 60 nautical miles.

157. The above regulations represent measures at sea and ashore sufficiently equivalent for all practical purposes, and probably embody or provide for regulations as applied to sealing on the high seas as stringent as would be admitted by any maritime power, whether directly or only potentially interested.

158. As an alternative method of effecting a compensatory adjustment of the stringency of measures of protection, it is possible that some advantages might be found in the adoption of a sliding scale of length for the season of sealing at sea, with a fixed width of zone of protection about the islands.

In this case it is believed that, in correspondence with a decrease of 10,000 seals killed upon the breeding islands, the length of the sealing season at sea might be curtailed by seven days, such curtailment to be applied either to the opening or closing time of the sealing season.

159. It may be objected to the principle involved in any correlative regulation of shore and sea sealing that it would be impossible in any particular year to make known the number fixed for killing on the islands in time to secure a corresponding regulation of pelagic sealing. As a matter of fact, however, if the condition of the breeding rookeries called for any change, it should be possible to fix this number with sufficient precision a year in advance, while, on the other hand, the general effect would be almost equally advantageous if the number killed on the islands in any one year were employed as the factor of regulation for pelagic sealing in the following year.

160. While a zone of protection has been spoken of as the best method of safely guarding the vicinity of the breeding islands, it is to be borne in mind that such an area might be defined for practical purposes as a rectangular area bounded by certain lines of latitude and longitude. Even in dense fog, and therefore comparatively calm weather, an arrested vessel could be anchored with a kedge and warp until the weather cleared, according to frequent custom. The special advantages of a concentric zone appear to be that it is more directly in conformity with the object in view, and that in fine weather the visibility or otherwise of the islands themselves might serve as a rough guide to sealers.

161. The restriction of the number of seals killed on the breeding islands, appropriate safeguards being provided, admits of very considerable precision and requires no special explanation. That the restriction of the number taken at sea may be accomplished practically and with all necessary certainty, and that the means of control available in the case of this branch of the sealing industry are sufficient, is clearly shown by the successful application of measures such as these here proposed, to the Jan-Mayen and Newfoundland hair-seal fisheries, as well as of those based on like principles, which are generally employed in protecting fish and game.

(C) Methods of Giving Effect to Regulations.

162. The means suited to secure the practical efficiency of regulations at sea are generally indicated by those adopted in the instances just cited. It is unnecessary to formulate these here in full detail, but the following suggestions are offered as pointing out those methods likely to prove most useful in the particular case under consideration:

(1) Statutory provisions should be made, declaring it unlawful to hunt or take fur seal during the close season by subjects or vessels of the respective powers.

(2) The time of commencement of the sealing season should be further regulated by the date of issuance of special customs clearances and of licenses for sealing, and preferably by the issuance of such clearances or licenses from certain specified ports only.

(3) As elsewhere explained, the regulation of the time of opening of the sealing season is the most important, and the closing of the season is practically brought about by the onset of rough weather in the early autumn. If, however, it be considered desirable to fix a precise date for the close of sea sealing in each year, this can be done, as in the case of the date of sealing under the Jan-Mayen convention.

(4) The liability for breach of regulations, of whatever kind, should be made to apply to the owner, to the master, or person in charge of any vessel, and to the hunters engaged on the vessel.

(5) The penalty imposed should be a fine (of which one-half should go to the informant), with possibly, in aggravated cases or second offenses, the forfeiture of the catch and of the vessel itself.

(6) To facilitate the supervision of the seal fishery and the execution of the regulations, all sealers might, in addition, be required to fly a distinctive flag, which might well be identical with or some color modification of that already adopted for the same purpose by the Japanese Government.

(D) Alternative Methods of Regulation.

163. Although the general scheme of measures above described appears to us, all things considered, to be the most appropriate to the actual circumstances, measures of other kinds have suggested themselves. Some of these, though perhaps less perfectly adapted to secure the fullest advantages, recommend themselves from their very simplicity and the ease with which they might be applied. Of such alternative methods of regulations, three may be specially referred to:

(1) *Entire prohibition of killing on one of the breeding islands, with suitable concurrent regulations at sea.*

164. The entire reservation and protection of one of the two larger islands of the Pribilof group, either St. Paul or St. George Island, might be assured; such island to be maintained as an undisturbed breeding place, upon which no seals shall be killed for any purpose. On the remaining island the number of seals killed for commercial purposes would remain wholly under the control of the Government of the United States.

In consideration of the guaranteed preservation of a breeding island with the purpose of insuring the continuance of the seal stock in the common interest, a zone of protected waters might be established about the Pribilof Islands, and pelagic sealing might be further controlled and restricted by means of a close season, including the early spring months, or by a protected area to the south of the Aleutian Islands, defined by parallels of latitude, such provisions at sea to have, as far as possible, quantivalent relation to those established on the breeding islands.

(2) Recurrent periods of rest.

165. This implies the provision of a period of rest or exemption of all seals from killing, both at sea and on shore, to extend over a complete year at such recurrent intervals as may be deemed necessary.

Such a period of rest might be fixed in advance for every fifth, or possibly as often as every fourth year, and be made to form a part of a general scheme imposing limitation of number of seals killed on the islands in intervening years, together with restriction by time or by area of pelagic sealing.

While proximately equal in effect on both shore and sea killing a period of rest of this kind would, in other respects, cause some inconvenience, by its interruption of the several industries, and this, though minimized by the fact that the date of occurrence of the year of rest would be known in advance, would not be wholly obviated by this circumstance.

(3) *Total prohibition of killing on the breeding islands, with concurrent strict regulation of pelagic sealing.*

166. While the circumstance that long usage may in a measure be considered as justifying the custom of killing fur seals on the breeding islands, many facts now known respecting the life history of the animal itself, with valid inferences drawn from the results of the disturbance of other animals upon their breeding places, as well as those made obvious by the new conditions which have arisen in consequence of the development of pelagic sealing, point to the conclusion that the breeding islands should, if possible, remain undisturbed and inviolate.

167. If this view should be admitted, and particularly if the United States and Russia, as the owners of the principal breeding islands of the North Pacific, should agree to cooperate in entirely prohibiting all killing of seals on these islands, and in guarding and protecting the breeding places upon them, it should be possible to obtain, in consideration of such care exercised in the common interest, an international assent to measures regulating sea sealing of any required degree of stringency, including certain special rights of supervision by the powers mentioned.

168. It might, for example, under such circumstances, be provided—

(1) That all sealing vessels should be registered, and should take out special licenses at one or other of certain specified ports, as, for instance, Victoria, Port Townsend, Honolulu, Hakodate, and Vladivostock.

(2) That such annual clearances or licenses be not issued before a given date (say 1st of May), and that certain license fees be exacted. Such license fees to be collected by the customs authorities of the licensing Government, and to be eventually transferred, in whole or in part, proportionately, to the Governments protecting the breeding islands, to go toward meeting the cost of this protection.

(3) That no vessel should seal in Bering Sea before some fixed date (say 1st of July) in each year, and that vessels intending to seal in Bering Sea should report either to the United States or to the Russian authorities on or after that date at named ports, such as Unalaska or Petropavlovsk.

(4) That all duly licensed sealing vessels should be required to fly a distinctive flag, and that any unlicensed vessel found engaged in sealing should be subject to certain penalties.

(5) That a zone of protected waters should be established about the breeding islands, within which no sealing should under any circumstances be permitted.

(E) INTERNATIONAL ACTION.

169. In the foregoing remarks on the measures available for the protection and preservation of the fur seal of the North Pacific, reference is made throughout especially to the eastern part of that ocean, including more particularly the area comprised in the range of those fur seals of which the summer haunts and breeding places are about or on the Pribilof Islands, and of which the winter home is found especially off the coast of British Columbia. It is evident, however, that the same remarks and recommendations apply equally to those fur seals which in summer center about the Commander Islands and in winter frequent the seas off the coast of Japan.

170. It may be stated, further, that no system of control can be considered as absolutely complete and effective which does not include under common regulations all parts of the North Pacific, and that the facility of execution of measures and their efficiency would, under any system of regulations, be much increased by the concurrent action of Great Britain, the United States, Russia, and Japan, as indicated in the message of the President of the United States in 1889. Apart from the fact that vessels prevented from sealing at given dates in certain areas might at these times frequent other waters in increased numbers, the circumstance that there is a

certain though not fully known interrelation and interchange of seals between the eastern and western breeding islands of Bering Sea points very clearly to the advisability of such cooperation in protection. (Report of British Bering Sea Commissioners, p. 25.)

The most casual observer will see at a glance that the commissioners' suggestions are all in favor of the pelagic sealer and his "industry," and against the United States and the seals. That the public at large may see this as I see it, I will briefly review a few of the most prominent points suggested.

The commissioners say:

The maximum number of seals to be taken on the Pribilof Islands to be fixed at 50,000.

That is to say, the United States must agree to reduce their catch on land one half, to begin with, and the suggestion, remember, was made long after it was known that the pelagic sealers had captured 78,000 seals in 1891.

They continue:

A zone of protected waters to be established, extending to a distance of 20 nautical miles from the islands.

As the largest catches are made at distances of from 80 to 200 miles from the islands, and as the commissioners were well aware of that fact when they made the suggestion, its worthlessness may be understood so far as the protection and safety of the seals go.

Again, they suggest:

A close season to be provided, extending from the 15th of September to the 1st of May in each year, during which all killing of seals shall be prohibited, with the additional provision that no sealing vessel shall enter Bering Sea before the 1st of July in each year.

As the killing season never did open on the islands till June, and always closed on or before August 10 (excepting the few seals killed from time to time for natives' food), and as it is from May to October that protection is absolutely necessary for the preservation of the seal herd; and as the pelagic sealer hardly ever enters Bering Sea before July it is difficult to see how the "suggestion" could benefit the United States or save the seals.

The next "suggestion" deserves careful attention, for it is the keynote of the whole superstructure raised by the commissioners, who say:

Respecting the compensatory feature of such specific regulations, it is believed that a just scale of equivalency as between shore and sea sealing would be found, and a complete check established against any undue diminution of seals, by adopting the following as a unit of compensatory regulation: For each decrease of 10,000 in the number fixed for killing on the islands, an increase of 10 nautical miles to be given to the width of protected waters about the islands. The minimum number to be fixed for killing on the islands to be 10,000, corresponding to a maximum width of protected waters of 60 nautical miles.

Here they make the pelagic sealer the senior partner in the fur-seal "industry," and the repressive part of the "suggestion" is intended for the United States only.

The situation at the start is to be something like this: The United States are to kill not to exceed 50,000 seals, and the pelagic sealer is not to approach the breeding islands nearer than 20 nautical miles. Then for every additional 10 miles we would remove the pelagic sealer we must reduce our catch on shore by 10,000, so that by the time he is 60 miles away our maximum catch is to be 10,000.

The first thought that suggests itself here is, What would happen were we to ask him for a protected zone of 70 miles from the seal islands? Logically, we would have to stop killing on the islands altogether and turn them into breeding grounds for the use of a class of sportsmen who are fond of the manly art of hunting gravid female seals and cutting out their unborn young.

Turning to the "alternative methods of regulation" suggested by the commissioners we find that they, too, were possessed of the same thought, for they "suggest:"

> Entire prohibition of killing on one of the breeding islands, with suitable concurrent regulations at sea. The entire reservation and protection of one of the two larger islands of the Pribilof group, either St. Paul or St. George Island, might be assured; such island to be maintained as an undisturbed breeding place, upon which no seals shall be killed for any purpose. On the remaining islands the number of seals killed for commercial purposes would remain wholly under the control of the Government of the United States.
>
> In consideration of the guaranteed preservation of a breeding island, with the purpose of insuring the continuance of the seal stock in the common interest, a zone of protected waters might be established about the Pribilof Islands, * * * such provisions at sea to have, as far as possible, quantivalent relation to those established on the breeding islands. (See section 164.)

Growing bolder and bolder as they proceed they finally come forward with a suggestion, which, for downright coolness, may well claim "first place" among all the cool propositions made in any age or country; it is nothing less than the "total prohibition of killing on the breeding islands, with concurrent strict regulation of pelagic sealing."

Here, at last, the mask is thrown off and the commissioners stand forth in their true character of "advocates" for the pelagic sealer and apologists for his horrible methods.

It does not take long to get at the meaning of the "suggestions" offered, for a careful reading shows at once the whole animus of the thing is to prevent the killing of seals on the seal islands, and to turn the whole herd over to the pelagic sealer.

Had the most heartless of all the pelagic sealers been given carte blanche to write suggestions, the adoption of which would inure to his own benefit, he could not improve on those of the British Bering Sea commissioners.

This may seem to be a hard saying, but, from the testimony given by the pelagic sealers themselves, it is well known that the killing of female seals anywhere is sure destruction to the herd; and the British commissioners have admitted it to be on more than one occasion.

Speaking of the indiscriminate killing of the seals at sea, they say:

> But it is unfortunately the case that at certain seasons considerable numbers of gravid females are thus killed, and this killing is deprecated by the better classes of the pelagic sealers themselves, not alone on grounds of humanity, but because they see clearly that it is unduly destructive to the industry in which their fortunes are embarked. (Report of British Bering Sea commissioners, section 633, p. 109.)

And yet the gentlemen who say so are the same men who have "suggested" the "total prohibition of killing on the breeding islands" and the turning over of the seals to indiscriminate slaughter.

The commissioners were instructed to ascertain:

> First. The actual facts as regards the alleged serious diminution of seal life on the Pribilof Islands, the date at which such diminution began, the rate of its progress, and any previous instance of a similar occurrence.
>
> Second. The causes of such diminution; whether, and to what extent, it is attributable—
>
> (a) To a migration of the seals to other rookeries.
> (b) To the method of killing pursued on the islands themselves.
> (c) To the increase of sealing upon the high seas, and the manner in which it is pursued.

And then they were admonished as follows:

I need scarcely remind you that your investigation should be carried on with strict impartiality, that you should neglect no sources of information which may be likely to assist you in arriving at a sound conclusion, and that 'great care should be taken to sift the evidence that is brought before you.

It is equally to the interest of all the Governments concerned in the sealing industry that it should be protected from all serious risk of extinction in consequence of the use of wasteful and injudicious methods. (British Bering Sea Commissioners' report, p. 2.)

To which they replied as follows:

To the Queen's Most Excellent Majesty:

May it please Your Majesty, we, Your Majesty's commissioners, appointed to undertake an inquiry into the condition of seal life and the precautions necessary for preventing the extermination of the fur-seal species in Bering Sea and other parts of the North Pacific Ocean, beg to submit the following report. * * *

Wherefore, in carrying out the terms of our commission, it has been our object to acquire and record the most complete information available, in order to promote, in the true interests of all concerned an equitable, impartial, and mutually satisfactory adjustment of the questions at issue. (British Bering Sea Commissioners' report, p. 3.)

When Mr. Elliott was urging the appointment of a joint commission, as the remedy for all our troubles on the seal islands, he addressed the Secretary of State as follows:

NOVEMBER 22, 1890.

MY DEAR MR. BLAINE: * * * We must take some of the best British representation up to the islands and let it see the wreck and ruin thereon.

I have no fear of the result; these Englishmen will return our friends, and work in harmony with us in the labor of saving these anomalous interests from their impending ruin.

I believe that subsequent events have shown him that his faith was misplaced, to say the least, unless we can fully appreciate the kindness with which they propose to prohibit all killing on the islands and assume the whole burden themselves.

Had they suggested the prohibition of all pelagic sealing and an even division between the nations interested of the burdens, expenses, and proceeds resulting from a strict and constant protection of the breeding islands there would be some semblance of justice and right as well as a desire to perpetuate the seals indefinitely; but the suggestion that the United States shall be forbidden to kill seals ashore and that the islands must be turned into brēeding grounds for the sake of the pelagic sealer is so repugnant to cōmmon sense and decency that were not the commissioners' report at my hand I should not believe they could have been guilty of making such a suggestion under any circumstances, but especially under the plea of protecting and perpetuating the fur seals.

Had they attempted to prove the wastefulness of present methods, or had they quoted the testimony of one honest and disinterested person to show that American management of the seals on the islands had ever been inimical to their increase and improvement, there would be some excuse for the suggestions offered, but it was beyond their power to produce testimony of that sort.

Therefore, I deem the remarks of the American counsel at Paris on this point as most just and opportune, and as they express my own views much better than my own feeble words oan possibly do it I quote them as follows:

We are reluctant to make any reference to motives; but where opinions are, as in this case, made evidence, the question of good faith is necessarily relevant. Why is it that these commissioners have chosen to disregard the plain dictates of reason and natural laws which they were bound to accept, and to recommend some cheap devices in their place, when they so clearly perceived those dictates? We are not permitted to think that this was in conscious violation of duty, if any other explanation is possible. The only apology we can find comes from the fact, clearly apparent upon nearly every page of their report, that the predominating interest which they conceived themselves bound to regard was not the preservation of the seals, but the protection of the Canadian sealers. This explanation at once accounts for all their extraordinary recommendations, and all their varying inconsistencies. Hence, every degree of restraint upon pelagic sealing is reluctantly conceded, and yielded only when it is compensated for, and more than compensated for, by an added restriction of the supply furnished to the market from the breeding islands. As the work of the pelagic sealers is on the one hand restricted in time or place, and thus discouraged, it is on the other stimulated by the certainty of a better market and a richer reward. So persistently and exclusively have they kept this policy before them as their main object, that an ideal has been formed in their minds which they openly avow, and to attain which is their constant effort. This ideal is that all taking of seals on land should be prohibited, and pelagic sealing be made the only lawful mode of capture.

They thus express themselves: "It has been pointed out, and we believe it to be probable, that if all killing of seals were prohibited on the breeding islands, and these were strictly protected and safe-guarded against encroachment of any kind, sealing at sea might be indefinitely continued without any notable diminution, in consequence of the self regulative tendency of this industry."

And suggesting, as the only objection to this policy which occurs to them, that it might be too much to expect of the United States to thus guard the islands and support a native population of 300 at its own expense, they continue: "It may be noted, however, that some such arrangement would offer, perhaps, the best and simplest solution of the present conflict of interests, for the citizens of the United States would still have equal rights with all others to take seals at sea, and in consequence of the proximity of their territory to the sealing grounds they would probably become the principal beneficiaries."

And they finally come to the conclusion that any taking of seals at the breeding places is an error for which there is no defense except long usage, and even that they regard as a doubtful apology. They say:

"While the circumstance that long usage may, in a measure, be considered as justifying the custom of killing fur seals on the breeding islands, many facts now known respecting the life history of the animal itself, with valid inferences drawn from the results of the disturbance of other animals upon their breeding places, as well as those made obvious by the new conditions which have arisen in consequence of the development of pelagic sealing, point to the conclusion that the breeding islands should, if possible, remain undisturbed and inviolate."

These references to the opinions expressed in the report of the commissioners of Great Britain, when taken together with the scheme recommended by them, leave no room for doubt that the defense of the Canadian sealers was from first to last, their predominating motive, and enable us to make for them the apology that they conceived that this was the duty with which they were especially charged. If this be the fact, it is easy to perceive how all their reasonings and recommendations should receive a color and character. We feel obliged to say that we can perceive no other ground upon which their action may be made consistent with good faith. (Argument of the United States, p. 209.)

*   *   *   *   *   *   *

The real conflict between the report of the British commissioners and the case of the United States seems to be as to the number of cows in a harem. The British commissioners assert that the number is unduly large of cows served by one bull; the United States produce credible and experienced witnesses to show that, on the contrary, the number of females is decreasing. A comparison is invited between the two statements and the quality of proof adduced in favor of each It is plain that the British commissioners could not admit the diminution in number of female seals without admitting that decrease to be wholly due to pelagic slaughter. They are therefore reduced to the necessity of insisting that there is a redundancy of females and a deficit of males on the islands. They are kind enough to admit, however, that "the sparing of females in a degree prevented, for the time being, the actual depletion of seals on the islands" (section 58). It is not probable that any reasonable person will take issue with them on that point. The intelligence and legislation of the civilized world, not to speak of humanity in its broad sense, have concurred that to spare the female was not the best but the only effective method of preventing depletion and eventual extermination.

Even if we should concede, for the sake of the argument and in direct disregard of the fact, that the diminution is due to the smaller number of males, we would venture to remind this high tribunal, if such a reminder were needed, that the pirates or poachers who pursue and slaughter the pregnant and nursing females are killing, by starvation in the one case, by the mother's death in the other, a large number of males. Even, according to their own showing, the British commissioners must realize that pelagic sealing is responsible, to some extent at least, for the decrease in the number of males, as well as of females. They may speak of this "industry," as they term it, and glorify it as requiring all the courage and skill which can be brought to bear on it (whatever that may mean). (Section 609.) They may contrast its "sportsmanlike" character with the "butchery" committed on the islands (section 610); but they can not fail to perceive that the mode of destruction, which principally deals with gravid females, necessarily strikes at the very foundation of life, and must eventually extinguish the race, because, as they mildly state it, it is unduly destructive (section 633).

The pelagic sealer not only kills or attempts to kill the males that he happens to meet, but prevents the birth of males to take their place. He often kills three with one discharge of his rifle, viz, the mother, the unborn young, and the pup at home; but he does it in a "sportsmanlike" manner, and he gives the sleeping animal a "fair sporting chance for its life." (Section 610.) In many cases he either misses his object or wounds it and loses it. So that there is by this manly process an utterly useless waste of life, in many cases a waste more or less appalling as the "sportsman" is more or less skillful. How destructive in reality this process is proven to be may be seen from the British commissioners' report under the head of "Proportion of seals lost" (p. 104, section 603). It must be a consolation to those disposed to extol this kind of sport that while nearly "all the pelagic sealers concur in the opinion that the fur seal is annually becoming more shy and wary at sea," it is certain that "the dexterity of the hunters has been increased pari passu with the wariness of the seals." (British commissioners' report, section 401.)

That the number of the seals has been diminished in recent years at a cumulative rate and that such diminution is the consequence of destruction by man is certified by the joint report of all the commissioners. That this human agency is pelagic sealing exclusively, and not the mode, manner, or extent of capture upon the breeding islands, is abundantly clear.

This follows necessarily from admitted facts. The fur seals being polygamous, and each male sufficient for from 30 to 50 females, and being able to secure to himself that number, it follows that there must be at all times a larger number of superfluous males, and the killing of them produces no permanent diminution of the number of the herd. On the other hand, the killing of a single breeding female necessarily reduces pro tanto the normal numbers.

An excessive killing of males might indeed tend toward a decrease if carried to such an extent as not to leave enough for the purpose of effectual impregnation of all the breeding females. The taking from these herds of 100,000 males would not, if that were the only draft allowed, be excessive. This is evident from many considerations.

(a) Those who, like the British commissioners, propose to allow pelagic sealing to such an extent as would involve the annual slaughter of at least 50,000 females in addition to a slaughter of 50,000 young males on the breeding islands can not certainly with the least consistency assert that the capture limited to 100,000 males would be excessive. Nor could they consistently assert this, even though the pelagic slaughter should be restricted (by some means which no one has yet suggested) to 10,000 females. It requires no argument to show that the destruction of even that number would be rapidly disastrous to the herds.

(b) And when we turn to the proofs, they are conclusive that prior to the practice upon any considerable scale of pelagic sealing the annual draft of 100,000 young males did not tend to a diminution of numbers.

(c) Of course, it is easily possible that the indiscriminate slaughter effected by pelagic sealing may soon so far reduce the birth rate as to make it difficult to obtain the annual draft of 100,000 young males. This draft, under such circumstances, would necessarily at once diminish the birth rate, for, the number of females being less, a less number of males would be required. The number of the whole herd might be rapidly diminished by the slaughter of females and the consequent diminution of the birth rate and still 100,000 males continue to be taken for a time without damage. How soon a point would be reached at which so large a draft of males from a constantly diminishing number of births would operate to produce an insufficiency of males is a problem which from want of precise knowledge of the relative numbers of the sexes it would be difficult to solve.

The British commissioners' report upon this subject is as follows:

"The systematic and persistent hunting and slaughter of the fur seal of the North Pacific, both on the shore and at sea, has naturally and inevitably given rise to certain changes in the habits and mode of life of that animal, which are of importance not only in themselves, but as indicating the effects of such pursuit and in showing in what particular this is injurious to seal life as a whole. Such changes doubtless began more than a century ago, and some of them may be traced in the historical precis elsewhere given (section 782 et seq.). It is unfortunately true, however, that the disturbance to the normal course of seal life has become even more serious in recent years, and that there is therefore no lack of material from which to study its character and effect even at the present time."

In the zeal of their advocacy on behalf of pelagic sealing and their denunciation of the methods in use on the islands the commissioners have experienced much and evident difficulty in framing their theory. If they admitted, in unqualified terms, a decrease in number, the obvious deduction from the concession would be that the unlimited slaughter of females must bear the blame and burden of such a result. If, on the other hand, they should assert that the number actually increased, this would only be consistent with an approval of the methods in use on the land. Between this Scylla and this Charybdis a way of escape can be found, and it was found. The ingenuity here displayed deserves full notice and acknowledgment. The joint report contains this statement:

"We find that since the Alaska purchase a marked diminution in the number of seals on and habitually resorting to the Pribilof Islands has taken place, that it has been cumulative in effect, and that it is the result of excessive killing by man."

Bearing in mind that the fur seals forming the object of this controversy have no other home on land than the Pribilof Islands, and that the British commissioners themselves concede that they, for the most part, breed on those islands; bearing in mind, too, that these gentlemen have not yet discovered any other summer habitat for the seals, it would seem that this declaration is equivalent, in its fair sense and meaning, to a statement that the fur seals that frequent the American coast and the Bering Sea have suffered a marked decrease.

Perhaps it was so intended by the British, as it was by the United States commissioners; but if so, the former gentlemen have lost sight of their original intention and have been led to nice distinctions, which we shall now examine.

That the seal, although "essentially pelagic" (section 26), has not yet learned to breed at sea is not denied, although to the vision of the commissioners the prospect of such a transformation or evolution is evidently not very remote. We must, in justice to them, quote one single passage, which admirably illustrates the complacency and self-confidence with which they wrest to their own purposes with unhesitating violence the laws of nature and the mysteries of ulterior evolution. If this quotation does not give a just idea of the imaginative powers of these officials nothing but a perusal of the whole of their work will do them justice:

"The changes in the habits and mode of life of the seals naturally divide themselves into two classes, which may be considered separately. The first and most direct and palpable of these is that shown in the increased shyness and wariness of the animal, which, though always pelagic in its nature, has been forced by circumstances to shun the land more than before, so that but for the necessity imposed upon it of seeking the shore at the season of birth of the young it might probably ere this have become entirely pelagic."

An animal "always pelagic," forced by circumstances to shun the land more than before, and which would become entirely pelagic long before this if it were not obliged to seek the shore for so trifling an object as giving birth to its young, deserves to be classed among the curiosities of nature. The difference between animals (now) always pelagic and those (in the future) entirely pelagic may not readily be understood without explanation not vouchsafed. How can they be always pelagic if they are obliged to seek the land or perish, and why is it reasonable to talk of the probability of their becoming something different from what they are when that conjecture is based upon nothing but reckless and grotesque assumption? Of course, this and other specimens of affront to common sense are merely gratuitous and pointless vagaries. But the thesis must be sustained, viz, that the seals are not even amphibious animals; their resort to land is a merely accidental necessity, and therefore the United States can no more claim a right to or possession in them than in other "essentially pelagic animals," such as the whale, the codfish, or the turbot.

If anything more were needed to emphasize the absurdity of this defiance of well-known facts and settled distinctions in the animal world we might still further cite the British commissioners on the subject of the seal pelage or shedding of hair. It seems that these pelagic animals were not endowed by nature with the proper skin to perform this function in their native element. Unless they can find a suitable place out of water they retain the old hair and disregard the laws which would compel an annual shedding. Lest this seem an exaggeration, read their report citing Mr. Grebnitsky: "During the 'stagey' or shedding season their pelage becomes too thin to afford a suitable protection from the water." (See section 202; also 281, 631, 632.)

It is hardly necessary to say that this theory, so gravely and seriously advanced, that the seal is naturally and essentially a pelagic animal, is utterly unsustained by evidence, is refuted by the language of the commissioners themselves, and disputed by elementary writers. It is only necessary to ascertain how naturalists define pelagic animals and then compare such definition with the known characteristics and rudimentary elements of seal life (see especially for this the books of Johns Hopkins University). Besides, the unanimous and unquestioned testimony of the agents for the Government and the lessees shows that the fur seals spend at least four months of the year on the Pribilof Islands.

Having found, with the American commissioners, a marked diminution in the number of seals on and habitually resorting to the Pribilof Islands, the British commissioners proceed to show that the seals are more numerous than ever. They have, no doubt, demonstrated this to their entire satisfaction on pages 72 and 73 of their report. Captain Warren they quote as saying that he noticed no diminution in the number of seals during the twenty years that he had been in that business, and, if any change at all, an increase (section 403). To the same effect Captain Leary, who says that in Bering Sea they were more numerous than he had ever seen them (section 403), while Mr. Milne, collector of customs at Victoria, reports, what others have said to him, that owners and masters do not entertain the slightest idea that the seals are scarce (section 403). What a tribute this must be to the management of the Pribilof Islands if, notwithstanding the conceded destruction of gravid and nursing females, these statements should be true. Capt. W. Cox took 1,000 seals in four days 100 miles to the westward of the Pribilof Islands (section 405). He found the seals much more plentiful in Bering Sea than he had ever seen them before. It would have added much to the interest of Captain Cox's statement if he had told us how many of these seals gave evidence of having left their pups at home.

The British commissioners multiply the evidence to show that the general experience as stated to them has been that seals were equally or more abundant at sea at the time of their extermination than they had been in former years. It is difficult to treat this with the respect that a report emanating from gentlemen of character and high official position should meet. Either the statement in the joint report is true and the assumption of an increase is untrue, or vice versa. In view of the evidence that these seals have no other home than the Pribilof Islands, it is plain, beyond the necessity of demonstration, that all the seals killed by Captain Cox and others in the Bering Sea were inhabitants of those islands, and the testimony only goes to show that the mothers do go out to sea a hundred miles or more, as is sworn to by the witnesses for the United States, and that it is while they are on the feeding grounds, or searching abroad for food, that they are captured by the Canadian poachers. If this is not so, then let the commissioners or those advocating their views tell us where these seals slaughtered by Captain Cox and others found their "summer habitat."

Any pretense that the seals are decreasing at home—i. e., where they live through the summer, and breed, and nurse, and shed their hair—and at the same time are increasing in the sea is simply an absurdity. It would have added much to the value of the testimony of all these masters if they had not sedulously avoided stating the sex of the animals that they killed.

There is one, and one explanation only, of this, and that explanation makes the stories above quoted plausible. The pelagic sealers were engaged in hunting nursing mothers on the feeding grounds. where those animals are found in large numbers. The decrease proved. and indeed admitted to exist (see joint report), had not yet been so great as to be manifest to those sealers who were so fortunate as to fall in with a number of females either intent upon finding the food necessary to produce a flow of milk or sleeping on the surface of the water after feeding.

And here we may note another illustration of the thesis and its advocacy. Having satisfied themselves that pelagic sealing rather operated to increase the supply of seals they remembered that the killing of young males was objectionable and likely to result in extermination, and thereupon discovered the fact that "a meeting of natives was held" at which the aborigines unanimously expressed the opinion that the seals had diminished and would continue to diminish from year to year (an opinion, too plain, we think, for argument), but they at once assign the reason, which is not the killing of many females, but the extraordinary fact that "all the male seals had been slaughtered without allowing any to come to maturity upon the breeding grounds." (Section 438.)

Having thus proved that the seals were in a flourishing condition of increase, and that they were decreasing in an alarming degree, the conclusion is reached that the decrease is on the land and the increase in the water:

"The general effect of these changes in the habits of the seals is to minimize the number to be seen at any one time on the breeding islands, while the average number to be found at sea, at least proportionately, though perhaps in face of a general decrease in the number of seals, not absolutely increased." (Section 445 of British Commissioners' Report.)

Would it be irrelevant to inquire what was the "summer habitat" of the numerous seals slaughtered by Captain Warren, Captain Leary, and Captain Cox? Were they not all of the Pribilof family? Did not the commissioners, who quoted Captain Cox to the effect that he had, no doubt in true sportsmanlike fashion, with a shotgun, killed 250 seals a day for four days, know that the enormous majority of these were nursing mothers whose pups were starving at home? (Argument of the United States. p. 288.)

PELAGIC SEALING, CLOSE SEASON, ETC.

That many honest and patriotic men have differed in their opinions about the true cause of the destruction and threatened total extinction of the Alaskan fur seal is not to be denied; for, unfortunately, the rival interests have been so many and so diverse, and the seal islands are so far beyond the reach and ken of the public, that it has been very difficult to get at the plain truth of the matter as it really exists. Above all the theories advanced, however, there are two facts which are most intimately connected with the discussion, which never should be lost sight of if we would understand the matter thoroughly.

First. That from 1835, when the Russians first prohibited the further killing of the female seals, to 1884, when the pelagic sealers became numerous and powerful, the seal herds grew and flourished and the rookeries expanded notwithstanding long drives and other barbarous methods continued until the United States purchased Alaska; and that from 1868 to 1886 an average annual killing of 100,000 young males was made before a sign of decrease or diminution appeared on the islands.

Second. In spite of all that has been said and reiterated against the lessees' management of the islands and the methods pursued for so many years in caring for the rookeries and the seals, in driving and killing, and the waste of seal life resulting therefrom, it must be admitted that under this same management (which has been the same, practically for twenty-five years), the seals increased steadily from 1868 to 1884, or until the pelagic sealers appeared in force in Bering Sea.

These are facts that have been proved beyond the possibility of a doubt, and although interested or meddlesome parties may and often do make wild charges and unreliable statements about bad management, bad methods, and barbarity in the driving and killing of the seals, there is not a shadow of truth in the stories, nor has any honest man who ever lived on the seal islands ever said or thought of anything of the sort.

The word "monopoly" is often used for the purpose of bringing odium on the seal question when facts are lacking, but the truth is that, despite all the wicked and idle insinuations thrown out in that way, the leasing of the seal islands to a responsible company was the best as well as the most prudent thing the Government could have done under the circumstances, as the result showed before the pelagic sealer appeared to interfere with the prosperity of the rookeries which had been fostered and built up by the wise management of the lessees.

As an answer to the fault-finder who proclaims the destruction of the seals through the mismanagement of a monopoly, I will quote from the island records the number of seals actually killed for their skins on the islands during the twenty years' lease of the Alaska Commercial Company, and also the number of skins which were rejected or lost out of all that were killed.

[Senate Ex. Doc. No. 107, Fifty-second Congress, second session, appendix.]

Total number of seals killed for their skins by the lessees from 1870 to 1889, both inclusive.

St. Paul	1,463,907
St. George	318,120
Total	1,782,027

Total number of skins rejected from same.

St. Paul	2,480
St. George	628
Total	3,108

In other words, for every 1,000 seals killed by the lessees, during their twenty years' lease, there was a loss of 1¾ skins.

As these figures were compiled by me, originally, from the books kept on both of the seal islands, I know they can not be denied or successfully contradicted, and I respectfully submit them, and the lesson they teach, to the most careful consideration of the Department.

Lest some critic may say I have not quoted all the figures, let me add right here that I am speaking of the large young males which were actually killed for their skins to make up the lessees, annual quota, and of those only.

That the natives killed, for food, 99,684 young male seals during the same twenty years (in addition to pups), and that 27,690 of the skins were rejected, is true; but the lessees are not blamable for that, for they had nothing whatever to do with it, and consequently I have counted only the skins of the seals killed during the regular sealing season and before the seals became "stagy."

Most of the seals killed for natives' food were taken during the "stagy" season, hence the rejection of so many of the skins.

This is why I have repeatedly advised in this and former reports that no killing for any purpose should be permitted during the "stagy" season.

The management of the seal islands, and the care bestowed on the seals by the lessees and their agents, are matters of history into which it is not necessary to enter, because the above figures show far more eloquently and conclusively than words of mine could that that must of necessity be a well-managed business which can make such a showing at the end of twenty years.

Only 7 rejected skins out of every 4,000 seals killed is a record for good and careful management that the lessees may very well be proud of, and it is a withering reply to all the idle story-tellers who have attempted from time to time to make the world believe that carelessness and brutality united in driving the seals hurriedly to the killing grounds, leaving hundreds dead on the road, and that bad management, corruption, and dishonesty reigned supreme on the seal islands.

In another part of this report I have given a table showing the number of seal skins actually recorded as sold as a result of pelagic sealing from 1868 to 1894, both inclusive, which shows the gradual increase of the catch from year to year as the sealing fleet increased in numbers and efficiency, until the 4,367 skins taken in 1868 have grown into 121,143 in 1894.

To further illustrate the growth of pelagic sealing and the havoc it has wrought on the seal herd I will now insert another table comparing the numbers taken on the Pribilof Islands with those taken on the open sea from 1890 to 1894, both inclusive.

Year.	Official, Pribilof Islands.	Official, pelagic catch, as entered in United States and Victoria (British Columbia) custom-houses.	As corrected by trade sales, adding skins shipped via Suez Canal.
1890	20,995	51,814	a 60,000
1891	13,482	69,788	a 75,000
1892	7,549	73,394	a 85,000
1893	7,500	78,083	109,669
1894	16,031	121,143	142,000
Total	65,557	394,222	471,669

a Estimated.

Those taken on the islands, it is hardly necessary to say, were young males—the surplus males of the herd, those taken at sea were taken indiscriminately, without regard to sex, and were mostly gravid females or nursing mothers whose young perished too.

The official figures for the Pribilof Islands catch are taken from the Treasury agents' annual reports on file in the Department. The official figures of the pelagic catch are based on the reports of the collectors of customs at San Francisco, Astoria, Port Townsend, and other ports in the United States, and at Victoria, British Columbia, and, for some of the years, from the London trade sales of pelagic skins.

There is every reason to believe that the real number of pelagic skins taken during the five years last named aggregate 500,000, and if we consider the loss sustained by the wounding and sinking of seals that are never secured, the numbers would run up to three-quarters of a million destroyed, lost to the United States, in five years.

But let us take only what are given officially as entered in United States and Victoria (British Columbia) custom-houses, 394,222, and allow that only 50 per cent of them were females, or, say, 200,000 mothers, one-half of whose pups were "cut out alive" and thrown overboard at sea, and the other half of whose pups starved to death on the rookeries, then the account would run thus:

```
Male seals killed......................................................... 194, 222
Gravid females........................................................... 100, 000
Pups "cut out of same"................................................... 100, 000
Mothers in milk.......................................................... 100, 000
Pups starved on rookeries................................................ 100, 000

    Total................................................................ 594, 222
```

But coming back once more to the bare official figures as given by the collectors of customs, what do they teach us?

They show on their face that the pelagic sealers are reaping the wealth of the seal herd while the United States are paying all the expenses; that during the existence of the modus vivendi, when it was agreed that all parties should cease killing seals until an impartial inquiry and investigation could be made, the United States lived up to the agreement, and the pelagic sealer increased his fleet and killed more seals than he ever killed before.

In 1890, on the discovery of the decrease on the rookeries, we immediately reduced our catch from the regular annual quota of 100,000 to 20,995, but the pelagic sealer continued on his cruise and captured in the whole North Pacific Ocean and Bering Sea 51,814 skins.

In 1891 our catch amounted to 13,482; the pelagic sealer, in spite of pledge, promise, law, and two armed fleets, captured nearly 70,000.

In 1892, with the modus vivendi thoroughly understood by our agents on the islands, we took 7,549 seals to feed the natives of the seal islands as per agreement with Great Britain, and the pelagic sealer, in defiance of all law, took 73,394.

In 1893, still abiding by the terms of the modus vivendi, we took 7,500, and the pelagic sealer took 109,000. (These figures include seals killed on the Asiatic side of the North Pacific Ocean.)

It was in 1893 the Tribunal of Arbitration met at Paris, and, after carefully reviewing the whole situation and the questions at issue, a decision was rendered and regulations suggested for the settlement of the Bering Sea question and for the protection of the fur seals.

The full text of the award will be found in the Appendix.

With the Tribunal of Arbitration, and the questions of national and international law decided by it, I have nothing to do; but with all that appertains to the practical side of the seal question and the measures which should be adopted for the absolute protection of the seals, I have to do, and I say, without the least hesitation, that the regulations adopted for that express purpose by the Tribunal of Arbitration are a failure.

That the two great nations directly interested in the questions laid before the tribunal were honestly anxious to have a definite and mutually satisfactory settlement is not to be doubted; that the questions at issue were fully and ably presented by counsel on both sides can not be disputed; that our own representatives were in full possession of all the facts and testimony, and that they had a thorough knowledge and grasp of the actual situation is shown by the able manner in which they presented their case and met the arguments of opposing counsel, and yet notwithstanding all this, regulations have been made professedly for the protection of the seals but practically for the benefit of the pelagic sealer.

No better proof of this could be given than the official figures already quoted for 1894—a total pelagic catch in the North Pacific Ocean and Bering Sea, from shore to shore, of 142,000 seals, while only 16,031 were killed on the Pribilof Islands from August, 1893, to August, 1894.

Let it be remembered, too, that out of a total of 95 vessels employed in pelagic sealing only 37 entered Bering Sea in 1894, and yet, in about five weeks, these 37 vessels killed over 7,000 seals more than were taken by the 95 vessels on the American side of the North Pacific Ocean, exclusive of Bering Sea, in four months, from January to April, inclusive.

That the regulations have already accomplished much good in the Pacific Ocean outside of Bering Sea is freely admitted; but so long as they allow the same seals to be killed in August in Bering Sea which they protected in May, June, and July in the Pacific Océan they can not be of permanent benefit to the herd as a whole. The fault is not the fault of the Tribunal of Arbitration nor of any of the American gentlemen in any way connected with it, for they very clearly showed that extermination would be the result of pelagic sealing in Bering Sea at any time from May to September, as the following extracts from argument of American counsel will show:

PELAGIC SEALING.

The British commissioners, in their report (section 132), say the coast catch is made from February to June, inclusive, five months, while the Bering Sea catch is taken during July, August, and part of September, or two months and a half.

For each of the 96 vessels engaged in the coast sealing, the average per month is 113, while the monthly average for each of the 86 vessels entering Bering Sea is 290.

It is at once apparent that sealing in Bering Sea is over twice as damaging to the seal herd as sealing in the North Pacific, and that in three years 8,000 more seals were taken in Bering Sea than along the coast in half the time by a fleet numbering ten vessels less than the coast fleet.

Certain witnesses examined by the United States give sufficient data to show the time occupied in sealing along the coast and that occupied in Bering Sea, also the catches made in each place, respectively, and in many instances the distance from the islands at which seals were taken. These have all been collated and arranged in the form of a table, an examination of which will show that they fully corroborate the statement that pelagic sealing is much more damaging in Bering Sea than in the North Pacific. The first four witnesses were examined at Victoria. The page references are to the United States case, Appendix, Vol. II.

Such data as these appearing in the above table can not be found in the depositions appearing in the British counter case. It is unfortunate that this important matter should have been left out of the British testimony.

This testimony further corroborates the statement of the British commissioners that the Bering Sea is not entered until about the 1st of July.

Of the pelagic sealers examined by the United States (United States case, Appendix, Vol. II, pp. 313–507, inclusive) 79 give testimony as to the time they entered Bering Sea. Of this number 68 entered the sea after June 20 and 61 entered between July 1 and July 15.

Of the 316 depositions taken by Great Britain and printed in the British counter case (Appendix, Vol. II) but 5 give the time of entering Bering Sea. One of these (Miner, p. 113) gives the time as "the latter part of June;" 2 (Hartiven, p. 112, and Figuera, p. 125) "early in July;" and the 2 others (Gaudin, p. 111, and Lutjens, p. 121), "July 20."

From the testimony stated above, it is evident why Great Britain failed to examine witnesses on this point, since the British commissioners proposed as a restrictive regulation that Bering Sea should not be entered before the 1st of July, and the British counsel, in presenting a scheme for regulations to the tribunal, incorporated the same suggestion therein. It scarcely seems possible, in face of the evidence that sealing does not usually begin in Bering Sea until July, that Great Britain's advisers can really believe that it would restrict pelagic sealing to prohibit the sealers from doing what they have never done, do not do, and never would do.

Of the sealers examined by the United States and Great Britain, 29 suggest a definite period for a close time. They are arranged below in the form of a table, showing the months in which they think pelagic sealing should be prohibited. The first 7 were examined by Great Britain, and their depositions are included in the British counter case. (Appendix, Vol. II.) The remainder were examined by the United States, and their statements appear in the United States case. (Appendix, Vol. II).

These men, being pelagic sealers, know what months sealing is injurious to the seal herd. If, therefore, the advice of all these witnesses were followed, every month in the year would be closed to pelagic sealing.

Tabulation of opinions of pelagic sealers, showing during what months protection is needed in Bering Sea.

	January	February	March	April	May	June	July	August	September	October	November	December	
Dishow													July.
O'Leary													July–August.
E. P. Miner													January–July.
Geo. Scott													July–September.
Lutjens													July–December.
Conners													July–September.
Moreau													July–September,
Anderson													April–August.
Andricius													January–August 15.
Ball													July–September.
Henri Brown													April–December.
Brennan													April–August.
Clausen													July–October.
Culler													April–November 15.
Franklyn													May–September.
Funcke													July–September 15.
Griffin													April–August.
Hannon													May–September 15.
Hannan													June 15–December.
Harrison													January–July 15.
Hansen													July–November.
Hoffman													June–July.
Johnson													July–December.
Kiernan													March–September.
Lawson													April–October.
Lenard													March–October.
A. McLean													July–August.
D. McLean													June 15–October.
Sundwall													July–October.

An examination of the foregoing table shows that as to some months all are substantially agreed that sealing should be prohibited if the seals are to be preserved. These months are July and August, the principal sealing months in Bering Sea.

All the 29 include July, except one, who thinks the close season should end on July 15.

Twenty-four, or four-fifths of the witnesses, include August, and 17 include September in their proposed close season.

On the facts above stated the United States claim that the following propositions have been demonstrated beyond refutation:

(1) That female seals 2 years old and over are preguant at all times when found in the waters of Bering Sea.

(2) That the nursing females are the only class of seals which feed to any extent while pelagic sealing is carried on in Bering Sea.

(3) That the nursing females are taken in large numbers over 50 miles from the islands.

(4) That the seal pups are not weaned until after the sealing season has closed in Bering Sea.

(5) That the killing of a nursing female in Bering Sea destroys at least two lives, namely, the female and the fetus; and it is an irresistible conclusion that the pup left upon the islands by the female killed also perishes (see paper directed particularly to that subject).

(6) That in point of numbers alone sealing in Bering Sea is over twice as destructive to seal life as sealing in the North Pacific.

(7) That the sealing season in Bering Sea comprises only the months of July, August, and a part of September.

(8) That all the sealers examined by the United States and Great Britain as to the months when sealing should be prohibited include July in the close season proposed, and nearly all include August.

(9) That to open Bering Sea during the months of July and August, with a protective zone of 20 miles about the Pribilof Islands, as proposed by Great Britain, would mean the extermination of the seal herd.

(10) That absolute prohibition of pelagic sealing at all times in the whole Bering Sea east of the 180 degrees meridian from Greenwich is necessary to preserve the Alaskan seals. (Notes for United States counsel, p. 10.)

REGULATIONS.

[Extract from Senator Morgan's opinion.]

I will now state, as I gather from all the evidence before us, what is the evil that these Governments have found to be so threatening to seal life in the Alaskan herd as to draw them into an agreement that it should be repressed by their concurrent action.

I will not attempt to examine again the details of the evidence so thoroughly presented and with such judicial impartiality by Mr. Justice Harlan. I can find no flaw or omission in his careful statements of the evidence, or in the conclusions that he drew from it as to matters of fact. I believe that he stated the exact truth of the situation, and I fully concur in his treatment of the subject and in the conclusions that he has reached.

The present situation, as I understand it, is as follows, as shown by a comparison of the Pribilof and pelagic catches:[1]

Year.	Pribilof Islands.	Total pelagic catch.
1890	21,234	51,655
1891	12,071	68,000
1892	7,500	73,394
1893	7,500	a 80,000
Total	48,305	273,049

a Estimated.

In 1889 the Pribilof catch was 102,617, which fell off to 21,234 in 1890, and this was all that the islands would yield of killable seals, leaving a deficit as compared with the previous year of 81,379 seals upon the islands. If this contrast in the number of seals that could be taken on the islands in 1889 and 1890 was due to the overkilling of males on the islands and not to pelagic sealing, the falling off of numbers would have been indicated in each of the six years prior to 1889. No one has asserted such a fact, and we know that a male seal must be of 6 years old before he is able to take up and maintain a harem on the rookeries. So that this falling off between 1889 and 1890, if it was due to an excessive killing of males, must have occurred at least as early as 1882. This is not true, and no one pretends that it is. The killing of 51,655 seals that the pelagic hunters got, and at least three-fold that number, including

[1] These figures, cited by Senator Morgan, include seals taken off the Asiatic coast of the North Pacific Ocean.

those that were lost, must have reached 300,000 seals that were destroyed. Of this number three-fourths were females, that are not killable seals on the islands and are not counted in the Pribilof catch.

The verification of this calculation is almost perfect in 1892, when the pelagic sealers took 73,000 seals, and in 1891, when they took 68,000. The close approximation of these figures shows that the loss of the seals on the islands was due to pelagic sealing and not to the want of virility in the bulls on the breeding grounds or to any other cause.

That the process which has actually depleted the seal herd in four years to the extent of 569,065 (273,000 of which were females) is an evil that requires to be remedied, for the sake of the protection and preservation of seal life, no one can doubt, as it seems to me. This progressive depletion of this herd of seals can not fail to destroy them very soon, and, in the meantime, to deprive the United States of all possible advantage and compensation derived from its efforts to save the species. What the United States has done, or omitted to do, to deserve treatment at the hands of this tribunal that will expose its lawful industries to ruin, its revenues to depletion, and its wards on the Pribilof Islands to the loss of their only valuable industry will be an inquiry that will seriously challenge the justice of such an award, in the estimate of the civilized world.

The evil to be provided against by this tribunal is, clearly, pelagic sealing with firearms.

If there is, or has been, any detriment to the seal herd from the treatment of the United States, on the islands, the facts on this subject were not unknown to Great Britain when the treaty was made and before ratifications were exchanged. This subject was not referred to in any of the correspondence between the Governments, and the treaty is silent as to this supposed mismanagement.

Will the tribunal, in such a case, make an objection to protecting and preserving the fur seals on the water because Great Britain has not thought it proper or necessary to call the methods into question, or the United States into account for its manner of dealing with that subject on land? True, if it can be shown that the depletion of the herd is due to that cause, and not to pelagic hunting, that is a just and proper inquiry. If it is due to both causes, this tribunal will deal with the pelagic evil, that is submitted to its consideration, and leave it to the nations concerned in the protection of seal life to deal with the evil on land.

If the United States are not so wise in caring for the seals on land as the pelagic hunters are in caring for them at sea, as seems to be asserted, they are quite as earnest in the wish to do so. They destroy no female seals, while the pelagic hunter never spares one. They do not fire upon the breeding rookeries when the seals are massed, many of them asleep, with double-barreled shotguns and buckshot cartridges. They do not kill indiscriminately all seals that come in sight.

The United States permit no female seals to be killed; while 75 per cent of those killed by the pelagic hunter are females heavy with young and almost helpless.

In that condition, as well as in accordance with a law of their nature, which is an important fact in connection with their domesticity, the female fur seals require a great deal of sleep. When asleep, they turn upon their backs, fold their flippers over their breasts, and curving their hind flippers upward, they form their bodies as a sort of boat, the spinal column representing the keel. They can only breathe the upper air; they can not, like a fish, extract air from the water.. After inhaling the air the nostrils close firmly together, and the air, heated by their bodies, expands and buoys them up. They seldom breathe oftener than once in fifteen minutes, and, when diving, they need not return to the surface for air oftener than every thirty minutes. We know nothing of their habits at night while in the ocean. On land they are so boisterous at night with their howlings that sleep would seem to be impossible, except from sheer exhaustion. They have not a keen vision, and the sunlight is painful to them, so that they leave the land and go to sea on days that are bright. This causes them to seek a summer home in a place where fogs and rains prevail. Yet they must have warmth. Nature has amply provided for this necessity by giving them a double coating of thick, strong hair, and of the thickest and finest fur that was ever bestowed upon any species of animals. It is as impervious to water as the down of an eider duck. The pups are born without this fur, and hence their aversion to swimming until it has grown out; and this detains them on land for four months, at least, during which period they can subsist only on the milk of the cow seals. While their vision is not keen, their auditory organs and sense of smell are exceedingly acute. They are attracted by sounds as few other animals are. In this faculty they make a close approach to the endowments of mankind. Sir John Thompson is amused at an account, read by Mr. Justice Harlan, of the seals being attracted in great numbers near to the shore at Hoy by the ringing of a church bell. In his credulous sport over this incident Sir John forgot that it is the personal observation of Mr. Low, one of the greatest naturalists who ever lived, the friend and companion of Cuvier, and is more than confirmed by M. Peron, whom France has honored in the most conspicuous way. His abilities as a naturalist, acquainted intimately with seal life, are as far in advance of those of Professor Elliott, from whom Lord Hannen quotes with much satisfaction, as Napoleon was in advance of the Sioux chieftain, Sitting Bull, as a military genius.

I will presently quote something further about fur seals from Mr. Peron.

I know Mr. Elliott, whom the British Government has dubbed "professor." I have respect for his character and sprightliness. He is a painter in water colors of no mean pretensions, but his use of color does not stop with his canvas. It enters into all he says, and makes him too vivid an enthusiast for a safe reliance on questions of measurements, statistics, and cold facts. Mr. Elliott was out on the Pribilof Islands on the 10th of July, 1890, taking field notes, which, to be of any value, should be free from all romantic conjecture. The following is one of his highly colored extracts from his report of that day:

"In company with Mr. Goff and Dr. Lutz I made my plotting of the breeding seals as they lay on the Reef and Garbotch to-day. Here at the very height of the breeding season, when the masses were most compact and uniform in their distribution in 1872-1874, I find the animals as they lay to-day, scattered over twice and thrice as much ground, as a rule, as the same number would occupy in 1872—scattered because the virile bulls are so few in number and the service which they render so delayed or impotent. In other words, the cows are restless; not being served when in heat, they seek other bulls by hauling out in green jagged points of massing (as is shown by the chart) up from their landing belts. This unnatural action of the cows, or rather unwonted movement, has caused the pups already to form small pods everywhere, even where the cows are most abundant, which shadows to me the truth of the fact that in five days or a week from date the scattering completely of the rookery organization will be thoroughly done. It did not take place until the 20th to the 25th of July, 1872. In 1872 these cows were promptly met with the service which they craved on the rookery ground. The scattering of these old bulls to-day over so large an area is due to extreme feebleness and combined in many cases to a recollection of no distant day when they had previously hauled thus far out on this very ground surrounded by bareness, though all is vacant and semi-grass grown under and around them now." (Dissenting opinions, Harlan and Morgan, pp. 106, 109.)

It is assumed throughout the report of the British commissioners that pelagic sealing is not necessarily destructive, and that, under regulation, the prosecution of it need not involve the extermination of the herds. This assumption and the evidence bearing upon it will be elsewhere particularly treated in what we may have to say upon the subject of regulations. It will there be shown that it is not only destructive in its tendency, but that, if permitted, it will complete the work of piratical extermination in a very short period of time. But so far as it is asserted that a restricted and regulated pelagic sealing is consistent with the moral laws of nature and should be allowed, the argument has a bearing upon the claim of the United States of a property interest, and should be briefly considered here. Let it be clearly understood, then, just what pelagic sealing is, however restricted or regulated. And we shall now describe it by those features of it which are not disputed or disputable.

We pass by the shocking cruelty and inhumanity, with its sickening details of bleating and crying offspring falling upon the decks from the bellies of mothers as they are ripped open, and of white milk flowing in streams mingled with blood. These enormities which, if attempted within the territory of a civilized State, would speedily be made the subjects of criminal punishment, are not relevant, or are less relevant, in the discussion of the mere question of property.

It is not contended that in pelagic sealing (1) there can be any selective killing, or (2) that a great excess of females over males is not slain, or (3) that a great number of victims perish from wounds without being recovered, or (4) that in most cases the females killed are not either heavy with young or nursing mothers, or (5) that each and every of these incidents can not be avoided by the selective killing which is practiced on the breeding islands. We do not stop to discuss the idle questions whether this form of slaughter will actually exterminate the herds or how long it may take to complete the destruction. It is enough for the present purpose to say that it is simple destruction. It is destructive because it does not make or aim to make its draft upon the increase, which consists of the superfluous males, but, by taking females, strikes directly at the stock, and strikes at the stock in the most damaging way, by destroying unborn and newly born pups, together with their mothers. Whoever undertakes to set up a moral right to prosecute this mode of slaughter on the ground that it will not necessarily result in complete destruction must maintain that while it may be against the law of nature to work complete destruction, it is yet lawful to destroy. But what the law of nature forbids is any destruction at all unless it is necessary. To destroy a little and to destroy much are the same crimes.

If there were even something less than a right, or rather some low degree of right—for nothing other than rights can be taken notice of here—some mere convenience, it might be worthy of consideration; but there is none. It can not even be said that pelagic sealing may furnish to the world a seal skin at a lower price. Nothing can be plainer than that it is the most expensive mode of capturing seals. It requires the expenditure of a vast sum in vessels, boats, appliances, and human labor, which is all unnecessary, because the entire increase can be reaped without them. This unnecessary expense is a charge upon the consumer, and must be reimbursed in the price he pays. In no way can pelagic sealing result in a cheapening of the product, except upon the assumption that the stock of seals is inexhaustible and that the amount of the pelagic catch is an addition to the total catch, which might be made on the land if capture were restricted to the land; and this assumption is admitted on all hands, and even by the commissioners of Great Britain, to be untrue.

If there were any evil, or inconvenience even, to be apprehended from a confinement of the capture of the seals to the breeding places, it might serve to arrest attention; but there is none. Much is said, indeed, in the report of the commissioners of Great Britain concerning a supposed monopoly which would thus be secured, as is pretended, to the lessees of the breeding islands, which would enable them to exact an excessive price for skins; but this notion is wholly erroneous. (Argument of the United States counsel, p. 98.)

The whole herd owes its existence not merely to the care and protection but to the forbearance of the United States Government within its exclusive jurisdiction. While the seals are upon the United States territory during the season of reproduction and nurture that Government might easily destroy the herd by killing them all, at a considerable immediate profit. From such a slaughter it is bound to refrain, if the only object is to preserve the animals long enough to enable them to be exterminated by foreigners at sea. If that is to be the result, it would be for the interest of the Government, and plainly within its right and powers, to avail itself at once of such present value as its property possesses, if the future product of it can not be preserved. Can there be more conclusive proof than this of such lawful possession and control as constitutes property, and alone produces and continues the exist-
· ence of the subject of it?

The justice and propriety of these propositions, their necessity to the general interests of mankind, and the foundation upon which they rest in the original principles from which rights of ownership are derived, have been clearly and forcibly pointed out by Mr. Carter. (Argument of the United States counsel, p. 134.)

Thus it will be seen that the danger menacing the seals in Bering Sea by hunting in July, August, and September was well understood by American counsel at Paris, and pointed out by them to the arbitrators with rare ability and conciseness.

As I write, the Congressional Record of December 12 is on my desk with a letter from Mr. Elliott in which he speaks very disparagingly of our agents and counsel at Paris, and of their lack of knowledge of the subject-matter before the Tribunal of Arbitration, thus:

At the time these articles of the Paris award were published immense stress was laid upon the fact that firearms were prohibited in Bering Sea by our agents, who declare that this prohibition would discourage and break up the business of the pelagic sealer. They were strangely ignorant of the truth in the matter, at least the lawyers were, and they had nobody on our side with them at Paris who really knew anything about the life and habits of the seals, who could teach them better.

That they were neither infallible nor omnipotent is freely admitted; that they may have made some mistakes may be true; but that they left behind them in America an equal number of men knowing one-half as much as what they knew about seals has not been, nor can it be, shown.

If mistakes have been made at all they were made when we first agreed to arbitrate the questions that have since been decided against us by the Tribunal of Arbitration; and it is now too late to enter into idle discussions, criminations, and recriminations as to who was right and who was wrong.

Having once put our case into the hands of the tribunal, we must abide by its decision until we can with honor and dignity, worthy our country, bring about other arrangements.

That the regulations, made in good faith, do not accomplish all that was expected of them is so patent to everyone that it needs no discussion here, and the proper steps ought to be taken as soon as possible to remedy their defects.

Of one thing we may rest assured, and that is that August and September—but August by all means—should have been included in the "close time," if the seals are to be saved from extinction.

It is in August the harems, or families, are broken up on the rookeries, and the mother seals go away from the islands to distances of from 80 to 200 miles after food and rest; and it is in August they sleep soundest and longest, after gorging themselves with the first full meal they have had time to secure since June. In August and September the weather is usually favorable in Bering Sea, and the pelagic hunter, having a license to work and nothing to fear, goes in among the sleeping mother seals and quietly spears them until his vessel is loaded with skins, and want of room for more admonishes him to stop.

That the seals have steadily decreased since 1884; that much of the decrease is due to the slaughter of the females by pelagic hunters; that the rookeries are in about the same depleted condition that they were in 1834, after a long period of female slaughter by the Russians; that the remedy applied then must be applied now if we would save the fur seals from total extinction, and build up and replenish the rookeries, are self-evident propositions and cannot be denied.

There is no time to debate mere questions of detail, and we are all agreed, I think, that on the absolute safety and continued protection of the female seal depends the perpetuation of the species. So well has this been understood and appreciated since 1835 that no female seal has been, knowingly, killed on the seal islands in Bering Sea for the past sixty years.

It makes but little difference now as to whose theory was the correct one when guessing was in order; nor does it matter much as to whether spears or shotguns are used in killing mother seals, or whether they are killed in the North Pacific Ocean or in Bering Sea; the only question worth considering in this matter of fur seals just now is "How can we prevent the killing of females?"

Fortunately the declarations made by the Tribunal of Arbitration suggest the most practicable way of solving the problem; and, with the consent of Great Britain, we can solve it immediately.

The Tribunal of Arbitration has declared that:

In view of the critical condition to which it appears certain that the race of fur seals is now reduced in consequence of circumstances not fully known, the arbitrators think fit to recommend both Governments to come to an understanding in order to prohibit any killing of fur seals, either on land or at sea, for a period of two or three years, or at least one year, subject to such exceptions as the two Governments might think proper to admit of.

Such a measure might be recurred to at occasional intervals if found beneficial.

In the spirit of that declaration, and being only too well aware of the present "critical condition of the race of fur seals," and fully appreciating the importance of immediate action, if they are to be saved from extinction, I respectfully offer the following suggestions:

(1) That the United States Government shall officially notify Great Britain of the failure of the "regulations" to adequately protect the seals from the destructive work of the pelagic sealer.

(2) That Great Britain shall be requested to join with the United States in establishing a modus vivendi until, jointly, they arrange to have the cooperation of both Russia and Japan in making regulations for the proper protection of the seal herds coming to the islands or territory of each.

(3) That during the time set apart for the modus vivendi no sealing vessels shall be cleared for sealing purposes, nor shall seals be taken anywhere in the North Pacific Ocean or in Bering Sea, excepting what may be taken by the Indians on the American and British Columbian

coasts for food, and by the natives of the Pribilof Islands for necessary food, fuel, and clothing, as was done during the last modus vivendi.

(4) That Congress shall be asked for an appropriation to defray the expenses of a commission of at least three competent and disinterested men, whose duty it shall be to visit all the seal islands and breeding rookeries in the Bering and Okhotsk seas, and any others whose seals range in either of those seas or in the North Pacific Ocean; to make a thorough investigation of fur-seal life, and to collect testimony bearing on the habits of the animal on land and at sea, and all data that it is possible to secure regarding the effect of driving and killing on land, and of pelagic sealing, and such other information as may be deemed necessary to a thorough understanding of the seal problem.

(5) That the said commission shall be appointed by the President of the United States, and that Great Britain, Russia, and Japan be invited to appoint similar bodies for similar purposes, who, at the completion of their joint investigations, shall jointly report the result thereof and suggest regulations for the proper and adequate protection of the fur seals on land and water.

In making these suggestions I have kept in view the fact that without concurrent action, which shall be mutually satisfactory to the nations directly interested, there can be no adequate protection given to the seals; for so long as pelagic sealers can operate freely in Japanese or Russian waters during a "close time" on the American side, and vice versa, the herds will eventually be exterminated.

The question has been asked, "Suppose Great Britain will not consent to a modus vivendi or a change in the regulations before the expiration of the five years' term established by the Tribunal of Arbitration; what then?"

It must be borne in mind that the regulations do not extend to the seal islands, nor have they anything whatever to do with our work thereon.

Let Congress at the present session repeal all laws which limit the numbers or designate the sex to be killed on the islands, and enact laws empowering the Secretary of the Treasury to kill without limit whenever it may appear that adequate protection to the herds has been sought for in vain. That this last resort is our right and our duty was plainly shown by the United States counsel at Paris, who said:

The whole herd owes its existence not merely to the care and protection, but to the forbearance of the United States Government within its exclusive jurisdiction. While the seals are upon the United States territory during the season of reproduction and nurture, that Government might easily destroy the herd by killing them all at a considerable immediate profit. From such a slaughter it is not bound to refrain if the only object is to preserve the animals long enough to enable them to be exterminated by foreigners at sea. If that is to be the result, it would be for the interest of the Government, and plainly within its right and powers, to avail itself at once of such present value as its property possesses if the future product of it can not be preserved. (Argument of the United States counsel, p. 134.)

And yet, while admitting our right, and asking for the enactment of a law conferring the authority to kill every seal on the Pribilof Islands, should the necessity arise to demand it, I abhor the thought of the possibility of such a dreadful contingency.

While it is well to be fully prepared, let us use all honorable means to avert it if possible.

Respectfully submitted.

JOSEPH MURRAY, *Special Agent.*

Hon. JOHN G. CARLISLE,
    Secretary of the Treasury.

APPENDIX.

PELAGIO SEALING.

Deposition of Milton Barnes, special employee of United States Treasury on St. Paul Island.

TERRITORY OF ALASKA,
St. Paul Island, ss:

I, Milton Barnes, being duly sworn according to law, depose and say as follows: 1 am a citizen of the United States, and when at home reside near Columbus, Ohio. Have been temporarily stationed during the last year on the island of St. Paul, one of the fur-seal or Pribilof group in Bering Sea, as a special employee of the United States Treasury Department on said island.

One day during the latter part of August or fore part of September last (exact date forgotten), Col. Joseph Murray, one of the Treasury agents, and myself, in company with the British commissioners, Sir George Baden Powell and Dr. Dawson, by boat visited one of the seal rookeries of that island known as Tolstoi or English Bay. On arriving there our attention was at once attracted by the excessive number of dead pups, whose carcasses lay scattered profusely over the breeding ground or sand beach bordering the rookery proper and extending into the border of the rookery itself. The strange sight occasioned much surmise at the time as to the probable cause of it. Some of the carcasses were in an advanced stage of decay, while others were of recent death, and their general appearance was that of having died of starvation. There were a few that still showed signs of life, bleating weakly and piteously, and gave every evidence of being in a starved condition, with no mother seals near or showing them any attention.

Dr. Dawson while on the ground took some views of the rookery with his kodak, but whether the views he took included the dead pups I could not say. Some days after this—can not state exact date—I drove with Mr. Fowler, an employee of the lessees, to what is known as Halfway Point, or Polavina rookery. Here the scene was repeated, but on a more extensive scale in point of numbers. The little carcasses were strewn so thickly over the sand as to make it difficult to walk over the ground without stepping on them. This condition of the rookeries in this regard was for some time a topic of conversation in the village by all parties, including the more intelligent ones among the natives, some of whom were with Mr. J. Stanley Brown in his work of surveying the island, and brought in reports from time to time of similar conditions at substantially all the rookeries around the island. It could not, of course, be well estimated as to the number thus found dead, but the most intelligent of the natives—chief of the village—told me that in his judgment there were not less than 20,000 dead pups on the various rookeries of the island, and others still dying. Dr. Akerly, the lessees' physician at the time, made an autopsy of some of the carcasses, and reported that he could find no traces of any diseased condition whatever, but there was an entire absence of food or any signs of nourishment in the stomach. Before Dr. Dawson left I called his attention to what Dr. Akerly had done, but whether he saw him on the subject I can not tell.

And further deponent saith not. MILTON BARNES.

Deposition of C. L. Hooper, captain, United States Revenue Marine.

DISTRICT OF COLUMBIA,
    City of Washington, ss:

Personally appeared before me, C. L. Hooper, who deposes and says:
From the investigations concerning seal life at sea, personally conducted by me, in the North Pacific during the months of March, April, May, and June; in Bering Sea during the month of August and part of September; in the vicinity of the Aleutian chain during the month of October and part of November, as well as from the experience obtained in six other cruises in Alaskan waters and in Bering Sea, I draw the following conclusions:

There were fewer seals to be seen in the water in the vicinity of the Pribilof Islands during the summer of 1892 than in 1891.

At least 75 per cent, and probably 80 or 90 per cent, of the seals in Bering Sea, outside of a narrow zone around the seal islands, are females, 75 per cent of which are nursing mothers and the remaining 25 per cent virgin cows too immature for bearing.

If barren cows exist at all they are rare. I have never known or heard of but one instance.

In Bering Sea mothers go long distances—as far as 200 miles from the islands—to feed, codfish furnishing the bulk of their food.

They sleep much in the water, are not timid, and are readily taken; and their death means the destruction of three lives—the mother, the fetus, and the pup—on the breeding grounds. The past season is the first in several years that such deaths among the pups have not occurred from this source.

At least 70 per cent, and probably 80 or 90 per cent, of any catch in Bering Sea will be females, either actually bearing or capable of bearing at no distant day. This is borne out by the character of the skins of the *Henrietta,* seized last summer for the violation of the modus vivendi. The captain informed me that nearly all the skins taken were those of male seals. Under my direction an examination was made of these skins by N. Hodgson, a man of experience, in whom I have entire confidence. The catch, as shown by the log and sealing book of this vessel, was made in Bering Sea and consists of 420 skins, 361 of which were found to be females, 33 males, and 26 those of seals too young to determine the sex.

For every 100 seals, the death of which results from pelagic hunting, not more than 65 or 75 skins are secured.

The female seals are widely distributed over the sea, and hence the establishment of zonal areas would be rendered impossible by climatic conditions.

There is a wide belt of 200 or 300 miles between the Commander and Pribilof groups of islands which are devoid of seals, and hence no commingling of the herds occur.

There is no foundation for the statement that during the summer months there are found in Bering Sea bodies of seals which are independent of, unattached to, or do not visit the Pribilof Islands.

The annual migration is caused by climatic conditions and feed supply.

The old bulls are the first to leave the islands, and most of them, together with many half bulls and large bachelors, remain in the waters of Bering Sea and off the coast of Alaska during the entire winter, individuals rarely being found south of the fifty-fifth parallel.

The major part of the herd, consisting of females and their pups and young males, begin to migrate about the end of October, and by January 1 all of them have begun their migration. These dates are somewhat earlier or later, according to the season.

Those that leave earliest go farthest south, arriving on the coast of California, and those leaving later reach the coast farther up. Their arrival is coincident with the coming of the smelt, herring, and eulachon, upon which they feed.

On reaching the coast their migration route is continually toward the islands, but following the general trend of the coast, the inner limit being about 25 miles offshore and the outer limit from 75 to 100.

As this migration progresses there is a bunching up of the herd, but the seals travel independently and not in bands or schools.

The migration route is from the Pribilof Islands through the passes across to the coast, up the coast and across the northern sweep of the North Pacific to the Aleutian Chain, and through the passes again to the islands.

There is no foundation in the statement that the Pribilof fur seals which migrate have a winter home off any coast. They appear at about the same time off a long line of coast, reaching from California to Washington. When they are so found they are known always to be moving northward up the coast.

The herd, by reason of hunting at sea, has steadily diminished, and such hunting will ultimately destroy the herd unless prohibited in the North Pacific and Bering Sea, for no matter how small the annual catch may be there is a possibility that the hunt will always be encouraged by the higher prices resulting from the decreasd catch, as in the case of the sea otter.

O. L. HOOPER.

Deposition of H. H. McIntyre, superintendent of the Pribilof Islands.

DISTRICT OF COLUMBIA,
        City of Washington, ss:

H. H. McIntyre, of West Randolph, Vt., being duly sworn, deposes and says:

I have stated in former depositions my connection with the sealeries of Alaska and opportunities for knowledge concerning them.

When the breeding male seals first arrive upon the islands in the spring they are much more timid and easily disturbed than at a later period, and might perhaps be then driven from their chosen places upon the rookeries, but at a later date, when their relation to their neighbors is fairly established and the cows begin to arrive, no amount of force will dislodge them, and they will die in defense of their harems rather than desert them.

In June, 1872, I carried a photographer's camera near the Reef rookery on St. Paul Island and while focusing the instrument, with my head under the black cloth, and the attention of my attendant was diverted, two old bulls made a savage assault upon me, which I avoided by dodging and running. The camera was left where I had placed it and could not be recovered until seal clubs had been sent for and one of the bulls killed and the other knocked down and stunned. The throwing of stones and noisy demonstrations had no effect whatever upon them. This experience only emphasized what I have observed on many occasions upon the islands. The female seals are more timid, and upon the near approach of man show signs of fear and generally move toward the water, but their flight is resisted by the bulls, and before impregnation they rarely succeed in escaping. After this occurs the discipline of the harem is relaxed and the females go and come at will.

I neither saw nor heard, in my twenty years' experience as superintendent of the sealeries, of any destruction of pups by reason of stampedes of seals. But I have occasionally witnessed the death of pups from being trampled upon by the old bulls during their battles for supremacy. This is, however, of rare occurrence. Even if stampedes occurred, the light bodies of the females, averaging only 80 or 90 pounds, would pass over a lot of pups without seriously injuring them.

Later in the season, after the old bulls have been superseded on the rookeries by the younger ones, the pups are already able to avoid being run over, and as a matter of fact the death of pups upon the rookeries from any cause whatever prior to the advent of pelagic sealers in Bering Sea was so rare as to occasion no comment.

It was not customary to drive from any points near enough to the breeding rookeries to cause stampedes, and even if this had been done I do not think any injury to the rookeries would have been occasioned by it. It might cause some of the cows to move away, but they would soon return again.

It is very difficult to determine the average number of females properly assignable to a single male, and difficult even to ascertain how many there are in any given family, because the boundaries of the groups are never well defined, and such as would be said by one observer to belong to a certain bull would be declared by another to be in a different harem. The surface of the ground mainly occupied as breeding rookeries is very irregular. Harems sometimes run together. Ledges, bowlders, and lava rocks hinder the uniform mapping of the family groups, and it is not difficult, therefore, to select certain spots and count a number of female seals which appear to be unattached to any male. On the other hand, there are often found full grown males upon the rookeries at all seasons with no families, and a still larger number with from one to five females each. Such variations have always occurred.

With our present knowledge of seal life, it is impossible to judge with any degree of accuracy how many females may safely be referred to a single male. But, by analogy, it is a very much larger number than has frequently been named as a fair average.

Horse breeders regard a healthy stallion as capable of serving from 40 to 50 mares in a single season; cattle breeders apportion at least 40 cows to a bull, and sheep raisers regard from 30 to 40 ewes as not too many for a single ram, and in the latter case, at least, the season of service is no longer than that permitted to the male seal. I think it would be safe to place an average of 40 or 50 seals to a harem as not excessive.

It is not unusual during the early years of the Alaska Commercial Company's lease to find exceptionally large harems containing from 50 to 100 females each, but we saw no reason to doubt that they were fully served by the male.

The erroneous idea seems to have gained lodgment that during the first decade of the lease a reserve of breeding seals was kept on certain rookeries, and that toward the end of this decade it became necessary to draw on these rookeries because killing 100,000 seals per annum had been too much of a drain upon the herd. This has no foundation in fact. In the early years of the lease the transportation facilities upon the islands, both by land and water, were very limited, and, as the Government agent in charge (Captain Bryant) did not object, we consulted our convenience and drove more frequently from near-by rookeries, but at all times worked the more distant rookeries more or less frequently, as appears by the seal island records. His successors in office theorized that all the rookeries ought to be worked in regular rotation, and so directed. We therefore increased our number of boats and mule teams in order to transport the skins from distant points, and complied with his orders. But we did not do this because of any scarcity of killable seals; no scarcity occurred until pelagic sealing had already made serious inroads. There was no such thing ever thought of upon the islands as reserves of seals," nor was any different practice pursued in respect to driving from year to year, except that all rookeries were worked more systematically after the first few years of the lease.

In the early years of the first lease a few of the bundles of seal skins shipped from the Pribilof Islands may have weighed as much as 60 pounds, but I would not undertake to say that I have seen any weighing as much. If there were any, the explanation is as follows: The skins in such bundles were those of small wigs, and such skins were bundled together so that the flesh sides should be covered completely and no overlapping edges left.

Excrement is voided by seals upon the rookeries as often, I think, as by other carnivorous animals. Those who assert the contrary apparently expect such discharges as they were accustomed to see in the track of the herbivora. The excrement of the seals is of very soft, often semifluid consistency, and in the porous soil or on the smooth rocks is easily brushed about by the trailing flippers of the seals and lost sight of. Their food is chiefly fish, which is highly organized and contains very little tissue that is not absorbed and assimilated. The excrement, therefore, is limited in quantity, even when the animal is full fed, and from its nature and surroundings easily overlooked.

H. H. McINTYRE.

HABITS, AND MANAGEMENT OF SEALS ON ROOKERIES, AND PELAGIC SEALING.

Deposition of L. A. Noyes, resident physician on the Pribilof Islands from 1880 to 1892.

ST. GEORGE ISLAND, PRIBILOF GROUP,
Alaska, *ss:*

L. A. Noyes, being duly sworn, deposes and says: I am a native American, and my home is in Randolph, Vt.; I am 52 years of age, and a physician by profession.

In 1880 I entered the service of the lessees of the Pribilof Islands as resident physician at the seal islands, and I have resided here continuously ever since, excepting an occasional visit to my home for a few months in winter, once or twice since 1880.

From June, 1880, to August, 1883, I was on St. George Island, and from 1883 to 1884 I was on St. Paul Island. I then returned to St. George, where I have resided ever since, except the vacations aforesaid.

I have given much time to the study of the Alaskan fur seal and its peculiar habits, and I have watched with care and solicitude the increase and the decline in numbers of the animal on the hauling grounds and rookeries, and also the methods followed by the lessees in taking the skins—the driving and killing of the young males of from

2 to 5 years old, and the salting, curing, bundling, and shipping the skins. I have likewise carefully observed and noted the coming of the seals in the spring, the hauling out at different times of the various ages and sexes, their disposition on the hauling grounds and rookeries, the formation of the "harem" or family, the breaking up of the harems, the scattering of the cows, and the general intermingling of the sexes in September, and finally the departure of the herd from the islands in November or later.

I have read most of all that has been written within the past quarter century on the fur-seal question, and I have listened to and taken part in many of the controversies indulged in by my associates and friends who have spent many years in the fur-seal industry, and whose practical experience, with all its details, gives weight and value to their assertions. It was I who, at the request of the United States Treasury agent in charge of the islands, measured all the rookeries and hauling grounds on St. George Island in 1887,[1] and I have kept the record of the climatic changes on St. George since the United States Government discontinued the meterological station at the Pribilof Islands.

In addition to my services as physician I have occasionally taught the school on St. George, and I have kept the books and accounts for many years for the lessees on the same island. I am thoroughly conversant with the orders issued by the general and local agents of the lessees to the native chiefs in regard to everything appertaining to the business of taking the annual catch and the care of the seals. I have been intimately acquainted with the Treasury agents who have had charge of the islands since 1880, and I acted as assistant agent myself during the temporary absence of the assistant special agent. I am quite familiar with the general and special orders and instructions issued from the Treasury Department from time to time to the special agents for the government of the natives and the care of the rookeries and seal herd; and I know those laws, rules, and regulations have been faithfully adhered to and fully enforced, published reports of transient visitors to the contrary notwithstanding.

The seal islands of St. Paul and St. George, geographically known as the Pribilof Islands, are situated in Bering Sea at about 170° west from Greenwich and 56° north latitude, and they are nearly 200 miles from the nearest land.

The climatic conditions in their immediate vicinity are so peculiar and their formation and situation are so unique that it is not hard to believe they were selected for a home and resting place by the Alaskan fur seal because of their adaptability to that purpose and to that only. The thermometer rarely goes higher than 60° or lower than zero, the average for a number of years being 35°.

In winter the islands are sometimes surrounded by broken ice, which comes from the north, and it will come and go with the tide and currents, generally from January to April, but occasionally remaining later, and again not appearing at all.

In June, July, and part of August the islands are enveloped for days at a time in dense fog, and a clear sunny day is of rare occurrence. The atmosphere is damp and cool, and the rain falls in a sort of fine mist which drenches one through before it is felt.

The islands are of volcanic origin, and the shores are rough, uneven lava rock, and broken rocks and bowlders of like formation. On this

---

[1] The measurements were made very imperfectly, and I never claimed anything but an approximate measurement. It was my opinion that the numbers were exaggerated, and I so stated at the time.—L. A. N.

rugged shore the Alaskan fur seals make their summer home; here they are born and reared for the first six months of their existence; here they come every spring as regular as time, and here they reproduce their species. The career of the fur-seal herd on these shores is not unlike that of any domesticated animal—it is simply a stock-breeding question.

Areas upon which it is agreeable for the females to breed are carefully reserved and set aside for that purpose.

Each year a sufficient number of breeding bulls are reserved for service on the rookeries. The utmost care is taken that the future of the herd is not jeopardized by the injury or death of a female.

So accustomed have the seals become to the presence of the natives that the timidity and shyness manifested in the ocean is not shown on the islands. In their infancy the pups will approach a native without fear, and later on they are readily handled and the sexes separated, should it be necessary to make a killing of pups for food. In the handling, management, and enlargement of the seal herd there is as much amenability to domestication as there is in a band of range cattle.

The male breeding seals, or bulls, begin to haul out on the breeding rookeries early in May, and they come in more and more rapidly as the month advances, and, selecting their respective stations, lie down and sleep almost continuously until within a few days of the coming of the females, or cows, when they assume a sitting posture, and set up a bellowing noise peculiar to themselves, which I suppose to be a "call" to the approaching herd of cows. It is at this time the bull appears at his best and in his most aggressive mood, and none but the physically strong and successful are allowed to remain within striking distance of the veterans.

The cows begin to haul out in June, and practically they are all on the breeding rookeries by July 15. Immediately on arriving they are taken possession of by the bulls, the strongest and most aggressive securing the greatest number and guarding with jealous care and increasing vigilance.

As a rule the pups are born soon after the cows reach the shore, though it occasionally happens that a cow will be two or three days on the rookery before bringing forth her young.

I think the pups are all born by July 22, and by the middle of August the cows have been fertilized for the next year, after which the harems are abandoned and the bulls begin to leave the islands, and the females and bachelors (or young males) intermingle indiscriminately on the rookeries. From the time the bulls haul out in May till they leave in September they neither eat nor drink, and their lean and lanky appearance in September is in striking contrast with their rotund form and sleek and glossy coats in May.

When the pup is born it is utterly helpless and dependent. It is not amphibious, and would drown if put into water. I have often watched the pups near the water's edge when in stormy weather the surf carried them off, and in every instance they drowned as soon as they went into deep water. The pup is entirely dependent on its dam for sustenance, and when it is a few days old she goes into the sea to feed, returning at intervals of a few hours at first, and gradually lengthening the time as the pups grow older and stronger, until she will be sometimes away for a whole week. During these journeys, in my opinion, she goes a distance of from 40 to 200 miles from the islands to feed, and it is at this time she falls a prey to the pelagic hunter.

Returned to the rookery, the cow goes straight to the spot where she left her pup, and it seems she instantly recognizes it by smelling; and it is equally certain that the pup can not recognize its dam. I have often seen pups attempt to suck cows promiscuously, yet no cow will suckle any pup but her own. When five or six weeks old the pups begin to run around and form bunches or "pods;" at seven to eight weeks old they try the water at the edge, where, after paddling in the shallows, they gradually learn to swim. After becoming expert swimmers they continue to show a preference for land, where they generally remain if not driven into the water by heavy rain or warm sunshine. They make no effort to secure sustenance of any sort beyond that furnished by their dams.

I have examined many pups at the food killings in November, and I never found anything but milk in their stomachs.

The young males or bachelors, whose skins are taken by the lessees, begin to haul out in May, and they continue to haul out until late in July, the older ones coming first and the younger ones later; and they herd by themselves during May, June, and July, because were they to approach the breeding grounds the bulls would drive them off or destroy them.

The bachelors of from 2 to 5 years old are the only seals driven or killed on the seal islands by anyone or for any purpose, and the sensational stories told of how they are tortured on the drive have no foundation in fact. When necessary to make a drive for skins from any given rookery, the local agent of the lessees informs the Treasury agent, and obtains his permission to make the drive. No seals are driven without the consent of the Treasury agent in charge of the island. All being ready, the native chief takes a squad of men to the hauling ground, where the seals are quietly surrounded without disturbing the breeding rookery, and they are then driven slowly along to the killing ground.

Since the improved methods of 1879 there are no drives of greater length than 2½ miles, and the majority of them do not exceed 1 mile. So carefully and so slowly are the drives made, the men driving are relieved every hour, because of the slow motion they get chilled on the road. Arrived at the killing grounds, the seals are driven out from the main body in "pods" of 20 or 30 at a time, and experienced men club and kill the desirable ones, and allow all that remain to return at their leisure to the adjacent waters. The most experienced men do the skinning, and after them come the women and children, who carry off the carcasses for food and the fat or blubber for winter fuel.

In accordance with instructions from the Department, the Treasury agent is always present at the killings, and he has full power and authority to interfere in all cases where there is cruelty practiced or attempted.

All seals killed by the lessees for skins are killed between June 1 and July 30, and generally the season closes on the 20th of July.

After the regular season closes, in July, the natives kill, weekly, for food, from 100 to 200 male seals whose skins are large enough to be accepted as part of the next year's quota; and it is during these food drives in August, September, and October that an occasional female is accidentally killed. Being mixed with the bachelors at this time, some females are driven and accidentally killed. The killing of a female is the greatest crime known on the seal islands and is never done intentionally. Of this I am most positive, for I know that every possible precaution has been taken to guard against it, and I believe there have not been 100 females killed on St. George Island since 1880, if I may except some killed by poachers who were driven off before they secured the skins of the seals they had killed.

Never since the islands have been American property has there been indiscriminate killing done upon them, nor has there been a desire on the part of anyone connected with them to injure or damage or waste seal life; on the contrary, everything has been done by the lessees, past and present, and by the United States, to foster and protect it, and to improve the methods of driving the seals, so that the herds might grow and thrive and increase, and perpetuate themselves indefinitely. Laws, rules, and regulations were made from time to time, prompted by experience, with a view to add to the value of the property and to abolish everything that was not beneficial and in strict accord with the most humane principles. To this end all long drives were prohibited and arrangements made by which the killing grounds have been brought as near the hauling grounds as is practicable without being injurious to the breeding rookeries.

Orders were issued by which the driving is regulated in such manner that no hauling grounds are molested or disturbed more than another, and, being taken in rotation, the seals are allowed several days rest between drives. The rules for driving are so strict, so rigidly enforced, and so faithfully carried out that I hardly know how they could be improved upon.

In my opinion the cows are the only seals that go into the sea to feed from the time they haul out in May till they leave the islands in November or December, and my opinion is based on the fact that the seals killed in May have plenty of food in their stomachs, mostly codfish, while those killed in July have no signs of anything like food in their stomachs.

Again, the males killed for food as the season advances are found to be poorer and poorer, and in all cases after July their stomachs are empty. I am convinced, therefore, that none but mother seals go into the sea to feed during the summer months, and this accounts for the sudden decrease in the herd after the sealing schooners became so numerous in Bering Sea about 1884. The decrease in the number of seals coming to the islands in the last three or four years became so manifest to everyone acquainted with the rookeries in earlier days that various theories have been advanced in an attempt to account for the cause of this sudden change, and the following are some of them: First, a dearth of bulls upon the breeding rookeries; second, impotency of bulls, caused by overdriving while they were young bachelors, and third, an epidemic among the seals.

The "dearth-of-bulls theory" has been thoroughly and impartially investigated without discovering a cow of 3 years old or over on the rookeries without a pup by her side at the proper time, and I am convinced that the virgin females coming onto the rookeries for the first time are the only ones to be found there without pups.

The investigation established the additional fact that hundreds of vigorous bulls were lying idle on the rookeries without cows and many others had to content themselves with only one or two.

The theory of "impotency of the bulls through overdriving" while young was also found to be untrue, and it was shown that after 1878 all long drives on both islands had been abolished, and instead of driving seals from 6 to 12 miles, as was done in Russian times, none were driven to exceed $2\frac{1}{2}$ miles.

It is also a well-known fact that none but the physically strong and aggressive bulls can hold a position on the rookeries and that a weak or an impotent animal has no desire to go there.

The epidemic theory was urged very strongly in 1891, when the rookeries were found covered with dead pups, but a careful and technical examination was made on several of the dead bodies without discovering a trace of organic disease, while starvation was so apparent that those who examined them decided that it was the true cause of their death. Had sickness or disease attacked the seal herd, it is only reasonable to suppose a few grown seals would be found dead where so many young ones had died so suddenly, but the most diligent search has failed to find a grown seal dead upon the islands from unknown causes.

From the discovery of the islands until the present time the flesh of the fur seal has been the daily meat ration of the natives and of the white people, and yet it is a fact that a tainted or diseased carcass has never been known.

In my opinion the solution of the problem is plain. It is the shotgun and rifle of the pelagic hunter which are so destructive to the cow seals as they go backward and forward to the fishing banks to supply the waste caused by giving nourishment to their young.

At this time they are destroyed by thousands and their young of but a few weeks old must necessarily die of starvation, for nature has provided no other means of subsistence for them at this time of life.

Unless the pelagic hunter is prevented from taking seals in Bering Sea and in the North Pacific, the Alaskan fur seal will soon cease to be of commercial value.

L. A. NOYES, M. D.

HABITS AND MANAGEMENT OF SEALS AND RULES OF FUR COMPANIES—PELAGIC SEALING.

Deposition of J. C. Redpath, agent of lessees on St. Paul Island.

ST. PAUL ISLAND, PRIBILOF GROUP,
Alaska, ss:

J. C. Redpath, being duly sworn, deposes and says: I am an American citizen, a native of Connecticut, and I am 48 years of age. At present I am a resident of St. Paul Island, Alaska. I have resided on the seal islands of St. George and St. Paul since my first coming to Alaska in 1875. My present occupation is that of local agent on St. Paul Island for the present lessees, the North American Commercial Company. I have a practical knowledge of, and am thoroughly conversant with, the habits and conditions of the fur seal as it exists on the Pribilof Islands of St. George and St. Paul, and also of the methods adopted and practiced in the taking of the skins, and of the several efforts made by the former and present lessees, as experience taught them, to increase the herd and to build up the rookeries and perpetuate seal life. I have had a personal experience of seventeen seasons on the killing grounds, in different situations, from that of seal clubber to foreman, several years of which I have been the resident local agent. My position as local agent has led me to make a careful study of the seal question, and it is my duty to report from time to time to the general agent of the lessees the result of my observations.

The Alaskan fur seal is a native of the Pribilof Islands, and, unless prevented, will return to these islands every year with the regularity of the season. All the peculiarities of nature that surround the Pribilof group of islands, such as low and even temperature, fog, mist, and perpetually clouded sky, seem to indicate their fitness and adaptability as a home for the Alaskan fur seal; and with an instinct bordering on reason they have selected these lonely and barren islands as the choicest spots of earth upon which to assemble and dwell together during their six months' stay on land; and annually they journey across thousands of miles of ocean and pass hundreds of islands, without pause or rest, until they come to the place of their birth. And it is a well-established fact that upon no other land in the world do the Alaskan fur seal haul out of water.

Early in May the bulls approach the islands and, after cautiously and carefully reconnoitering the surroundings, haul out and select their situations on the rookeries, where they patiently await the coming of the cows. When they first appear upon the rookeries the bulls are fat and sleek and very aggressive, but after a stay of from three to four months, without food, they crawl away from the rookeries in a very lean condition. In my opinion the bull seal returns to the spot he occupied the preceding years, and I know of several instances, where he could be distinguished by the loss of an eye or a flipper, in which he actually did return for a series of years to the same spot.

The mother seals or cows commence to haul about June 10, and nearly all of them are on the rookeries by July 15, and I believe they bring forth their young almost immediately after reaching their places on the rookeries. When the pup is from four to six days old the mother goes into the water for food and, as time passes, her stay becomes longer, until finally she will be away from her pup for several days at a time, and sometimes for a whole week. During these longer migrations she often goes 200 miles from the rookery, and I have been informed by men who were engaged in the trade of pelagic hunting that they had taken "mothers in milk" at a distance of 200 miles from the seal islands.

No cow will nurse any pup but her own, and I have often watched the pups attempt to suck cows, but they were always driven off; and this fact convinces me that the cow recognizes her own pup and that the pup does not know its dam. At birth and for several weeks after, the pup is utterly helpless and entirely dependent on its dam for sustenance; and should anything prevent her return during this period it dies on the rookery. This has been demonstrated beyond a doubt since the sealing vessels have operated largely in Bering Sea during the months of July, August, and September, and which, killing the cows at the feeding grounds, left the pups to die on the islands.

At about 5 weeks old the pups begin to run about and congregate in bunches or "pods," and at 6 to 8 weeks old they go into the shallow water and gradually learn to swim.

They are not amphibious when born, nor can they swim for several weeks thereafter, and were they put into the water would perish beyond a doubt, as has been well established by the drowning of pups caught by the surf in stormy weather. After learning to swim the pups still draw their sustenance from the cows, and I have noticed at the annual killing of pups for food in November that their stomachs were always full of milk, and nothing else, although the cows had left the island some days before. I have no knowledge of the pups obtaining sustenance of any kind except that furnished by the cows; nor have I ever seen anything but milk in a dead pup's stomach. The young males from 2 to 5 years old, whose skins are taken by the lessees, begin to haul out on land in May, and they continue to haul out till July. They herd by themselves during the months of May, June, and July, and they do this because, during the breeding season, they dare not approach the breeding rookeries, or the bulls would destroy them. Being thus debarred from a position on the breeding rookeries or from intermingling with the cows, they herd together on the hauling grounds, where they are easily approached and surrounded by the natives, who drive them to the killing grounds without disturbing the breeding rookeries.

Young males killed in May and June, when examined are found to be in prime condition, and their stomachs are filled with fish—principally codfish—but those killed later in the season are found to be poor and lean and their stomachs empty; which shows that the males rarely leave the islands for food during the summer months.

Statute law forbids the killing of the female seal, and nature regulates the matter so that there is no danger of their being driven or killed during the regular killing season, which takes place in June and July, when all the "killing for skins" is done; and after all my experience here I am free to say that a small fraction of 1 per cent would represent all the females killed on the islands since they became the property of the United States.

The compact family arrangement so tenaciously adhered to during the breeding season becomes relaxed in August, and the females scatter, and a few of them mix with the young males, and when the natives make a drive for food it occasionally happens that a female will accompany the males, and sometimes one or two may be accidentally killed. I use the word "accidentally" advisedly, because there is no good reason why the natives or the lessees should kill a female designedly, as the skin is of no more use or value (if so much), nor its flesh as good for food as is that of the male. And, excepting accidents, it is a fact that no female seals are, or ever were, killed on the Pribilof Islands since American rules and regulations were established there.

The regular killing season for the skins under the lease begins June 1 and ends practically the last of July; and during this period the first-class Alaskan fur-seal skins are taken. The seals are driven from the hauling to the killing grounds by experienced natives under the orders of the native chief, and the constant aim and object of all concerned is to exercise the greatest care in driving, so that the animals may not be injured or abused in any manner. As the regulations require the lessees to pay for every skin taken from seals killed by the orders of their local agents, and as the skin of an overheated seal is valueless, it is only reasonable to suppose that they would be the last men living to encourage and allow their employees to overdrive or in any manner injure the seals. I know that the orders given to me as local agent were always of the most positive and emphatic kind on this point, and they were always obeyed to the letter. Instead of overdriving or neglecting the seals, the lessees have endeavored to do everything in their power to shorten the distance between the hauling and killing grounds, or between the hauling grounds and the salt house.

Before the Alaska Commercial Company leased the seal islands in 1870 it was a common practice to drive seals from Northeast Point to the village of St. Paul Island, a distance of 12 miles, and from Zapadnie to the village of St. George Island, a distance of 6 miles, across a very rough and rugged country.

From Halfway Point and from Zapadnie on St. Paul Island seals were driven respectively 5 and 6 miles. When the Alaska Commercial Company took control of the islands the drive from Northeast Point was prohibited, and a salt house and other necessary buildings erected within 2 miles of the killing ground, and all the skins taken there were salted and stored and shipped from Northeast Point. In 1879 a killing ground was made and a salt house built at Halfway Point, within 2 miles of the hauling grounds, and all skins taken at the Point are salted there. At Zapadnie the same year a killing ground was made within a mile of the hauling ground, and the skins taken there are taken to the village salt house in boats, or when the weather is unfavorable by team and wagon.

Since 1878 there has not been a drive made on St. Paul Island to exceed 2 miles. At Zapadnie, St. George, a salt house was built about 1875 and the 6-mile drive prohibited and a trail made at great expense across the island, over which the skins are taken on pack saddles to the village. Since 1874 no seals have been driven on St. George Island to exceed 2½ miles.

Although the seals are comparatively tame after being on the land for a short time and do not get scared so easily as is commonly supposed, the rules and regulations of the Treasury Department are very strict on the question of absolute protection to the seals on the islands, and the Treasury agents have always most rigidly enforced them.

It is unlawful to fire a gun on the islands from the time the first seal appears in the spring until the last one leaves at the end of the season; and in order to properly enforce this law the firearms are taken from the natives and locked up in the Government house in care of the Treasury agents.

No person is allowed to go near a rookery unless by special order of the Treasury agent, and, when driving from the hauling grounds, the natives are forbidden to smoke or make any unusual noise, or to do anything that might disturb or frighten the seals. All driving is done when the weather is cool and moist, and when the condition of the weather demands it the drives are made in the cool of the night, and in no case are seals driven at a higher rate of speed than about half a mile an hour. So carefully is the driving done that it has been found necessary to divide the native drivers into several "watches," which relieve each other on the road, because the pace being so slow the men get cold.

From 1875 to 1883 it was no uncommon thing for the lessees to take the annual quota of 100,000 skins between June 1 and July 20, and yet there was no sign of any decrease, but rather an expansion of most of the rookeries.

I do not pretend to be able to say how many seals there are, or ever were, on the rookeries; nor do I believe anybody else can tell; for the rookeries are so broken and filled with rocks it is impossible to estimate the number of seals upon them with any approach to accuracy. The lines of expansion and contraction are plain enough, and can be seen and understood by the whole community.

Until 1884 sealing schooners were seen but very seldom near the islands or in Bering Sea, and the few seals taken by the hunters who raided the rookeries occasionally are too paltry to be seriously considered, because the raids were so few, and the facilities for taking many seals oft so utterly insignificant. In 1884 the sealing schooners became numerous. I believe there were about 30 in the sea that year, and they have increased very rapidly every year since, until now they are said to be about 120. As the schooners increased the seals decreased, and the lines of contraction on the rookeries were noticed to draw nearer and nearer to the beach, and the killable seals became fewer in numbers and harder to find. In 1886 the decrease was so plain that the natives and all the agents on the islands saw it and were startled, and theories of all sorts were advanced in an attempt to account for a cause.

A dearth of bulls on the breeding rookeries was a pet theory of one or two transient visitors, but it only needed a thorough investigation of the condition of the rookeries to convince the most skeptical that there were plenty of bulls, and to spare, and that hardly a cow could be found on the rookeries without a pup at her side.

For five years I have given this particular subject my most earnest attention, and every succeeding year's experience has convinced me that there is not, and never was, a dearth of bulls. The theory of impotency of the young bulls because of overdriving when young is not worthy of consideration by any sane or honest man who has ever seen a bull seal on a breeding rookery; and as I have already answered the question of overdriving, I will only add here that no young bull goes upon a breeding rookery until he is able to fight his way in, and an impotent bull has no desire to fight, nor could he win a position on the rookery were he to attempt it. The man is not alive who ever saw a 6 or 7 year old bull seal impotent.

Another theory, equally untrue, was that an epidemic had seized the herd, but investigations of the closest kind have never revealed the death on the islands of a full-grown seal from unknown causes. Let it be remembered that the flesh of the seal is the staple diet of the natives, and that it is eaten daily by most of the white employees as well; and yet it is true that a sign of taint or disease has never been found on a seal carcass in the memory of man. It was not until so many thousands of dead pups were found upon the rookeries that the problem was solved. The truth is that when the cows go out to the feeding grounds to feed they are shot and killed by the pelagic hunter, and the pups, deprived of sustenance, die upon the rookeries. Excepting a few pups killed by the surf occasionally, it has been demonstrated that all the pups found dead are poor and starved, and when examined their stomachs are found to be without a sign of food of any sort. In 1891 the rookeries on St. Paul Island were covered in places with dead pups, all of which had every symptom of having died of hunger, and on opening several of them the stomachs were found to be empty.

The resident physician, Dr. Akerly, examined many of them and found in every instance that starvation was the cause of death. The lowest estimate made at the time, placing the number of dead pups on the rookeries at 25,000, is not too high

It has been said that man can do nothing to facilitate the propagation of the fur seal. My experience does not support this. The reservation of females and the killing of the surplus males, so that each bull can have a reasonable number of cows, is more advantageous to the growth of the rookeries than when in a state of nature bulls killed each other in their efforts to secure a single cow.

The same care can be and is exercised in the handling and management of the seal herd as is bestowed by a ranchman upon his bands of ranging stock, and is productive of like results. The seals have become so accustomed to the natives that the presence of the latter does not disturb them. The pups are easily handled by the natives, and formerly, when used as an article of food, thousands of pups were actually picked up and examined, in accordance with Government requirement, to avoid the killing of a female. So easily are the seals controlled that, when a drive of bachelors is made to the killing grounds, a guard of two or three small boys is sufficient to keep them from straying, and from the general band any number from one upward can be readily cut out. It is possible in the future, as it has been in the past, to reserve unmolested suitable areas to serve as breeding grounds; to set aside each year a proper number of young males for future service upon the rookeries, and by the application of the ordinary stock-breeding principles not only to perpetuate, but to rapidly increase, the seal herd.

To one who has spent so many years among the seals as I have, and who has taken so much interest in them, it does appear to be wrong that they should be allowed to be so ruthlessly and indiscriminately slaughtered by pelagic hunters, who secure only about one-fourth of all they kill. There is no doubt in my mind that unless immediate protection be given to the Alaskan fur seal the species will be practically destroyed in a very few years; and in order to protect them pelagic hunting must be absolutely prohibited.

The foregoing is substantially the same testimony that I gave to the commissioners who visited the islands in 1891.

J. C. REDPATH.

PRIBILOF ROOKERIES.

Deposition of Charles J. Goff, Treasury agent in charge of Pribilof Islands.

DISTRICT OF COLUMBIA,
    City of Washington, ss:

Charles J. Goff, of Clarksburg, W. Va., being duly sworn, deposes and says: I am 45 years of age; during the years 1889 and 1890 I occupied the position of special Treasury agent in charge of the Pribilof Islands. I was located on St. Paul Island, only visiting St. George Island occasionally. About the 1st of June, 1889, I arrived on St. Paul Island, and remained there until October 12, 1889, when I returned to San Francisco for the winter. Again went to the islands in 1890, arriving there about the last week in May, and remaining until August 12, 1890. Since that time I have never been on the islands. My principal observations as to seal life upon the islands were confined to St. Paul Island, as I only visited St. George Island occasionally.

During my first year on the islands the Alaska Commercial Company was the lessee thereof, and during my second year the North American Commercial Company. In 1889 I made careful observations of the rookeries on St. Paul Island and marked out the areas covered by the breeding grounds; in 1890 I examined these lines made by me the former year and found a very great shrinkage in the spaces covered by breeding seals.

In 1889 it was quite difficult for the lessees to obtain their full quota of 100,000 skins. So difficult was it, in fact, that in order to turn off a sufficient number of 4 and 5 year old males from the hauling grounds for breeding purposes in the future the lessees were compelled to take about 50,000 skins of seals of 1 or 2 years of age. I at once reported this fact to the Secretary of the Treasury, and advised the taking of a less number of skins the following year. Pursuant to such report the Government fixed upon the number to be taken at 60,000, and further ordered that all killing of seals on the islands should stop after the 20th day of July. I was further ordered that I should notify the natives upon the Aleutian Islands that all killing of seals while coming from or going to the seal islands was prohibited. These rules and regulations went into effect in 1890, and pursuant thereto I posted notices for the natives at various points along the Aleutian chain, and saw that the orders in relation to the time of killing and number allowed to be killed were executed upon the islands. As a result of the enforcement of these regulations the lessees were unable to take more than 21,238 seals of the killable age of from 1 to 5 years during the season of 1890, so great had been the decrease of seal life in one year, and it would have been impossible to obtain 60,000 skins even if the time had been unrestricted.

The Table A appended to this affidavit[1] shows how great had been the decrease on St. Paul Island hauling grounds, bearing in mind the fact that the driving and killing was done by the same persons as in former years, and was as diligently carried on, the weather being as favorable as in 1889 for seal driving. I believe that the sole cause of the decrease is pelagic sealing, which from reliable information I understand to have increased greatly since 1884 or 1885. Another fact I have gained from reliable sources is that the great majority of seals taken in the open sea are pregnant or females in milk. It is an unquestionable fact that the killing of these females destroys the pups they are carrying or nursing. The result is that this destruction of pups takes about equally from the male and female increase of the herd, and when so many male pups are killed in this manner, besides the 100,000 taken on the islands, it necessarily affects the number of killable seals. In 1889 this drain upon male seal life showed itself on the islands, and this, in my opinion, accounts for the necessity of the lessees taking so many young seals that year to fill out their quota.

As soon as the effects of the pelagic sealing were noticed by me upon the islands I reported the same, and the Government at once took steps to limit the killing upon the islands, so that the rookeries might have an opportunity to increase their numbers to their former condition; but it will be impossible to repair the depletion if pelagic sealing continues. I have no doubt, as I reported, that the taking of 100,000 skins in 1889 affected the male life on the islands, and cut into the reserve of male seals necessary to preserve annually for breeding purposes in the future, but this fact did not become evident until it was too late to repair the fault that year. Except for the numbers destroyed by pelagic sealing in the years previous to 1889 the hauling grounds would not have been so depleted, and the taking of 100,000 male seals would not have impaired the reserve for breeding purposes or diminished to any extent the seal life on the Pribilof Islands. Even in this diminished state of the rookeries in 1889 I carefully observed that in the majority of cases the 4 and 5 year old males were allowed to drop out of a "drive" before the bachelors had been driven any distance from the hauling grounds. These seals were let go for the sole purpose of supplying sufficient future breeders.

A few seals are injured by redriving (often conflicted with overdriving and sometimes so called), but the number so injured is inconsiderable and could have no appreciable effect upon seal life through destroying the virility of the male. The decrease, caused by pelagic sealing, compelled whatever injurious redriving has taken place on the islands, as it was often necessary to drive every two or three days from the same hauling grounds, which caused many seals let go in a former "drive" to be driven over again before thoroughly rested. If a "drive" was made only once a week from a certain hauling ground, as had been the case before pelagic sealing grew to such enormous proportions and depleted the rookeries, there would be no damage at all resulting from redriving.

In my opinion pelagic sealing is the cause of redriving on the islands, the depletion of the rookeries, and promises to soon make the Alaska

---

[1] See "Island Records," Appendix.

fur-seal herd a thing of the past. If continued as it is to-day, even if killing on the islands was absolutely forbidden, the herd will in a few years be exterminated. I am, therefore, of the opinion that pelagic sealing should be absolutely prohibited both in Bering Sea and the North Pacific Ocean. If this is done and a few years are allowed the seal herd to recover from the enormous slaughter of the past seven years the Pribilof Islands will produce their 100,000 skins as heretofore for an indefinite period.

I hereby append to and make a part of this affidavit a table, marked A,[1] giving the number of seals killed each day on the island of St. Paul during the years 1889 and 1890 up to the 20th day of July.

*   *   *   *   *   *   *

CHARLES J. GOFF.

MANAGEMENT OF SEAL KILLING, AND PELAGIC SEALING.

Deposition of Abial P. Loud, special assistant Treasury agent on Pribilof Islands.

DISTRICT OF COLUMBIA,
    City of Washington, ss:

Abial P. Loud, being duly sworn, deposes and says: I am a resident of Hampden, Me., and am 55 years of age. On April 4, 1885, I was appointed special assistant Treasury agent for the seal islands, and immediately started for the islands, arriving at the island of St. Paul on May 28 or 30. Spent that season on St. Paul Island, and returned for the winter to the States, leaving the islands on the 18th of August. Went back again next spring, arriving there in latter part of May, and remained until August, 1887, on St. Paul Island. Spent the season of 1888 and 1889 on St. George Island, returning in the fall of 1889 to the States. In 1889 I spent some time in the fall on St. Paul Island. On whichever island I was located I always kept careful watch and made frequent examinations of the rookeries during this entire period. During the time from 1885 to 1889 there was a very marked decrease in the size of the breeding grounds on St. Paul Island, and from 1887 to 1889 I also noticed a great decrease in the areas covered by the rookeries on St. George Island.

In his reports of 1886 and 1887 George R. Tingle, special Treasury agent in charge of the seal islands, reported having measured the rookeries on the islands, and that the seals had largely increased in number, giving the increase at about 2,000,000. From this report I dissented at the time, as I was unable to see any increase, but, on the contrary, a perceptible decrease, in the rookeries. I expressed my views to many on the islands, and all agreed that there had been no increase in seal life. I do not think that there was a single person on the island except Mr. Tingle who thought there had been an increase, or, in fact, that there had not been a decrease in seal life. The measurements of the rookeries on which Mr. Tingle relied were made with a common rope by ignorant natives while the seals were absent from the islands, the grounds covered by them being designated by Mr. Tingle from memory. Even if these measurements had been correct, which was impossible, I

---

[1] See "Island Records," Appendix.

do not believe it is possible to calculate even approximately the number
of seals upon the rookeries because of the broken nature of the ground
and the irregular outlines of the breeding grounds. While I was on
the islands I attended nearly every drive of the bachelor seals from the
hauling grounds to the killing grounds, and these drives were conducted
by the natives with great care, and no seals were killed by overdriving,
plenty of time being always given them to rest and cool off. A few
were smothered by the seals climbing over each other when wet; but
the number was very inconsiderable, being a fraction of 1 per cent of
those driven, and did not to any extent affect the seal life on the islands.
The greatest care was always taken to avoid overdriving both by the
Government officers and employees of the lessees.

During my experience (and I was on the killing ground at every killing that took place while I was on the islands) I never saw a male seal
which had been injured by being redriven several times from the same
hauling ground. I am convinced that while I was there there was not
a single case in which the virility of a male seal was destroyed or
impaired in the slightest degree by driving, redriving, or overdriving,
and I took particular notice of the condition of the males during each
drive. The males old enough for service on the breeding grounds were
always allowed to return to the hauling ground from a drive, and I am
satisfied a sufficient number of males was always reserved for future
breeding purposes. A suggestion was made to the Secretary of the
Treasury in the fall of 1885 that some old bulls should be killed, but
the Secretary declined to permit such animals to be destroyed. I am
convinced that the decrease in the rookeries was caused entirely by open-
sea sealing. As I was not present on the islands in the fall of 1885, I
am unable to make a statement as to the number of dead pups on the
rookeries in that year, but in 1886 I saw a large number of dead pups
lying about. These pups were very much emaciated, and evidently had
been starved to death. I account for this by the killing of the mothers
by open-sea sealers before the pups were weaned, and because a mother
will not suckle any pup except her own.

In 1887 the number of dead pups was much larger than in 1886. In
1888 there was a less number than in 1887 or in 1889, owing, as I believe,
to a decrease of seals killed in Bering Sea that year, but in 1889 the
increase again showed itself. I believe the number of dead pups
increased in about the same ratio as the number of seals taken in Bering
Sea by pelagic sealers. While I was on the island there were not more
than three or four raids on the rookeries to my knowledge, and I think
that the destruction to seal life by raiding rookeries is a small part of
1 per cent as compared with the numbers taken by killing in the water.
Another fact in connection with open-sea sealing is that the great
majority of seals killed are females, and that a great part of the females
are pregnant or in milk. The milking females are most all killed while
visiting the feeding grounds, which are distant 40 or 60 miles, or even
farther from the islands. The female necessarily feeds so she can supply nourishment for her young, while the males during the summer
seldom leave the islands. This accounts for the large number of females
killed in Bering Sea. In July, 1887, I captured the poaching schooner
Angel Dolly while she was hovering about the islands. I examined the
seal skins she had on board, and about 80 per cent were skins of females.
In 1888 or 1889 I examined something like 5,000 skins at Unalaska
which had been taken from schooners engaged in pelagic sealing in
Bering Sea, and at least 80 to 85 per cent were skins of females.

I have conversed with the captains of several marauding schooners and others who were employed in pelagic sealing have informed me that they usually use rifles in shooting seals in the water. Some, however, use shotguns, but to no great extent. From these conversations I should judge they did not secure more than one-half of the seals killed, and this, I think, is a large estimate of the number secured. I am of the opinion that the Pribilof seal herd should be protected both in Bering Sea and the North Pacific Ocean. If an imaginary line were drawn about the islands, 30 or 40 miles distant therefrom, within which sealing would be prohibited, this would be little protection to seal life, for all the poachers whom I interviewed acknowledged that they could get more seals in the water near the fishing banks, 30, 40, or more miles from the islands, than in the immediate vicinity thereof, and the hunters on the schooners always complained if they got much nearer than 40 miles of the islands. I am certain that even if sealing were prohibited entirely upon the islands the seal herd would in a short time be exterminated by pelagic sealing, if permitted, because the females—that is, the producers—are the seals principally killed by open-sea sealing.

ABIAL P. LOUD.

PELAGIC SEALING—MANAGEMENT.

Deposition of Kerrick Artomanoff, native chief, resident of St. Paul Island.

ALASKA, UNITED STATES OF AMERICA,
St. Paul Island, Pribilof Group, ss:

Kerrick Artomanoff, being duly sworn, deposes and says: I am a native Aleut and reside on St. Paul Island, Pribilof group, Alaska. I was born at Northeast Point, on St. Paul Island, and am 67 years of age. I have worked on the sealing grounds for the last fifty years and am well acquainted with the methods adopted by the Russian and American Governments in taking of fur-seal skins and in protecting and preserving the herds on the island. In 1870, when the Alaska Commercial Company obtained a lease of the islands, I was made chief, and held the position for seventeen years.

It was my duty as chief to take charge of and conduct the drives with my people from the hauling to the killing grounds. The methods used by the Alaska Commercial Company and the American Government for the care and preservation of the seals were much better than those used by the Russian Government. In old Russian times we used to drive seals from Northeast Point to the village, a distance of nearly 13 miles, and we used to drive 5 or 6 miles from other hauling grounds; but when the Americans got the islands they soon after shortened all the drives to less than 3 miles.

From 1870 to 1884 the seals were swarming on the hauling grounds and the rookeries, and for many years they spread out more and more. All of a sudden, in 1884, we noticed there were not so many seals, and they have been decreasing very rapidly ever since. My people wondered why this was so, and no one could tell why until we learned that hunters in schooners were shooting and destroying them in the sea. Then we knew what the trouble was, for we knew the seals they killed and destroyed must be cows, for mostly all the males remain on or near the islands until they go away in the fall or fore part of the winter. We also noticed dead pups on the rookeries that had been starved to death. These young pups have increased from year to year since 1887, and in 1891 the rookeries were covered with dead pups. In my sixty-seven years' residence on the islands I never before saw anything like it. None of our people have ever known of any sickness among the pups or seals, and have never seen any dead pups on the rookeries, except a few killed by the old bulls when fighting, or by drowning when the surf washed them off. If they had not killed the seals in the sea there would be as many on the rookeries as there was ten years ago. There was not one-fourth as many seals in 1891 as there was in 1880.

The fur seal goes away from the island in the fall or winter and he returns in May or June; and I believe he will haul up in the same place each year, for I particularly noticed some that I could tell that hauled up in the same place for a number of years; and when we make drives, those we do not kill, but let go into the water, are all back where we took them from in a few hours. The pups are born between the middle of June and the middle of July, and can not swim until they are 6 or 7 weeks old; and if born in the water they would die. I have seen the surf wash some of the young pups into the sea, and they drowned in a very short time. In four or five days after it is born the mother seal leaves her pup and goes away in the water to feed, and when the pup is 2 or 3 weeks old the mother often stays away for five or six days at a time. The mother seals know their own pups by smelling them, and no seal will allow any but her own pup to suck her. When the pups grow to be 6 or 8 weeks old they form in "pods" and work down to the shore, and they try the water at the edge until they learn to swim. They will remain on the island until November, and, if not too cold, will stay till December. I have seen them swimming around the island late in January. All the seals when they leave the islands go off south, but I think they would stay around here all winter if the weather was not so cold.

When they come back to the islands they come from the south, and I think they come from the North Pacific Ocean over the same track that they went. The females go upon the rookeries as soon as they arrive here, but the yearlings, males and females, herd together. I think they stay in the water most of the time the first year, but after that they come regularly to the hauling grounds and rookeries, but do not come as early in the season as they do after they are 2 years old. Male seals from 2 to 6 years old do not go on the breeding rookeries, but haul out by themselves. The female seal gives birth to but one pup every year, and she has her first pup when she is 3 years old. The male seal establishes himself on the breeding rookery in May or June, when he is 7 or 8 years old, and he fights for his cows and does not leave the place he has selected until August or September. Our people like the meat of the seal, and we eat no other meat so long as we can get it.

The pup seals are our chicken meat, and we used to be allowed to kill 3,000 or 4,000 male pups every year in November; but the Government agent forbade us to kill any in 1891, and said we should not be allowed to kill any more, and he gave us other meat in place of pup meat, but we do not like any other meat as well as the pup-seal meat. We understand the danger there is in the seals being all killed off, and that we will have no way of earning our living. There is not one of us but what believes if they had not killed them off by shooting them in the water there would be as many seals on the island now as there was in 1880, and we could go on forever taking 100,000 seals on the two islands. But if they get less as fast as they have in the last five or six years, there will be none left in a little while.

KERRICK ARTOMANOFF.

Deposition of Daniel Webster.

ALASKA, UNITED STATES,
    St. George Island, Pribilof Group, ss:
    Daniel Webster, being duly sworn, deposes and says: I am 60 years of age, and am a resident of Oakland, Cal.; my occupation is that of local agent for the North American Commercial Company, and at present I am stationed on St. George Island, of the Pribilof group, Alaska; I have been in Alaskan waters every year but two since I was 14 years of age. I first went to Bering Sea in 1845, on a whaling voyage, and annually visited these waters in that pursuit until 1868, at which time the purchase and transfer of Alaska was made to the United States; since that time I have been engaged in taking of fur seals for their skins. In 1870 I entered the employ of the lessees of the Pribilof Islands, and have been so engaged ever since, and for the last thirteen years have been the company's local agent on St. George Island, and during the sealing season have, a part of the time, gone to St. Paul Island and took charge of the killing at Northeast Point, which is known to be the largest fur-seal rookery in the world. For ten years prior to 1878 I resided most of the time at Northeast Point, having landed and taken seals there in 1868. I have had twenty-four years' experience in the fur-seal industry as it exists in the waters of the North Pacific and Bering Sea, and have made a very careful study of the habits and conditions of this useful animal. During this period it has been my duty as a trusted employee of the lessees to observe and report, each year, the condition of the rookeries. My instructions were explicit and emphatic to never permit, under any circumstances, any practices to obtain that would result in injury to the herds. These instructions have been faithfully carried out by myself and other employees of the lessees of the Islands, and the laws and regulations governing the perpetuation of seal life have been rigidly enforced by all the Government agents in charge of the islands.

    In my twenty-three years' experience as a whaler in Bering Sea and the North Pacific, during which time I visited every part of the coast surrounding these waters, and my subsequent twenty-four years' experience on the seal islands in Bering and Okhotsk seas, I have never known or heard of any place where the Alaskan fur seals breed except on the Pribilof group in Bering Sea. These islands are isolated and seem to possess the necessary climatic conditions to make them the favorite breeding grounds of the Alaskan fur seals, and it is here they congregate during the summer months of each year to bring forth and rear their young. Leaving the islands late in the fall or in early winter, on account of the inclemency of the weather, they journey southward through the passes of the Aleutian Archipelago to the coast of California, Oregon, and Washington, and, gradually working their way back to Bering Sea, they again come up on the rookeries soon after the ice disappears from the shores of the islands; and my observation leads me to believe that they select, as near as possible, the places they occupied the year before. The young seals are born on the breeding rookeries in June and July. The head constitutes the greater part of this animal at this time, and they are clumsy and awkward in all their movements, and if swept into the water by accident or otherwise would perish from inability to swim—a fact that I have often observed, and one which is well known to all who have paid any attention to the subject. Practically, they remain in this helpless condition, though taking on fat rapidly, until they are from 6 to 7 weeks old, when they commence to go into shallow water, and, after repeated trials, learn to swim; but even then they spend most of their time on land until they leave the islands late in November. During the first few weeks after their birth they are not amphibious, and land is a necessity to their existence. The mother seals go out to sea to feed soon after giving birth to their young, and return at intervals of from a few hours to several days to suckle and nourish their young.

The mother seal readily distinguishes her own offspring from that of others, nor will she permit the young of any other seal to suckle her. I have noticed in the killing of young seals (pups) for food in November that their stomachs were full of milk, although, apparently, the mothers had not been on the islands for several days previous. I have observed that the male seals taken in the forepart of the season, or within a few days after their arrival at the islands, are fat and their stomachs contain quantities of undigested fish (mostly cod), while the stomachs of these killed in the latter part of the season are empty; and they diminish in flesh until they leave the islands late in the season.

I am of the opinion that while the female often goes long distances to feed while giving nourishment to her young, the male seals of 2 years old and over seldom, if ever, leave the islands for that purpose until they start on their migration southward. When the seals are on the breeding grounds they are not easily frightened unless they are too nearly approached, and even then they will go but a short distance if the cause of their fright becomes stationary.

It is impossible to estimate with any sort of accuracy the number of seals on the Pribilof Islands, because of the seals being constantly in motion, and because the breeding grounds are so covered with broken rocks of all sizes that the density varies. I think all estimates heretofore made are unreliable, and in the case of Elliott and others who have endeavored to make a census of seal life, the numbers are, in my opinion, exaggerated. Measurements of the breeding grounds, however, show an increase or decrease of the number of seals, because the harems are always crowded together as closely as the nature of the ground and temper of the old bulls will permit. My observation has been that there was an expansion of the rookeries from 1870 up to at least 1879, which fact I attribute to the careful management of the islands by the United States Government. In the year 1880 I thought I began to notice a falling off from the number of seals on Northeast Point rookery, but this decrease was so very slight that probably it would not have been observed by one less familiar with seal life and its conditions than I; but I could not discover or learn that it showed itself on any of the other rookeries. In 1884 and 1885 I noticed a decrease, and it became so marked in 1886 that everyone on the islands saw it. This marked decrease in 1886 showed itself on all the rookeries on both islands.

Until 1887 or 1888, however, the decrease was not felt in obtaining skins, at which time the standard was lowered from 6 and 7 pounds to 5 and 4½ pounds. The hauling grounds of Northeast Point kept up the standard longer than the other rookeries, because, as I believe, the latter rookeries had felt the drain of open-sea sealing during 1885 and 1886 more than Northeast Point, the cows from the other rookeries having gone to the southward to feed, where the majority of the sealing schooners were engaged in taking seal. There was never while I have been upon the island any scarcity of vigorous bulls, there always being a sufficient number to fertilize all the cows coming to the islands. It was always borne in mind by those on the islands that a sufficient number of males must be preserved for breeding purposes, and this accounts partly for the lowering of the standard weight of skins in 1888. The season of 1891 showed that male seals had certainly been in sufficient number the year before, because the pups on the rookeries were as many as should be for the number of cows landing, the ratio being the same as in former years. Then, too, there was a surplus of vigorous bulls in 1891 who could obtain no cows. Every care is taken in driving the seals from the hauling to the killing grounds, and during the regular killing season of June and July there are no females driven, because at this season they are on the breeding rookeries and do not intermingle with the young males. If occasionally one does happen to be in the drive great care is taken not to injure her; the law prohibiting the killing of the female seal is well understood by the natives, and they are thoroughly in sympathy with it. Even were I to request them to kill a female seal they would refuse to do it, and would immediately report me to the Government agent. I have known an occasional one to be killed by accident during the food drives late in the season, when the males and females intermingle on the hauling grounds, but the clubber was always severely rebuked by the chief for his carelessness, as well as by the Government and company officers.

My observation is that the number of female seals killed on the islands from all causes is too insignificantly small to be noticed. The longest drives made on St. George Island are from Starry Arteel and Great Eastern rookeries, and they are less than 3 miles long. Drives from these rookeries require from four to six hours, according to the weather. At Zapadnie rookery, on St. George, the drive to the killing grounds is less than a mile, the seals are now being killed there instead of being driven across the island as they were prior to 1878, when it took three days to make the journey. There is now a salt house at Zapadnie, at which the skins are salted as soon as taken. The killing grounds on both islands are all situated within a very short distance from the shore, and seals not suitable to be killed, or that are turned out for any cause, immediately go into the water, and, after sporting around for an hour or two, they return to the hauling grounds, and to all appearances they are as unconcerned and careless of the presence of man as they were before they were driven to the killing grounds. I have often observed that the seals when on the islands do not take fright easily at the presence of man; and the natives go among them with impunity. They will go into a herd of seals on the hauling grounds and quietly separate them into as many divisions and subdivisions as is necessary before driving them to the killing grounds. At the killing grounds they are again divided into bunches or "pods" of 20 or 30 each more readily than the same number of domestic animals could be handled under the same circumstances.

The bulls on the rookeries will not only stand their ground against the approach of man, but will become the aggressors if disturbed. Pups are tame and very playful when young, and previous to 1891, when it was the practice to kill 3,000 or 4,000 for natives' food in November, thousands of them were picked up and handled to determine sex, for only the males were allowed to be killed. Hair seal and seal lions haul out on the islands and are seldom disturbed, yet they will plunge into the water at once should they discover anyone upon their rookeries. But it is not so with the fur seal. They seem at home on the rookeries and hauling grounds, and they show a degree of domestication seldom found among similar animals. At Northeast Point rookery, on St. Paul Island, the longest drive is 2 miles. In former times the Russians used to drive from this rookery to St. Paul village, a distance of 12½ miles. Seals turned away from the killing grounds return to the rookery from which they were driven; therefore a male seal is not redriven day after day, because a hauling ground is always given several days' rest before being driven from again. I never saw or heard of the generative organs of a male seal being injured by driving or by redriving, and if such a thing had taken place, even in exceptional cases, the natives would have noticed and reported it, which they never did. I have seen a seal's flippers made sore by driving, but I never saw one that was seriously injured by driving. I do not believe that a male seal's powers of reproduction were ever affected by driving or redriving.

The bulls maintain their positions on the rookeries from the time they arrive till the cows come by most bloody battles, and after the cows commence arriving they are continually contending for their possessions. During these conflicts they are often seriously wounded, and their exertions are far more violent than any effort made by a young male during a drive. Then, too, the male seal must have great vitality to remain on the rookeries for three months without eating or drinking and with little sleep. In spite of this drain on his vital force he is able to fertilize all the cows which he can get possession of, and a barren cow is a rarity. I believe that a bull can serve one hundred or more cows, and it is an absurdity to think that an animal possessing such remarkable vigor could be made impotent by being driven or redriven when a bachelor. An impotent bull would have neither the inclination or vigor to maintain himself on the rookeries against the fierce and vigorous possessors of harems. The only bulls hauling up away from the breeding rookeries are those whose extreme old age and long service have made them impotent and useless, and I have never seen or heard tell of anything that would make an exception to this rule. The methods employed in taking the skins are, in my opinion, the best that can be adopted. The killing grounds are situated as near the rookeries and hauling grounds as is possible without having the breeders or bachelors disturbed by the smell of blood or putrefaction, and most stringent regulations have always been enforced to prevent disturbing or frightening the breeding seals.

I am convinced that if open-sea sealing had never been indulged in to the extent it has since 1885, or perhaps a year or two earlier, 100,000 male skins could have been taken annually forever from the Pribilof Islands without decreasing the seal herd below its normal size and condition. The cause of the decrease which has taken place can be accounted for only by open sea sealing; for, until that means of destruction to seal life grew to be of such proportions as to alarm those interested in the seals, the seal herd increased, and since that time the decrease of the number of seals has been proportionate to the increase in the number of those engaged in open-sea sealing. The majority of seals killed in the water are females, and all the females killed in Bering Sea are mothers who have left their pups on the rookeries and gone some distance from the islands in search of food. The death of every such mother seal at sea means the death of her pup on shore, because it is absolutely and entirely dependent on her for its daily sustenance. I never heard of any disease among the seal herd, nor of an epidemic of any sort or at any time in the history of the islands. I do not remember the precise date of the first successful raid upon the rookeries by sealing schooners, but I do know that for the past ten years there have been many such raids attempted, and a few of them successfully carried out, and that as the number of schooners increased around the islands, the attempted raids increased in proportion, and it has been deemed necessary to keep armed guards near the rookeries to repel such attacks. Although a few of the raids were successful, and a few hundred seals killed and carried off from time to time during the past ten years, the aggregate of all the seals thus destroyed is too small to be mentioned when considering the cause of the sudden decline of seal life on the Pribilof Islands.

Twenty-four years of my life have been devoted to the sealing industry in all of its details as it is pursued upon the Pribilof Islands, and it is but natural that I should become deeply interested in the subject of the seal life. My experience has been practical rather than theoretical. I have seen the herds grow and multiply under careful management until their numbers were millions, as was the case in 1880. From 1884 to 1891 I saw their numbers decline, under the same careful management, until in the latter year there was not more than one-fourth of their numbers coming to the islands. In my judgment there is but one cause for that decline and the present condition of the rookeries, and that is the shotgun and the rifle of the pelagic hunter, and it is my opinion that if the lessees had not taken a seal on the islands for the last ten years we would still find the breeding grounds in about the same condition as they are to-day, so destructive to seal life are the methods adopted by these hunters. I believe the number they secure is small, as compared with the number they destroy. Were it males only that they killed the damage would be temporary, but it is mostly females that they kill in the open waters, and it is plain to anyone familiar with this animal that extermination must soon follow unless some restrictive measures are adopted without delay.

The foregoing is substantially the same statement that I made to the commissioners who visited the islands in 1891.

DANIEL WEBSTER.

PELAGIC SEALING AND PRIBILOF ROOKERIES.

Deposition of Washington C. Coulson, United States Revenue Marine, in command of the Rush.

STATE OF CALIFORNIA,
City and County of San Francisco, ss:

Washington C. Coulson, having been duly sworn, deposes and says: I am captain in the United States Revenue Cutter Service. At present I am in command of the United States revenue cutter *Rush*. I was attached to the United States revenue cutter *Lincoln*, under the command of Capt. C. M. Scammon, during the year 1870, from June until the close of the year as a third lieutenant, and have been an officer in the revenue service ever since. In the month of that year that I was in the Bering Sea and at the seal islands of St. Paul and St. George. I went on shore at both islands and observed the seals and seal life, the method of killing, etc. I noticed particularly the great number of seal, which were estimated by those competent to judge that at least 5,000,000 and possibly 6,000,000, were in sight on the different rookeries. To me it seemed as though the hillside and hauling grounds were literally alive, so great was the number of seals. At St. George Island, though the seals were never in as great numbers nor were there so many hauling places, the seals were very plentiful. At this time and for several year thereafter pelagic sealing did not take place to any extent and the animals were not diverted from their usual paths of travel. All fire arms were forbidden and never have been used on these islands in the killing and taking of seals. In fact, unusual noise even on the ships at anchor near these islands is avoided.

Visiting the rookeries is not permitted only on certain conditions, and anything that might frighten the seals avoided. The seals are never killed in or near the rookeries, but are driven a short distance inland, to grounds especially set apart for this work. I do not see how it is possible to conduct the scaling process with greater care or judgment. Under the direction of Mr. Redpath, on St. Paul, and Mr. Webster, on St. George islands—men who have superintended this work for many years—the natives do the driving, and the killing is performed under the supervision of the Government agents. The natives understand just how much fatigue can be endured by the seals, and the kind of weather suitable for driving and killing; no greater precaution in that regard can be taken. The evidence of this is in the small percentage of animals injured or overheated in these drives. I do not believe the animals are much frightened or disturbed by the process of selecting the drives from the rookeries, nor do I think it has a tendency to scare the animals away from the islands.

During the seasons of 1890 and 1891 I was in command of the revenue cutter *Rush* in Bering Sea and cruised extensively in those waters around the seal islands and the Aleutian group. In the season of 1890 I visited the islands of St. Paul and St. George in the months of July, August, and September, and had ample and frequent opportunities of observing the seal life as compared with 1870. I was astonished at the reduced numbers of seals and the extent of bare ground on the rookeries in 1890 as compared with that of 1870, and which in that year was alive with seal life. In 1890 the North American Commercial Company were unable to kill seals of suitable size to make their quota of 60,000 allowed by their lease, and, in my opinion, had they been permitted to take 50,000 in 1891, they could not have secured that number if they had killed every bachelor seal with a merchantable skin on both islands, so great was the diminution in the number of animals found there.

I arrived with my command at St. Paul Island June 7, 1891; at that date very few seals had arrived and but a small number had been killed for fresh food. On the 12th of June, 1891, we were at St. George Island and found a few seals had been taken there, also for food, the number of seals arriving not being enough to warrant the killing of any great number. During that year I was at and around both these islands every month from and including June until the 1st day of December (excepting October), and at no time were there as many seals in sight as in 1890. I assert this from actual observation, and it is my opinion we will find less this year; and should pelagic sealing in the North Pacific and Bering Sea continue, it is only a question of a very few years when seal in these seas, and especially at the seal islands, will be a thing of the past, for they are being rapidly destroyed by the killing of females in the open sea.

As to the percentage of seals lost in pelagic sealing where the use of firearms is employed, I am not able to state of my own observation, but from conversations with those engaged in the business I am of the opinion that the number secured is small compared with those lost in attempts to secure them. No mention was ever made of any unusual number of dead pups upon the rookeries having been noticed at any time prior to my visit in 1870, but when I again visited the islands in 1890 I found it a subject of much solicitude by those interested in the perpetuation, and in 1891 it had assumed such proportions as to cause serious alarm. The natives making the drives first discovered this trouble, then special agents took note, and later on I think almost every-one who was allowed to visit the rookeries could not close their eyes or nostrils to the great numbers of dead pups to be seen on all sides. In company with Special Agent Murray, Captain Hooper, and Engineer Brerton, of the *Corwin*, I visited the Reef and Garbotch rookeries, St. Paul Island, in August, 1891, and saw one of the most pitiable sights that I have ever witnessed. Thousands of dead and dying pups were scattered over the rookeries, while the shores were lined with emaciated, hungry little fellows, with their eyes turned toward the sea uttering plaintive cries for their mothers, which were destined never to return. Numbers of them were opened, their stomachs examined, and the fact revealed that starvation was the cause of death, no organic disease being apparent.

The greatest number of seals taken by hunters in 1891 was to the westward and northwestward of St. Paul Island, and the largest number of dead pups were found that year in rookeries situated on the western side of the island. This fact alone goes a great way, in my opinion, to confirm the theory that the loss of the mothers was the cause of mortality among the young.

After the mother seals have given birth to their young on the islands they go to the water to feed and bathe, and I have observed them not only around the islands, but from 80 to 100 miles out at sea.

In different years the feeding grounds or the location where the greater number of seals are taken by poachers seem to differ; in other words, the seals frequently change feeding grounds. For instance, in 1887, the greatest number of seals were taken by poachers between Unamak, Akutan Passes, and the seal islands, and to the southwestward and east-ward, in many cases from 50 to 150 miles distant from the seal islands. In the season of 1890 to the southward and westward, also to north-west and northeast of the islands, showing that the seals had been scat-tered. The season of 1891 the greatest number were taken to northward or westward of St. Paul, and at various distances, from 25 to 150 miles away.

On my cruise to St. Matthews and Unamak Island we did not discover any seal within 25 or 30 miles of those islands, nor do I know of or believe that the seals haul out upon land in any of the American waters of Bering Sea except at the Pribilof Islands. If the seal life is to be preserved for commercial purposes the seals must be protected, not only in the Bering Sea, but in the waters along the Pacific Coast from the Aleutian Passes to the Columbia River.

WASH. C. COULSON,
Captain, United States Revenue Marine.

Deposition of Thomas F. Morgan, agent of lessees of Pribilof and Commander islands.

STATE OF CONNECTICUT,
  New London County, ss:

Thomas F. Morgan, being duly sworn, deposes and says: I am the person described in and who verified two certain affidavits on the 5th day of April, 1892, before Sevellon A. Brown, notary public, in rela-tion to the habits, management, etc., of the fur seals.

The harems on the Pribilof Islands have at all times varied very much in size. In the years when I was on the islands, between 1874 and 1887, it was always possible to find individual harems with 50 or perhaps 80 females, while others would only have 4 or 5 females, notwithstanding the average harem would perhaps contain from 15 to 30 females. Large harems, though in smaller numbers, continued to exist even in the years 1885 and 1887, when, as I have already stated in a former affidavit, the number of females began to decrease.

While I was on the islands there was no such thing known as disturbing breeders or stampeding the rookeries. The herd is driven from the rookery, is kept away from filth as much as possible, for the reason that the skins which are taken, if clean, take salt better, cure in better condition, and bring better prices. Filth, grease, and oil make skins come out of kench flat, and such skins are classed as low when sold. Mud spoils the salt for quick work, so the cleaner the skins are, the better. As the rejected seals are only to be got away from the killing ground, the quickest way to the sea is the route chosen, and they often pass over decaying carcasses, but not of necessity, as they are allowed to choose their own gait and route to the sea. They do not seem to object to this any more than to the filth caused by the excrement and decaying placentas on the breeding grounds.

I was on the Commander Islands in 1891 as agent of the Russian Seal Skin Company. I never heard anyone state that barren females (I mean females without young) were noticed there, and I don't believe that any person whose opinion would be entitled to consideration noticed this fact. It soon would have become a matter of common knowledge on the islands if there had been any number of adult females without young. The only sure way to determine whether an adult female is barren is to examine her as to whether she is giving milk or is dry. As the young seals do not follow the mother continuously, the fact of seeing females without pups with them does not prove that they have not pups somewhere on the breeding grounds, and no person having any knowledge of rookery life could draw such an inference, and claim that the females were therefore barren.

While on the Pribilof Islands I don't know that I ever saw a sterile female seal. It is impossible to recognize the same seal from year to year unless, as in the case of a few old bulls which have large scars, a torn lip, a white blind eye, the nose split, or some unnatural mark. And, although I have seen old females without milk, very fat, associating with the young males, I could not say that they had not been fertilized, and, not having an offspring to care for, were associating with the males until the season arrived for the herd to leave. At one time the suggestion was made that it would be a good plan to kill these females. I denied that it was possible for anyone to know that they would not bear young, and that if the killing of one female was authorized it would open the way to do great injury to the herd. For, when it became desirable to market a large number of skins, the clubber would see large numbers of females unfit for breeding.

It is difficult to discover fresh excrement on the rookeries, for the seals' flippers soon wipe out the evidence looked for. Still I have often seen it. In color it is orange, light yellow to almost colorless, and in consistency soft, almost liquid. At times it is very offensive, and at others nearly odorless. But the soil of the breeding ground is impregnated with it, which gives to the rookery a most disagreeable odor that is increased by the decaying placentas.

I am quoted by the British commissioners (section 825 of their report) to show that in 1884 au irregularity in the habits of the seals took place at the Pribilof Islands. This irregularity consisted in the following: In previous years the seals that arrived in June furnished nearly all 8-pound skins and over; very few of these seals were let go or rejected, and when any were rejected it was principally because they were too large. But this year the 2-year-old seals commenced to land much earlier, and the run of large half bulls arrived in more scattered bunches, just as if the herd had been turned back in places and hurried ahead in others, thus hurrying the smaller seals, so that they came on with the head of the flock, and turning back some of the large seals which formerly had arrived later. No irregularity was observed in the habits of the female seals.

THOMAS F. MORGAN.

Deposition of James G. Swan, former inspector of customs, employee of Indian Bureau and of Fish Commission of United States.

STATE OF WASHINGTON,
    Jefferson County, ss:

James G. Swan, having been duly sworn, deposes and says: I am 74 years old, a resident of Port Townsend, Wash., and by occupation a lawyer. I am also United States commissioner, Hawaiian consul, commissioner for the State of Oregon, and a notary public. I came to the Pacific Coast in 1850 and to Port Townsend in 1859, where I have since held my residence the greater part of the time to the present date. From 1862 to 1866 I was employed in the Indian Bureau of the Interior Department and stationed at Neah Bay, and again from 1878 to 1881 I was inspector of customs at the same place. In 1883 I also visited there in the employ of the Fish Commissioner.

In 1880, at the request of the late Professor Baird, of the Smithsonian Institute at Washington, I made a careful study of the habits of the fur seal (*Callorhinus ursinus*) found in the vicinity of Cape Flattery and the Strait of Juan de Fuca, and the result of my observation is embodied in the Tenth United States Census (report of United States Fish and Fisheries, sec. 5, vol. 2, p. 293. Fur seal of Cape Flattery and Vicinity) and in the report of the United States Fish Commission. (Bulletin United States Fish Commission, vol. 3, pp. 201-207.)

The observations upon which these reports are based were mostly confined to the immediate vicinity of Cape Flattery, and I had at that time no opportunity for extended inquiry as to the pelagic habits of the animals. The natural history of the seal herd of the Pribilof Islands, when upon or in the immediate vicinity of the land, had been minutely, and, I have no doubt, accurately, described by H. W. Elliott in his monograph published in 1875. There had been up to that date no series of observations nor good evidence on which to base the hypothesis that the Pribilof herd and the large mass of seals annually seen on the latitude of Cape Flattery were identical. On the contrary, there seemed then to be many evidences that some other rookeries than those of the Pribilof Islands were located at some point on the Oregon, Washington, or British Columbia coast. Young seals were occasionally found by the Indians upon or near the beaches, and pregnant females were often captured by them so heavy with pup, and apparently so near their full term of pregnancy, as to warrant the belief that the young must be either born in the water upon bunches of kelp or upon the rocks and beaches on or near the coast. Young seals were often brought to the

Indian villages, and the testimony of both Indian and white hunters at that time pointed strongly to the conclusion that the breeding grounds of the animals with which we were familiar could not be far distant. I have myself seen the black pups in the water when they appeared to be but a few weeks old, and others have assured me that a considerable number were found from time to time swimming with their mothers. This phenomenon being of constant occurrence year after year, and in the absence of a wider range of observations, we were naturally confirmed by them in the conclusion to which I have above referred.

In recent years it has been demonstrated by the large catches obtained off the coast by pelagic hunters, and by the testimony of a great number of people whose attention has been directed to the matter, that the herd of seals, of which we saw only a very limited proportion from the Neah Bay station, is a very large one; and it now seems beyond a doubt that the comparatively few authentic cases in which pups were seen upon or in the vicinity of the coast were anomalous, for it is reasonable to suppose that in so large a mass of pregnant females an occasional one would be prematurely overtaken by the pains of the parturition, and that the offspring brought forth under favorable conditions, as upon a bunch of kelp or some rock, should survive at least a few days and be brought in and kept by the Indians, as I have occasionally seen them. I have also seen at the villages late in the season, in the hands of the Indian boys, live pups which had been recently removed from their speared mothers, and whose vitality was such that they continued to live for several days; but it is a well-known fact that young mammalia may be born several days, or possibly even a month or two, before full term and still survive. It is possible, too, that as a source of error the hunters may have mistaken gray pups whose coats had been darkened by wetting, or those a few months old, born the preceding summer, for the so-called black pups.

At the Neah Bay station large bull seals are seldom seen, and the major part of those killed are pregnant females, having in them small fetuses early in the season—say about January or February—and later full-grown young. From all the evidence I am able to gather, I believe the different classes of seals remain apart when upon the British Columbia coast, and old bulls and immature young males being chiefly found at a considerable distance from the land, while the pregnant females and young males travel close along the shore, and are frequently seen in limited numbers in the straits and inlets.

In the light of investigation and research had since the date of my observations, the most of which were made more than ten years ago, I am satisfied that the mass of the herd from which the British Columbia or Victoria catch is obtained are born neither in the water nor upon the land in the vicinity where they are caught, and it appears most probable from the routes upon which they are followed and the location in which they are found by pelagic hunters between March and August that they originate in, migrate from, and annually return to Bering Sea.

It has been stated in print that I said I had seen pups born on the kelp in the water. This is a gross misrepresentation. I merely said that it had been reported to me that such birth had been witnessed, and quoted as my authority Capt. E. H. McAlmond, of the schooner *Champion* (p. 203, vol. 1, of United States Fish Commission's report).

Pelagic sealing was carried on by the Indians at Neah Bay long before I first went among them, but they were then, and until within a few years, provided only with their canoes, spears, and other native implements, constituting the necessary outfit for an aboriginal seal hunter. The destruction wrought by them upon the seal herd was, compared with the vast number of which it was composed, very slight, and did little harm to anyone, while the result to the Indians was then and is still of great importance. Now pelagic seal hunting is carried on in quite a different manner. Numerous expeditions are fitted out in well-equipped vessels, some of them under both steam and sail, manned by whites and Indians, and armed with guns and spears. I am informed and believe that the herd has greatly decreased within the last two or three years, and that if pelagic sealing is not soon checked the herd will be driven hither and thither and so decimated as to render it commercially valueless. This would be a great wrong to the Indians, who are dependent to a great measure upon the seals for a livelihood, as well as needless, wanton waste, which civilized nations ought not to permit. It can not be denied that the natives, who have utilized the seal fisheries adjacent to their settlements from their earliest history and profited by them, deserve some consideration. I believe that in order to preserve the rookeries upon the islands and build them up to their former productiveness it is only necessary to restrict pelagic sealing to the coast south of 54° 40' and confine it to the use of the primitive methods formerly employed by the natives.

JAMES G. SWAN.

Deposition of Joseph Stanley-Brown, Treasury agent.

DISTRICT OF COLUMBIA,
  City of Washington, ss:

Joseph Stanley-Brown, being duly sworn, deposes and says: I am 37 years of age; am a citizen of the United States; reside at Mentor, Ohio, and am by profession a geologist.

I spent the entire season of 1891 upon the Pribilof Islands, and during the summer of 1892 again visited them and spent the period between June 9 and August 14 upon the islands of St. Paul and St. George in continuation of my investigations concerning seal life. This season, in addition to the continuous general examination of all the rookeries and the plottings of the breeding-ground areas upon charts, certain special stations were selected at points within easy reach of the village and daily visits made thereto. This method of work gave me an excellent opportunity to make comparisons between the breeding areas of 1891 and those of 1892.

As the result of my observations during the past season, it is my opinion that there was no increase among the females—the producing class—but on the contrary that there was a perceptible falling off. This decrease was the more noticeable at points on the rookeries where the smaller groups of breeding seals are to be found.

There was so little driving during the season of 1892 that an excellent opportunity was given to observe life upon the hauling grounds, several of which were not disturbed during the entire season. There seemed to be a slight increase of the young bachelor seals, although this may have been more apparent than real from the fact that being unmolested they accumulated in large bands.

It is quite certain that the normal habit of the holluschickie is to remain most of their time upon shore, and if left to themselves spend more time there than in the water. I have kept a close daily watch upon groups of young males, the members of which did not go into the water for a week or ten days at a time.

H. Doc. 92, pt. 2——8

Any statement to the effect that the occasional occurrence of large harems indicates a decrease in the available number of virile males, and hence deterioration of the rookeries, should be received with great caution if not entirely ignored. The bulls play only a secondary part in the formation of harems. It is the cow which takes the initiative. She is in the water beyond the reach or control of the male and can select her own point of landing. Her manner on coming ashore is readily distinguished from that of the young males which continuously play along the sea margin of the breeding grounds. She comes out of the water, carefully noses or smells the rocks here or there like a dog, and then makes her way to the bull of her own selecting. In this incipient stage of her career on shore there is but little interference on the part of the male, but once well away from the water and near the bull she has chosen, he approaches her, manifests his pleasure, and greetings are exchanged. She then joins the other cows and as soon as dry lies down and goes comfortably to sleep. I have seen this selective power exercised repeatedly, and the result is that one bull will be especially favored while those within 15 or 20 feet will be ignored.

The size of the harems, therefore, has of itself but little to do with the question of lack of virile males, but indicates only the selective power of the females. If 100 bulls represented the necessary supply of virile males we might, by reason of this fact, find 10 bulls with very large harems, 10 with still less, 50 with a reasonable number, 20 with a few, and 10 with none. An onlooker would not, therefore, be justified in stating that by reason of these few large harems there is a lack of virile males.

In the very nature of things it seems impossible that any method other than this one of selection on the part of the female could ever have existed.

Large harems are frequently due to topographic conditions, the configuration of the land being such that the females can only reach the breeding grounds through narrow passageways between the rocks, and around the terminations of which they collect.

Harems often coalesce; then boundaries become indefinite, and when their size and position make them too large for control, cows pass to the rear and are appropriated by the bulls there.

When once the female is located, the bull exercises rigid control and permits no leaving of the lands until she has been served. I never saw a harem so large that the vigilance of the bull in this respect was ever relaxed. His consorts may escape to another harem, but they are never permitted to go to sea until an inspection convinces the bull that they are entitled to do so. No intelligent observer would be so bold as to assert that during the season of 1892 there was not an abundance of males of complete virility, despite the occurrence of occasional large harems. The accompanying photographs[1] show that even at the height of the season, and just previous to the disintegration of the breeding grounds, there were unsupplied with cows old males which had taken their stand and from which I was unable to drive them with stones.

I should have been extremely glad to have been able to note a great many more of these large harems, but the work of the pelagic hunter among the females has been so effective that the average size of the harems is growing smaller and smaller, while the number of the idle bulls is steadily increasing. The rookeries of the Pribilof Islands will never be destroyed by superabundance of large harems.

_____
[1] Not furnished.

I arrived on the islands this year a few days after the coming of the first cows, and by selecting a small harem composed of seals, the arrival of which I have seen, and giving it daily observation, I was able to satisfy myself that females begin to go into the water from fourteen to seventeen days after first landing. On first entering the sea they make a straight line for the outer waters, and as far as the eye can follow them they seem still to be traveling. The first cows to arrive are the first to depart in search of food, and by the first week in July the cows are coming and going with such frequency as to be readily seen at any time. The accompanying photograph[1] (taken on July 8, 1892, from the same position but one day earlier than the one of last year which faces page 13 of volume 2, of the case) shows pups, the mothers of which are at sea.

The fact that the coat of the cow assumes from residence on the shore a rusty or sunburned aspect gives a ready means of observing her movements. The rustiness is quickly lost by life in the sea.

The movements of females can also to a certain extent be well observed by their appearance after giving birth to their pups—after fasting and after gorging themselves with food. After the birth of the pup, and after remaining upon the rookeries even for a few days when the period of coming from and going into the water has been entered upon, the mother has a very decidedly gaunt appearance, in strong contrast to the plumpness of pregnancy or full feeding. After feeding at sea they come ashore again well rounded up. So marked is this that I have been repeatedly misled by mothers in such a condition, mistaking them for pregnant cows, and have discovered my error by seeing her call her pup and suckle it. If I had any doubt in my mind as to cows feeding at sea it was dispelled by an examination of three cows I shot at Northeast Point on July 25, 1892. Two "sunburnt" cows were first killed, and their stomachs were found to be empty. Another was shot just as she came ashore and her stomach was gorged with half-digested codfish, which was identified by Mr. Townsend, an expert of the United States Fish Commission. A dissection was made of this seal, and the udder—which extends, as a broad, thick sheet, thinning out toward the edges, over the entire abdominal portion of the cow and well up to the fore flippers—was so charged with milk that on removing the skin the milk freely flowed out in all directions, and previous to skinning it was possible with but little effort to extract a sufficient amount to enable me to determine its taste and consistency. A large supply of food is necessary to furnish such an abundant amount of milk. I have no doubt that a well-developed mother seal could yield between a pint and a quart of milk in the first twenty-four hours after landing from a feeding expedition, and with such rich fountains to draw upon it is no wonder that the voracious pups increase during their residence upon the island not less than four times their weight at birth. And it is equally certain that without such a constant supply of nourishment they could not make such a rapid growth as they do.

The presence of excrementitious matter upon the breeding rookeries is recognized both by sight and smell. It is of a yellowish color, and though much of it is excreted, it is of such a liquid consistency that it is quickly rubbed into and mingled with the soil, and thereafter its existence can only be noticed through the discoloration of the soil and the offensive odor. The latter is readily detected at a distance of miles, when the wind is completely impregnated with it. The odor bears no

---

[1] Not furnished.

resemblance to that which arises from the bodies of a large number of assembled animals.

The quantity of excrementitious matter present is influenced by the nature of their diet, which, being fish, is largely assimilated, while in their coming and going much of it may be deposited in the water, to say nothing of drenching from rain, to which the rookeries (many of which are solid rock) are subjected.

On the hauling grounds, on the other hand, it is almost impossible to detect such matter, either through its presence, the appearance of the soil, or its odor. This is a well-known fact to anyone who has even casually inspected such hauling grounds as Middle Hill, parts of Zapadnie, western end of English Bay, western end of North Rookery, Starry Arteel, Great East Rookery, and others.

This difference between the breeding grounds and the true hauling grounds is explained by the fact that the former are occupied by nursing females, which are constantly feeding, while the latter are frequented chiefly by young males, which take but little food during the summer. This abstention from food on their part is further indicated by the fact that, with exceptions now and then observed on the killing grounds, they grow thinner and thinner as the season advances.

The pup at birth is received by the mother with an affectionate regard that is unmistakable; a sound not unlike that made by an ewe, but not so loud, can be heard, and care is exercised by the mother for the pup's protection. I have repeatedly seen a mother, when her offspring was still so young as to be helpless, remove it beyond the reach of the surf, or gently lift it from a hole between the bowlders into which it had fallen. I have seen them often place the udder in the most available position for the pup to suck, and move themselves sufficiently close for it to be within easy reach. After an absence in the sea, the mother invariably calls to her young repeatedly, and manifests pleasure on finding it. Later on the pup is able to recognize its mother, and as the female will suckle only her own pup the pleasure and contentment which the meeting gives both is evident to the most careless observer.

Dead pups were as conspicuous in their infrequency in 1892 as by their numerousness in 1891. In no instance was there to be noted an unusual number of dead pups, except on the breeding grounds of Tolstoi, the position, character, and size of which gave prominence to the carcasses. Here the mortality, while in no way approaching that of the previous season, was still beyond the normal, as indicated by the deaths upon the other breeding grounds.

Any surreptitious killing of the mothers can not be charged with it, for such killing either there or anywhere else on the island would have become the gossip of the village and readily detected by the attempt to dispose of the skins. There are no hauling grounds so close to the breeding areas that the driving of the young males could cause consternation among the females during the breeding season. Stampedes or disturbances can not account for it, for not only are the breeding grounds in this particular case of Tolstoi one-fourth of a mile away from the hauling grounds, namely, at Middle Hill (the nearest point to that breeding ground from which seals were driven in 1891 and 1892), but it would be practically impossible to stampede this breeding ground by any disturbing cause save of such magnitude as to be the subject of common knowledge on the islands, and I know that no cause for such a commotion occurred.

Seals will stand a large amount of annoyance before leaving their harems, or, indeed, being permitted to do so by the bulls, and the man does not live who can stampede rookery bulls. No smoke of vessels or presence of ships ever cause the stampede of an entire breeding ground. Such things have been reported but no one has ever seen it, and it would require persistent effort to accomplish such a result. I have had cause to send natives on several occasions entirely across a rookery, and no stampede ensued. I have thrown eggshells filled with blue paint at female seals, for the purpose of marking them, until rocks and seals were a mass of blue color, but with no disturbing effect. In the prosecution of my investigations I have shot females with a noiseless rifle upon a small detached breeding ground, have crawled in and dragged out the seals killed without causing the other mothers to recede more than 20 feet, and in fifteen minutes thereafter the breeding grounds presented their wonted appearance.

After two seasons' observation I unhesitatingly state that I do not believe there has ever been breeding grounds stampeded in such a wholesale manner as to cause the death of pups. If such occurred in 1891 and 1892 it is certainly extraordinary that only the starvelings met death.

The true explanation of the deaths upon Tolstoi this year is not readily found, and must be sought in local causes other than those indicated above, and I am confident that to none of those causes can be justly attributed the dead pups of 1891 and 1892. The following explanation, based upon my acquaintance with the facts, is offered in a tentative way:

A glance at the map will show that the location and topographic character of this rookery have no counterpart elsewhere on the island. The rookeries upon which deaths are infrequent are those which are narrow and upon the rear of which are precipitous bluffs that prevent the wandering of pups backward. The larger part of Tolstoi, as will be seen from the map, extends far back and has great lateral dimensions. Much of it is composed of drifting sands and it has rather a steep inclination down to the sea. The shore is an open one, and the surf, either gentle or violent, is almost constantly present. As the time for learning to swim approaches the pups find it easy to come down the incline. They congregate in large numbers upon the sandy shore and begin their swimming lessons. This is at a period when they are still immature and not very strong. The buffeting of the waves exhausts them and coming ashore they either wander off, or struggling a certain distance up the incline, made more difficult of ascent by the loose sand of which it is composed, lie down to rest and sleep, and are overlooked by their mothers returning from the sea. I have seen mother seals go up the entire incline seeking their pups.

I find nothing in the history of dead pups upon the island this year which does not confirm my belief that the great mortality of the season of 1891 was due to pelagic sealing in Bering Sea. Had it not been so, there is no reason why the deaths in 1892 should not have been as widely distributed as they were the previous year.

During the past summer particular care was taken to have the drives conducted in the same manner as in previous years, in order that the effect of driving upon the young males might be noted.

From June 10 (the day after my arrival) to the close of the season, on August 9, there were eleven drives made, the longest one being from Middle Hill, about 2 miles from the village killing ground. With two exceptions, no drives were made from the same hauling grounds except at intervals of two weeks. As the killing this year was limited to 7,500, there could be but few seals taken each week, and this necessitated turning back to the water, about 200 yards distant from the killing ground, from 75 to 85 per cent of those driven up, and gave an excellent opportunity to observe the effect of driving upon large bands of seals.

In driving it is true that if the weather is unfavorable a few may die en route, or in anticipation of their death are clubbed, skinned, and their pelts added to the quota. It is also true that sometimes there are manifestations of weariness and exhaustion among the driven seals; that driving causes some excitement; that occasionally smothering occurs, and that there are other episodes happening on and about the killing field which are necessarily incident to and must always form part of the killing of seals on land, and which are likely to obscure the judgment of the observer or be allowed to assume undue prominence in his mind. But the chief question is the potency of these episodes as destructive agents. To what extent do they occur and to what extent do they effect the herd at large are the points to be fairly considered; and their consideration must not be influenced by an exaggeration due to the sensibilities of the observer. Care should be and is at all times exercised to avoid needless waste; but after giving the greatest prominence possible to the injurious methods which are alleged to have been employed at different times since the American occupancy of the islands, my observations lead me to believe that the loss of life from the causes indicated above would be but a fraction of 1 per cent of the seals driven; and I also believe that it can not, with any show of justice, be made to account for or play other than a very insignificant part in the diminution of seal life. After my observations of two seasons I can not believe that creatures which in their maturity possess sufficient vitality to live for eighty or ninety days without food or water, and in which their fetal life can be cut from the mother and still live for days, are as bachelor seals injured in their virility or to any extent disabled physically by the driving to which they are subjected on the Pribilof Islands.

JOSEPH STANLEY-BROWN.

DEAD PUPS.   .

Deposition of J. C. S. Akerly, surgeon United States Revenue Marine, and resident surgeon on St. Paul Island.

STATE OF CALIFORNIA,
    City and County of San Francisco, ss:

J. C. S. Akerly, Ph. B., M. D., having been duly sworn, deposes and says: I am a graduate of the University of California, 1882, and a graduate of the Cooper Medical College, 1885. From June to August 18, 1891, I was surgeon on the revenue-marine steamer *Corwin*. From August 18 to November 24, 1891, I was resident physician on St. Paul Island, one of the Pribilof or seal islands. I am at present a practicing physician at Oakland, Cal. During my stay on the islands I made frequent visits to the different seal rookeries. One thing which attracted my attention was the immense number of dead young seals; another was the presence of quite a number of young seals on all the rookeries in an emaciated and apparently very weak condition. I was requested by the Government agent to examine some of the carcasses for the purpose of determining the cause or causes of their death. I visited and walked over all the rookeries. On all, dead seals were to be found in immense numbers. Their number was more apparent on those rookeries such as Tolstoi and Halfway Point, the water sides of which were on smooth ground, and the eye could glide over patches of ground hundreds of feet in extent which were thickly strewn with carcasses.

Where the water side of the rookeries, as at Northeast Point and the Reef (south of the village), was on rocky ground, the immense number of dead was not so apparent, but a closer examination showed that the dead were there in equally great number scattered among the rocks. In some localities the ground was so thickly strewn with the dead that one had to pick his way carefully in order to avoid stepping on the carcasses. The great mass of dead in all cases was within a short distance of the water's edge. The patches of dead would commence at the water's edge and stretch in a wide swath up into the rookery. Amongst the immense masses of dead were seldom to be found the carcasses of full-grown seals, but the carcasses were those of pups or young seals born that year. I can give no idea of the exact number of dead, but I believe that they could only be numbered by the thousands on each rookery. Along the water's edge, and scattered amongst the dead, were quite a number of live pups, which were in an emaciated condition. Many had hardly the strength to drag themselves out of one's way, thus contrasting strongly, both in appearance and actions, with the plump condition and active, aggressive conduct of the healthy appearing pups.

The majority of the pups, like all healthy nursing animals, were plump and fairly rolling in fat. I have watched the female seals draw up out of the water, each pick out its pup from the hundreds of young seals sporting near the water's edge, and with them scramble to a clear spot on the rookery, and lying down give them suck. Although I saw pups nursing in a great many cases, yet I never saw one of the sickly looking pups receiving attention from the female. They seemed to be deserted.

The cause of the great mortality among the seal pups seemed to me to have ceased to act in great part before my first visits to the rookeries, for subsequent visits did not show as great an increase in the masses of dead as I would have expected had the causes still been in active operation. It seemed to me that there were fewer sickly looking pups at each subsequent visit. This grew to be more and more the case as the season advanced. When I visited the rookeries for the purpose of examining the dead bodies it was with extreme difficulty that carcasses could be found fresh enough to permit of a satisfactory examination. I examined a large number of carcasses. All showed an absence of fatty tissue between the skin and muscular tissue. The omentum in all cases was destitute of fat. These are the positions where fat is usually present in all animals. Well-nourished young animals always have a large amount of fat in these localities. The few carcasses which were found in a fair state of preservation were examined more thoroughly. The stomachs were found empty and contracted, but presented no evidence of disease. The intestines were empty, save in a few cases, where small amounts of fecal matter were found in the large intestines. A careful examination of the intestines failed to discover any evidence of disease. The heart, lungs, liver, and kidneys were in a healthy condition.

Such is the evidence on which I have founded my opinion that the cause of the great mortality during 1891 among the young seals on St. Paul Island, Bering Sea, was caused by the deprivation of mothers' milk. The result of my investigation is that there was great mortality exclusively among nursing seals. Second, the cause of this mortality seemed to have been abated pari passu with the abatement of sea sealing. Third, the presence of emaciated, sickly looking pups which were apparently deserted by their mothers. Fourth, the plump, healthy appearance of all the pups I saw nursing. Fifth, the emaciated condition of the dead. Sixth, the absence of food in the stomachs and their contracted condition. Seventh, the absence of digested food in the intestines. Eighth, the absence of even fecal matter, save in small amounts in a few cases. Ninth, the absence of structural changes in the viscera or other parts of the bodies to account for the death

                                    J. C. S. AKERLY, Ph. B., M. D.

Deposition of Henry W. Elliott.

CITY OF WASHINGTON,
     District of Columbia, ss:

Henry W. Elliott, being duly sworn, deposes and says: I am a resident of Cleveland, Ohio, where I was born; am 46 years of age, and am a citizen of the United States.

I first visited the Pribilof Islands in April, 1872, under the joint appointment of the United States Treasury Department and of the Smithsonian Institution, and resided thereon until August, 1873. In 1874 I made another prolonged visit under the authority of a special act of Congress. I visited the islands again briefly in 1876, and during May, June, July, and August, under authority of a special act of Congress, in 1890. During each visit I carefully studied the seal life on these islands, and investigated the habits of the fur seals. In these years I also visited the various islands in and around Bering Sea, the leading ports and inhabited places on the mainland and islands of Alaska in the Pacific Ocean, as also the ports of British Columbia and the United States; witnessed the methods of pelagic sealing, conversed with many pelagic seal hunters, shipmasters, and fur traders, and sought in all possible ways to acquaint myself fully with seal life and the taking of seals.

CLIMATIC CONDITIONS OF PRIBILOF ISLANDS.

The Pribilof Islands possess a peculiar climate. There are but two seasons, winter and summer; the former begins with November and ends with April, the mean temperature being 20° to 26° F. above zero; summer brings only a slight elevation in the temperature, between 15° or 20°, so that the mean temperature of that season is 40° to 46°. With the opening of the summer, about the 1st of May, a cold, moist fog settles down upon these islands, and is ever present until the latter part of October. It is doubtless to this remarkably damp and sunless atmosphere, together with the isolation of these islands, and the fact that from their formation they are rapidly drained, that the seals seek these islands to breed; in fact, it is necessary that such a sunless and moist climate with a low temperature should exist for this species of fur seal when on land, and it becomes highly important that they should be so protected as to make their chosen home as free from unnecessary molestation as possible. It is quite certain that the seal herd which perennially frequents the Pribilof Islands has no other terrestrial haunt, and now never lands, even temporarily, on any other terra firma in or bounding the Pacific Ocean or Bering Sea.

When all the climatic, topographical, and other facts are considered, which are so remarkably favorable to seal life on the Pribilof Islands, and which, with the exception of the Commander Islands of Russia, can not be found anywhere else in the Northern Pacific or Bering Sea, the reasons are plain why these islands have been selected by the fur seals for their breeding resorts, since reproduction of their kind can not be effected in the sea.

My personal observation and study of seal life during the past twenty years have led me to the certain conclusion that all the herd of fur seals (*Callorhinus ursinus*) which now make their annual migration from and back to the Pribilof Islands (described hereafter) were all born in June and July (annually) upon the Pribilof Islands, pass the first four months of their existence on these islands, nursing at irregular intervals, learning to swim, and in shedding their fetal coats of black hair for their seagoing jackets of hair and fur, leave in November, and annually return there to spend from four to six months of each year. In my published observations of 1872 and 1874 I thought it possible there might be some commingling of the Pribilof seals with the seal herd of the Russian Islands, but from my subsequent study of their migrations and of the varietal differences in the herds in the two localities, it is now very clear to me that they never mingle on the islands, each herd keeping to its own side of the ocean and annually resorting to its own fixed breeding grounds.

ARRIVAL OF THE BULLS.

Between the 1st and 5th of May a few of the adult males (bulls) may be found upon the breeding grounds on the Pribilof Islands, but many of them may be seen swimming a short distance from the shore for several days before landing. The method of landing is to come collectively to these rookeries which they occupied the former season, but whether a bull always takes up the same position or strives to do so I was unable to gather sufficient data to determine, my opinion being to the contrary. After landing, the bulls fight furiously for positions upon the rookeries, the place of advantage being nearest the sea.

FASTING ON THE ROOKERIES.

All the bulls, from the time they have established themselves upon the breeding grounds, do not leave them for a single instant, night or day, nor do they until the end of the breeding season, which closes some time between the 1st and 10th of August as a rule. The bulls therefore for the space of three or four months abstain entirely from food of any kind or water. When they do return to the water they are greatly emaciated and lack life and activity. But the females, directly to the contrary, feed at frequent intervals during the suckling period, and at the end of the season are as sleek and fat as when they first hauled out.

ARRIVAL OF THE COWS.

The cows, or females, begin to come up from the sea during the fore part of June, and after continual battles between the rival bulls are finally settled upon the rookeries. All the females of 2 years of age or older "haul up" on the breeding rookeries, whether they are pregnant or not, and during the period from June until the middle of August they may be found coming and going almost continuously to and from the rookeries, except a few barren cows, which I will mention hereafter. The pregnant cows land upon the islands from instinctive knowledge that their period of gestation, which is about twelve months, lacking only a few days, has come to an end. As the pups (the young seals) can not be born in the water, the female's instinct causes her to seek the land, upon which her young is brought forth, sometimes in a few hours, but usually in a day or two, after landing.

AGE OF SEALS.

The bulls on the rookeries are at least 6 years of age, that being about the time when they attain their growth, the age of puberty being probably about 5 years. The remainder of the male seals, being those younger and less powerful, called "bachelors," I will refer to later. The cows probably reach their growth between 4 and 5 years, but give birth to their first pup when 3, so that cows 2 years old are found upon the breeding grounds; they are the nubiles.

ORGANIZATION OF THE ROOKERIES.

As the cows haul up on the shores they are met by the bulls, who coax and urge them toward their own position on the rookeries. During this process the most bitter fights occurred between the bulls for possession of the cows in 1872–1874; those nearest the water being the most advantageously located, obtained the greatest number for their harems, sometimes having as many as 40 or 50 cows in their possession, while those farther inland could obtain sometimes only 2 or 3; it was very difficult to fix the average number of cows to a harem in 1872–1874, but I estimated it at about 15 or 20.

PELAGIC COITION IMPOSSIBLE.

In the act of coition on the breeding rookeries I have noticed the fact that no effective coition took place until the cow was brought up to or laid against an inequality of the rookery or fragment of the rock; that, in spite of the bulk of the male being so great and resting upon the female as she lies upon her belly, the orgasms are so rapid and violent that she is shoved forward unless some obstruction holds her in place. This fact is, I believe, sufficient to satisfy anyone who carefully considers the matter that it is a physical impossibility for these seals to copulate in the water. In my opinion there is no conceivable position in which effectual coition can take place in the water. I also observed that the period of connection in the early part of the season lasted from eight to fourteen minutes, and in the latter part, when the bull was not as vigorous, from four to six minutes.

THE PUPS.

Immediately after birth the seal pup begins to move about and to nurse, which it often does to gorging itself. It weighs but 3 or 4 pounds when born and is only about 12 inches long. The female after bringing forth her young goes frequently to and from the water, to feed and bathe. On returning from the sea she will recognize the cry of her young though ten thousand pups are bleating at once, and will immediately go to it. The pups themselves do not recognize their own mothers—a fact I ascertained by careful observation. The mother, however, will not permit any pup but her own to suckle, and will fight off any which attempt it.

YOUNG SEALS LEARNING TO SWIM.

The pup when born can not swim. If he is thrown a rod or two into the water, his head, which is heavy, will immediately sink, and his posterior parts rise to the surface. Suffocation is only a matter of a few minutes. Until he is almost six weeks old a pup can not live in the water. He then begins to try the water, never going intentionally beyond his depth; soon he becomes bolder and strikes out, using at first only his flippers; then he grows more and more expert, until finally the sea alongshore is his frequent abiding place. The young seal, therefore, up to the time it learns to swim, is a land animal—in no way a full-fledged amphibian; and it requires four months of suckling by its mother on the land before it becomes able to shift for itself and is abandoned by its parents.

BARREN FEMALES.

Whenever a female ceases to breed or is barren she hauls up with the bachelors, and no longer goes on the breeding grounds; she, however, can be easily distinguished, and whenever one became mixed in a drive the natives pointed her out to me in 1872–1874. The whole number of barren cows was then very inconsiderable.

UNATTACHED MALES.

Behind the harems there were always a number of idle and vigorous bulls in 1872–1874, who were unable to obtain any consorts, but they had to do severe battle to maintain their position at all.

DISORGANIZATION OF THE ROOKERIES.

Between the 20th of July, when the rutting season closes, and the 5th or 8th of August, the harems have changed from their methodical compact disposition on the rookeries. The old bulls begin to leave; the pups are gathered into pods or groups. The cows, pups, and idle bulls before mentioned now take possession of the rookeries in a disordered manner, together with a large contingent of the bachelor seals, who have not thus far been permitted to land on the breeding grounds by the other males. By the middle of August three-fourths of the cows spend the greater part of their time in the water, only coming on shore at irregular intervals to nurse their young. The food of the fur seals is mainly fish, squids and crustaceans, and mothers, while nursing their young, I am satisfied, go great distances in Bering Sea for this food—50, 100, and even 200 miles away from the Pribilof Islands for that subsistence.

SWIMMING OF SEALS.

I am unable to state positively how rapidly a seal can swim, but I have seen squads of young bachelors follow the revenue cutter, *Reliance*, upon which I was, swimming alongside and around the vessel for hours, when she was moving at the rate of 14 knots an hour. My opinion is that the bachelors and those cows which are not heavy with pups can travel through the water from 18 to 20 miles an hour for many consecutive hours without pausing to rest.

HOLLUSCHICKIE, OR BACHELOR SEALS.

The male fur seals under the age of 6 years are not allowed to land upon the breeding grounds by the older and stronger males, and so are compelled to herd by themselves. These seals are called holluschickie or bachelors, and the places which they occupy on land are called hauling grounds, in contradistinction to the breeding rookeries. It is from this class of seals that the killable seals are selected.

The hauling grounds are located on the low, free beaches not occupied by the breeding grounds, or else inland behind the harems. In
the latter case, lanes are left between the harems by the old bulls for
the bachelors to pass to and from the sea. In 1872 I noticed one of
these lanes on the Polavina rookery and the one at Tolstoi and the
two at the Reef rookery, but when I returned in 1874 the lanes had
been entirely closed up. But the other locations on unoccupied beaches
are the most favored hauling grounds. The bachelors when on land
can be readily separated into their several classes as to age by the
color of their coats and sizes.

Only the bachelor seals of from 2 to 5 years of age have been killed
by the lessees of the islands. No female has been or is allowed to be
taken; a few have been killed by accident. A number of seals are
driven from the hauling grounds to the killing grounds after being
separated from the rest by the natives. They can be driven safely at
the speed of half a mile an hour, providing the weather is reasonably
wet and cold. On arriving at the killing grounds they are killed with
clubs and their skins removed. During my visit to the islands, in 1890,
I was led to the conclusion that some unnecessary loss of life had been
occasioned by excessive driving, and that the methods of culling the
herd must be abolished; but this loss, which is bad enough, bears no
comparison in its injurious effect upon the herd to that loss by reason
of indiscriminate slaughter which is inflicted upon the fur-seal herd
unchecked by pelagic hunting. Of this I will speak later. Besides,
the injurious effect of excessive driving can be easily corrected. It
was stopped in 1890, and has been still further restricted since on the
islands.

A bull when full grown weighs between 400 and 500 pounds, sometimes even 600, and measures from 6 to 7 feet in length. The female
weighs from 70 to 120 pounds, and measures 4 to $4\frac{1}{2}$ feet in length.
The bachelors, over 1 year and up to 5 years old, weigh from 50 to 200
pounds, and are from 4 to $5\frac{1}{2}$ or 6 feet in length.

About the 1st of November the great mass of the cows and bachelors
begin to depart, and the pups following from the islands, going southward, the old bulls having nearly all preceded them in September and
October. Some, however, remain as long as the ice and snow will permit, and when the winters are mild and little ice is about the islands,
which occasionally occurs, fur seals are seen there until late in January
in small numbers, a few hundreds at the most.

To this, my affidavit, I append a track chart[1] of the path traveled by
the Pribilof fur-seal herd in the North Pacific Ocean from the time it
leaves the seal islands and Bering Sea in the late autumn until it
reenters Bering Sea in June or 4th to 10th of July following. From
records kept at Unalaska and Umnak for the last eighty years, and
from other information, I believe it to be a fact, well settled, that the

---

[1] "Not furnished."

fur seals regularly pass out from the waters of Bering Sea into the North Pacific by the middle or end of November as a body; that these animals do not turn to the eastward and up by the peninsular and Kadiak coast, but keep directly south till lost to view.

From ship captains who have sailed during the last twenty years between San Francisco and Puget Sound, I have learned that while making out from San Francisco from the Sound, a long westerly reach, they have seen large numbers of fur seals 800 or more miles at sea in January or late December moving toward the California Coast. Early in January the first stragglers begin to appear off the California Coast and by the middle of February the main body of the herd arrives simultaneously off between Santa Barbara and Cape Mendocino. From this point the progress of the herd northward is indicated on the chart hereto attached.[1] The fact of this annual migration of the Pribilof fur-seal herd and the route thereof is stated from knowledge derived from my own study in the field, and from the testimony of those traders and mariners who responded to my inquiries at Unalaska, Umak, Sannak, Belcovskie, Kadiak, Nuchek, Yakutat, Sitka, Fort Simpson, Victoria, Port Townsend, and Astoria.

THE HERD VISIT ONLY THE PLACE OF THEIR BIRTH.

From all the facts that have come to my knowledge in relation to the annual migration of the fur-seal herd, and also from information carefully gathered, I am convinced and believe that the Pribilof herd of fur seals now never land upon any other coast or islands save the Pribilof group, the land of their birth. At no time along the coast does the herd approach nearer than gunshot of the shore, and is often 100 to 200 miles distant therefrom.

GROWTH OF PELAGIC SEALING.

When I first visited Alaskan waters in 1865–66, and again in 1872, pelagic sealing was almost unknown, except by Indians in canoes along the North Pacific Coast and the catch was small, from 5,000 to 10,000 annually. In 1885 it began to assume larger proportions, for white men then embarked, and in 1886 the number of vessels engaged with white crews in pelagic sealing was 17; the number in 1890 was 42, and in 1891, 86 known craft; and probably 10 or 12 more clearing for "whaling and trading," where, in fact, they intended to seal.

The distinctive effects of open-water killing on the seal herd may be better understood by examining the manner in which pelagic sealing is now carried on.

MANNER OF PELAGIC SEALING.

A sealing schooner is seldom over 80 or under 40 tons measurement, employing 15 or 20 men. The vessel sails well into the track of the migrating herd of fur seals. Each boat, to the number of 7 or 8, is manned with 2 men, one of whom rows; the other sits in the bow with his shotgun or rifle and gaff-pole. The boat also contains a small keg of water, some provisions, ammunition locker, skinning knives, and an extra pair of oars and sail. These boats are let down over the side of the vessel, and row out one after the other to the windward, taking up positions just so far from each other as to be in hail of the one next to them toward the schooner; in this way they can cover 6, 7, or 8 miles, and the furthermost may be out of sight of the schooner.

---

[1] "Not furnished."

When the boats have taken their position the oarsman just keeps the boat's nose to the wind, and the hunter keeps a lookout for seals.

A fur seal, when discovered by the hunter in the open ocean, is either sleeping or feeding, and so the only classification by these hunters is "feeders" or "sleepers." It is an absolute impossibility for the hunter to determine the sex or the age of any fur seal when in the water, until it is dragged into the boat.

In swimming the seal is always submerged several feet below the surface. The seal also devours its food beneath the water. It is, however, compelled to come up every three, five, or fifteen minutes to breathe, rising head and shoulders above the water for a second or two. If the seal rises very near the hunter's boat it will dive again too quickly to be shot at, but if it raises 30, 50 or 100 yards from the boat, it will pause a moment—long enough for the hunter to shoot at it.

If the seal is not hit or is wounded it at once dives and can never be secured; if it is killed by the shot it sinks, and unless the boat is moved up in a minute or two to the spot where the animal sank the carcass will be invisible from the surface. If, however, the seal happens to be wounded so as to be stunned or dazed, it will flounder on the surface of the water until secured. Except, therefore, in the last peculiar manner of wounding, the seal hunter never knows whether he has missed, wounded, or killed the seal. Provided, however, the boat can be rowed immediately to the spot where the seal was, which depends on the accuracy of fixing the spot—necessarily a most difficult matter—the hunter may perceive the sinking body, if the seal was killed, some 4, 6, or 8 feet below the surface. In that case he reaches down with his gaff and fastens on to the carcass and drags it up to the boat. Seals wounded either fatally or slightly are never found. They instantly dive and swim away, to perish sooner or later.

THE WASTE OF LIFE.

A hunter takes, say, 200 cartridges when he leaves the schooner in the morning, and after perhaps sixteen hours' work returns to the vessel with all these expended. If for these he can show 10 or 12 skins it is considered a good day's work. The pelagic hunter certainly kills and fatally wounds a very large number of animals which he never secures the bodies of, the number hit and secured depending very largely upon the retrieving skill of the hunter. From conversations I have had with pelagic hunters, I am of the opinion that a large majority of them do not get one out of every five that they shoot at within and beyond a range of 50 yards. At 30 to 50 yards' distance they are almost sure to hit them if they use buckshot. No hunter who uses a gun can tell the exact number he secures, as compared with the number he kills or fatally wounds. He can not possibly tell the truth, even if he wants to do so. He usually blazes away at every seal that rises within range to a hundred yards or even farther.

The Indian hunters accompanying a sealing schooner generally use a toggle-headed spear, fastened to the canoe by a line which they use. After a storm the seals sleep more than at any other time, and it is then the Indian hunters are let down in their canoes and paddle off to the windward, the hunter standing or squatting in the bow, spear in hand, looking for the protruding nose of a sleeping seal. When a "sleeper" is seen, the canoe is silently paddled as near the animal as possible, the spear is thrown, and if the seal is struck she is dragged into the canoe by the line. An Indian hunter secures nearly every seal he strikes; but it is also indiscriminate slaughter, as he can not distinguish the age or sex of the "sleeper" before striking it.

PROHIBITION IN BERING SEA AND NORTH PACIFIC NECESSARY.

After carefully examining the situation, actual records, and trustworthy testimony of men engaged in scaling, with whom I have conversed, and also from knowledge of the migratory habit and peculiar circumstances of seal life, I am of the opinion that unchecked pelagic sealing is sure, speedy destruction of the Pribilof herd of fur seals; that if allowed to continue, and the fleet increases in number of vessels and increased skill of hunters, even though the present modus vivendi should remain in force, it would result in the utter commercial ruin of the herd; that in order to preserve the seals from complete destruction, as a commercial factor, it is necessary that pelagic sealing should not only be prohibited in Bering Sea, but also in the North Pacific, from the 1st of May until the end of October, annually. The pelagic hunters to-day kill at least 90 per cent cows, the great majority being with young, nearly ready for delivery, in the Pacific Ocean.

As the physical conditions are such that it is utterly impossible to discriminate in matters of sex or age when shooting or spearing in the water, it is evident that pelagic sealing can not be regulated in the slightest degree beyond its complete prohibition within certain limits. A zone or belt of 30 or even more miles about the Pribilof Islands will be entirely ineffective. No pelagic sealing can be permitted in Bering Sea with safety to the preservation of the herd, and the prohibition should extend into the North Pacific to a period sufficiently early (at least by the 1st of May) in the season to protect in great measure the pregnant female seals as they pass along up the coast.

The visit which I made to the Pribilof Islands in 1890 satisfied me that a very great decrease had taken place in the seal herd which annually resorts to those islands. My observations in 1872, 1874, and 1876 led me to the conclusion that, provided matters were conducted in the seal islands as they were then, 100,000 male seals under 5 years of age might be safely taken each year without injury to the regular birth rates or natural increase of the herd, provided no abnormal cause of destruction occurred. But in 1890 I found an entirely different condition of affairs existing. This decrease I attribute in the greatest measure to the pelagic sealing above mentioned. Its effect has been so great that there is demanded, in my opinion, a cessation of all killing on the islands, except for the necessities of the natives for a few years, as well as the permanent prohibition of pelagic sealing, as already indicated, thus giving an opportunity for the herd to reestablish itself approximately to its normal conditions. When the killing is again permitted on the islands for commercial purposes the regulations of the Treasury Department can be rigidly enforced, overdriving can easily be prevented, and the present killing of pups by the natives for food should be prohibited, at least until the herd shall have reached the form and condition which I found during 1872–1876.

With such regulations in force, and with pelagic sealing discontinued, it may be confidently anticipated that within a few years this species, so valuable to the human race, will be restored to a condition which will render it valuable once again to the commerce of the civilized world; and this restoration will prove enduring.

HENRY W. ELLIOTT.

Subscribed and sworn to before me, a notary public in and for the District of Columbia, this 13th day of April, 1892.

[L. S.]   SEVELLON A. BROWN.

UNITED STATES COAST AND GEODETIC SURVEY,
Steamer McArthur, December 9, 1892.

Hon. JOHN W. FOSTER,
Secretary of State, Washington, D. C.

SIR: I have the honor to forward the affidavit desired, and will forward the duplicate to-morrow.

We anchored off Sechat village at 3.30 p. m. April 20. Our native chief came alongside and was requested to come on board in the morning and bring with him some of the chief men of the village. He had planned to go hunting wild geese, which were flying at the time; so I promised him $3 or $5 for his loss of time and to accompany us to other villages. We took their testimony in the morning of the 21st, and ran to another village, anchoring at 10.45 a. m., took testimony and left at 1.20. Anchored off Uchielet at 2.40 and left at 4.25. Anchored off Taylor Island at 7.20 p. m. and left for Port Townsend at 10 p. m.

Two to three dollars were given to each head chief and one dollar each to the others for their loss of time and witness fee after testimony was given. All that was requested of them was to answer the questions truthfully. The white storekeeper was on board but a few minutes, and was invited to take a glass of beer or liquor. The priest dined on board, and, I believe, took a drink and some claret wine. We were not long enough in any one place to intoxicate anyone if we had been foolish enough to do so. I sincerely believe they would give the same testimony to an English party at any time.

There were four commissioned officers of the Navy present during the testimony, and as many of the witnesses could speak and understand English, all were satisfied of their truthfulness.

Very respectfully,

W. P. RAY,
Lieutenant, United States Navy, Commanding.

Deposition of W. P. Ray.

STATE OF CALIFORNIA:

W. P. Ray, being duly sworn, deposes and says: I am an officer in the United States Navy, holding the grade of lieutenant. Under instructions from Washington I went from Port Townsend to Barclay Sound, on the west coast of Vancouver Island, April 19, 1892, in the steamer *McArthur*, of the United States Coast Survey. I returned to Port Townsend three days later. The object of my visit to Barclay Sound was to procure information in the form of affidavits as to the habits of the fur seals, to be used in the pending arbitration with Great Britain. I visited the various points in that vicinity inhabited by the Indians, and took the testimony of a number of these people and of the priest of the village. For greater convenience I took the testimony aboard the steamer, and I agreed to pay, and did pay, each witness a reasonable sum for attendance, which sum did not exceed the usual fee allowed a witness in a court of justice. The total amount disbursed by me in obtaining the testimony of these witnesses was $35, which amount was distributed among 15 men. It was made up partly of the amount paid to each individual witness on account of his attendance on board my vessel, partly of sums paid out to men who undertook to ascertain the whereabouts of certain of the witnesses and secure their attendance as above. Each witness received a plug of tobacco. No other gratuity of any sort was dispensed.

At no time during my stay at Barclay Sound was any intoxicating liquor dispensed to any native witness, nor was any witness under the influence of liquor at the time when I took his testimony. We were not more than two and a half hours at any one village.

The testimony which I obtained was given in every instance willingly and cheerfully. Neither the witness fees nor the gratuities above mentioned formed any part of the consideration for the giving of this testimony, and I firmly believe the same statements will be made to anyone visiting the place for information at any time.

Just before leaving, Chief Charlie, chief of police, stated that he and his people had given food, clothing, shelter, and protection to many shipwrecked Americans, and he requested blue cloth enough to make a uniform suit, as he could not procure any there. It was given to him as a slight acknowledgment of his kindness to our people in distress. Value, $10.

W. P. RAY,
Lieutenant, United States Navy,
Commanding Coast Survey Steamer McArthur.

Sworn to before me this 9th day of December, 1892.

[SEAL.]   A. S. MACDONALD.
Notary Public in and for Alameda County, State of California.

Deposition of C. L. Hooper.

DISTRICT OF COLUMBIA,
      City of Washington, ss:

Personally appeared before me C. L. Hooper, who, being duly sworn, deposes and says: I am 50 years of age; a resident of Oakland, Cal., and am an officer in the United States Revenue-Marine Service, holding the grade of captain, and commanding the United States revenue-steamer *Corwin.*

In obedience to instructions from the Secretary of the Treasury, I cruised in the North Pacific Ocean from March 9 to May 16, 1892, for the purpose of investigating the habits of the fur seal when at sea.. During these investigations I had occasion to take the depositions of a number of natives and white men familiar with the subject.

During a portion of September, all of October, and a portion of November these investigations were continued in the vicinity of the Aleutian Archipelago, and a number of depositions were taken also from the natives of the Aleut villages situated thereon.

No depositions were taken by me from the natives of Vancouver Island, nor from the natives from any other localities except as previously indicated.

In no instance was liquor in any form given by me, or by anyone on my vessel, to any affiant; no affiant was under the influence of liquor when his statement was made; no undue influence of any sort or description was used; no gratuities were given; only such witness fees were paid as would be a fair compensation for loss of time when such loss of time actually occurred, and the testimony obtained was given freely and willingly.

Two hundred and eighty depositions were taken, and the aggregate fees paid was $69.50.

C. L. HOOPER.

Subscribed and sworn to before me this 13th day of December, 1892.

[SEAL.]   SEVELLON A. BROWN.

H. Doc. 92, pt. 2——9

Deposition of William H. Williams.

DISTRICT OF COLUMBIA,
    City of Washington, ss:

Personally appeared before me William H. Williams, who being duly sworn, deposes and says: I reside at Wellington, Ohio; I am 56 years of age, and am United States Treasury agent in charge of the Pribilof Islands.

I have seen several newspaper articles in which I am charged with having "suborned Indian testimony," with employing "unfair means" in obtaining evidence from Indians, and that conclusive proof of this misconduct has been procured by Major Sherwood of the Dominion police.

The facts in connection with the procuring of these depositions are as follows: During the summer of 1892 I had occasion, in accordance with instructions from the Secretary of the Treasury, to take the depositions of certain natives concerning the subjects of seal life and sealing at sea.   The Indians from whom I took depositions were the Makah Indians at the Makah Agency, two Nitnat Indians at the same place, and the natives on the Pribilof Islands.   No depositions were taken by me from any other natives, and I was never at Barclay Sound, on the west of Vancouver Island, or on the west coast of British Columbia, In taking depositions from the Makah Indians the only sum of money paid was $2.50, which was given by me to Chestoqua Peterson, son of the chief, for his services as interpreter for two and one-half days.   On the Pribilof Islands the sum of $5 was paid to Simeon Melivedof, a native and school teacher on the island, for four days' services as a copyist.   These were the only sums of money paid by me to Indians or to natives, or to anyone in Alaska.

In no instance was any liquor given to an affiant by me, nor by any one either directly or indirectly associated with me; nor was any affiant under the influence of liquor when his deposition was made or verified; and no undue influence of any sort or description was employed.   No gratuities in any form were given.   The testimony obtained was not only freely and willingly given, in all instances, but often voluntarily. This was especially true of the two Nitnat Indians.

In the case of the natives at the Makah Agency, the depositions were taken in the office of the Indian agent, Glynn, and under his personal knowledge.   He is a radical in his opposition to the giving of intoxicants to natives, and had anyone attempted to offer one of the Indians liquor he would have been at once ejected from the agency.

                                                WM. H. WILLIAMS.

Subscribed and sworn to before me this 20th day of December, 1892.
    [SEAL.]   CHAS. S. HUGHES, *Notary Public.*

Additional deposition of William H. Williams.

DISTRICT OF COLUMBIA,
    City of Washington, ss:

Personally appeared before me William H. Williams, who, being duly sworn, deposes and says: I reside at Wellington, Ohio; am 56 years of age, and am United States Treasury agent in charge of the Pribilof Islands.

During the summer of 1892 I had occasion, in accordance with instructions from the Secretary of the Treasury, to take the depositions of certain white men concerning the subject of sealing at sea. The depositions were taken in Victoria from ship captains, seamen, boat pullers and steerers, seal hunters, and others interested in sealing, among them the vice-president of the Sealers' Association. All depositions were taken and verified before the United States consul, Myers, at Victoria. This was the only place in which I took depositions in British Columbia. In no instance was any liquor given by me to an affiant; nor was any affiant under the influence of liquor when his deposition was made or verified; and no undue influence of any sort or description was employed. No gratuities were given. The testimony obtained was, in all instances, not only given freely and willingly, but often voluntarily. The usual witness fees (in this case ranging from $1 to $3) were paid, and only in three instances was the latter sum given, the usual price being $2.

WM. H. WILLIAMS.

Subscribed and sworn to before me this 20th day of December, 1892.

[SEAL.] CHAS. S. HUGHES, *Notary Public.*

Deposition of Joseph Murray.

DISTRICT OF COLUMBIA,
City of Washington, ss :

Personally appeared before me Joseph Murray, who, being duly sworn, deposes and says: I reside at Fort Collins, Colo.; I am 50 years of age, and am the first assistant Treasury agent at the Pribilof Islands.

In obedience to instructions from the Secretary of the Treasury, I accompanied the Fish Commission steamer *Albatross* on the cruise made by that vessel during the month of April, 1892, and took depositions from the natives of Cooks Inlet and Prince William Sound. I also took depositions in Kodiak, Victoria, Port Townsend, and Seattle from white men. In no instance was any liquor given to an affiant; nor was any affiant under the influence of liquor when his statement was made or verified; and no undue influence of any sort or description was employed. No compensation whatsoever was given by the Government to any native or other person for any purpose, and the testimony obtained was in all instances given freely and willingly.

JOSEPH MURRAY.

Subscribed and sworn to before me this 21st day of December, 1892.

[SEAL.] JOSEPH A. KINSLEY, *Notary Public.*

DESTRUCTION OF FEMALE SEALS.

Testimony of American furriers.

Relative to matter of depletion of seal herds of the Pribilof Islands, this most deplorable fact is due in our opinion in great part, if not entirely, to the action of sealers in the indiscriminate killing of these animals while in transit to and from these islands for breeding purposes, the females being killed in much greater proportionate numbers, owing to their less aggressive nature and their being less able to escape. While on their way to these islands the bow (female) seal is in a condition of pregnancy, the period of gestation ending shortly after their landing. If intercepted and killed while in this condition the loss is obvious. (G. G. Gunther's Sons.)

At that time (1865) he made his purchases from the Indians on the western coast of the American continent, who offered to him only the skins of female seals; that the price he originally paid for them was as low as 50 cents per skin; that he offered the Indians a much higher price for male skins, and was told by them that the male seals could not be caught, and that many Indians whom he has personally seen kill seals, and from whom he has bought skins, have told him that male seals and the young cows were too active to be caught, and that it was only the female seals heavy with young which they could catch. The males, for instance, as deponent was told by the seal hunters, come up to the surface of the water after diving often as much as a mile from the place they went down, whereas the females can, when pregnant, hardly dive at all.

Deponent says that, from his own observation of live seals during many years, and from his personal inspection of the skins, he knows the difference between the skin of a female seal and a male seal to be very marked, and that the two are easily distinguishable. The skin of a female shows the marks of the breasts, about which there is no fur. The belly of the female seal is barren of fur also, whereas on a male the fur is thick and evenly distributed. The female seal has a much narrower head than the male seal, and this difference is apparent in the skins; also that the differences between the male and female seals' skins are marked; that there is now and always has been a difference in the price of the two from 300 to 500 per cent. For example, at the last sales in London, on the 22d day of February, 1892, there were sold 30,000 female skins at a price of 40 shillings apiece, and 13,000 male seals at a price of 130 shillings apiece, on an average.

Second. That from the year 1864 down to the present day deponent or his firm have been large purchasers of seal skins on the western coast of America from the Indians and residents on the British coast; and deponent believes that he has handled nearly three-fourths of the catch from that time down to the present. That during the whole of this period he has purchased from 30,000 to 40,000 seal skins a year, and that he has personally inspected and physically handled the most of the skins so bought by him or his firm.

That from the year 1880 he has been in the habit of buying skins from American and English vessels, engaged in what is now known as poaching, and that he has personally inspected every cargo bought, and seen unloaded from the poaching vessels, and subsequently seen and superintended the unpacking of the same in his own warehouse; that the most of the skins mentioned as purchased by him have been bought from the poaching vessels, and that of the skins so bought from the vessels known as poachers deponent says that at least 90 per cent of the total number of skins were those of female seals, and that the skins of male seals found among those cargoes were the skins of very small animals, not exceeding 2 years of age; and, further, that the age of the seal may be told accurately from the size of its skin.

Third. That the skins bought at Victoria from the poaching vessels are shipped by him largely to the firm of C. M. Lampson & Co., in London, who are the largest sellers of skins in the world, and the agents of deponent's firm; that he has been through the establishment of C. M. Lampson & Co., in London, very frequently; that he has frequently heard stated by the superintendent thereof that the great majority of the skins received by them from what is called the "Northwest catch"—that is, the northwest coast of Victoria—are the skins of seals caught by vessels in the open Pacific or the Bering Sea, and that a large proportion of said skins, amounting to at least 90 per cent, were in his, the said superintendent's, judgment obviously the skins of female seals.

Fourth. That deponent has frequently requested the captains of the poaching vessels sailing from Victoria and other ports to obtain the skins of male seals, and stated that he would give twice as much money, or even more, for such skins than he would pay for the skins of females. Each and all of the captains so approached laughed at the idea of catching male seals in the open sea, and said that it was impossible for them to do it, and that they could not catch male seals unless they could get upon the islands, which, except once in a long while, they were unable to do, in consequence of the restrictions imposed by the United States Government; because, they said, the males were more active, and could outswim any boat which their several vessels had, and that it was only the female seals who were heavy with young which could be caught. Among the captains of vessels with whom deponent has talked, and who have stated to him that they were unable to catch anything but females, are the following: Captain Cathcart, an American, now about 75 years of age, who commanded the schooner *San Diego*, and who subsequently commanded other vessels; Capt. Harry Harmson, Capt. George W. Littlejohn, Capt. A. Carlson, Gustav Sundvall, and others, whose names he does not now remember. (Herman Liebes.)

I find in handling the skins taken in Bering Sea that the teats of those from the cow seals are much larger and much more developed than from the ones taken in the North Pacific before they have given birth to their young; and the fur on the belly of the former is thinner and poorer than on the latter, as a result, I suppose, of the heat and distention of the udder consequent upon giving milk. (Isaac Liebes.)

In my examination of skins offered for sale by sealing schooners I found that over 90 per cent were skins taken from females. The sides of the female skins are swollen, and are wider on the belly than those of the males. The teats are very discernible on the females, and it can be plainly seen where the young have been suckling. The head of the females is also much narrower. (Sidney Liebes.)

I have read the affidavit of John J. Phelan, verified the 18th day of June, 1892. I was present at the examination of seal skins therein referred to. While Phelan inspected all of these seal skins, I assisted him in the inspection of about three-fourths of them. I know that of those inspected jointly none were improperly classed as the skins of female animals. (Chas. E. McClennen.)

I was visiting in San Francisco in the winter of 1890–91, and I worked in a fur store during several months of my stay there, and I was called on to handle and inspect thousands of the skins taken by schooners in Bering Sea, and they were nearly all cow seal skins. (Anton Melovedoff.)

In buying the catch of schooners engaged in the sealing business I have observed that fully 50 per cent of them were females, and had either given birth to their young or were heavy in pup when killed, which was easily observed by the width of the skin of the belly and the small head and development of the teat. (R. H. Sternfels.)

The first consignment was placed in cold storage at the Central Stores in New York City. A short time since I consented, at the request of the United States Government, that this consignment be examined, in

order to determine how many female skins it contained. To perform
the examination I detailed John J. Phelan. This man has been in the
employ of my father or of myself since the year 1868. I regard him
as one of the most competent, trustworthy men in our service. I have
read an affidavit verified by him on the 18th of June. I agree entirely
with what he says concerning his experience in the handling and dressing of skins, and from what I know of his character and ability I
believe that everything stated by him in this affidavit is correct. (Geo.
H. Treadwell.)

It is true that the Northwest Coast catches have of late years placed
upon the market a certain number of good skins which could be purchased at prices far below those for which the skins of the Alaska
catch were sold. But I realize that this can not continue to be the
case, for it is a matter of common knowledge among furriers that
these Northwest Coast catches are composed mainly of the skins of
female animals, and I understand that the killing of the seals is rapidly
impairing the value of the herd. (Samuel Ullmann.)

I have for many years personally examined numerous shipments of
Northwest Coast skins purchased at Victoria. I have had such experience in handling fur-seal skins as enables me, readily in most cases,
but always upon careful examination, to distinguish a female skin from
a male skin, and I know it to be a fact that a very large proportion of
the skins in such shipments are those taken from female animals. It is
also true that a large number of skins in many of these shipments are
rendered almost valueless through the numerous bullet holes which they
contain. (Samuel Ullmann.)

I have observed that by far the larger portion of skins purchased by
me were taken from female seals. Not less than eight out of every ten
were from cows with pup or in milk. (C. T. Wagner.)

During the past two years I have handled large numbers of Northwest Coast skins (i. e., skins of animals taken in the Pacific Ocean or
in Bering Sea). I have assorted all of them, and in doing so have
specially noticed the fact that a very large proportion were skins of
female animals. To determine this fact in the case of dressed skins I
see whether there are any teat holes. I never call a skin a female skin
unless I can find two such holes on either side. These holes can be
easily distinguished from bullet or buckshot holes, of which there are
generally a great number in Northwest Coast skins. In the case of a
shot hole it is always evident that the surrounding fur has been abruptly
cut off, while around the edge of a teat hole the fur gradually shortens
as it reaches the edge and naturally ceases to grow at the edge. I have
just looked over an original case of 90 dressed and dyed Northwest
Coast fur-seal skins, which have been lately received from London, and
were still under seal placed on them in London. I found that of these
90 skins 9 only were those of male animals. (Wm. Wiepert.)

Deponent further says that the skins of the Northwest catch are
almost entirely the skins of females; that the skins of males and the
skins of females may be as readily distinguished from each other as
the skins of the different sexes of any other animals when seen before
being dyed and dressed, and that the skins of this
catch are almost exclusively females is that the male seal is much more
active and much more able to escape from the boats engaged in this
manner of hunting than the female seal, and that a large number of
the female seals included in the Northwest catch are of animals heavy

with young. A large number of females are also caught on their way from and to the Pribilof Islands and their feeding grounds before and after the delivery of their young on those islands. (C. A. Williams.)

A statement is attached thereto,* prepared by deponent, giving his estimate of the number of female seals killed by pelagic hunting in the past twenty-one years. (C. A. Williams.)

That for the last fifteen years he has had consigned to him by fur sealers from 8,000 to 10,000 seal skins annually, for the purpose of dressing and dyeing the same; that about 50 per cent. of the skins so received by him came from London in casks marked as they are catalogued by C. M. Lampson & Co., and are the skins belonging to what is known as the Northwest catch; and deponent is informed and believes that the Northwest catch, as the term is used in the trade, means the skins of seals caught in the open sea and not upon the islands. Another reason for this belief is the fact that all of the skins of the Northwest catch contain marks showing that the animal has been killed by bullets or buckshot, the skins being pierced by the shot, whereas the skins killed on the American and Russian islands are killed on land by clubs and are not pierced.

That of the skins of the Northwest catch coming into his hands for treatment probably all are the skins of the female seal, and that the same can be distinguished from the skins of the male seal by reason of the breasts and of the thinness of the fur around the same and upon the belly, most of the females being killed while they are bearing their young, and the fur therefore being stretched and thinner over that part of the body; and also for the further reason that the head of the female seal is much narrower than that of the male seal, and that this point of difference is obvious in the skins of the two classes; that of the total number of skins received by him about 25 per cent are the skins of the Alaska and Copper catches; that all the skins of the Alaska catch are male seals, and an overwhelming proportion of the Copper catch are likewise male skins; that the remainder of the skins sent to deponent for dressing and dyeing, as aforesaid, are received by him through the house of Herman Liebes & Co., of San Francisco, and others, the majority, however, from Herman Liebes & Co. The skins received from the latter sources are from each of the three catches known to the trade as the Copper, Alaska, and Northwest catches, although the major part thereof belong to what is known as the Northwest catch, and are, as in the case of the skins received from London of that catch, all skins of the female seal. (Jos. D. Williams.)

In examining and purchasing seal skins from schooners in their raw state I have observed that 90 per cent of their catch are females. I know that to be a fact, because the heads of the females are smaller, the bellies larger, and the teats can be plainly seen. The teats show more plainly when the skin is dressed and dyed. In examining the skins taken by sealing schooners I have found most of them perforated with shot, making them much less valuable thereby; formerly more of them used to be killed with a rifle, which did not injure the skin as much. (Maurice Windmiller.)

The destruction of the seals in the North Pacific Ocean, as well as in the Bering Sea, is largely confined to females. This fact can not be disputed successfully. I made an examination of the reports of the

---

* Not furnished.

gentlemen who handled the North Pacific collection, up to and including the year 1889, and all agreed that the skins were nearly all from females. It may not be out of place to explain that the smaller value of the female seal, especially after the birth of her pup, is in a measure due to the wearing of the fur around the teats. The amount of merchantable fur being reduced to that extent, makes it necessary for the handlers of skins to observe carefully whether pelts are male or female, as well as their general condition. They make a complete classification, and being experts in their business are not likely to make mistakes. (Theo. T. Williams.)

PELAGIC SEALING.

Deposition of Maurice Windmiller, furrier, San Francisco.

STATE OF CALIFORNIA,
     City and County of San Francisco, ss:

Maurice Windmiller, having been duly sworn, deposes and says: My age is 46; I reside in San Francisco; my occupation is that of a furrier. I have been engaged in the fur business all my life, and my father was a furrier before me. I am an expert in dressed and undressed, raw, and made-up furs, and also a dealer and manufacturer in the same. I have bought and examined large numbers of fur-seal skins during the last twelve years caught by sealing schooners both on the American and Russian side of the North Pacific and Bering Sea, and I can easily distinguish one from the other.

The Russian seal is a smaller seal, and the fur is not as close as the fur of the Alaska seal, nor as good quality. They are an entirely different herd from those on the American side, and their skins have peculiar characteristics by which it is not difficult to separate them. In examining and purchasing seal skins from the schooners in their raw state I have observed that 90 per cent of their catch are females. I know that to be a fact because the heads of the females are smaller, the bellies larger, and the teats can be plainly seen. The teats show more plainly when the skin is dressed and dyed. In examining the seals taken by sealing schooners I have found most of them perforated with shot, making them much less valuable thereby. Formerly more of them used to be killed with a rifle, which did not injure the skins as much.

MAURICE WINDMILLER.

GENERAL SEAL-SKIN INDUSTRY.

Deposition of Joseph D. Williams, furrier, New York.

STATE OF NEW YORK,
     City and County of New York, ss:

Joseph D. Williams, being duly sworn, says that he is 74 years of age, a citizen of the United States, and a resident of Brooklyn, in the State of New York; that he has been engaged in the business of dressing and dyeing fur-seal skins continuously for fifteen years past, and prior to that time, at intervals during the whole time he has been engaged in business, during a period of some fifty-odd years, he has dressed and dyed seal skins, and that his father was engaged in the same business before him; that for the last fifteen years he has had consigned to him by fur dealers from 8,000 to 10,000 seal skins annually for the purpose of dressing and dyeing the same; that about 50 per cent of the skins so received by him came from London in casks marked as they are catalogued by C. M. Lampson & Co., and are the skins belonging to what is known as the Northwest catch; and deponent is informed and believes that the Northwest catch, as the term is used in the trade, means the skins of seals caught in the open sea, and not upon the islands. Another reason for this belief is the fact that all of the skins of the Northwest catch contain marks showing that the animal had been killed by bullets or buckshot, the skins being pierced by the shot, whereas the skins killed on the American and Russian islands are killed on land by clubs and are not pierced.

That of the skins of the Northwest catch coming into his hands for treatment probably all are the skins of the female seal, and that the same can be distinguished from the skins of the male seal by reason of the breasts and of the thinness of the fur around the same and upon the belly, most of the female seals being killed while they are bearing their young, and the fur therefore being stretched and thinner over that part of the body; and also for the further reason that the head of the female seal is much narrower than that of the male seal, and that this point of difference is obvious in the skins of the two classes. That of the total number of the skins received by him about 25 per cent are the skins of the Alaska and Copper catch. That all the skins of the Alaska catch are male seals, and an overwhelming proportion of the Copper catch are likewise male skins. That the remainder of the skins sent to deponent for dressing and dyeing as aforesaid are received by him through the house of Herman Liebes & Co., of San Francisco, and others, the majority, however, from Herman Liebes & Co. The skins received from the latter sources are from each of the three catches known to the trade as the Copper, Alaska, and Northwest catch, although the major part thereof belong to what is known as the Northwest catch, and are, as in the case of the skins received from London of that catch, all skins of the female seal.

JOSEPH D. WILLIAMS.

DESTRUCTION OF FEMALE SEALS.

Testimony of British furriers.

I can also tell by examining a skin whether it has been taken from a female or a male. I have examined and sorted a great many thousand skins taken from sealing schooners, and have observed that they are nearly all females, a few being old bulls and yearlings. A female seal has a smaller head and a larger belly when with young than a male seal, and the fur on the belly when with young is much thinner, and the fur on the belly part where the teats are, in consequence of being worn, is not worth much, and has to be cut off after being dyed. (George Bantle.)

The skins of the male and female animal are readily distinguishable from each other in the adult stage by reason of the difference in the shape of the heads. That the Copper and Alaska skins are almost exclusively the skins of the male animal, and the skins of the Northwest catch are at least 80 per cent of the skins of the female animal. That prior to and in preparation for making this deposition deponent says he carefully looked through two large lots of skins now in his warehouse for the especial purpose of estimating the percentage of female skins found among the Northwest catch, and he believes the above estimate to be accurate. That the skins in the Northwest catch are also pierced with shot and spear marks, in consequence of having been killed in the open water instead of upon land by club. (H. S. Bevington.)

And in the same way deponent thinks, from his own personal experience in handling skins, that he would have no difficulty whatever in separating the skins of the Northwest catch from the skins of the Alaska catch by reason of the fact that they are the skins almost exclusively of females, and also that the fur upon the bearing female seals is much thinner than upon the skin of the male seals; the skin of the animal while pregnant being extended and the fur extended over a large area. (Alfred Fraser.)

That the said firm can distinguish very readily the source of production of the skins when the latter are in their undressed state; that for several years besides the skins of the regular companies, such as the Alaska Company (American concessionaire) and the Copper Company (Russian concessionaire), the said firm has bought quantities of skins called Northwest Coast, Victoria, etc. That these skins are those of animals caught in the open sea by persons who apparently derive therefrom large profits, and nearly three-quarters of them are those of females and pups, these probably being less difficult to take than males; that these animals are taken by being shot. That the seals taken by the Alaska and Copper companies are males; the destruction of which is much less prejudicial to the preservation of the race, and which furnish the best skins, these being much finer and more furnished with down; that they are killed on the islands with clubs. That every animal killed by ball or shot bears the traces of such slaughter, which marks greatly depreciate the value of the skin. (Emin Hertz.)

An essential point of difference between the skins of the Northwest catch and the skins of the Alaska and Copper Island catches consists in the fact that most of the Northwest skins are the skins of the female seal, while the Copper and Alaska skins are of the male seal. Deponent has made no computation or examination which would enable him to say specifically what proportion of the Northwest catch are the skins of the female seal, but it is the fact that the great majority, deponent would say 75 to 80 per cent, of the skins of this catch are the skins of the female animal. The skins of the female seal, for instance, show the marks of the breast, and the fur on the belly is thinner, and the whole of the fur is also finer, lower in pile; that is, the fibers composing the fur are shorter than in the case of the male seal. Another means of distinguishing the female skins from the skins of the male lies in the fact that the skins of the female are narrower at the head and tail and proportionately wider in the belly than the skins of the male seal. Another means of distinguishing the seals of the Northwest catch from those of the Copper Island and Alaska catches consists in the fact that nearly all the skins of this catch have holes in them, which deponent understands is caused by the fact that the seals from which they are taken have been shot or speared in the open sea, and not—as is the case with the seals from which the skins of Copper Island and Alaska catches are taken and killed—with clubs upon land. (Walter E. Martin.)

Both the Copper Island skins and the Alaska skins are almost exclusively the skins of the male seals, and the difference between the skin of a male seal and a female seal of adult age can be as readily seen as between the skins of different sexes of other animals. That the Northwest skins are, in turn distinguishable from the Copper Island and Alaska skins, first by reason of the fact that a very large proportion of the adult skins are obviously the skins of female animals; second, because they are all pierced with a spear or harpoon or shot, in consequence of being killed in open sea, and not, as in the case of Copper Island and Alaska skins, being killed upon land by clubs; third, because the Northwest skins are cured upon vessels by the crews of which they are killed, upon which there are not the same facilities for flaying or salting the skins as there are upon land, where the Copper and Alaska skins are flayed and salted. The Japanese skins, which, I think, are now included in the Northwest catch, are distinguishable from the other skins of the Northwest catch by being yellower in color, having a much shorter pile, because they are salted with fine salt, and have plenty of blubber on the pelt. That the skins purchased by deponent's firm are handed over by it to what are called dressers and dyers, for the purpose of being dressed and dyed. (Henry Poland.)

That the differences in the skins of the adult male and the adult female seals are as marked as the difference between the skins of the two sexes of the other animals, and that in the Northwest catch from 85 to 90 per cent of the skins are of the female animal. Deponent does not mean to state that these figures are mathematically accurate, but they are, in his judgment, approximately exact. (Geo. Rice.)

I should estimate the proportion of female skins included within the Northwest catch at at least 75 per cent, and I should not be surprised at, nor be inclined to contradict, an estimate of upward of 90 per cent. My sorter, who actually handles the skins, estimates the number of female skins in the Northwest catch at 90 per cent. One means of distinguishing the skins of the Northwest catch from those of the other catches is the fact that they are pierced with shot or spear holes, having been killed in the open sea, and not, as in the case of the Copper and Alaska catches, killed upon land with clubs. (William C. B. Stamp.)

The number of Japanese skins averages, deponent should say, about 6,000 a year, although there is a good deal of fluctuation in the quantity from year to year, and deponent says that, like the other skins included in the Northwest catch, they are principally the skins of female seals, not easily distinguishable from the skins taken from the herds frequenting the eastern part of the Pacific Ocean and Bering Sea, except by reason of their being principally speared instead of shot. The most essential difference between the Northwest skins and the Alaska and Copper catches is that the Northwest skins, so far as they are skins of adult seals, are almost exclusively the skins of female seals, and are nearly always pierced with shot, bullet, or spear holes. The skins of the adult female seals may be as readily distinguishable from the skins of the adult male as the skins of the different sexes of other animals; that practically the whole of the adult Northwest-catch seals were the skins of female seals, but the skins of the younger animals included within this Northwest catch, of which we have at times considerable numbers, are much more difficult to separate into male and female skins, and I am not prepared to say that I could distinguish the male from the female skins of young animals. A certain percentage of young animals is found among the consignments received by us at the beginning of each season, which, we understand, and are informed, are the skins of seals caught in the Pacific Ocean off the west coast of America, but a much smaller percentage of such small skins is found among the consignment later in the season, which we are informed are of seals caught in the Bering Sea. (Emil Teichmann.)

From C. M. Lampson & Co. to C. A. Williams, August 22, 1889.

LONDON, 64 QUEEN STREET, E. C.,
August 22, 1889.

DEAR SIR: We beg to acknowledge receipt of your favor of the 10th instant, inclosing draft of a paper to be submitted to Congress on merchant marine and fisheries.

We have read the paper with a great deal of interest and consider that it places the matter in a thoroughly impartial way before its readers. It has been so carefully prepared and goes into all details so fully that we can add but little to it. There are, however, one or two points to which we beg to draw your attention, and which you will find marked in red ink on the paper.

When speaking of the supply of fur skins we would suggest mentioning the following localities:

Cape of Good Hope.—From some islands off this cape, under the protection of the Cape Government, a yearly supply of from 5,000 to 8,000 skins is derived. All these skins come to the London market, part of them being sold at public auction, the remainder being dressed and dyed for account of the owners.

Japan.—The supply from this source has varied very much of late years, amounting sometimes to 15,000 skins a year, at others to only 5,000. Last year, we understand, the Japanese Government passed stringent laws prohibiting the killing and importation of seals, with the view of protecting seal life and encouraging rookeries, and the consequence has been that this year very few skins have come forward.

Vancouver Island.—For many years past, indeed long before the formation of the Alaska Company, regular supplies of fur seals in the salted and parchment state have come to the London market, killed mostly off Cape Flattery. The quantity, we should say, has averaged at least 10,000 per annum. This catch takes place in the months of March and April, and we believe that the animals from which these skins are derived are the females of the Alaska seals, just the same as those caught in the Bering Sea.

Had this quantity been materially increased, we feel sure that the breeding on the Pribilof Islands would have suffered before now; but fortunately the catch must necessarily be a limited one, owing to the stormy time of the year at which it is made and the dangerous coast, where the seals only for a short time are found. It must, however, be evident that if these animals are followed into the Bering Sea and hunted down in a calm sea in the quietest months of the year, a practically unlimited quantity of females might be taken, and, as you say, it would be only a few years till the Alaska seal was a thing of the past.

C. M. LAMPSON & CO.

C. A. WILLIAMS, Esq., *New London.*

TESTIMONY RELATING TO THE GENERAL SEAL-SKIN INDUSTRY IN GREAT BRITAIN.

Deposition of H. S. Bevington, head of the firm of Bevington & Morris, furriers, London.

H. S. Bevington, M. A., being duly sworn, doth depose and say: That he is 40 years of age and a subject of Her Britannic Majesty, and is the head of the firm of Bevington & Morris, doing business as fur merchants and manufacturers at 28 Cannon street, in the city of London; that his said firm was founded in the year 1726, and has been continued in the same family during the whole of these years down to the present time, and has been engaged during the whole of the period since 1726 in the same business, dealing in furs and leather; that deponent has been in the business ever since the year 1873. During the whole of the period since that date his said firm has been in the habit of buying fur-seal skins, and he knows from his general knowledge of the business that prior to that time they were in the habit of buying seal skins ever since they became an article of commerce; that deponent has personally handled many thousands of skins of the fur seal, and by reason of that fact and of his experience in his business, has a general knowledge of the history of the fur-seal skin business and a general and precise knowledge of the several kinds of skins which now and for many years last past have come upon the London market; that since deponent has been in business skins coming upon the London market have been principally divided into three classes, known as the Alaska catch, the Copper catch, and the Northwest catch. Small supplies have also been received from the Southern Sea, and Lobos Islands, Falkland Islands, and Cape Horn, but the skins arriving from these last-mentioned localities make no figure in the market; that what is known as the Alaska catch consists of skins of seals which are killed upon the Pribilof Islands, in the Bering Sea, and the Copper catch of skins which are killed upon the Copper and Bering islands, in Russian waters.

That the Northwest skins consist of skins taken from animals which are caught in the open Pacific Ocean off the coast of British Columbia or in the Bering Sea; that the differences between the three several sorts of skins last mentioned are so marked as to enable any person skilled in the business or accustomed to handle the same to readily distinguish the skins of one catch from those of another, especially in bulk, and it is the fact that when they reach the market the skins of each class come separately and are not found mingled with those belonging to the other classes. The skins of the Copper Island catch are distinguished from the skins of the Alaska and Northwest catch, which two last-mentioned classes of skins appear to be nearly allied to each other, and are of the same general character, by reason of the fact that in their raw state the Copper skins are lighter in color than either of the other two, and in the dried state there is a marked difference in the appearance of the fur of the Copper and the other two classes of skins. This difference is difficult to describe to a person unaccustomed to handle skins, but it is nevertheless clear and distinct to an expert, and may be generally described by saying that the Copper skins are of a close, short, and shiny fur, particularly down by the flank, to a greater extent than the Alaska and Northwest skins. The skins of the male and female animal are readily distinguished from each other in the adult stage by reason of the difference in the shape of the heads;

that the Copper and Alaska skins are almost exclusively the skins of the male animal, and the skins of the Northwest catch are at least 80 per cent of the skins of the female animal; that prior to and in prepa- ration for making this deposition, deponent says, he carefully looked through two large lots of skins now in his warehouse for the special purpose of estimating the percentage of female skins found among the Northwest catch, and he believes the above estimate to be accurate.

That the skins in the Northwest catch are also pierced with shot and spear marks, in consequence of having been killed in the open water instead of upon land by clubs; that the business of dealing in fur-seal skins in the city of London has become an established and important industry. Deponent is informed that practically all the seal skins in the world are sold in London, and the number runs up in the year to between 100,000 and 200,000, averaging considerably over 150,000 a year. These skins are sold for the most part either by the firm of C. M. Lampson & Co., through their brokers, Goad, Rigg & Co., or by the firm of Culverwell, Brooks & Co. At the auction sales, which are advertised two or three times in the year by these firms, skins are bought by dealers from all over the world, who are present either in person or by proxy. The next stage in the industry is the dressing and dyeing of the furs, and practically the whole of these fur-seal skins sold in London are dressed and dyed in that city. The principal firms engaged in that business are C. W. Martin & Sons and George Rice. Deponent's own firm dress a small number of skins, and have dressed in one year as many as 23,000, and formerly dyed large numbers of skins, but do not now dye skins, as the secret of the present fashionable color is now in the hands of other firms. After having been dressed and dyed, the skins of the fur seal are then passed into the hands of fur merchants, by whom, in turn, they are passed to furriers and dra- pers and retail dealers generally. Deponent estimates the total num- ber of persons engaged in one way or another, directly or indirectly, in the fur-seal industry in the city of London to be at least 2,000 or 3,000, many of whom are skilled laborers, all receiving high wages.

That a large amount of capital is also invested in the business in the city of London, and the precise value of the industry can be estimated by reckoning the amount expended in the various processes which deponent has enumerated upon each skin. For instance, after the skins arrive at the London market they are sold at the sales at prices which in the year 1890 averaged, say, 80 shillings apiece. The commissions on the selling of the goods, including warehousing, insurance, and so forth, deponent believes, amount to 6 per cent of the price obtained. That the amount paid for dressing, dyeing, and machining each skin aver- ages, say, 16 shillings. These processes take together about four or five months. The next expenditure upon the skin is, say, an average of 5 shillings at least for each skin for cutting up, and that thereafter there will be an average of at least from 3 to 4 shillings per skin expended in quilting, lining, and making up the jackets or other garments, showing a total expenditure upon each skin for labor alone, in the city of Lon- don, of 25 shillings in addition to the percentage paid for brokerage, before the processes of manufacture began, and the most of this money is actually paid out in wages.

Deponent says that in the above estimates he has given the bottom figures, and that the amount actually expended upon the skins in the city of London undoubtedly averages a larger sum. This would make on an average of 200,000 skins a year, which is not excessive, a total expenditure annually in the city of London of £250,000, minus the amounts paid for cutting and making up in respect to the skins sent to the United States.

Deponent further says that the preservation of the seal herds and the continued supply of fur-seal skins, which furthermore, it is important should be constant and regular in supply, is absolutely necessary to the maintenance of this industry. Deponent says that the reason for this opinion is shown in the history of last season's business. For instance, at the October sale the prices of skins were very high, as a short supply was expected. The skins purchased at that sale were then put into the hands of the dressers and dyers, where they would be retained, as above stated, in process of treatment four or five months. During this interval it appeared that instead of there being a short supply the poaching vessels had caught a large number of skins, 50,000 or 60,000, which, being unexpectedly plumped on the market, brought the price down so that there was a loss of perhaps 25 shillings per skin on the skins bought at the October sales; and deponent further says that it is of course obvious that the business can not be maintained unless the herds are preserved from the destruction which has overtaken the South Sea herds, which formerly existed in such large numbers, and so important has the seal-skin business become that if the herds were exterminated deponent says it would hardly be worth while to remain in the fur business.

Deponent says while he does not wish to express any opinion upon the matters which are in controversy, that nevertheless, looking at the question of preserving the seals from a natural-history point of view alone, and having no regard whatever to the rights of any individuals or nations, but looking at the matter simply from the point of view of how best to preserve the seals, he has no hesitation in saying that the best way to accomplish that object would be to prohibit absolutely the killing of all seals except upon the islands, and furthermore to limit the killing of seals on the islands to the male species at particular times, and to limit the numbers of the males to be so killed. If, however, the rights of individuals are to be considered, and sealing in the open sea is to be allowed, then deponent thinks that the number of vessels to be sent out by each country ought to be limited, and the number of seals which may be caught by each vessel should be specified.

Deponent says that one reason why he thinks the killing of seals in the open sea should be prohibited, and all killing limited to the islands, is because deponent is of the opinion that when seals are killed in the open sea a large number must be killed which are not recovered, and consequently that the herds must suffer much greater loss than is measured by the skins of the seals caught or coming to market.

Deponent further says that one reason for this opinion is that he has had some small experience in shooting hair seals in the Scilly Islands, and has himself personally killed hair seals at a distance of 40 or 50 yards which sank before he could reach them. Hair seals are of the same general family as fur seals, and he has no doubt that the same thing occurs and must occur when the fur seals are killed on the open sea.

HERBERT SHELLEY BEVINGTON, M. A.

Deposition of Alfred Fraser, member of firm of C. M. Lampson & Co., furriers, London.

STATE OF NEW YORK,
    City and County of New York, ss:

Alfred Fraser, being duly sworn, says:

(1) That he is a subject of Her Britannic Majesty and is 52 years of age and resides in the city of Brooklyn, in the State of New York. That he is a member of the firm of C. M. Lampson & Co., of London, and has been a member of said firm for about thirteen years; prior to that time he was in the employ of said firm and took an active part in the management of the business of said firm in London. That the business of C. M. Lampson & Co., is that of merchants, engaged principally in the business of selling fur skins on commission. That for about twenty-four years the firm of C. M. Lampson & Co. have sold the great majority of the whole number of seal skins sold in all the markets of the world. That while he was engaged in the management of the business of said firm in London he had personal knowledge of the character of the various seal skins sold by the said firm, from his personal inspection of the same in their warehouse and from the physical handling of the same by him. That many hundred thousands of the skins sold by C. M. Lampson & Co. have physically passed through his hands, and that since his residence in this country he has, as a member of said firm, had a general and detailed knowledge of the character and extent of the business of said firm, although since his residence in the city of New York he has not physically handled the skins disposed of by his firm.

That during the last year or two a large number of skins have been sold in London by the firm of Culverwell, Brooks & Co., and that said firm, as deponent is informed and believes, have secured the consignment of skins to them during the period aforesaid by advancing to the owners of vessels engaged in what is now known as pelagic sealing sums of money, which is stated to be $15 per skin, as against shipments from Victoria of such skins.

(2) That the seal skins which have been sold in London from time to time since deponent first began business have been obtained from sources and were known in the market as—

(*a*) The South Sea skins, being the skins of seals principally caught on the South Shetland Islands, South George Islands, and Sandwich Land. That many years ago large numbers of seals were caught upon these islands, but in consequence of the fact that no restrictions were imposed on the killing of said seals, they were practically exterminated, and no seal skins appeared in the market from those localities for many years. That about twenty years ago these islands were again visited, and for five seasons a considerable catch was made, amounting, during the whole five seasons, to about 30,000 or 40,000 skins. Among the skins found in this catch were those of the oldest males and the smallest pups, thus showing, in the judgment of deponent, that every seal of every kind was killed that could be reached. That in consequence thereof the rookeries on these islands were then completely exhausted. Once or twice thereafter they were visited without result, no seals being found, and about five years ago they were again revisited and only 36 skins were obtained. Deponent is informed that all the South Sea skins were obtained by killing seals upon the islands above mentioned, and that it is obviously everywhere much easier to kill seals upon the land than in the water; and, in the judgment of the deponent, the seals of the above-mentioned islands were thus entirely exterminated because of the entire absence of any protection or of any restriction of any kind whatever upon the number, age, or sex of seals killed, and not merely, as deponent understands has been claimed by some authorities, because they were killed on land instead of in open sea, which moreover, in that locality, deponent is informed, is practically impossible, by reason of the roughness of the sea and weather.

(b) A considerable number of seal skins were formerly obtained upon the Falkland Islands; how many deponent is not able to state.

(c) That a certain number of seals were also caught at Cape Horn, and that more or less are still taken in that vicinity, though the whole number has been very greatly reduced.

(d) That at the present time and for many years last past the skins coming to the market and which are known to commerce have come from the following sources: By far the most important are the Northern Pacific skins, which are known to the trade under the following titles:

The Alaska catch, which are the skins of seals caught on the Pribilof Islands, situated in Bering Sea. For many years past the whole of the skins caught upon these islands have been sold by deponent's firm, and a statement of the number of skins so sold in each year is appended hereto and marked Exhibit A, showing the aggregate of such skins sold from the year 1870 to the year 1891, inclusive, as 1,877,977.

The Copper catch, being the skins of seals caught upon what are known as the Commander Islands, being the islands known as Copper and Bering islands. All the skins so caught have been sold by deponent's firm in the city of London, and the total number of such Copper catch from the year 1872 to 1892 appears upon the statement which is hereto annexed and marked Exhibit B, showing the total so sold during such years of 768,096 skins.

The Northwest catch, being the skins of seals caught in the open sea either of the Pacific Ocean or the Bering Sea. These skins were originally caught exclusively by the Indians and by residents of the colony of Victoria and along the coast of the British possessions. A statement of the total number of the catch from the year 1868 to 1884, inclusive, is appended hereto and marked Exhibit C, showing a total of 153,348. That statement is divided into three heads: First, the salted Northwest coast skins; second, the dried Northwest coast skins, both of which were mainly sold through deponent's firm in London; and third, salted Northwest coast skins, dressed and dyed in London, but not sold there. It will be noticed that in the years 1871 and 1872 an unusually large proportion of dried skins appeared to have been marketed. Those skins were purchased in this year from the American-Russian Company and sold when the Americans took possession. For the years 1871 and 1872, therefore, the surplus skins over the average for the other years should be rejected in a computation of the general average of seals killed during the years from 1868 to 1884, inclusive.

From the year 1885 to the year 1891 the number of skins included in the Northwest catch enormously increased, and a statement of such skins is hereto annexed and marked Exhibit D, showing a total of 331,962, and is divided, like the statement marked Exhibit C, into three heads: The salted Northwest coast skins, the dried Northwest coast skins, and the salted skins dressed and dyed in London but not sold there. The majority of the first two classes were, as in the previous case, sold by deponent's firm. The great majority of these skins appearing in the last-mentioned statement are the skins caught by vessels sent out from the Canadian provinces; many also by vessels sent out from San Francisco, Port Townsend, and Seattle, and a few from vessels sent out from Yokohama; the majority, however, are supposed to have been caught by vessels sent out from British harbors. A large number of the skins included in Exhibit D have been consigned to C. M. Lampson & Co. by the firm of Herman Liebes & Co., of San Francisco. In estimating the total number of the Northwest catch it should also be mentioned that something like 30,000 skins belonging to that catch have been dressed and dyed in the United States, which have not gone to London at all.

(e) Besides the Alaska, Copper, and Northwest skins there are also a certain number of skins arriving in London known as the Lobos Island skins, although the same are not handled by the firm of C. M. Lampson & Co., but the total number of which, from the year 1872 to the year 1891, inclusive, is, as appears from the catalogues of sales, 247,777. The Lobos Island skins are those of seals killed on the Lobos Island, belonging to the Republic of Uruguay; and deponent is informed and believes that there is no open-sea sealing in the vicinity of such island, and that the animals are protected on the island as they are on the Russian and Pribilof islands, by prohibition from the killing of females and limiting the number of males killed in each year. A statement of the seals killed on Lobos Island is hereto annexed and marked Exhibit E, from which it appears that there is a regular annual supply obtained from that source, which shows no diminution.

(f) There are also a certain number of skins sold in London obtained from rookeries at or near the Cape of Good Hope, the exact number of which deponent is not able to state, but which, he is informed, shows a steady yield.

The statements marked A, B, C, D, and E, hereunto appended, have been carefully prepared by me personally, and the figures therein stated have been compiled by me from the several sale catalogues of C. M. Lampson & Co., and others from my private books which I kept during all the years covered by the statements, and I am sure that these statements are substantially accurate and truly state the respective numbers of the skins caught and sold which they purport to state.

(3) The great majority of the skins sold from the Northwest catch are the skins of female seals. Deponent is not able to state exactly what proportion of such skins are the skins of females, but estimates it to be at least 85 per cent, and the skins of females are readily distinguishable from those of the males by reason of the fact that on the breast and on the belly of the bearing female there is comparatively little fur, whereas on the skins of the male seals the fur is evenly distributed; and also by reason of the fact that the female seal has a narrow head and the male seal a broad head and neck; and the skins of this catch are also distinguishable from the Alaska and Copper catch by reason of the fact that seals are killed by bullets or buckshot or speared, and not, as on the Pribilof and Commander islands, by clubs. Marks of such bullets or buckshot or spears are clearly discernable in the skins, and there is a marked difference in the commercial value of the female skins and of the male skins. This fact, that the Northwest skins are so largely the skins of females, is further evidenced by the fact that in many of the early sales of such skins they are classified in deponent's books as the skins of females.

(4) Deponent further says, that in his judgment the absolute prohibition of pelagic sealing, i. e., the killing of seals in the open sea, whether in the North Pacific or the Bering Sea, is necessary to the preservation of the seal herds now surviving, by reason of the fact that most of the females so killed are heavy with young, and that necessarily the increase of the species is diminished by their killing. And further, from the fact that a large number of females are killed in the Bering Sea while on

the search for food after the birth of their young, and that in consequence thereof the pups die for want of nourishment. Deponent has no personal knowledge of the truth of this statement, but he has information in respect of the same from persons who have been on the Pribilof Islands, and he believes the same to be true. Deponent further says that this opinion is based upon the assumption that the present restriction imposed by Russia and the United States on the killing of seals on their respective islands are to be maintained, otherwise it would be necessary to impose such restrictions as well as to prohibit pelagic sealing in order to preserve the herds.

(5) Deponent is further of the opinion, from his long observation and handling of the skins of the several catches, that the skins of the Alaska and Copper catches are readily distinguishable from each other, and that the herds from which such skins are obtained do not in fact intermingle with each other, because the skins classified under the head of Copper catch are not found among the consignments of skins received from the Alaska catch, and vice versa.

(6) Deponent further says that the distinction between the skins of the several catches is so marked, that in his judgment he would, for instance, have had no difficulty had there been included among 100,000 skins in Alaska catch 1,000 skins of the Copper catch, in distinguishing the 1,000 Copper skins and separating them from the 99,000 Alaska skins, or that any other person with equal or less experience in the handling of skins would be equally able to distinguish them. And in the same way deponent thinks, from his own personal experience in handling skins, that he would have no difficulty whatever in separating the skins of the Northwest catch and the Alaska catch, by reason of the fact that they are the skins almost exclusively of females, and also that the fur upon the bearing female seals is much thinner than upon the skin of the male seals, the skin of the animal while pregnant being distended and the fur extended over a large area.

(7) Deponent says that the number of persons who are employed in the handling, dressing, dyeing, cutting, and manufacturing of seal skins in the city of London is about 2,000, many of whom are skilled laborers earning as high as £3 or £4 a week. Deponent estimates the amount paid in the city of London for wages in the preparation of fur-seal skins for a manufacturer's use, and excluding the wages of manufacturers' employees, prior to the beginning of the pelagic sealing in 1885, at about £100,000 per annum; and deponent further says that in his judgment if this pelagic sealing be not prohibited, it is but a question of a few years, probably not more than three, when the industry will cease by reason of the extermination of the seals in the same way in which they have been exterminated on the South Sea Islands, by reason of no restrictions being imposed upon their killing.

ALFRED FRASER.

EXHIBIT A.

Salted Alaska fur seal skins sold in London.

Year.	Skins.	Year.	Skins.	Year.	Skins.	Year.	Skins.	Year.	Skins.
1870	9, 965	1875	99, 634	1880	100, 161	1885	99, 719	1890	{ 20, 994
1871	100, 896	1876	90, 267	1881	99, 921	1886	99, 910		{ a 4, 158
1872	96, 283	1877	75, 410	1882	100, 100	1887	99, 940	1891	13, 473
1873	101, 248	1878	99, 911	1883	75, 914	1888	100, 000		
1874	90, 150	1879	100, 036	1884	99, 887	1889	100, 000	Total..	1, 877, 977

a Food skins.

EXHIBIT B.

Salted Copper Island fur-seal skins sold in London.

Year.	Skins.	Year.	Skins.	Year.	Skins.	Year.	Skins.	Year.	Skins.
1872......	7,182	1877	25,380	1882	39,111	1887	54,584	1892	30,678
1873	21,614	1878	19,000	1883	36,500	1888	46,333		
1874	30,349	1879	28,211	1884	26,675	1889	47,416	Total..	768,096
1875	34,479	1880	38,885	1885	48,929	1890	95,486		
1876	33,298	1881	45,209	1886	41,752	1891	17,025		

EXHIBIT C.

Salted Northwest coast fur-seal skins sold in London prior to pelagic sealing in Bering Sea.

Year.	Skins.	Year.	Skins.	Year.	Skins.	Year.	Skins.	Year.	Skins.
1872......	1,029	1875	1,646	1878	264	1881	9,997	1884	9,242
1873		1876	2,042	1879	12,212	1882	11,717		
1874	4,949	1877		1880	8,939	1883	2,319	Total..	64,366

Salted Northwest coast fur-seal skins dressed and dyed in London (but not sold there) taken prior to pelagic sealing in Bering Sea.

Year.	Skins.	Year.	Skins.	Year.	Skins.	Year.	Skins.	Year.	Skins.
1872......	699	1875	578	1878	2,434	1881	5,890	1884	9,242
1873	40	1876	1,062	1879	2,397	1882	11,727		
1874	122	1877	772	1880	4,562	1883	2,319	Total..	46,215

Dry Southwest coast fur-seal skins sold in London prior to pelagic sealing in Bering Sea.

Year.	Skins.	Year.	Skins.	Year.	Skins.	Year.	Skins.	Year.	Skins.
1868	2,141	1872	14,584	1876	903	1880		1884	785
1869	1,671	1873	891	1877	1,173	1881	686		
1870	684	1874	2,772	1878	912	1882	321	Total..	42,767
1871	12,495	1875	1,351	1879	918	1883	390		

Of the skins sold in 1871 and 1872 a very large proportion were the accumulation of the Russian-American Company, and sold by them after the purchase of Alaska by the United States.

RECAPITULATION.

	Years.	Skins.
Salted skins sold in London..	1872–1884	64,366
Salted skins dressed and dyed in London...................................	1872–1884	46,215
Dry skins sold in London...	1868–1884	42,767
Grand total..		153,348

EXHIBIT D.

Dry Northwest coast fur-seal skins sold in London after commencement of pelagic sealing in Bering Sea.

Year.	Skins.	Year.	Skins.	Year.	Skins.
1885	1,520	1888	1,252	1891	1,083
1886	979	1889	228		
1887	2,843	1890	699	Total...............	8,604

Salted Northwest coast fur-seal skins dressed and dyed in London (but not sold there) taken after the commencement of pelagic sealing in Bering Sea.

Year.	Skins.	Year.	Skins.
1885	16, 667	1889	2, 017
1886	15, 087		
1887	3, 589	Total	39, 200
1888	1, 930		

In addition to above, it is estimated that from 25,000 to 30,000 skins have been dressed and dyed in the United States.—E. T. R., jr., notary public.

Salted Northwest coast fur-seal skins sold in London after commencement of pelagic sealing in Bering Sea.

Year.	Skins.	Year.	Skins.
1885	2, 078	1890	38, 315
1886	17, 909	1891	54, 180
1887	36, 907	1892	28, 298
1888	36, 818		
1889	39, 503	Total	254, 068

RECAPITULATION.

	Years.	Skins.
Dry skins sold in London	1885–1891	8, 004
Salted skins dressed and dyed in London, but not sold there	1885–1889	39, 200
Salted skins dressed and dyed in United States, estimated	1885–1889	30, 000
Salted skins sold in London	1885–1892	254, 068
Grand total		331, 962

EXHIBIT E.

Salted Lobos Island fur-seal skins sold in London.

Year.	Skins.	Year.	Skins.	Year.	Skins.
1873	6, 956	1881	13, 569	1889	8, 755
1874	8, 509	1882	13, 200	1890	18, 541
1875	8, 179	1883	12, 861	1891	15, 834
1876	11, 353	1884	16, 258	1892 a	4, 800
1877	13, 066	1885	10, 953		
1878	12, 301	1886	13, 667	Total	247, 777
1879	12, 295	1887	11, 068		
1880	14, 805	1888	20, 747		

a To date.

Additional deposition of Alfred Fraser, member of the firm of C. M. Lampson & Co., furriers, London.

STATE OF NEW YORK,
        City and County of New York, ss:

Alfred Fraser, being duly sworn, says: I am a member of the firm of C. M. Lampson & Co., of London, and the person described in and who verified an affidavit on the 1st day of April, 1892, relating to the fur-seal industry. The tables hereto annexed, marked A, B, C, D, E, and F, have been prepared by me from the printed catalogues of public auction sales in London of fur-seal skins, and also from my private memoranda, and from knowledge and information of the fur-seal industry, I believe them to be correct in every particular. Said tables state all of the salted fur-seal skins of the Alaska, Copper, Northwest coast, and Lobos catches, which, according to the said catalogues and memoranda, were sold at public auction in London between the years 1868 and 1891, together with the average price per skin obtained during each of said years for the aforesaid skins.

ALFRED FRASER.

———

EXHIBIT A.

Salted Alaska fur-seal skins sold in London from 1870 to 1891.

Year.	Skins.	Year.	Skins.	Year.	Skins.	Year.	Skins.	Year.	Skins.
1870	0, 905	1875	90, 634	1880	100, 161	1885	99, 719	'800	{ 20, 994
1871	100, 896	1876	90, 267	1881	99, 921	1886	90, 910		{ 4, 158
1872	06, 283	1877	75, 410	1882	100, 100	1887	99, 940	1891	13, 473
1873	101, 248	1878	99, 911	1883	75, 914	1888	100, 000		
1874	90, 150	1879	100, 036	1884	99, 887	1889	100, 000	Total..	1, 877, 977

EXHIBIT B.

Salted Copper Island fur-seal skins sold in London in the years 1870 to 1892.

Year.	Skins.	Year.	Skins.	Year.	Skins.	Year.	Skins.	Year.	Skins.
1870	12, 030	1875	34, 479	1880	38, 885	1885	48, 929	1890	95, 486
1871	9, 522	1876	33, 298	1881	45, 209	1886	41, 752	1891	17, 025
1872	7, 182	1877	25, 380	1882	39, 111	1887	54, 584	1892	30, 678
1873	21, 614	1878	19, 000	1883	36, 500	1888	46, 333		
1874	30, 349	1879	28, 211	1884	26, 675	1889	47, 416	Total..	789, 648

EXHIBIT C.

Salted Northwest coast fur-seal skins sold in London prior to pelagic sealing in Bering Sea.

Year.	Skins.	Year.	Skins.	Year.	Skins.	Year.	Skins.	Year.	Skins.
1872	1, 029	1875	1, 646	1878	264	1881	9, 997	1884	9, 242
1873	....	1876	2, 042	1879	12, 212	1882	11, 717		
1874	4, 949	1877	....	1880	8, 939	1883	2, 319	Total..	64, 366

EXHIBIT D.

Salted Northwest coast fur-seal skins sold in London after commencement of pelagic sealing in Bering Sea.

Year.	Skins.	Year.	Skins.
1885..	2, 078	1890..	38, 315
1886..	17, 909	1891..	54, 180
1887..	36, 907	1892 *a*..	28, 298
1888..	36, 818		
1889..	39, 563	Total............................	254, 068

a To March 25.

EXHIBIT E.

Salted Lobos Island fur-seal skins sold in London.

Year.	Skins.	Year.	Skins.	Year.	Skins.
1873	6,596	1881	13,569	1889	8,755
1874	8,509	1882	13,200	1890	18,541
1875	8,179	1883	12,861	1891	15,834
1876	11,353	1884	16,258	1892 a	4,800
1877	13,066	1885	10,953		
1878	12,301	1886	13,667	Total	247,777
1879	12,295	1887	11,068		
1880	14,865	1888	20,747		

a To date.

EXHIBIT F.

Salted Alaska fur-seal skins sold in London in the years 1868-1871 taken prior to the leasing of the Pribilof Islands.

Year.	Skins.	Year.	Skins.
1868	28,220	1871	20,111
1869	121,820		
1870	110,511	Total	280,662

The following table, prepared by Hutchinson, Kohl, Philipeus & Co., of San Francisco, lessees of the right to take fur seals upon the Commander and Robben islands, shows the number of seal skins secured annually from these respective islands from 1871 to 1891:

Year.	Commander Islands.	Robben Island.	Total.	Year.	Commander Islands.	Robben Island.	Total.
1871	3,614		3,614	1883	26,650	2,049	28,699
1872	29,356		29,356	1884	49,444	3,819	53,263
1873	27,710	2,694	30,404	1885	41,737	1,838	43,575
1874	28,886	2,414	31,300	1886	54,591		54,591
1875	33,152	3,127	36,279	1887	46,347		46,347
1876	25,432	1,528	26,960	1888	47,362		47,362
1877	18,584	2,949	21,533	1889	52,859		52,850
1878	28,198	3,142	31,340	1890	53,780		53,780
1879	38,749	4,002	42,750	1891	5,800		5,800
1880	45,174	3,330	48,504				
1881	39,314	4,207	43,521	Total			776,467
1882	40,514	4,106	44,620				

Table of annual seal-skin supply compiled from table of London trade sales as given by Emil Teichmann.

Year.	Lobos Island.	Cape Horn.	Northwest catch.	Alaska catch.	Copper catch.	Total.
1870			684	9,965		10,649
1871			12,405	100,896		113,301
1872			16,303	96,283	7,182	119,768
1873	6,956		981	101,248	21,614	130,749
1874	8,507		7,843	90,150	30,349	136,851
1875	8,170		3,575	99,634	34,497	145,867
1876	11,353	6,306	4,097	90,267	33,298	145,321
1877	13,066	7,631	1,945	75,410	25,380	123,432
1878	12,301	18,227	3,607	99,911	19,000	143,046
1879	12,295	12,180	15,527	100,036	28,211	168,249
1880	14,386	17,562	13,501	100,161	38,885	184,045
1881	13,569	13,164	16,573	9,094	45,209	188,436
1882	13,200	11,711	23,207	100,100	39,111	187,329
1883	12,861	4,655	9,544	75,914	36,500	139,474
1884	16,258	6,743	20,142	99,887	26,057	169,705
1885	10,953	8,404	20,265	90,719	48,920	183,270
1886	13,667	909	33,975	90,910	41,752	190,213
1887	11,068	2,762	43,330	99,940	54,584	211,693
1888	20,747	4,403	40,000	100,000	46,333	211,483
1889	8,755	3,021	41,808	100,000	47,416	201,000

*Table of percentages of annual seal-skin supply compiled from table of London trade sales
as given by Emil Teichmann.*

Year.	Lobos Island.	Cape Horn.	Northwest catch.	Alaska catch.	Copper catch.	Total.
1870			0.0620	0.9380		1.0000
1871			.110	.890		1.0000
1872			.136	.8047	0.051	1.0000
1873			.1362	.7743	.059	1.0000
1874	0.0532		.0072	.6830	.1653	1.0000
1875	.0560		.0246	.6204	.2364	1.0000
1876	.0782	0.0440	.0282	.6113	.2143	1.0000
1877	.1054	.0618	.0158	.698	.2578	1.0000
1878	.0831	.0575	.00251	.5944	.1363	1.0000
1879	.0730	.0722	.0927	.813	.1677	1.0000
1880	.0804	.0946	.0730	.5417	.2103	1.0000
1881	.0720	.0697	.0825	.5307	.2451	1.0000
1882	.0703	.0624	.1233	.5343	.2097	1.0000
1883	.0923	.0334	.0685	.5442	.2616	1.0000
1884	.0950	.0332	.1187	.5821	.1631	1.0000
1885	.5540	.0196	.113	.5447	.2684	1.0000
1886	.0718	.0047	.1795	.5307	.2143	1.0000
1887	.0521	.0133	.2047	.4721	.2578	1.0000
1888	.0981	.0207	.1894	.4728	.2190	1.0000
1889	.0435	.0156	.2075	.4975	.2359	1.0000

CRUISE OF THE LOUIS OLSEN IN THE BERING SEA.

BY A. B. ALEXANDER.

On the 25th of May, at Seattle, I met Capt. E. P. Miner, master of
the American sealing schooner *Harry Dennis*, who, on the same day,
had arrived from Japan, his vessel having been wrecked on that coast.
At the time of meeting him he was endeavoring to charter another
vessel, and hoped to be in the Bering Sea by the 1st of August. Informing him that the United States Commissioner of Fish and Fisheries
was anxious that I should make a cruise with him should he succeed
in getting a suitable vessel, he freely consented, and informed me that
if he found out in time he would let me know by mail.

On the morning of the 28th I sailed in the *City of Topeka* for Sitka,
and from there took passage in the *Crescent City* for Unalaska. Soon
after arriving there I joined the *Albatross*, and remained by her until
the evening of the 29th of July, when I joined the sealing schooner
Louis Olsen, of Astoria, Oreg., Captain Guillams, master, who did not
for a moment hesitate about giving me a passage. My reason for
joining the *Olsen* was on account of not having heard from Captain
Miner as to whether he had succeeded in obtaining a vessel; I was
also informed by several sealing captains that he did not get a vessel,
and in consequence would not be in the sea. The time having arrived
when pelagic sealing was about to commence, I was glad to accept, as
I thought, the only opportunity which would be offered for the season.

The next day, in latitude 54° 38′ north, longitude 167° 04′ west, we
saw our first seals, 20 in number, 12 of which were "sleepers." Seals
when sleeping are by sealers always called by the above name. In the
afternoon we saw 6 seals about half a mile from the vessel playing in
a bunch of seaweed. The sea at the time was perfectly smooth with a
light air stirring. Two hunters and myself started out in a boat to
watch them and see how near we could approach without disturbing
them. We soon learned that they were unusually tame, as we approached
near enough to touch one with a spear pole which was in the boat. They
showed little signs of fear, notwithstanding that we were within 30 feet
of them for fully five minutes. Diving under the seaweed and sud-

denly thrusting their heads up through it seemed to afford them great pleasure. Rolling over and over in the seaweed, their flippers becoming tangled in it, was also a pleasant enjoyment. They paid but little attention to us and seemed almost indifferent as to how near we approached so long as we did so quietly. This caused the hunters to exclaim several times, "If we only had a gun we could kill them all." Under the circumstances it was but natural that a gun should be the uppermost thought in their minds.

Early in the spring, both on the Northwest coast and off the coast of Japan, seals are sometimes found which evince little signs of fear, but after one day's shooting on the ground they become very wild and mistrustful, and, like a crow and some land animals, seem to scent a gun in the air. On this particular occasion a kodak camera would have given good results—it was one opportunity of a thousand.

The following day, 31st, seals were plentiful. The wind being light during the previous night, our position had changed but little.

On August 1, at a very early hour, the spears were brought forth and the seal on them broken. While this was going on many remarks were made in regard to the first day's trial. Some of the hunters were already discouraged and were confident that they were only wasting time by attempting to use spears; the thought of being obliged to adopt the primitive weapon of the Siwash was indeed humiliating to them. A few on board felt more hopeful and were willing to give the spear a fair trial. Ever since leaving Unalaska the hunters had been practicing at throwing the spear pole. Every piece of floating seaweed or other object which came within range had been a target. On several occasions the boats had been lowered and a supply of chips and small pieces of wood taken along. These were thrown ahead of the boat as targets to throw at. It was soon found that an object that could be easily hit at a distance of 25 or 30 feet from the vessel was not so easily reached from a boat, as the smallest wave would cause her to move just enough to cause the pole to go wide of its mark. A day's practice throwing from the boats had the effect of teaching the hunters the various ways of holding the spear to make more sure of its hitting the mark under the many conditions of sea and wind. All this time spear throwing had been carried on with lifeless objects for a mark. The opportunity was about to present itself to exhibit skill in throwing at something that if missed the first time would not be likely to remain stationary long enough to give the marksman another trial. A cool head and steady nerves would be the special requirements to insure a successful day's hunt should seals be plentiful.

The 1st day of August did not prove a success, so far as sealing was concerned, the weather being too foggy to send out the boats. Scattering seals were observed all day, but they were all "travelers;" that is, they were all moving in various directions. Our noon position was latitude 56° 11' north, longitude 172° 01' west. The next day seals were less numerous. None were observed in the forenoon; in the afternoon 12 were seen; all but one were traveling to the westward. This individual was asleep; a boat was quickly lowered and the hunter on watch was rowed toward it. Before the boat had covered half the distance the seal showed signs of waking, and shortly after, becoming aware of approaching danger, it disappeared. Our noon position on this day was latitude 57° 21' north, longitude 173° 46' west. Seals here were not so plentiful as they were farther south. In the afternoon we hove to and caught two cod in 65 fathoms of water. No more seal life was observed until the afternoon of the following day, when two sleep-

ing seals were sighted, latitude 57° 50' north, longitude 173° 48' west.
Five boats were sent out. They returned at 5 p. m., having taken
no seals, although 8 had been seen, but they were all "travelers."
Heavy fog and strong indications of the wind breezing up fresh caused
the boats to return earlier than they otherwise would. In the evening,
the vessel being hove to, several seals came close alongside. They
seemed to be very curious to know what we were. All the spears on
board were repeatedly thrown at them, but they had the good sense to
keep just out of range. Whistling had the effect of enticing them close
aboard, but the sight of a spear or two being aimed at them would cause
them, without any apparent effort, to increase their distance by 20 or
more feet.

The first seal captured by the *Olsen* was on August 4, in latitude 57°
50' north, longitude 173° 48' west, the same position recorded on the
previous day. At 8.45 a. m. sail was made, and at 10.45 the boats were
lowered, two "sleepers" having been seen. The weather being foggy
the boats were soon lost to sight. The vessel was hove to, it being
much easier for the boats to keep the bearing of the vessel than for
the vessel to keep track of the boats.

As soon as the boats had left a hand line was put over in 70 fathoms
of water. An hour's fishing resulted in 18 cod. Their average weight,
as near as could be judged, was 12 pounds; the largest weighed not far
from 30 pounds. All but two of the cod were in a healthy condition.
These two had sores on their backs about the size of a half dollar, which
had eaten nearly to the backbone. This instance is mentioned here on
account of the part of the fish where the sores were. In both cases
they were situated near the neck, directly over the vertebra; they were
as round and smooth as if cut with a knife.

Notwithstanding that the fog did not lift during the day, the boats
remained out until 9 p. m. The result of the day's hunt was 12 seals—
4 males and 8 females. One of the seals had previously been speared
in one of its flippers, as it was nearly severed from its body, showing
that the seal must have had a hard struggle to free itself. The largest
number of seals caught for this first day's successful hunt was taken
by two boats, they bringing in five each; two other boats captured one
each, and the remaining two boats brought in nothing. The hunters
in these boats, on learning that 12 seals had been captured, indulged in
strong language at their nonsuccess.

About 50 seals had been observed from the boats, the most of which
were awake. Only an occasional individual had been seen during the
day from the vessel, the fog being too dense to see more than an eighth
of a mile.

In no single instance was the first seal speared at captured; it was
only after repeated attempts by each hunter that one was hit. The
excitement caused by the desire on the part of each to be the first
to capture a seal, combined with the inexperience of throwing the
primitive weapon, was no doubt the chief cause of the poor results.
The eight females captured were all nursing seals; but little food was
found in their stomachs, and that was too much digested to tell what
it consisted of; it was, however, placed in alcohol.

In the evening, after the seals had been skinned and everything made
snug for the night, each hunter told his experience during the day,
which, as may be supposed, was more entertaining than instructive.

On the 5th the wind and weather were not suitable for scaling; a
very fresh southeast wind prevailed, and in order to keep our present
position the vessel was hove to under easy sail. A large number of

seals was noticed; they were frequently seen playing about, sometimes on the crest of a wave, and then in the hollow of a sea. They seemingly had no fixed course, but would swim in one direction a half a mile or so, return and go in an opposite way. It is more than likely, had the wind been blowing a strong gale, they would all have been bound in one direction. The wind continued fresh, with a rough, choppy sea, until the following noon, when the fog which had come in during the night lifted and the wind suddenly subsided into a calm. The boats were put in readiness and sent out for an afternoon's hunt. Considering the state of the sea and the time of starting, a fair afternoon's work was done, 19 seals being landed on deck by 8.30 p. m. Fifteen of the number were cows and 4 males. Only 6 had food in their stomachs. Every hunter reported seals numerous, about half of the number being asleep. They slept in bunches of 6 and 8, and when aroused from their slumber were very tame, but owing to the inexperience of the hunters with spears in a comparatively rough sea, the successful throws were few and far between. Had the hunters been provided with shotguns instead of spears, it is pretty safe to say that a hundred or more seals would have been nearer the day's catch. To be compelled to see seals escape that could easily have been killed with a shotgun brought forth from both hunters and boat crews loud and imprecatory language upon the heads of all those who were instrumental in prohibiting the use of firearms in the Bering Sea. These men had not been used to seeing their prey get away so easily, and to them the sight was more than exasperating. During the absence of the boats a large number of traveling seals had been seen from the vessel and also an occasional "sleeper." One of the latter was observed close aboard a little on our lee. It evidently was sleeping soundly, for neither the slatting of the sails nor the blowing of the fog horn had the effect of awaking it, and it was only when the scent of the vessel reached its nostrils that it showed signs of life. After being fully aroused it did not exhibit any great signs of alarm, but played about not far off for some time. It seldom happens that a seal will show such indifference to its surroundings as this one. The captain and mate said they had never in all their experience seen a seal so tame. The general opinion on board was that it was due to there being no firearms used or hunting allowed in the Bering Sea for the past few years that caused the seals thus far observed to show so little fear of man.

The highest catch for any one single day was taken on the 7th. The day commenced with a gentle breeze from the south, and a smooth sea. A light fog hung low over the water which prevented the boats from being seen more than 20 yards. At 8 a. m. the last boat shoved off, and they were not seen again until evening. Noon position: Latitude, 58° 30' north; longitude, 173° 56' west. In the forenoon hand-line fishing was carried on. The depth of water here was 60 fathoms. Six good-sized cod were caught in quick succession; 2 males and 4 females. Their stomachs were well filled with food. In the stomach of a large female was found an octopus; it had been recently swallowed, as its skin showed no discoloration. Cod were abundant; we could have filled the decks in a day's fishing with a single line. The abundance of cod may have been the cause of seals being plentiful in this region. In the evening the boats all returned nearly at the same time, bringing in 34 seals, 30 of which were females. Twenty-four of the number had food in their stomachs. The material, however, was finely masticated, and hard to identify, but a portion of it looked very much like the flesh of cod. If a portion of the food was cod, the question arises, did the seals dive to the bottom in 60 fathoms of water and bring their prey to the surface? As a rule cod are found very close to the bottom, especially in deep water; in shallow places they are sometimes found nearer the surface. It is not probable that seals in this region found an abundance of cod or even scattering ones near the surface. Just how deep a seal can dive and secure food is a mooted question. Mr. Henry Elliott gives them credit of being able to dive to profound depths. The writer has conversed with a good many sealers on the subject, but has never been able to gather any reliable information. Sealers as a rule are not a very observing class of men, for the reason that their interest is all centered in the commercial side of the question, and think little of the habits or other peculiarities of seal life. The most satisfactory evidence the writer ever had that seals are deep divers was two years ago on the Fairweather Ground, a large bank off the coast of Alaska, while on a cruise in the revenue-cutter *Corwin*. We were about to return to the ship at the end of a successful afternoon's hunt, when a large bull suddenly came up close to our canoe, not over 30 feet away, with a very large red rockfish in its mouth, which it immediately proceeded to devour. The fish was alive and could be plainly seen struggling in the seal's mouth. Our position at the time was some 75 or 80 miles offshore from Yakutat Bay. We had no means of ascertaining the depth of the water, but it could not have been much less than 100 fathoms. Red rockfish is also a species that generally swims close to the bottom, although like cod it is possible that they sometimes feed near the surface. The writer does not maintain that seals can go to the bottom in 100 fathoms of water, but thinks they can dive much deeper than is generally supposed.

All the hunters on this day reported seals plentiful, but could find very few asleep. Had the sun been shining it is safe to say that the majority of those with food in their stomachs would have slept during a greater part of the day, for, as a rule, seals with full stomachs sleep when the sun is out, the air warm, and the sea smooth or comparatively so. Their time of sleeping, however, is not always when conditions are favorable, for after a gale of long duration they are frequently seen asleep when the air is cold and the sea uncommonly high. At such times seals are completely exhausted. It is not an infrequent sight during the winter and spring months, at the end of a long and heavy gale, to see seals sleeping soundly in a snowstorm, with that portion of the body out of the water covered with snow. In consequence of the seals on this day being restless, a great many of the 34 taken were what is known to sealers as "finners," that is, seals about half asleep, rolling about and scratching themselves. Sometimes "finners" are hard to approach, and at other times very easy. A restless one will try very hard to take a nap, but just as he gets comfortably fixed something disturbs him; holding its head up he will take a look all around, as if danger was scented in the air. These are hard to capture with spears.

Indians seldom pay any attention to moving seals when hunting with spears; they think it a waste of time. White hunters, when they can find no sleeping seals, frequently give chase to "finners" and "travelers," and in many cases are rewarded for their trouble. The hunters on the *Olsen* soon found that few seals would be taken on certain days if they only selected sleeping ones. Many haphazard throws were made at swimming and finning seals, the majority of which were failures, but enough good shots were made to make the experiment a paying one.

For several days seals had been observed chasing some kind of fish, and during this last day's hunt they were quite plentiful. Only a single individual would be seen; it would dart first in one direction and then in another, and occasionally would make a desperate leap out of water. Presently a seal would be noticed not far off swimming as rapidly and in as many different directions as the fish. On the day in question, two seals were speared, just as they came to the surface, each with one of these fish in its mouth. The seals did not relinquish their hold when speared, but kept a firm grip until knocked on the head. The specimens proved to be Alaskan pollock. In both cases the specimens of fish secured were brought up by large males; one was somewhere between 8 and 10 years of age.

The two following days, 8th and 9th, the weather was too boisterous for sealing; wind southeast and every indication of a gale. We lay to under the foresail in order to keep as near our present position as possible. A heavy sea set in from the westward, but the wind did not increase above a strong breeze. Scattering seals were about each day, all traveling to the westward. From observation we learned that during stormy weather seals traveled in an opposite direction to the wind. In a gale it will be found that seals are far more numerous on the lee side of the Pribilof Islands than to the windward of them. When the wind is heavy and the sea rough seals as a rule travel from the seal islands directly to leeward or nearly so. Just how much the wind changes the course of the main body of seals would be hard to say, but so far as our investigations extended, in connection with the traveling herd which came under our notice, we are inclined to think that seals within 100 miles of the seal islands, bound to the feeding grounds, will in most cases seek the grounds to the leeward of the group. Seals in a gale take every advantage of wind and sea. It is necessary that they should, for there is evidently a limit to their endurance.

On the morning of the 10th light winds prevailed, but a choppy sea, combined with a long rolling swell from the west-southwest, rendered it unfit for sealing, although scattering seals had been noticed. Two days of idleness had made everybody on board anxious to get out in the boats. In the afternoon the wind fell to a calm, and the boats were put over in latitude 58° 27' north, longitude 172° 46' west, and remained out until evening, bringing in only three seals. Very few were seen from the boats, although they covered considerable ground during the day. Seals were equally scarce in the vicinity of the vessel; only six were observed. One of these, more bold than the rest, kept circling around the vessel, coming nearer each time. Finding that it was inclined to be inquisitive, it was encouraged to make further investigations as to what we were by continual whistling by those on board. It was finally enticed alongside and captured, the spear passing through one of its hind flippers. A series of photographs showing all the different positions the seal was in during its struggle for liberty would have been valuable. It fought bravely for life while in the water, but on being hauled on board its power was greatly lessened. It did not, however, give up without a desperate struggle to regain its liberty. At one period of the fight it drove everybody from the main deck, and it was only when a noose was thrown over its neck and its head hauled down to a ring bolt that terms of peace could be made, which was by knocking in on the head. The catch of seals for the day was 4—3 females and 1 male; the total catch to date 69—13 males and 56 females.

The next day (11th) the boats made an early start. Everything looked favorable for a good day's hunt, the wind being light and the sea smooth, two things which are almost indispensable in seal hunting.

No seals had been noticed during the morning, but it does not necessarily follow that because none are observed from the vessel they are
not about, for frequently it happens that good catches are made when
not a seal has been seen from the vessel. This was not one of those
exceptional days; 13 was the catch—3 males and 10 cows. Seals had
been comparatively plentiful, but were not inclined to sleep and were
too wild to approach. A piece of an Alaskan pollock was brought in
by one of the hunters, it having been taken from a seal's mouth in the
same manner as the two others previously described. The shape of an
Alaskan pollock would indicate it to be a fish that could easily escape
from a seal. It may be, however, that seals do not select a single fish,
but give chase to a body of them after the manner of whales, sword-
fish, and sharks, and out of many succeed in capturing one or more.
The reason for seals seen on this day being so wild could be accounted
for only in one way, they having had little to eat. The stomachs of
those taken fully corroborated this theory. A series of trials were
made for bottom fish, but with negative results; we seemed to be
drifting over barren ground. The noon position on this day was 57°
42' 38'' north latitude; 172° 52' west longitude.

Our pleasant weather was about to be broken for a considerable
length of time, for on the 12th the day began with a gale from the
southeast, accompanied by a heavy sea. Lay hove to under single-
reefed foresail and trysail. In the afternoon spoke with the schooner
Teresa, of San Francisco; also saw the schooner *Kate*, of Victoria,
British Columbia, a short distance away. Seals frequently seen all
through the day. In the early part of the night the wind increased to
a heavy gale, and in the latter part the wind decreased in force and
hauled to the west-southwest. A heavy sea kept up all day. In the
forenoon a vessel was sighted low on the horizon. An occasional seal
observed; phalaropes numerous.

August 14: Weather pleasant, but wind fresh from the westward.
In the evening boarded the schooner *Fawn*, of Victoria, British Columbia. She reported losing a boat and three men on the 11th. (They
were afterwards picked up.) The *Fawn* had an Indian crew and had
taken 20 skins in the sea. This news gave our hunters considerable
encouragement. Position, latitude 57° 37' north; longitude 173° 14'
west.

August 15: Pleasant weather in the early part of the day, but very
squally in the latter part; sea rough. Latitude 57° 11' north; longitude 173° 09' west.

August 16: Variable weather; clear in the morning, thick and squally
in the afternoon; sea very rough. But few seals seen. Noon position,
latitude 57° 04' north; longitude 172° 30' west.

August 17: At 7 a. m. made sail and ran to the southward; wind
northwest and fresh, gradually decreasing to a light breeze in afternoon. A heavy fog came in later in the day. Position, latitude 56° 54'
north; longitude 172° 45' west. Continued on our course until 8 a. m.
the next day, at which time saw a seal "finning" close by. A boat was
quickly manned and started in pursuit, but the seal was on the alert
and soon increased the distance between itself and enemy. Shortly
after this a "sleeper" was noticed not far off on the weather bow.
Another boat was hoisted out, which was silently rowed toward the
coveted prize. No attempt was made by the hunter to throw the spear
until the boat was within 20 feet of it. It was easily captured. When
opened its stomach was found to be well filled with food, which no
doubt was the cause of its sleeping so soundly.

Later in the day all the boats went out, but returned at the end of three hours with only one seal. The sea was smooth and but little wind stirring, but the air grew suddenly chilly and the sky very cloudy, which practically put an end to the chances of seals sleeping for the day. On this particular occasion the hunters were very much disgusted on account of not having shotguns. They claimed that with guns the day's catch at the least calculation would have been between 60 and 70 seals, instead of the small number of two. On a day like this, when seals showed no inclination to sleep, shotguns in the hands of skillful hunters would have done very destructive work to the seal herd, for experienced hunters kill nearly if not quite as many traveling seals during the course of a season as sleeping ones. In the early history of pelagic sealing hunters sought sleeping seals only, but they have learned the movements of the seal so thoroughly that traveling and finning seals are almost as desirable as sleeping ones.

Hand-line fishing was carried on from the vessel in 60 fathoms of water. In one hour 10 cod were caught, their average weight being about 9 pounds. It was estimated that the largest would weigh 30 pounds, the smallest 4 pounds. In their stomachs were found small starfish, prawns, squid, medusæ, and a quantity of decomposed fish, all of which was saved.

Unfortunately this was our last day's hunt. From this time on we had stormy weather and heavy gales. Eighty-four seals had been taken, 16 males and 68 females. All the females were nursing cows, except one, which was a yearling. The last seal caught by the *Olsen* was taken in latitude 56° 05′ north, longitude 172° 17′ west.

Early in the morning of the 19th the weather was pleasant, with indications of its being a suitable day for sealing, but shortly after the wind began to freshen from the southeast, gradually increasing in force and hauling to the westward. Lay to under a double-reefed foresail; heavy squalls at times. Noon position, latitude 55° 39′ north, longitude 172° 12′ west.

August 20: Heavy gale from the northwest; very high sea running. Ran before the wind for three hours, hoping to run out of the heaviest part of the gale, but no perceptible difference was felt. Lay hove to until 10 p. m., at which time again kept off before the wind and ran until 10 a. m. the next day. About this time saw several seals, and soon after ran close to a bunch of seals, five in number, all huddled together. It was evident that they were well tired out, or else they would not have been asleep in such weather. Position, latitude 54° 38′ north, longitude 168° 01′ west. In the afternoon sighted several vessels.

On the 22d bore away for Unimak Pass; wind north-northwest and blowing a gale, followed by a heavy sea. On the morning of the 23d sighted the lower part of Akutan Island, the top of it being enveloped in a heavy fog. All through the day seals were plentiful, many of which were asleep. During the past few days enough seals had been seen to cause a vessel to lay by and wait until the weather should moderate. The captain thought that bad weather had set in for the fall, and accordingly had made up his mind to go home. A mistake was made in this decision, for after we had left the sea and were on our way home good catches were being made by all the vessels that remained.

At 6 o'clock in the evening we had left Unimak Pass behind us, and were standing on an east by south course. The next day, when about 75 miles from the pass, saw a sleeping seal, and 10 miles farther on saw two more. When about 200 miles offshore salmon were noticed jumping. They were so near that we could hardly mistake the species. Whales were also plentiful.

For the first two or three days after leaving the sea the weather was pleasant, but during the greater part of the voyage home heavy gales from the westward prevailed, which made the captain all the more confident that no mistake had been made by leaving so early.

On the evening of the 6th of September we arrived at Victoria, having been twelve days on the voyage home.

The writer was very kindly treated by the captain, officers, and crew of the *Olsen*, every effort being made by them to lend assistance and collect such material as was desired. Had the *Olsen* been among seals under favorable circumstances, as many vessels were, the writer could, with the assistance of the kindly disposed crew, have gathered considerable material; but we were one of the unfortunate ones. It was subsequently learned that during the time we were having exceedingly stormy weather—often hove to in a gale—many vessels of the fleet that were several degrees farther south were having pleasant weather and getting good catches every day.

Seals taken in the Bering Sea by the schooner Louis Olsen, 1894.

Date.	Position.		Number.	Male.	Female.
	North latitude.	West longitude.			
	° ′ ″	° ′ ″			
Aug. 4	57 50 00	173 48 00	12	4	8
6	58 30 00	173 56 00	19	4	15
7	58 30 00	173 50 00	34	4	30
10	58 27 00	172 46 00	4	1	3
11	57 42 38	172 52 00	13	3	10
18	56 05 00	172 17 00	2	....	2
Total			84	16	68

[Statistics compiled by H. H. McIntyre, 1889.]

Seal skins landed at Victoria from Bering Sea, as shown by the Victoria custom-house records.

Schooner.	1881.	1882.	1883.	1834.	1885.	1886.	1887.?	1888.	1889.
San Diego (American)	193	327	908	980	1,726	....	1,187	....	....
Alex. and Otter (American)				1,700	....	....	....	....	....
Mary Ellen (British)				1,409	1,773	3,550	2,130	700	....
Vanderbilt (American)					1,244	1,420	1,349	....	....
City of San Diego (American)					1,953	1,600	1,187	....	....
Lookout (American)					1,100	....	....	....	....
Favorite (British)					1,385	3,492	1,887	....	1,700
Annie (American)						182	507	1,040	....
Therese (American, now British)						2,000	900	....	650
Sylvia Handy (American)						1,700	440?	....	614
Helen Blum (American)						(?)	536	....	....
Dolphin (British, now American, J. G. Swan)						2,200	S'zed.	....	....
Alfred Adams (British, now the Lily)						1,455	....	....	....
Black Diamond (British)						328	990	765	55
Pathfinder (British)						1,700	2,377	800	50
Sierra (British)						1,000	....	....	....
Active (British)						1,338	....	....	....
Annie Beck (British)						1,142	S'zed.	....	....
W. P. Sayward (British)						1,600	S'zed.	....	1,600
Grace (British, now the J. H. Lewis)						1,700	S'zed.	....	....
Penelope (British)						194	1,292	1,054	1,850
Mountain Chief (British)						630	624	780	....
Mary Taylor (British)							1,000	....	....
Kate (British)							1,625	....	911
Triumph (British)							500	2,470	00
Lottie Fairfield (British)					...		2,507	....	....
Ada (British)							S'zed.	....	....

Seal skins landed at Victoria from Bering Sea, etc.—Continued.

Schooner.	1881.	1882.	1883.	1884.	1885.	1886.	1887.?	1888.	1889.
Juniata (British)................................								1,030	......
Annie C. Moore (British).....................								715	1,300
Viva (British).....................................								2,069	2,180
Maggie Mc (British)............................								1,424	1,290
Adele (German)...................................						1,350?		450	1,600
Webster (American).............................								520	
Olsen (American).................................								500	
Walter A. Rich (American)....................								400	
Allie J. Alger (American).....................						400?		380	284
Newton (American) (Venture)................									239
J. G. Swan (American).........................									60
Henry Dennis (American).....................									700
Sapphire (British)...............................									1,629
Lilly (British)....................................									74
Ariel (British)....................................									1,316
Minnie (British)..................................									521
Beatrice (British)................................									700
San Jose (American)............................									700
Lily L (American)................................									800
Mollie Adams (American)......................									1,537
Bessie Reuter (American)......................									550
Challenge (American)..........................									96
Total.........................	193	327	908	4,089	9,181	27,240	22,331	15,097	23,066
Skins seized by the United States, approximate...............							2,000	12,000a		2,500
Grand total......................	193	327	908	4,089	9,181	29,240	34,331	15,097	25,566

a Actual number of skins seized 11,618 or 11,902. See page 337 United States counter case.—J. S. B.
NOTE.—The interrogation point (?) following figures in 1887 column indicates doubt as to the correctness of the report.

Number of Victoria and Northwest Coast fur-seal skins sent to market from 1881 to 1889, inclusive.

Year.	Bering Sea skins.	North Pacific skins.	Total.
1881	193	16,380	16,573
1882	327	22,880	23,207
1883	908	8,186	9,094
1884	4,089	16,053	20,142
1885	9,181	11,184	20,365
1886	29,240	4,735	33,975
1887	34,331	8,908	43,239
1888	15,097	24,801	39,898
1889	25,566	20,580	46,146

The above totals are believed to be very nearly correct, having been compiled from the London catalogues of sales, but the numbers caught, respectively, in Bering Sea and the North Pacific are not definitely known. The catch of the North Pacific has been ascertained by deducting the number reported from the Victoria custom-house records as having been taken in Bering Sea from the total number sold in London. It will be noticed that nearly in proportion as the Bering Sea catch increased, that of the Northern Pacific decreased; and, that while the total catch of 1888, following the seizures and repression of 1887, was not very materially less, the proportion taken from Bering Sea was much smaller than in the preceding and following year.

The inference may be clearly drawn that to the extent to which illicit sealing is suppressed in Bering Sea, it will be more active in the North Pacific, and that the simple closure of the former body of water against marauders will do little toward the effective protection of seal life.

H. Doc. 92, pt. 2——11

Value of Victoria, British Columbia, sealing vessels, estimated by A. R. Milne, esq., surveyor of the port of Victoria, and T. T. Williams, of San Francisco, August, 1889.

Schooner.	Owner's name.	Milne's valuation.	Williams's valuation.	Tonnage.	Crew. White.	Crew. Indian.
Mary Taylor	Carne, Munsie & Co	$8,000	$4,500	43	5	22
Pathfinder	....do	10,000	10,000	66	22	......
Viva	....do	12,000	12,000	92	22	......
Mary Ellen	D. McLean	8,000	6,000	63	22	......
Triumph No. 1	R. C. Baker & Co	14,000	14,000	98	30	......
Favorite	C. Spring	8,000	6,000	80	7	30
Kate	....do	7,000	7,000	58	5	30
Aurora	Not known	8,000	8,000	41	4	24
Minnie	H. Jacobson	8,500	8,500	46	4	30
Sapphire	Marvin & Co	15,000	15,000	124	6	40
Winifred	McDolau	2,500	2,500	13	5	10
Blk Diamond	A. Frank	9,500	5,000	82	5	36
Lily	....do	8,500	5,000	69	5	36
Penelope	Gray & Moses	10,000	10,000	70	20	......
Maggie Mac	Dodd & Co	10,000	7,000	71	21	......
W. P. Sayward	Lundberg & Co	8,000	8,000	60	5	28
Juniata	Hall & Geopel	6,000	3,000	40	5	28
Annie C. Moore	Moore & Hackett	15,000	15,000	118	22	......
Theresa	Babbington & Co	10,000	10,000	63	20	......
Ariel	Buckman	9,000	9,000	90	21	......
Mountain Chief	Indians	3,000	1,000	23	......	20
Wanderer	Paxton & Co	3,000	1,000	16	3	20
Triumph No. 2	Muir Bros	3,000	1,750	15	3	15
Lotetia	Indians	2,000	100	28	......	20
Total		a 200,500	b 173,350	1,464	261	389

a Actual result, $198,000.   b Actual result, $171,350.

Slight errors appear to have been made in footing the above, but the totals are as reported respectively by Milne and Williams.

The above estimates include cost of outfit for a season's cruise, comprising boats, guns, spears, ammunition, provisions, etc.

The schooner *Araunah*, formerly belonging to the Victoria sealing fleet, was seized by the Russian Government July 1, 1888.

Mr. Milne estimates the cost of a sealing venture as follows:

Wages of crews and hunters per vessel	$7,000
Insurance, 7 per cent of $8,000	560
Provisions, salt, ammunition, etc	3,000
Total per vessel, average	10,560

He also estimates the annual average catch at 2,000 skins per vessel, but as a matter of fact the average has been for Victoria vessels during the last four years only about 1,288 skins per vessel annually.

Mr. Williams estimates the expense of a sealing trip as follows:

For five boats	$500
Five Marlin rifles, at $35	175
Five shotguns, at $35	175
Two extra guns	70
Salt for skins	200
Five thousand rounds of ammunition	125
Insurance, one-third of a year	175
Captain's wages four months	400
Ten men at $35, and 5 at $20 per month	1,800
Paid hunters, 1,600 skins at $2 per skin	3,200
Provisions, 20 men 4 months at $8 per month	640
Total per vessel, average	7,460

The annual average price paid for seal skins in Victoria from 1881 to 1889, inclusive, is as follows:

Per skin.		Per skin.		Per skin.
1881................. $9. 25	1884................. $7. 75	1887................. $5. 50
1882................. 8. 00	1885................. 7. 50	1888................. 5. 62
1883................. 10. 00	1886................. 7. 65	1889................. 6. 50

General average, $7.53 per skin.

It will be observed that the price of Victoria and Northwest coast skins has decreased. This has resulted from the fact that it was found by the London dressers that the skins of seals taken indiscriminately, chiefly from females, in the water, did not compare favorably with those taken from carefully selected young males on the islands.

On the basis of the foregoing figures, the value of the fur-seal trade, as conducted by the Canadians, is surprisingly small. Their annual catch at present prices is worth about $125,000, and the highest estimated value of the tonnage engaged is only $200,000—amounts incomparably small in proportion to the loss that would be sustained by the United States and England in case the seal fisheries were broken up, as will inevitably result if the Canadian manner of killing is continued.

The following is extracted from the report of United States Consul Stevens, of Victoria, British Columbia, to the Department of State, in June, 1889:

Since the beginning of the present decade the hunting of the fur seal has been vigorously pursued from this port. There are some 21 vessels, varying from 26 to 126 tons register (an aggregate tonnage of 1,737 tons), employing 458 men, and valued at about $126,000, engaged in hunting the fur seal. These vessels, some of them having small steam power, leave here about the 1st of January and proceed southward, returning in May and landing the skins, taking some of them as far south as San Diego, Cal., and along the coast up. They again leave for the north, going as far as the Bering Sea, returning in September. The total catch for 1888 amounted to 26,720 skins, much smaller than for recent previous years. Of these, 14,987 were reported as "the Bering Sea collection," the distinctive name given to those taken far north, in the neighborhood of the Aleutian Islands, and claimed to be finer furs than any other.

These skins are sold here in bundles, salted to preserve them, and they may be kept many months in that condition without injury. Ordinarily sales are made at so much per skin for the lot; sometimes, however, they are sold in assortments of males, females, and pups, the average price for the latter being $6 per skin. They are shipped from here to London, where they are dressed and dyed, paying a duty when they reach the United States, as they mostly do, of 30 per cent on their then value of about $22.50 per skin.

During these years (1886–87) some eight of these vessels were seized in the northern waters by the United States revenue cutters for violation of the law of July, 1870, "to prevent extermination of fur-bearing animals." No seizures were made in 1888.

Victoria and Northwest coast fur-seal skins sold and dressed in London.

[Compiled by Mr. Alfred Fraser, of the house of Messrs. C. M. Lampson & Co.]

Year.	Dry skins sold.	Salted skins sold.	Dressed for owners.	Total.
1868	2,141			2,141
1869	1,671			1,671
1870	684			684
1871	12,495			12,495
1872	14,584	1,029	699	16,312
1873	891		40	931
1874	2,772	4,949	122	7,843
1875	1,351	1,646	578	3,575
1876	993	2,042	1,062	4,097
1877	1,173		772	1,945
1878	912	264	2,434	3,610
1879	918	12,212	2,397	15,527
1880		8,939	4,562	13,501
1881	686	9,997	5,890	16,573
1882	321	11,727	11,159	23,207
1883	390	2,319	6,385	9,094
1884	785	9,242	10,115	a 20,142
1885	1,520	2,078	16,667	b 20,365
1886	979	17,909	15,087	c 33,975
1887	2,843	36,907	3,589	d 43,239
1888	1,252	36,816	1,930	e 39,998
1889				46,146
Total	**49,361**	**158,076**	**83,488**	**337,071**

a Retained in United States (estimate).
b + 3,000 = 23,365.   c + 3,000 = 36,975.   d + 3,000 = 46,239.   e + 3,000 = 42,000.

NOTE.—Indians dried the skins.

During the past four years, say from 1885 to 1888, about 3,000 Bering Sea and Victoria skins have been annually dressed and dyed in the United States, and must be added to the above.

The large number of dry skins sold in 1871–72 doubtless consisted in part of the stock of the Russian-American Company taken before the cession of the Territory, and held in their warehouse at Sitka.

Adding to the above grand total.. 337,071
Skins dressed in the United States, as above stated.......................... 12,000
                                                                                          _____
                                                                                          349,071
And deducting those from the Russian-American Company's stock of 1867.. 24,000
Makes total killing in the waters of the North Pacific and Bering Sea, from _____
1868 to 1889... 325,071

That this number should be considerably more than doubled to represent the total illegitimate destruction of seal life has been so frequently repeated and so thoroughly proved as to need no further proof or demonstration.

It is worthy of note that of the above 325,071 skins, 203,865 have been taken within the last six years in constantly increasing numbers, except during the year 1888 following the seizures and repression of 1887.

Sealing vessels (schooners) fitted out in 1889 under the American flag.

Sylvia Handy.	San Jose.	Lottie.	Caroline.
Allie J. Alger.	Lily L.	Mary Deleo.	Adele (German).
J. G. Swan.	Mollie Adams.	O. S. Fowler.	Marie de las Cruzes(?)
Venture.	Bessie Reuter.	City of San Diego.	Alexander.
Henry Dennis.	Challenge.	Adonia.	Webster.

Decrease in size of Pribilof Island seal skins, 1885–1889.

[Compiled by Mr. Alfred Fraser, of Messrs. C. M. Lampson & Co., of London.]

	Average weight.	1885.	1886.	1887.	1888.	1889.
	Lbs. oz.					
Large middlings	19 0	149	133	29	2	....
Middlings	15 0	1,811	1,173	696	177	220
Middlings and smalls	12 2	5,300	4,875	2,254	2,318	2,133
Smalls	9 13	20,664	13,318	8,950	9,298	7,020
Large pups	8 4	34,270	28,578	23,178	18,305	11,040
Middling pups	7 0	25,207	30,910	35,591	36,669	26,476
Small pups	6 0	10,684	17,045	24,814	29,239	33,859
Extra small pups	5 4	1,291	3,857	4,426	3,962	18,728
Extra extra small pups	4 0	....	....	....	....	521
Total	....	99,376	99,889	99,938	99,970	99,997

The lessees of the seal islands have been unable during the last three years to secure the most desirable sizes of skins, owing to diminished number of seals, the result of illicit killing by marauders.

The decrease in the size of skins taken by lessees is in proportion to the increase of numbers caught by the marauders.

LOSS OF FEMALE SEALS.

British and American testimony.

[Extract from letter of Sir George Baden-Powell, published in the London Times November 30, 1889.]

As a matter of fact the Canadian sealers take very few, if any, seals close to the islands. Their main catch is made far out at sea, and is almost entirely composed of females. (Case of the United States, p. 200, and Senate Ex. Doc. No. 55, Fifty-second Congress, first session, p. 96.)

[Extract from letter of Rear-Admiral Hotham, of the British navy, to Admiralty.]

WARSPITE, *at Esquimalt, September 10, 1890.*

I have to request you will bring to the notice of the Lords Commissioners of Admiralty this letter with reference to my telegram of the 8th instant.

I personally saw the masters of the sealing schooners named below, and obtained from them the information here reported: Capt. C. Cox, schooner *Sapphire;* Captain Petit, schooner *Mary Taylor;* Captain Hackett, schooner *Annie Seymour;* Capt. W. Cox, schooner *Triumph.*

*   *   *   *   *   *   *

They also mentioned that two-thirds of their catch consisted of female seals, but that after the 1st of July very few indeed were captured "in pup," and that when sealing outside the Bering Sea, round the coast, on the way up (where this year the heaviest catches were made), they acknowledged that the seals "in pup" were frequently captured. (Extract from Vol. III, Appendix to Case of Great Britain, cited in United States, No. 1, 1891, p. 17.)

There were killed this year so far from 40,000 to 50,000 fur seals, which have been taken by schooners from San Francisco and Victoria. The greater number were killed in Bering Sea, and were nearly all cows or female seals. This enormous catch, with the increase which will take place when the vessels fitting up every year are ready, will, I am afraid, soon deplete our fur-seal fishery, and it is a great pity such a valuable industry could not in some way be protected. (Extract from reports of the department of fisheries of Canada, 1886, by Thomas Mowat, inspector of fisheries for British Columbia, p. 268. Cited in British Case, Appendix, Vol. III, p. 173; United States No. 1890.)

The majority of our hunters contend that there are over 7 per cent of pups in the entire catch of fur seals on the coast, while in Bering Sea the catch does not exceed 1 per cent. But they can not deny the fact that 60 per cent of the entire catch of Bering Sea is made up of female seals. (Extract from reports of the department of fisheries, Canada, 1888, p. 241, by Thomas Mowat, inspector of fisheries for British Columbia. Cited in United States Case, p. 201.)

Niel Bonde, of Victoria, sealer. (Case of United States, Appendix, Vol. II, p. 315.)

Bonde has been out four years on sealing schooners from Victoria, namely, from 1887 to 1890, inclusive. He says:

The seals caught along the coast after the 1st of April were mostly pregnant females, and those caught in Bering Sea were females that had given birth to their young. I often noticed the milk flowing out of their breasts when being skinned and have seen live pups cut out of their mothers and live around on the decks for a week.

Cross-examination by the British Government (see British Counter Case, Vol. II, p. 94):

That on each of said vessels [namely, the four he had served on] I have had more or less to do with skinning the seals, and would say that about 60 per cent on the coast were females and about 5C per cent in Bering Sea. I distinguished the male skin from the female by the absence of teats.

Christ Clausen, of Victoria, master mariner (Case of United States, Appendix, Vol. II, p. 319):

Acted as mate in 1889. Was navigator on schooner *Minnie* in 1890.
My catch that year was 2,600, of which about 2,000 were caught in Bering Sea. Acted as navigator on same vessel in 1891.
The seals we catch along the coast are nearly all pregnant females. It is seldom we capture an old bull, and what males we get are usually young ones. I have frequently seen cow seals cut open and the unborn pups cut out of them, and they would live for several days. This is a frequent occurrence. It is my experience that fully 85 per cent of the seals I took in Bering Sea were females that had given birth to their pups, and their teats would be full of milk. I have caught seals of this kind from 100 to 150 miles away from the Pribilof Islands.

E. M. Greenleaf, of Victoria, master mariner (Case of the United States, Appendix, Vol. II, p. 324):

Since then (1882) I have been interested in the sealing business, and am well acquainted with it and the men engaged in it and the methods employed. I am acquainted with the hunters and masters who sail from this port, and board all incoming and outgoing vessels of that class. These men all acknowledge that nearly all the seals taken off the Pacific Coast are females, and that they are nearly all with young.

*   *   *   *   *

I have also learned by conversation with Bering Sea hunters that they kill seal cows 20 to 200 miles from the breeding grounds and that these cows had recently given birth to young. I have observed in the skins that the size of the teats show either an advanced state of pregnancy or of recent delivery of young.

Arthur Griffin, of Victoria, sealer (Case of United States, Appendix, Vol. II, p. 325):

He went sealing in 1890.

Began sealing off the northern coast of California, following the sealing herd northward, capturing about 700 seals in the North Pacific Ocean, two-thirds of which were females with pups; the balance were young seals, both male and female. We entered Bering Sea July 13 through Unimak Pass and captured between 900 and 1,000 seals therein, most of which were females in milk.

Of the following year, he says:

Wo captured between 900 and 1,000 on the coast, most all of which were females with pups. We entered the sea July 12 through Unimak Pass and captured about 800 seals in those waters, about 90 per cent of which were females in milk from 20 to 100 miles from the rookeries.

James Harrison, of Victoria, sealer (Case of the United States, Appendix, Vol. II, p. 326):

We commenced sealing right off the coast; went as far south as the California Coast, and then hunted north to the west coast of Vancouver Islands. Caught 500 skins during the season; almost all of them were pregnant females. Out of 100 seals taken about 90 per cent would be females with young pups in them. I can't tell a male from a female while in the water at a distance. On an average, I think the hunters will save about one out of three that they kill, but they wound many more that escape and die afterwards. We entered Bering Sea about the 1st of June, and caught about 200 seals in those waters. They were mostly mothers that had given birth to their young and were around the fishing banks feeding. The hunters used shotguns and rifles. In Bering Sea we killed both males and females, but I do not know the proportion of one to the other.

James Hayward, of Victoria, sealer (Case of the United States, Appendix, Vol. II, p. 327):
He went out sealing in 1887, 1888, 1890, and 1891. His vessels appear to have made large catches. He makes the following statement:

Most of the seals killed on the coast are pregnant females, while those we killed in Bering Sea after the 1st of July were females that had given birth to their young on the seal islands and come out into the sea to feed. Have caught them 150 miles off from the shore of the seal islands, and have skinned them when their breasts were full of milk. Seals travel and go a long way to feed.

Alfred Dardean, of Victoria, sealer (Case of United States, Appendix, Vol. II, p. 322):
He went sealing in 1890.

We caught over 900 skins before entering the sea and our whole catch that year was 2,159 skins. Of the seals that were caught off the coast fully 90 per cent out of every hundred had young pups in them. The boats would bring the seals killed on board the vessel and we would take the young pups out and skin them. If the pup is a good, nice one we would skin it and keep it for ourselves. I had 8 such skins myself. Four out of five, if caught in May or June, would be alive when we cut them out of the mothers. One of them we kept for pretty near three weeks alive on deck by feeding it on condensed milk. One of the men finally killed it because it cried so pitifully. We only got 3 seals with pups in them in Bering Sea. Most all of them were females and had given birth to their young on the islands, and the milk would run out of the teats on the deck when we would skin them. We caught female seals in milk more than 100 miles off the Pribilof Islands.

Morris Moss, furrier, and vice-president Sealers' Association of Victoria (Case of United States, Appendix, Vol. II, p. 341):
He has bought from 10,000 to 20,000 seal skins per annum.

I believe the majority of seals captured by white hunters in Bering Sea are females in search of food.

J. Johnson, of Victoria, sealer and sailing master (Case of the United States, Appendix, Vol. II, p. 331):
Has spent six years of his life sealing, and been captain of four different schooners.

A large majority of the seals taken on the coast are cows with pup. A few young males are taken, the ages ranging from 1 to 5 years. Once in a while an old bull is taken in the North Pacific Ocean. I use no discrimination in killing seals, but kill everything that comes near the boat in the shape of a seal. The majority of the seals killed in Bering Sea are females. I have killed female seals 75 miles from the islands that were full of milk.

Victor Jacobson, of Victoria, sealer (Case of the United States, Appendix, Vol. II, p. 328):

He is a British subject. Has been engaged in sealing for eleven years, ten years as a master. He is now master and owner of schooner *Mary Ellen* and owner of schooner *Minnie.*

The female seals go through the passes from the Pacific Ocean into Bering Sea between June 25 and July 15. Females killed previous to this time I found with pup, but none with pups after that latter date. I have killed female seals taken by me that three in five are females and nearly all with pup.

Cross-examination by the British Government (See British Counter Case, Appendix, Vol. II, p. 83):

My experience has been that about three out of five seals taken on the coast are females, and about the same in Bering Sea.

Edwin P. Porter, of Victoria, sealer (Case of the United States, Appendix, Vol. II, p. 346):

My experience in four years' sealing is that nearly all the seals taken along the coast are pregnant females, and it is seldom that one of them is caught that has not a young pup in her. In the fore part of the season the pup is small, but in May and June, when they are taken off the Queen Charlotte and Kodiak islands, the unborn pup is quite large, and we frequently take them out of the mothers alive. I have kept some of them alive for six weeks, that were cut out of their mothers, by feeding them condensed milk. The seals we capture in Bering Sea were fully 80 per cent females that had given birth to their young. A fact that I often noticed was that their teats would be full of milk when I skinned them, and I have seen them killed from 20 to 100 miles from the seal islands.

Charles Peterson, of Victoria, sealer (Case of the United States, Appendix, Vol. II, p. 345):

We entered Bering Sea about the 15th of August, through the Unimak Pass, and captured therein 1,404 seals, most of which were cows in milk. On that voyage we caught female seals in milk over 80 miles from the rookeries where they had left their young. I have seen the deck almost flooded with milk while we were skinning the seals. Ninety per cent of all the seals we captured in the water were female seals.

Robert H. McManus, of Victoria, journalist (Case of the United States, Appendix, Vol. II, p. 337):

Tuesday, August 25, rain in morning. Boats and canoe out at half past 9 o'clock; out all day (returning to dinner). Result: First boat, 2 seals reported; wounded and lost 5; seals said to be shy and wary, and not so numerous as formerly; attention called to cow seal being skinned (which I had taken for a young bull). The snow-white milk running down blood-stained deck was a sickening sight. Indian canoe, 1 seal. Total, 3 seals, 2 mediums, and 1 cow.

Wednesday, August 26, cloudy morning. Seals floating round schooner. Boats and canoe out all day. Result: First boat, 1 seal; second boat, none; Indian canoe, 10 seals; total, 11 seals; 8 cows in milk and 3 medium. Skipper in first boat blamed the powder. Second boat said it was too heavy and clumsy for the work. Skipper reported having wounded and lost 7, and the men in second boat 9—16 in all. Skipper said seals not so numerous as formerly, more shy; also blamed the powder. Evidently a great deal of shooting and very few seals to correspond.

Saturday, August 29, ship's cook brought down from deck a large cow seal at 40 yards rise. Boats and canoe out all day. Fine, clear, balmy weather. Akutan Island in sight. Result: First boat, 3 seals; second boat, 3 seals; cook, from deck, 1; Indian canoe, 10; total catch, 17 seals, greater proportion cows in milk. Horrid sight; could not stay the ordeal out till all were flayed. A large number reported as wounded and lost. According to appearances, slaughter and indiscriminate.

Sunday, August 30. Result of hunt: First boat, 2 seals; second boat, 1; Indian canoe, 7; total, 10 seals, 7 of which were cows in milk. Several, as usual, reported wounded and lost by the boats. The great superiority of the Indian spear evident.

The British commissioners, in their report, express the following views with regard to pelagic sealing, which views should be specially noted in connection with the foregoing descriptions of how gravid nursing females are killed:

633. By the pelagic sealers and by Indian hunters along the coast, fur seals of both sexes are killed, and, indeed, it would be unreasonable, under the circumstances, to expect that a distinction should be made in this respect, any more than that the angler should discriminate between the sexes of the fish he may hook.

610. The accusation of butchery laid against those who take the seals on shore can not be brought against this pelagic method of killing the seal, which is really hunting as distinguished from slaughter, and in which the animal has what may be described as a fair chance for its life.

Capt. C. L. Hooper, of the United States revenue marine (United States Counter Case, p. 214):

Captain Hooper made extensive official investigations in regard to seal life on the Pribilof Islands, in Bering Sea, and the North Pacific Ocean in 1891 and 1892. In the course of these investigations he captured, between July 24 and August 31, 1892, 41 seals in Bering Sea. He made no efforts to secure large numbers or all that he saw. The 41 seals were composed of the following classes: Old males, 1; young males, 11; nursing cows, 22; virgin cows, 7. He says:

Since leaving San Francisco on March 9 the *Corwin* has steamed 16,200 miles, and 8,713 miles since the date of my reporting for duty, as part of the Bering Sea fleet. Of this distance, 5,567 miles were steamed in Bering Sea.

* * * * *

I find in general, as one of the results of my investigations, that more than two-thirds of the seals taken are now having young or capable of bearing them at no distant day; that it is impossible to discriminate as to age or sex of seals while in the water, except in the case of young pups and old bulls; that even under the most favorable conditions a large percentage is lost by sinking or wounding; and that by reason of the tameness of the nursing cows, which form the larger part of the seals sent, pelagic hunting in Bering Sea is peculiarly destructive and unless stopped will wholly exterminate the already greatly depleted herds.

I do not believe that it is possible to indicate any zonal limit in Bering Sea beyond which pelagic sealing could be carried on and at the same time preserve the seals from complete annihilation. Further, I wish to renew a statement contained in a former report made to the Secretary of the Treasury, that, unless supplemented with protection in the Pacific Ocean, no amount of protection in Bering Sea will preserve the herds.

Capt. L. G. Shepard, United States revenue marine (Case of the United States, Appendix, Vol. II, p. 187):

I am 45 years of age, a resident of Washington, D. C., and am captain in the United States Revenue-Marine Service, chief of division revenue marine, Treasury Department. In command of the revenue steamer *Rush*, I made three cruises to Bering Sea in the years 1887, 1888, and 1889 for the purpose of enforcing existing law for protection of seal life in Alaska and the waters thereof. I hereby append to and make a part of this affidavit a table, marked A, giving the names of the vessels seized by me in Bering Sea while violating the law of the United States in relation to the taking of fur-bearing animals.

* * * * * * *

I examined the skins taken from sealing vessels seized in 1887 and 1889, over 12,000 skins, and of these at least two-thirds were the skins of females. Of the females taken in the Pacific Ocean, and early in the season in Bering Sea, nearly all are heavy with young, and the death of the female necessarily causes the death of the unborn pup seal; in fact, I have seen on nearly every vessel seized the pelts of unborn pups which had been taken from their mothers. Of the females taken in Bering Sea nearly all are in milk, and I have seen the milk come from the carcasses of dead females lying on the decks of sealing vessels which were more than 100 miles from the Pribilof Islands. From this fact, and from the further fact that I have seen seals in the water over 150 miles from the islands during the summer, I am convinced that the female, after giving birth to her young on the rookeries, goes at least 150 miles, in many cases, from the islands in search of food. It is impossible to distinguish a male from a female seal in the water, except in the case of a very old bull, when his size distinguishes him Therefore, open-sea sealing is entirely indiscriminate as to sex or age.

Capt. Daniel McLane (Case of the United States, Appendix, Vol. II, p. 443):

Captain McLane has been engaged in pelagic sealing for eleven years as master of vessels and deposes in part as follows:

Q. Of what sex are the seals taken by you, or usually killed by hunting vessels in the North Pacific and Bering Sea?—A. Females.

Q. What percentage of them are cows? Suppose you catch 100 seals, how many males would you have among them?—A. About 10.

Q. What percentage of the cows taken are with pup?—A. The females are mostly all with pup; that is, up until the 1st of July.

Q. Have you noticed any decrease in the quantity of animals in the last few years?—A. Yes, sir.

Q. To what do you attribute the cause?—A. Killing off the females.

Q. If sealing continues as heretofore, is there any danger of exterminating them?—A. Yes, sir; they will all be exterminated in three years, and there will be no more sealing.

Q. Do you think it is absolutely necessary to protect the cows in the Bering Sea?—A. Yes, sir.

James Kiernan, of San Francisco, sealing captain (Case of the United States, Appendix, Vol. II, p. 449):

My experience has been that the sex of the seals usually killed by hunters employed on vessels under my command, both in the ocean and Bering Sea, were cows. I should say that not less than 80 per cent of those caught each year were of that sex. I have observed that those killed in the North Pacific were mostly female carrying their young, and were generally caught while asleep on the water, while those taken in the Bering Sea were nearly all mother seals in milk, that had left their young and were in search of food.

*   *   *   *   *   *   *

The mother does not leave the rookery in search of food until she has dropped her young and become pregnant again, hence when she has been slain it means the loss of three, as the young pup will unquestionably die for lack of sustenance.

Michael White, of San Francisco, sealing captain (Case of the United States, Appendix, Vol. II, p. 489):

I am 50 years of age. I reside at East Oakland. My occupation is master mariner, and I have been so engaged for twenty-seven years, off and on. I have been engaged in seal hunting during the years 1885, 1887, 1888, and 1889 in the North Pacific and Bering Sea. I first went out in 1885 in the schooner City of San Diego, chartered by myself and others, and my catch for that year was between 2,300 and 2,400 seals. Of that number about 1,900 were caught in Bering Sea. There were but very few vessels sealing at that time. In 1886 I was master of the schooner Terese, sailing from San Francisco on the 2d day of February, and commenced capturing seals on the coast of California, and followed them from that date north into Bering Sea. We caught them from 5 to 60 miles off the coast. I entered Bering Sea on the 6th day of June, 1886, and previous to that time had caught about 880 seals. Then I sealed in Bering Sea from that time to the 28th of August; caught about 2,200 more, the whole catch being 3,000 for the year.

In 1887 I was master of the schooner Lottie Fairchild, sailing from San Francisco on or about the 17th day of March, and worked northward to Bering Sea, and captured 883 seals. I then entered Bering Sea about the 6th of July, cruising there until the 29th day of August, and took 2,517 seals more, the whole catch being 3,400 for the year.

In 1888 I took the schooner Undaunted on a fishing and sealing voyage, leaving here on or about the 20th of March, and cruised in the North Pacific up to the island of Kodiak, capturing 400 seals up to the 7th day of June. I did not enter Bering Sea that year. I did the same in 1889, my trip being the same as in 1888, and my capture of seals was about the same. I then quit sealing, and I am now engaged in trading with the Gilbert and Marshall islands in the South Pacific Sea.

In my captures off the coast between here and Sitka 90 per cent of my catch were females, but off the coast of Unimak Pass there was a somewhat smaller percentage of females, and nearly all the females were cows heavy with pup, and, in some instances, the period of gestation was so near at hand that I have frequently taken the live pup from the mother's womb.

I never paid any particular attention as to the exact number of or proportion of each sex killed in Bering Sea, but I do know that the larger portion of them were females and were mothers giving milk. I have never hunted within 15 miles of the

Pribilof Islands, but I have often killed seals in milk at distances of not less than 100 to 200 miles from these islands. From my knowledge and experience in the business it is my conviction that within the last few years, since the sealers have become so numerous in the Pacific and Bering Sea, that not more than one out of three is secured. Our purpose and practice was to take all seals we could get, regardless of their age or sex, without any discrimination whatever.

M. A. Healy:

My own observation and the information obtained from seal hunters convince me that fully 90 per cent of the seals found swimming in Bering Sea during the breeding season are females in search of food, and the slaughter results in the destruction of her young by starvation. I firmly believe that the fur-seal industry at the Pribilof Islands can be saved from destruction only by a total prohibition against killing seals, not only in the waters of Bering Sea, but also during their annual immigration northward in the Pacific Ocean.

This conclusion is based upon the well-known fact that the mother seals are slaughtered by the thousands in the North Pacific while on their way to the islands to give birth to their young, and extinction must necessarily come to any species of animal where the female is continually hunted and killed during the period required for gestation and rearing of her young. As now practiced, there is no respite to the female seal from the relentless pursuit of the seal hunters, for the schooners close their season with the departure of the seals from the northern sea and then return home, refit immediately, and start out upon a new voyage in February or March, commencing upon the coast of California, Oregon, and Washington, following the seals northward as the season advances into Bering Sea.

Captain Coulson says:

In company with Special Agent Murray, Captain Hooper, and Engineer Brerton, of the *Corwin*, I visited the Reef and Garbotch rookeries, St. Paul Island, in August, 1891, and saw one of the most pitiable sights that I have ever witnessed. Thousands of dead and dying pups were scattered over the rookeries, while the shores were lined with emaciated, hungry little fellows, with their eyes turned toward the sea, uttering plaintive cries for their mothers, which were destined never to return. Numbers of them were opened, their stomachs examined, and the fact revealed that starvation was the cause of death, no organic disease being apparent.

The great number of seals taken by hunters in 1891 was to the westward and northwestward of St. Paul Island, and the largest number of dead were found that year in rookeries situated on the west side of the island. This fact alone goes a great way, in my opinion, to confirm the theory that the loss of the mothers was the cause of mortality among the young.

After the mother seals have given birth to their young on the islands they go to the water to feed and bathe, and I have observed them, not only around the island, but from 80 to 100 miles out at sea.

In different years the feeding grounds or the location where the greater number of seals are taken by poachers seem to differ; in other words, the seals frequently change feeding grounds. For instant, in 1887 the greatest number of seals were taken by poachers between Unimak and Akutan passes and the seal islands, and to the southwestward of St. George Island. In 1889 the catching was largely done to the southward and eastward, in many cases from 50 to 150 miles distant from the seal islands. In the season of 1890, to the southward and westward, also to northwest and northeast of the islands, showing that the seals have been scattered. The season of 1891, the greatest number were taken to northward and westward of St. Paul, and at various distances from 25 to 150 miles away.

Mr. Redpath:

The Alaskan fur seal is a native of the Pribilof Islands, and, unless prevented, will return to those islands every year with the regularity of the seasons. All the peculiarities of nature that surround the Pribilof group of islands, such as low and even temperature, fog, mist, and perpetually clouded sky, seem to indicate their fitness and adaptability as a home for the Alaskan fur seal; and, with an instinct bordering on reason, they have selected these lonely and barren islands as the choicest spots of earth upon which to assemble and dwell together during their six months' stay on land; and annually they journey across thousands of miles of ocean and pass hundreds of islands, without pause or rest, until they come to the place of their birth. And it is a well-established fact that upon no other land in the world do the Alaskan fur seal haul out of water.

J. C. Redpath says:

No cow will nurse any pup but her own, and I have often watched the pups attempt to suck cows, but they were always driven off, and this fact convinces me that the cow recognizes her own pup but that the pup does not know its dam. At birth, and for several weeks after, the pup is utterly helpless and entirely dependent upon its dam for sustenance, and should anything prevent her return during this period, it dies on the rookery. This has been demonstrated beyond a doubt since the sealing vessels have operated largely in Bering Sea during the months of July, August, and September, and which, killing the cows at the feeding grounds, left the pups to die on the islands.

At about 5 weeks old the pups begin to run about and congregate in bunches or "pods," and at 6 to 8 weeks old they go into the shallow water and gradually learn to swim. They are not amphibious when born, nor can they swim for several weeks thereafter, and were they put into the water would perish beyond a doubt, as has been well established by the drowning of pups caught by the surf in stormy weather. After learning to swim the pups still draw sustenance from the cows, and I have noticed at the annual killing of pups for food in November that their stomachs were always full of milk and nothing else, although the cows had left the islands some days before. I have no knowledge of the pups obtaining sustenance of any kind except that furnished by the cows, nor have I ever seen anything but milk in a dead pup's stomach.

Karp Buterin says:

Schooners kill cows, pups die, and seals are gone. Some men tell me last year, "Karp, seals are sick." I know seals are not sick; I never seen a sick seal, and I eat seal meat every day of my life. No big seals die unless we club them; only pups die for food after the cows are shot at sea. When we used to kill pups for food in November, they were always full of milk; the pups that die on the rookeries have no milk. The cows go into the sea to feed after the pups are born, and the schooner men shoot them all the time.

Captain Carthcut says:

About 80 per cent of the seals I caught in Bering Sea were mothers in milk, and were feeding around the fishing banks just north of the Aleutian Islands, and I got most of my seals from 50 to 250 miles from the seal islands. I don't think I ever sealed within 25 miles of the Pribilof Islands. They are very tame after giving birth to their young, and are easily approached by the hunters. When the females leave the islands to feed, they will go very fast to the fishing banks, and after they get their food they will go to sleep on the waters. That is the hunter's great chance. I think we secure more in proportion to the number killed than we did in the North Pacific. I hunted with shotgun and rifle, but mostly with shotgun. Seals were not nearly as numerous in 1887 as they were in 1877, and it is my belief that the decrease in numbers is due to the hunting and killing of female seals in the water. I do not think it possible for seals to exist for any length of time if the present slaughter continues. The killing of the female means death to her born or unborn pup, and it is not reasonable to expect that this immense drain on the herds can be continued without a very rapid decrease in their numbers, and which practically means extermination within a very few years.

Christ Clausen says:

The Indian hunters, when they use spears, saved nearly every one they struck. It is my observation and experience that an Indian hunter, or a white hunter, unless very expert, will kill and destroy many times more than he will save if he uses firearms. It is our object to take them when asleep on the water, and any attempt to capture a breaching seal generally ends in failure. The seals we catch along the coast are nearly all pregnant females. It is seldom we capture an old bull, and what males we get are usually young ones. I have frequently seen cow seals cut open and unborn pups cut out of them, and they would live for several days. This is a frequent occurrence. It is my experience that fully 85 per cent of the seals I took in Bering Sea were females and had given birth to their pups, and their teats would be full of milk. I have caught seals of this kind 100 to 150 miles from Pribilof Islands. It is my opinion that spears should be used in hunting seals, and if they are to be kept from extermination the shotgun should be discarded.

George Dishow says:

I use a shotgun exclusively for taking seals. Old hunters lose but very few seals, but beginners lose a great many. I use the Parker shotgun. A large proportion of all seals taken are females with pup. I never examined them as to sex. But very few old bulls are taken, only five being taken out of a total of 900 seals taken by my schooner. Use no discrimination in killing seal, but shoot everything that comes near the boat in the shape of a seal. Hunters shoot seal in the most exposed part of the body. Have never known any pups to be born in the water, nor on the land on the coast of Alaska anywhere outside of the Pribilof Islands. Have never known fur seal to haul up on the land anywhere on the coast except on the Pribilof Islands. Most of the seals taken in Bering Sea are females. Have taken them 70 miles from the islands that were full of milk. I think a closed season should be established for breeding seal from January 1 to August 15 in the North Pacific Ocean and Bering Sea.

George Fairchild says:

Most of them were cows, nearly all of which had pups in them. We took some of the pups alive out of the bodies of the females. We entered Bering Sea May 25, and we got 704 seals in there, the greater quantity of which were females with their breasts full of milk, a fact which I know by reason of having seen the milk flow on the deck when they were being skinned. We had five boats on board, each boat having a hunter, boat puller, and steerer. We used shotguns and rifles. We got one out of every five or six that we killed or wounded. We wounded a great many that we did not get. We caught them from 10 to 50 miles off the seal islands.

Norman Hodgson says:

I do not think it possible for fur seals to breed or copulate in water at sea, and never saw nor heard of the action taking place on a batch of floating kelp. I have never seen a young fur-seal pup of the same season's birth in the water at sea nor on a patch of floating kelp, and, in fact, never knew of their being born anywhere save on a rookery. I have, however, cut open a gravid cow and taken the young one from its mother's womb alive and crying. I do not believe it possible for a fur seal to be successfully raised unless born and nursed on a rookery. I have seen fur seals resting on patches of floating kelp at sea, but do not believe they ever haul up for breeding purposes anywhere except on rookeries.

Captain Tanner, lieutenant-commander in the United States Navy, makes a deposition which is entitled to particular consideration. The following is a short extract:

Seals killed in Bering Sea after the birth of pups are largely mother seals, and the farther they are found from the islands the greater the percentage will be. The reason for this seeming paradox is very simple. The young males, having no family responsibilities, can afford to hunt nearer home, where food can be found if sufficient time is devoted to the search. The mother does not leave her young except when necessity compels her to seek food for its sustenance. She can not afford to waste time on feeding grounds already occupied by younger and more active feeders, hence she makes the best of her way to richer fields farther away, gorges herself with food, then seeks rest and a quiet nap on the surface. Under these circumstances she sleeps soundly, and becomes an easy victim to the watchful hunter.

A double waste occurs when the mother seal is killed, as the pups will surely starve to death. A mother seal will give sustenance to no pup but her own. I saw sad evidences of this waste on St. Paul last season, where large numbers of pups were lying about the rookeries, where they had died of starvation.

DECREASE OF SEALS.

EXCESSIVE KILLING THE ADMITTED CAUSE.

We find that since the Alaska purchase a marked diminution in the number of seals on and habitually resorting to the Pribilof Islands has taken place; that it has been cumulative in effect, and that it is the result of excessive killing by man. (Joint report of United States and British Bering Sea commissioners.)

PELAGIC SEALING THE SOLE CAUSE.

Opinions of American commissioners.

Having answered the first of the two queries relating to conditions of seal life at the present time, the second becomes important. It is: Has the decrease in number been confined to any particular class of seals, or is it most notable in any class or classes? In answer to this, it is our opinion that the diminution in numbers began and continues to be most notable in female seals. (Report of American commissioners.)

As a matter of fact, there is sufficient evidence to convince us that by far the greater part of the seals taken at sea are females; indeed, we have yet to meet with any evidence to the contrary. The statements of those who have had occasion to examine the catch of pelagic sealers might be quoted to almost any extent to the effect that at least 80 per cent of the seals thus taken are females. On one occasion we examined a pile of skins picked out at random, and which we have every reason to believe was a part of a pelagic catch, and found them nearly all females. When the sealers themselves are not influenced by the feeling that they are testifying against their own interests, they give similar testimony. The master of the sealing schooner *J. G. Swan* declared that in the catch of 1890, when he secured several hundred seals, the proportion of females to males was about four to one, and on one occasion in a lot of 60 seals, as a matter of curiosity, he counted the number of females with young, finding 47. (Report of American commissioners.)

The decrease in the number of seals is the result of the evil effects of pelagic sealing. (Report of American commissioners.)

Opinion of Dr. Allen.

From the foregoing summary it is evident that the decline in the number of the killable seals at the Pribilof rookeries and the immense decrease in the total number of seals on the Pribilof Islands are not due to any change in the management of the seal herd at the islands, but to the direct and unquestionably deleterious effects of pelagic sealing. At the islands the killing is regulated with reference to the number of killable seals on the rookeries, the designated quota is limited to nonbreeding young males, and every seal killed is utilized. The killing, as thus regulated, does not impair the productiveness of the rookeries. In pelagic sealing the slaughter is indiscriminate and unlimited, and a large proportion of the seals killed are lost. The catch also consists almost wholly of breeding females, which at the time of capture are either heavy with young or have young on the rookeries depending upon them for sustenance. Thus two or more seals are destroyed to every one utilized, and nearly all are drawn from the class on which the very existence of the seal herd depends. (Article by Dr. J. A. Allen.)

Opinions of experts.

I have always taken a great interest in the sealing industry and felt a great desire to have them protected from destruction, and I say without hesitation that the great decrease in the number now annually arriving at the seal islands is due entirely to the killing of female seals by pelagic sealers. (George Adams.)

From my general knowledge of natural history, from my study of the habits of seals, as well as from the opportunities I have had to acquaint myself with the sources of destruction which are at work, I firmly believe that pelagic sealing would not only account for the diminution of the seal herd, but if continued the seals will inevitably be commercially destroyed. (A. B. Alexander.)

I believe there has been a great decrease of seals on the islands since I left there, and this is no doubt due to pelagic sealing. (James Armstrong.)

My people wondered why this was so, and no one could tell until we learned that hunters in schooners were shooting and destroying them in the sea. Then we knew what the trouble was, for we knew the seals they killed and destroyed must be cows, for most all the males remain on or near the islands until they go away in the fall or fore part of the winter. We also noticed dead pups on the rookeries that had been starved to death. If they had not killed the seals in the sea there would be as many on the rookeries as there was ten years ago. There was not more than one fourth as many seals in 1891 as there was in 1880. We understand the danger there is in the seals being all killed off and that we will have no way of earning our living. There is not one of us but what believes if they had not killed them off by shooting them in the water there would be as many seals on the islands now as there was in 1880, and we could go on forever taking 100,000 seals on the two islands; but if they get less as fast as they have in the last five or six years there will be none left in a little while. (Kerrick Artomanoff.)

Upon examining the Bering Sea catch for 1891, as based upon the records of the Victoria custom-house, I ascertained that nearly 30,000 seals had been taken by the British fleet alone in Bering Sea during the summer of 1891. When there is added to this the catch of the American vessels, the dead pups upon the rookeries, and allowances made for those that are killed and not recovered, we have a catch which will not only nearly reach in numbers the quota of male seals allowed to be taken upon the islands in years gone by, but we have a catch in the securing of which destruction has fallen most heavily upon the producing females. This is borne out by a further fact. The young bachelor seals can lie idly on the hauling grounds and through the peculiarities of their physical economy sustain life with a small supply of food, but the cows must range the ocean in search of nourishment that they may meet the demands made upon them by their young. That seals go a great distance from the islands I know from personal observation, for we saw them 120 miles to the northward of the island on the way to Nunival. That the females outnumber the males ten to one is well known, otherwise the hauling ground would present such an array of killable seals that there would be no necessity for the Government to suspend the annual quota. It inevitably follows that the females are the class most preyed upon in Bering Sea. No class of animals which bring forth but a single offspring annually can long sustain itself against the destruction of the producers. As a result of my investigation I believe that the destruction of females was carried to the point, in about 1885, where the birth rate could not keep up the necessary supply of mothers, and that the equilibrium being once destroyed and the drain upon the producing class increasing from year to year from that date, the present depleted condition of the rookeries has resulted directly therefrom. (J. Stanley-Brown.)

When we first noticed that the seals on the rookeries were not so many as they used to be, we did not know what was wrong, but by and by we found that plenty of schooners came into the sea and shot seals, and we often found bullets and shot in seals when we were skinning them. And then we found plenty of dead pups on the rookeries, more and more every year, until last year (1891), when there were so many the rookeries were covered with them, and when the doctor (Akerly) opened some of them there was no milk or food in their stomachs. Then we all knew the cows had been shot when they went into the sea to feed, and the pups died because they had nothing to eat. Plenty of schooners came first about eight or nine years ago, and more and more every year since, and the seals get less and less ever since schooners came; and my people kept saying "No cows," "No cows." First the cows get less and then the bachelors get less, and the company agent he says "kill smaller seals," and we kill some whose skins weigh only 4½ pounds, instead of 7 pounds, same as they always got. Then we could not get enough of seals and at last we could hardly get enough for meat. Schooners kill cows, pups die, and seals are gone. (Karp Buterin.)

The cause of this decrease I believe to be due to the promiscuous killing of the seals by hunters in the open sea and the disturbance caused by their presence in destroying the mother seals and scattering the herds. (James H. Douglass.)

I know of no other cause for the decrease than that of the killing of the cows at sea by the pelagic hunters, which I believe must be prohibited if the Alaskan fur seal is to be saved from total destruction. (C. L. Fowler.)

In my opinion pelagic sealing is the cause of redriving on the islands, the depletion of the rookeries, and promises to soon make the Alaska fur-seal herd a thing of the past. If continued as it is to-day, even if killing on the islands was absolutely forbidden, the herd will in a few years be exterminated. (Charles J. Goff.)

. During my visit to the islands of St. Paul and St. George for the last twenty years I have carefully noticed that those islands were visited by great herds of fur seals during the breeding season, and that, although 100,000 male seals were taken annually at the islands by the lessees, no perceptible diminution in their numbers was noticeable until within the past few years, when the killing of seals in the open sea on the part of fishing vessels became prevalent, since which time there has been a very perceptible diminution in the number of seals seen in the water of Bering Sea and hauling grounds on the islands. This decrease has become alarmingly sudden in the last three or four years, due, I believe, to the ruthless and indiscriminate methods of destruction employed by vessels in taking female seals in the open sea. (Capt. M. A. Healey.)

I made the conditions of seal life a careful study for years, and I am firmly of the opinion their decrease in number on the Pribilof Islands is due wholly and entirely to hunting and killing them in the open sea. (W. S. Hereford.)

When in 1886 we all saw the decrease of seals upon the hauling grounds and rookeries, we asked each other what was the cause of it, but when we learned that white men were shooting seals in the water with guns we knew what was the matter; we knew that if they killed seals in the water that they must be nearly all females that were going out to feed, for the males stay on the islands until they get ready to go away in the fall or winter. It was among the cows we first noticed the decrease, and as we never kill the cows on the islands, we knew they must be killing them in the water. (Aggei Kushen.)

There can be no question, in my opinion, about the ultimate result to the rookeries of marine sealing. If it is continued as it has been for the last two or three years, the seals will be so nearly wiped out of existence in a short time as to leave nothing to quarrel about, and an article of commerce that has afforded a vast amount of comfort and satisfaction to a large class of wearers and a large income to both American and British merchants will be a thing of the past. (Isaac Liebes.)

I am convinced that the decrease in the rookeries was caused entirely by open-sea sealing. (Abial P. Loud.)

There were no destructive agencies at work upon the island that would not have left the rookeries in better condition in 1890 than they were in 1870. Until the effects of the true agent of destruction began to be manifest there was an excess of male life on the islands sufficient to permit of an annual catch of 100,000 seals for an indefinite period without jeopardizing the rookeries. If it be remembered that the seals taken in the waters by hunters are chiefly females, that their young die with them, and that all of those killed are not secured, and if then an examination be made of the pelagic skins actually sold during the past twenty years, the real source of the depletion of the rookeries will be found. In my judgment such depletion was caused by pelagic sealing, and that it grew greater from year to year, as the number of so-called poaching schooners increased, and that its effects began to manifest themselves about 1885 or 1886. The depletion on both hauling and breeding grounds is accounted for by the fact that the catch of said sealers consists of at least 85 per cent cows. Said cows, when taken in the North Pacific, are in the majority of cases with pups, and in Bering Sea are so-called milking females. Whenever a milking cow is killed, her pup on the rookeries dies of starvation. In support of this fact last stated, the number of dead pups during the last four years I was upon the islands increased annually. The effect of the comparatively few raids upon the rookeries themselves, while injurious, bear but a small ratio to the enormous damage done by the pelagic hunting. Those in charge of the islands did not, when the decrease on the rookeries commenced, know exclusively the cause thereof. My opinion then was that it was caused by pelagic sealing, but I had been informed and believed that the United States Government intended to seize all such poaching vessels. Relying upon such information I authorized the taking of seals as before. The proper protection of seal life was not fully carried out in Bering Sea and the North Pacific by reason of England's interference, and the rookeries were thus depleted. (H. H. McIntyre.)

From statements made by personal acquaintances and friends, I became aware of a rapid decrease in seal life in Alaska, and reports of pelagic sealing, as made public through the press, combined with previous personal affairs as existing prior to 1882, leaves no possible doubt as to the cause of such decrease of seals. Pelagic sealing as practiced prior to the year 1882 had no apparent effect upon seal life, and even when to this was added the taking of a definite number year after year under lease from the United States Government, there was still a constant increase of seals observed. I am, therefore, fully confirmed in the belief that the decrease in their numbers is due solely to the indiscriminate killing at sea of all ages, regardless of sex, as practiced since 1884. (H. W. McIntyre.)

The seals have rapidly decreased since sealing vessels appeared, but before the inroads of these seal hunters there was no trouble in obtaining the full quota of the best grades of skins, as the herds previous to that time had been noticeably increasing. (John Malowansky.)

Q. To what do you attribute the decrease in the number of seals on the rookeries?—A. To the great number of cows killed by poachers, and consequently less pups are born on the rookeries.

Q. How do you know that cows have been killed by poachers?—A. I have handled and seen a great number of skins captured by the revenuc cutters from the poaching vessels, and there were very few male skins among them; also have seen among them a great number of unborn pups. Twice upon the rookeries I have seen cows killed and left there by the poachers. (Anton Melovedoff.)

I know of no other explanation than this: The cows are shot and killed when they go into the sea to feed and the pups die on the rookeries. This, I think, is the true solution of the vexed question, "What has become of the seals?" (Anton Melovedoff.)

Since 1883, however, there is said to have occurred a very material diminution of the seal life on the Pribilof Islands, due, as it is claimed, to a large and indiscriminate slaughter of these animals in the waters of Bering Sea and the Pacific Ocean. The cause assigned for this loss is undoubtedly the true one. If no other proof were forthcoming in relation to it, the large display of dead pups on the rookeries would in itself furnish all the evidence required. Such diminution could not, in my opinion, be the result of the yearly slaughter of skins. It is shown that an appreciable expansion of the rookeries took place after twelve or fourteen years of such slaughter, and I think this fact conclusively demonstrates that the number of seals which the law permitted to be killed each year was not greater than the known conditions of the seal's life would safely warrant. (J. M. Morton.)

From the experience gained, and observations made, during three killing seasons; from the information gleaned from men who have devoted their lives to the practical side of the seal question, and from the books and reports in the Government offices on the islands, I am able to say that, in my opinion, there is only one great cause of the decrease of the fur seal, and that is the killing of the females by pelagic hunting. (Joseph Murray.)

I believe this decrease is owing to the large number of vessels engaged in hunting the fur seal at sea, and the indiscriminate methods employed by these sealing vessels in taking skins. (Arthur Newman.)

The practice of pelagic seal hunting was followed by the Northwest Coast Indians from their earliest history, but amounted to so little as to be inappreciable on the islands. Even after white hunters engaged in it in a limited way our losses from this source were attributed to the marine enemies of the seals, and was so far overcome by the good management of the islands as to permit the growth of the herd to continue so long as it was limited to a few vessels and confined to the vicinity of the Oregon, Washington, and British Columbian coasts. But even before any considerable slaughter had taken place in the waters of Bering Sea, as early as 1882, it was noticed that the rookeries had stopped expanding, though they were treated in every way as they always had been. An examination of the London Catalogue of seal-skin sales shows that the "Victoria catch" already aggregated a very considerable number of skins, and now brings home the conviction that pelagic sealing, when confined almost wholly to the Pacific, is still a very dangerous enemy of seal life on the islands. After 1886 the force of pelagic hunters was greatly augmented, and became more and more aggressive, until they appeared in alarming numbers in Bering Sea in 1884 and 1885. In 1887 we were forced to commence taking smaller skins in order to obtain our quota and preserve enough breeding bulls. In 1888 they were still smaller, while in 1889 more than half of them were such as we would not have killed in former years; and we called the attention of the Treasury Department to the evident diminution of seal life, and recommended that fewer seals be killed in future. There can be no question as to the cause of the diminution. It is the direct result of pelagic sealing, and the same destruction, if continued a few years longer, will entirely dissipate any commercial value in the rookeries, if it does not, indeed, annihilate them. (Gustave Niebaum.)

In my opinion the solution of the problem is plain. It is the shotgun and the rifle of the pelagic hunter which are so destructive to the cow seals as they go backward and forward to the fishing banks to supply the waste caused by giving nourishment to their young. At this time they are destroyed by thousands, and their young of but a few weeks old must necessarily die of starvation, for nature has provided no other means of subsistence for them at this time of life. (L. A. Noyes.)

Q. How do you account for it?—A. By the numbers, principally females, that are killed in the waters by marauders. (J. C. Redpath.)

I saw no diminution of seal life during my three years on the island. The outlines of the rookeries remained just about the same from year to year. I was told at the time that there had formerly been a large increase, and did not then understand why it did not continue, as every condition seemed favorable for it. There were, apparently, an abundance of bulls for service; every cow seemed to have a pup, and all were healthy and in good condition. No females were killed, and in the natural order of growth there ought to have been at this time a constantly increasing area covered with breeding rookeries. Yet such was not the case. The explanation of the matter came later, when we fairly awoke to the fact that our animals were being slaughtered by tens of thousands in the North Pacific. I knew in a commercial way from our sales catalogue that a very large number of "Victoria skins," as they were called, were being sent to market, and that this number grew constantly larger; but I did not then know, as I now do, that each skin sold represented a waste of two or three and perhaps even four or five seals to obtain it. Nor was any attention given to the now well-known fact that these animals were a part of our herd as wrongfully stolen from us, I believe, as my cattle would be if driven in and appropriated from the highway when lawfully feeding. (Leon Sloss.)

Since my residence on the Pribilof Islands I have kept a very careful watch of the progress of the events there, and have interviewed a great many connected with the seal industry. I am of the conviction that the reported decrease in seal life on these islands can be attributed to no other cause save pelagic sealing. While I was located at St. George Island in 1881 pelagic sealing was then and previous to that time had been of very little consequence, having very slight effect upon seal life. Not more than four or five vessels were engaged in pelagic sealing in 1881 in the waters of Bering Sea, and prior to that time a still fewer number were so engaged. But since 1881 this industry has grown yearly until now about a hundred vessels are destroying the seals in great numbers, and, as I am informed and believe, the great majority of those killed are females. Then, too, large numbers are killed in this way which are never recovered nor reported. (W. B. Taylor.)

Scarcity of seal can be attributed to no other cause than pelagic hunting and the indiscriminate shooting of seals in the open sea, both in the North Pacific and Bering Sea. (John C. Tolman.)

I am sure the decrease is caused by the killing of female seals in the open sea, and that if their destruction by the indiscriminate killing in the open sea is permitted to continue it will only be a very short time until the herd is destroyed. (Charles T. Wagner.)

I have no doubt that it is caused by the killing of female seals in the water, and, if continued, will certainly end in their extermination. (M. L. Washburn.)

I am convinced that if open-sea sealing had never been indulged in to the extent it has since 1885, or perhaps a year or two earlier, 100,000 male skins could have been taken annually forever from the Pribilof Islands without decreasing the seal herd below its normal size and condition. The cause of the decrease which has taken place can be accounted for only by open-sea sealing; for, until that means of destruction to seal life grew to be of such proportions as to alarm those interested in the seals, the seal herd increased, and since that time the decrease of the number of seals has been proportionate to the increase in the number of those engaged in open-sea sealing. From 1884 to 1891 I saw their numbers decline, under the same careful management, until in the·latter year there was not more than one-fourth of their numbers coming to the islands. In my judgment there is but one cause for that decline and the present condition of the rookeries, and that is the shotgun and rifle of the pelagic hunter, and it is my opinion that if the lessees had not taken a seal on the islands for the last ten years we would still find the breeding grounds in about the same condition as they are to-day, so destructive to seal life are the methods adopted by these hunters. (Daniel Webster.)

Deponent, by reason of his experience in the business, his observation, conversations with those physically engaged in catching and curing skins, and the custody of herds on the islands, feels justified in expressing the opinion that the numbers of the seal herds have, since the introduction of the open-sea sealing on a large scale, suffered serious diminution. The killing of large numbers of females heavy with young can not, in deponent's knowledge, but have that effect. (C. A. Williams.)

I made careful inquiry of the people on the islands, both native and white, and of those who were or had been employed as masters or mates on sealing vessels, and others interested one way or another in the capture of fur seals for food or for profit, and failed to find any of them but who admitted that the number of seals in Bering Sea was much less now than a few years since, and nearly all of them gave it as their opinion that the decrease in number was due to pelagic hunting, or, as they more frequently expressed it, the killing of females in the water. (W. H. Williams.)

DECREASE OF THE ALASKAN SEAL HERD.

After 1882 they seemed to stay about the same, as far as the number of breeders was concerned, as long as I was there. (John Armstrong.)

I ascertained by questioning those who had years of continuous experience with the seals that up to the year 1882 there was an annual expansion of the boundaries of the breeding grounds; that this was followed by a period of stagnation, which in turn was followed by a marked decadence from about 1885–86 down to the present time. (J. Stanley-Brown.)

I am unable to state whether the seals increased or not during my residence on St. Paul, but they certainly did not decrease, except, perhaps, there was a slight decrease in 1884. In all my conversations with the natives—which were, of course, a great many—they never spoke of the seals being on the decrease, as they certainly would have done if such had been the case. (H. A. Glidden.)

While on St. Paul I do not think the number of seals increased, and in the last year (1884) I think there was a slight decrease. (J. H. Moulton.)

Upon the Commander Islands, as I have already said, the increase in seal life was constant for many years, but in 1890 we noticed a decided disturbance in the rookeries and a considerable decrease in their population. This we subsequently attributed, when the facts were ascertained, to pelagic sealing in the adjacent waters. (Gustave Niebaum.)

I noticed during this period no perceptible increase in the breeding rookeries on St. George. (B. F. Scribner, Treasury agent.)

ON PRIBILOF ISLANDS.

In 1882 there was no scarcity of killable seals. The men drove up as many every day as they could handle, and those selected for killing comprised only the choicest ones. (W. C. Allis.)

There seemed to be also a large surplus of full-grown bulls for rookery service, and enough escaped from the slaughter ground to keep the number good as the old ones passed the age of usefulness. I do not believe the condition of the rookeries nor the manner of driving and killing the seals at this time could have been improved. It was perfect in every respect, and the lessees, employees, and natives, as well as the seals, all appeared to be and were, I believe, contented and happy. In 1886 the conditions had somewhat changed. The natives complained that big seals were growing scarcer; that there were many dead pups on the rookeries, and the superintendent intimated to me that he did not like the outlook as compared with a few years previous, and said he thought either the number killed or the size of the animals taken for their skins would have to be reduced of killable seals, and the work went on as during my first year (1882) in the service. But the trouble of which they complained grew more serious in the following years, and I think it was in 1888 the superintendent told the bosses they must kill less large seals and more "yellow bellies," or 2-year-olds. In 1889 a very large proportion of the catch was made up of this class. It was then perfectly apparent to everybody, myself included, that the rookeries were "going to the bad" and that a smaller number must inevitably be killed the following year. (W. C. Allis.)

The aggregate size of the areas formerly occupied is at least four times as great as that of the present rookeries. (Report of American Bering Sea commissioners.)

I have noticed a great decrease in the numbers of the fur seals since 1887, both on the rookeries of St. Paul Island, which are much shrunken in the area covered by seals, and in the waters of the Pacific and Bering Sea. On the rookeries, ground formerly hauled over by seals is now grown up with a scattering of recent growth. (C. H. Anderson.)

The skins taken prior to 1886 weighed from 6 to 10 pounds each, averaging about 8 pounds per skin; but I understand from those who remained there on duty that much smaller ones were afterwards taken, because the large seals had become scarce and were needed for rookery service. (John Armstrong.)

From 1870 to 1884 the seals were swarming on the hauling grounds and the rookeries, and for many years they spread out more and more. All of a sudden, in 1884, we noticed there was not so many seals, and they have been decreasing very rapidly ever since. (Kerrick Artomanoff.)

There are not nearly as many seals on the coast as there were two or three years ago. (Johnny Baronovitch.)

There are certain physical as well as historical sources of information upon the island from which the relation of the present to the past condition of the rookeries can be very clearly made out.

(1) Not only upon, but immediately to the rear of, the area at present occupied by the breeding seals occur fragments of basalt whose angles have been rounded and polished by the flippers of seals. Among these latter rocks grass is found growing to an extent proportionate to their distances from the present breeding grounds, and further, the soil shows no recent disturbance by the seals. This rounding of the bowlders of the abandoned areas was not due to the impingement of sand driven by the wind. No geologist would be willing to risk his reputation by asserting that this rounding came from any such agency. The distinction between the result of sand-blast action and seals' flippers is very marked.

(2) A careful examination among the roots of the grass will often show the former presence of seal by the peculiar appearance of the soil, due to the excrementa of the seal and the occurrence of a thin mat of seal hair. The attention of Dr. George M. Dawson was called to such a felt of hair upon the summit of Hutchinson Hill, and both he and Dr. C. Hart Merriam collected specimens of it from among the grass roots at that locality.

(3) At the rear of the rookeries there is usually an area of mixed vegetation—an area the boundary of which is sharply defined, and between which and the present breeding grounds occurs a zone of grass of only a single variety. In the immediate vicinity of the present breeding grounds only scanty bunches are to be seen. These gradually coalesce as the line of mixed vegetation is approached. The explanation of this is that the seals were formerly so abundant as to destroy the normal mixed vegetation at the rear of the breeding grounds, and that the decrease of the seals has been followed by the encroachment of the uniform variety of grass.

(4) The statements made to me by competent observers who have lived upon the islands for years all agree that the shrinkage in the breeding area has been rapid during the past five or six years.

After observing the habits of seals for a season, I unhesitatingly assert that to satisfactorily account for the disturbance to vegetable life over areas whose extent is visible even to the most careless and prejudiced of observers, would require the presence of from two to three times the amount of seal life which is now to be found upon the islands. That there has been enormous decrease in the seals there can be no question. (J. Stanley-Brown.)

Have observed carefully the areas occupied by the seals on the rookeries and hauling-out grounds, especially at Northeast Point and the Reef, on St. Paul Island, in 1884, 1885, 1886, and 1891, and on both rookeries the areas formerly occupied by seals have greatly decreased, so much so that at first appearance it seemed, in 1891, as if the hauling-out grounds had been entirely deserted. Subsequent examination disclosed the fact that this was not strictly true, there still being a small number of male seals left on the grounds. Have also observed that the seals are much more scattered on the breeding rookeries than in former years (1884, 1885, 1886); also that the number of seals in the water has proportionately decreased, and that they have grown very much more shy and difficult to approach. Without presuming to be absolutely correct, would estimate the number of seals present at St. Paul Island during the year 1891 to about 10 per cent of the number there in former years of observation—1884, 1885, 1886. (John C. Cantwell.)

I did not notice any falling off in the size of the rookeries from the landmarks to which they came when I first saw them during the first two years I was on the island, and all agreed, in discussing the matter, that the seals had never been more numerous than they were; but in the following years, and particularly in 1888 and 1889, no other opinion was heard than that the animals had greatly diminished, and in this opinion I fully coincided. (Henry N. Clark.)

During the seasons of 1890 and 1891, I was in command of the revenue cutter *Rush*, in Bering Sea, and cruised extensively in those waters around the seal islands and the Aleutian group. In the season of 1890 I visited the islands of St. Paul and St. George, in the months of July, August, and September, and had ample and frequent opportunities of observing the seal life as compared with 1870. I was astonished at the reduced numbers of seals and the extent of bare ground on the rookeries once teeming with seal life. In 1890 the North American Commercial Company were unable to kill seals of suitable size to make their quota of 60,000 allowed by their lease, and in my opinion, had they been permitted to take 50,000 in 1891, they could not have secured that number if they had killed every bachelor seal with a merchantable skin on both islands, so great was the diminution in the number of animals found there. (W. C. Coulson.)

I arrived with my command at St. Paul Island June 7, 1891. At that date very few seals had arrived, and but a small number had been killed for fresh food. On the 12th of June, 1891, we were at St. George Island and found a few seals had been taken there, also for food, the number of seals arriving not being enough to warrant the killing of any great number. During that year I was at and around both these islands every month from and including June until the 1st day of December (excepting October), and at no time were there as many seals in sight as in 1890. I assert this from actual observation, and it is my opinion we will find less this year. (W. C. Coulson.)

During my annual cruising in Bering Sea and to and from the Pribilof Islands I have carefully noted the number and appearance of seals in the water and on the breeding rookeries from the deck of my vessel, and have also repeatedly visited the hauling grounds from year to year, and it was about 1884 and 1885 that bare spots began to appear on the rookeries, so much so that myself and the other officers often spoke of it and discussed the causes therefor. The decrease in number of seals both on the Pribilof Islands and in the waters of Bering Sea and North Pacific has been very rapid since 1885, especially so in the last three or four years, and it is my opinion that there is not now more than one-third of the number of seals in these waters and on the islands that there were ten years ago. (Leander Cox.)

During my last visits to the islands I observed a very marked diminution in the number of seals thereon as contrasted with the herd on the rookeries five or six years previously. I am familiar with the area and topography of the various rookeries on the islands, and have observed that spaces formerly occupied by seal herds are now vacant and parts of them covered with grass. This diminution was particularly noticeable in 1887 and 1888, the last two years' visit to the islands. (James H. Douglas.)

For many years prior to 1890 I have observed the rookeries from my ship and also from the islands. The first decrease in the number appearing on the rookeries and in the surrounding sea that I particularly noticed was in the summer of 1884, and it has become more marked from year to year since. For the last three or four years their disappearance has been very marked. In October, 1890, I made a trip from Unalaska to St. Michaels. When about 20 miles south of St. George we commenced to watch for seals passing the Zapadnie rookery close inshore along the west end of St. George Island to Otter Island and Seal Island rock; thence to Northeast Point, about a mile and a half offshore. When we started, I requested the officers to keep a sharp lookout and to report if they saw any seals in the water. I was on deck most of the time myself also, and we only saw two seals in the whole run; whereas ten years ago, when on a similar voyage, seals were so plentiful that it was impossible to count them. From my long observation I do not think there are as many seals by two-thirds now annually arriving on the islands or in those waters as there were ten years ago, when I first commenced to notice that they were decreasing. By this statement I mean to say that only one-third as many are now to be seen as formerly. (M. C. Erskine.)

Seals have decreased in numbers very rapidly in the last few years, and to anyone who saw the breeding rookeries as I did in 1880 the change is most wonderful. (C. L. Fowler.)

It was on the breeding rookeries and among the cows that I first began to notice the decrease in seal life, and I do not think there were more than one-fourth as many cows on the breeding rookeries in 1891 as were there in 1887. (C. L. Fowler.)

I have been a resident of the seal islands for the past ten years; formerly assistant agent of the Alaska Commercial Company, now agent of the North American Company, and during that time was engaged in the taking of seals. I have listened to the testimony of J. C. Redpath, as above, and fully concur in all that he has said concerning seal life, with the exception that the number of seals on the islands this season are, in my judgment, not more than one-fourth of what they were in 1887. (C. L. Fowler.)

In those days (from 1869 to 1882 or 1883) we used to get plenty of seals on the Zoltoi sands near the Reef rookery, and now there are none there. It was in 1884 that I first noticed a decrease in the seals, and it has been a steady and a very rapid decrease ever since 1886, so that at present there are not one-fourth as many seals on the island as there was every year from 1869 to 1883. (John Fratis.)

In 1889 I made careful observations of the rookeries on St. Paul Island and marked out the areas covered by the breeding grounds; in 1890 I examined these lines made by me the former year, and found a very great shrinkage in the spaces covered by breeding seals. In 1889 it was quite difficult for the lessees to obtain their full quota of 100,000 skins; so difficult was it in fact, that in order to turn off a sufficient number of 4 and 5 years old males from the hauling grounds for breeding purposes in the future, the lessees were compelled to take about 50,000 skins of seals of 1 or 2 years of age. I at once reported this fact to the Secretary of the Treasury, and advised the taking of a less number of skins the following year. Pursuant to such report the Government fixed the number to be taken as 60,000, and further ordered that all killing of seals upon the islands should stop after the 20th day of July. I was further ordered to notify the natives upon the Aleutian Islands that all killing of seals while coming from or going to the seal islands was prohibited. These rules and regulations went into effect in 1890, and pursuant thereto I posted notices for the natives at various points along the Aleutian chain, and saw that the orders in relation to the time of killing and number allowed to be killed were executed upon the islands. As a result of the enforcement of these regulations, the lessees were unable to take more than 21,238 seals of the killable age of from 1 to 5 years during the season of 1890, so great had been the decrease of seal life in one year, and it would have been impossible to obtain 60,000 skins even if the time had been unrestricted. (Charles J. Goff.)

The Table A, appended to this affidavit, shows how great has been the decrease on St. Paul Island's hauling grounds, bearing in mind the fact that the driving and killing were done by the same persons as in former years, and were as diligently carried on, the weather being as favorable as in 1889 for seal driving. I believe that the sole causes of the decrease is pelagic sealing, which, from reliable information, I understand to have increased greatly since 1884 or 1885. Another fact I have gained from reliable sources is that the great majority of the seals taken in the open sea are pregnant females or females in milk. It is an unquestionable fact that the killing of these females destroys the pups they are carrying or nursing. The result is, that this destruction of pups takes about equally from the male and female increase of the herd, and when so many male pups are killed in this manner, besides the 100,000 taken on the islands, it necessarily affects the number of killable seals. In 1889 this drain upon male seal life showed itself on the islands, and this, in my opinion, accounts for the necessity of the lessees taking so many young seals that year to fill out their quota. As soon as the effects of pelagic sealing were noticed by me upon the islands I reported the same, and the Government at once took steps to limit the killing upon the islands, so that the rookeries might have an opportunity to increase their numbers to their former condition; but it will be impossible to repair the depletion if pelagic sealing continues. I have no doubt, as I reported, that the taking of 100,000 skins in 1889 affected the male life on the islands and cut into the reserve of male seals necessary to preserve annually for breeding purposes in the future, but this fact did not become evident until it was too late to repair the fault that year. Except for the numbers destroyed by pelagic sealing in the years previous to 1889 the hauling grounds would not have been so depleted, and the taking of 100,000 male seals would not have impaired the reserve for breeding purposes or diminished to any extent the seal life on the Pribilof Islands. Even in this diminished state of the rookeries in 1889 I carefully observed that in the majority of cases the 4 and 5 year old males were allowed to drop out of a "drive" before the bachelors had been driven any distance from the hauling grounds. These seals were let go for the sole purpose of supplying sufficient future breeders. (Charles J. Goff.)

I believe there has been a great decrease in the numbers of the fur-seal species. I do not believe that there are now one-tenth as many fur seals frequenting the Pribilof Islands as there were ten years ago.

Nine or ten years ago, when lying off the Pribilof Islands in the fall, the young seals used to play in the water about the vessels in large numbers; in going to the westward in the month of May many seals were always to be seen between Unalaska and the Four Mountain islands. In midsummer, when making passages between Unalaska and the Pribilof Islands, used to see large bodies of fur seals feeding; they were invariably to be met with most numerously about 60 miles northwest true from Unalaska, and from there up to and from the feeding grounds. When last I visited the rookeries, three years ago, in 1889, I noticed a great shrinkage in the area covered by seals on the rookeries. (Charles J. Hague.)

In 1886 and 1887 there appeared to be enough seals, and the men were kept pretty steadily at work after the first few days of the season until the catch was completed. Good-sized skins were taken in these years, and there was no trouble in getting them, but large seals grew very scarce on the island in 1888, and still more so in the three following years. I am sure the size of the rookeries on St. Paul Island and the number of seals on them in 1891 were less than one-half their size and number in 1886. (Alex. Hansson.)

Coincident with the increase of hunting seals in the sea there was an increase in the death rate of pup seals on the rookeries; also a perceptible diminution of female seals. As hunting increased it became self-evident, even to the most casual observer, that the rookeries were becoming devastated. It is positively a fact that there are not near as many seals occupying the rookeries now, at the present time, as there were when I first saw the islands. The vacant spaces on the breeding and hauling grounds have increased in size from year to year since 1884, and have been very noticeable for the last four or five years. When I first went to the seal islands the seals were actually increasing in numbers instead of diminishing. Two facts presented themselves to me later on: First, seals were arriving each year in diminished numbers; second, at the same time that the female seals were decreasing in numbers the number of dead pups on the rookeries were increasing. The indiscriminate slaughter of seals in the water has so depleted their number that the company is at present unable to get their quota of skins on the island as allowed per contract with the Government, and is restricted to such an insignificant number that it is not enough to supply food to the native population of the islands. It is an indisputable fact that large portions of the breeding rookeries and hauling grounds are bare, where but a few years ago nothing but the happy, noisy, and snarling seal families could be seen. * * * The driving rookeries also necessarily have suffered, as witness the difference in the catch, a drop from 100,000 to about 20,000 in 1890. (W. S. Hereford.)

I have been employed on the seal islands since 1882, have resided upon them continuously for ten years, and have a personal knowledge of the seal life as it exists on the islands and in the waters surrounding them. There was less than one-third the number of seals on the islands last year than in 1882. The decrease in the number of seals coming to the islands was first noticed and talked about two or three years after I first came to live here; and since 1887 the decrease has been very rapid. A careful inspection of the rookeries each returning season since 1887 showed that the cows were getting less and less, although it was a rare thing to find a cow seal that did not have a pup at her side. (Edward Hughes.)

Ten or twelve years ago the rookeries and sea were full of seals, but now there are not a great many. We used to kill 85,000 in less than two months' time on St. Paul Island, and our people earned plenty of money to buy everything they wanted, and in the winter we killed 2,000 or 3,000 male pups for food and clothing. Now we are not allowed to kill any more pups, and only 7,500 male seals for food, and the people are very much worried to know what is to become of them and their children. (Jacob Kotchootten.)

I remember the first time I noticed a decrease of seals on the rookeries, about seven or eight years ago, and the seals have become fewer every year since. We used to kill 85,000 seals on St. Paul Island in less than sixty days' time until 1890, when they became so few we could not take more than about one-fourth of that number in the same length of time. (Nicoli Krukoff.)

All our people know the seals are getting scarcer every year, and we think it is because of the schooners coming in and shooting the cows in the sea. (Nicoli Krukoff.)

About 1885 a decrease was observed, and that decrease has become more marked every year from 1885 to the present time. (Aggei Kushen.)

There are not one-fourth as many seals now as there were in 1882, and our people are very much alarmed to know what is to become of them after the seals are killed off. If the seals decrease as fast as they have during the past five or six years there will be none left in a very short time for us to live upon. (Aggei Kushen.)

During the time from 1885 to 1889 there was a very marked decrease in the size of the breeding grounds on St. Paul Island, and from 1887 to 1889 I also noticed a great decrease in the areas covered by the rookeries on St. George Island. (Abial P. Loud.)

In his report of 1886 and 1887 George R. Tingle, special Treasury agent in charge of the seal islands, reported having measured the rookeries on the islands, and that the seals had largely increased in number, giving the increase at about 2,000,000. From this report I dissented at the time, as I was unable to see any increase, but, on the contrary, a perceptible decrease, in the rookeries. I expressed my views to many on the islands and all agreed that there had been no increase in the seal life. The measurements of the rookeries on which Mr. Tingle relied were made with a common rope by ignorant natives while the seals were absent from the islands, the grounds covered by them being designated by Mr. Tingle from memory. (Abial P. Loud.)

During the three years following 1882, namely, 1883, 1884, and 1885, I was not upon the islands. Upon my return in 1886 I noticed a slight shrinkage in the breeding areas, but am unable to indicate the year of the period of my absence in which the decrease of breeding seals began. From the year 1886 to 1889, inclusive, my observation was continuous, and there was a greater decrease of the seals for each succeeding year of that period in a cumulative ratio, proportionate to the number of seals killed by the pelagic sealers. (H. H. McIntyre.)

In 1886 I again assumed personal direction of the work upon the islands, and continued in charge to and including 1889. And now, for the first time in my experience, there was difficulty in securing such skins as was wanted. The trouble was not particularly marked in 1886, but increased from year to year to an alarming extent, until in 1889, in order to secure the full quota and at the same time turn back to the rookeries such breeding bulls as they seemed to absolutely need, we were forced to take fully 50 per cent of animals under size, which ought to have been allowed one or two years more growth. Concerning this matter I reported to the Alaska Commercial Company, under date of July 16, 1889, as follows: "The contrast between the present condition of seal life and that of the first decade of the lease is so marked that the most inexpert can not fail to notice it. Just when the change commenced I am unable from personal observation to say, for as you will remember I was in ill health and unable to visit the islands in 1883, 1884, and 1885. I left the rookeries in 1882 in their fullest and best condition and found them in 1886 already showing slight falling off, and experienced that year for the first time some difficulty in securing just the class of animals in every case that we desired. We, however, obtained the full catch in that and the two following years, finishing the work from the 24th to the 27th of July, but were obliged, particularly in 1888, to content ourselves with smaller skins than we had heretofore taken. This was in part due to the necessity of turning back to the rookeries many half-grown bulls, owing to the notable scarcity of breeding males. I should have been glad to have ordered them killed instead, but under your instructions to see that the best interests were conserved, thought best to reject them. The result of killing from year to year a large and increasing number of small animals is very apparent. We are simply drawing in advance upon the stock that should be kept over for another year's growth." (H. H. McIntyre.)

Q. How does the number of seals on the rookeries this year compare with the number five years ago?—A. The number now is about one-fourth of what they were then. (Noen Mandregin.)

In 1887 I began to notice a diminution in the number of seals arriving at the islands, which was due to the indiscriminate killing by sealing vessels in the open sea, some 50 or 60 miles distant. While we still obtain about the usual number of skins, many more are taken from the younger animals than formerly, and are somewhat inferior in quality. (John Malowansky.)

From 1885, which was about the time the sealers appeared in the waters, the decrease in seal life was rapid, and the natives commenced saying "no females," "no females," until now we are confronted with depleted rookeries and probable extermination. (John Malowansky.)

Q. Have you noticed any perceptible difference in the number of seals on rookeries from one year to another?—A. Yes.
Q. About how much less is the number of seals during the past year than they were six years ago?—A. The number of seals this year is about one-fourth of what they were six years ago, and about one-half of what they were last year.
Q. In what way do you form your above opinion as to the relative number of seals on the rookeries?—A. By the fact that many spaces on the rookeries which were formerly crowded are now not occupied at all. (Anton Melovedoff.)

About 1886 I noticed that the lines of former years were not filled with cows, and every succeeding year since then has shown a more marked decrease. In 1889 the bachelors were so few on the hauling grounds that the standard weight of skins was lowered to 5 pounds, and hundreds were taken at only 4 pounds in order to fill the quota of 100,000. (A. Melovedoff.)

Until the schooners came into Bering Sea the rookeries were always well filled, and many of them had grown steadily for years, when it was no uncommon thing for the lessees to take the quota of 85,000 seals on St. Paul Island between June 1 and 20 of each year. After 1884, when the original two or three sealing vessels had grown to be a well-organized fleet, we found a steady decrease of seals on all the rookeries, and we found it difficult to secure the quota of skins, and in 1889 the lessees had to lower the standard of weight lower than ever before in the history of the island. (Simeon Melovidov.)

From the year 1874 till 1885 we were able to get from St. George and St. Paul islands 100,000 male seals within the period known as the sealing season of six weeks, from the 10th of June to the 1st of August, and still leave a large percentage of marketable seals. In 1885, and in every year thereafter until I left in 1887, there was a marked decrease in the number of marketable skins that could be obtained in each year during the sealing season. We were able, down to the last year (1887), to get our total catch of 100,000 seals, but in order to get that number we had to take what in previous years we would have rejected, namely, undersized skins, i. e., the skins of young seals. Prior to 1887 we had endeavored to take no skins weighing less than 8 pounds, but in order to make up our quota in the last-mentioned year we had to take skins weighing as little as 6½ pounds to the number of several thousands. (T. F. Morgan.)

In the years 1885, 1886, and 1887 my attention was attracted not only to a diminution in the number of killable seals appearing on the island, but to a decrease in the females as well. Up to the year 1884 the breeding space in the rookeries had increased, and from that year down to 1887, when I left the island, the acreage covered by the rookeries which were occupied by seals constantly diminished. (T. F. Morgan.)

My attention was called to the decrease of seals and the depletion of the rookeries at an early date after my arrival. I attempted to study the habits and conditions and to note the numbers of seal on the several rookeries and hauling grounds. The natives and employees of the Alaska Commercial Company were unanimous in their opinion that the seal had been decreasing steadily and rapidly since 1884. I reported the fact to Agent Goff, who had found similar conditions existing on St. Paul, and he so reported to the Department, and suggested that not more than 60,000 seals should be taken in any one season in future. In pursuance of instructions from Agent Goff, I left St. George Island on the 19th of July, 1890, and landed on St. Paul Island on the 20th of the same month, and remained there until August, 1891. During the month of July, 1890, I walked over the rookeries and hauling grounds of St. Paul Island, and Agent Goff pointed out to me the lines to which in former years the seals hauled, and the large areas which they covered; and then he called my attention to the small strip covered by seals on that date, which was smaller than the year previous. Agent Goff stopped the killing of seals by the lessees on and after the 20th of July, 1890, because of the depleted condition of the hauling grounds; and I fully concurred in his order and action. I spent the sealing season of 1891 on St. Paul Island, and pursuant to instructions of Agent Williams, I gave my time and special attention to the study of the condition of the rookeries, both the breeding and hauling grounds. I visited the rookeries daily from the 7th to the 22d of July—during the period when the rookeries are fullest and at their best—and I carefully noted their condition and the number of seals; the number of cows to the family, and the number of idle, vigorous bulls upon each rookery. (Joseph Murray.)

Upon my first visit to the rookeries and hauling grounds of the island of St. Paul my attention was attracted to the evidences of recent and remote occupancy by the seals. Marked differences were noticeable in the appearance of vegetation on large areas formerly occupied as breeding and hauling grounds, while near the water's edge, more recently occupied, the ground was entirely bare of vegetation, enabling one to trace the gradual decrease of areas occupied during the last six to eight years. My examination of the rookeries on St. Paul and St. George during the years 1890, 1891, and 1892 enabled me to trace the yearly decreasing area occupied by the fur seals on these islands. Aside from the evidences of deserted rookeries and hauling grounds shown by native inhabitants of each island, the grounds occupied in former years were now deserted and grass grown. The silent witness of the deserted rookeries confirms the testimony of the resident agents of the lessees of the islands and of the native inhabitants that the number of seals began to decrease with the advent of pelagic sealing, and that the yearly decrease has been in proportion with the yearly increase in the number of vessels engaged in that enterprise. (S. R. Nettleton.)

The decrease in the number of seals coming to the islands in the last three or four years became so manifest to everyone acquainted with the rookeries in earlier days that various theories have been advanced in an attempt to account for the cause of this sudden change, and the fol lowing are some of them: (1) "A dearth of bulls upon the breeding rookeries;" (2) "Impotency of bulls caused by overdriving while they were young bachelors," and (3) "An epidemic among the seals." (L. A. Noyes.)

Q. Have you noted any perceptible difference in the number of seals on the rookeries from one year to another? If so, what changes have you observed?—A. Within the last four or five years I have observed a decided decrease in the number of seals on the rookeries.

Q. In what proportion have the seals decreased within the time mentioned?—A. As far as my judgment goes, I should say at least one-half. (J. C. Redpath.)

As the schooners increased the seals decreased, and the lines of contraction on the rookeries were noticed to draw nearer and nearer to the beach, and the killable seals became fewer in numbers and harder to find. In 1886 the decrease was so plain that the natives and all the agents on the islands saw it and were startled, and theories of all sorts were advanced in an attempt to account for a cause. (J. C. Redpath.)

I had no difficulty in getting the size and weight of skins as ordered, nor had my predecessors in the office, up to and including 1884. The casks in which we packed them for shipment were made by the same man for many years, and were always of uniform size. In 1885 the casks averaged about 47½ skins each, and in 1886 they averaged about 50⅔ skins each, as shown by the records in our office. After this date the number increased, and in 1888 they averaged about 55⅘ skins per cask, and in 1889 averaged about 60 skins per cask. These latter were not such skins as we wanted, but the superintendent on the islands reported that they were the best he could get. (Leon Sloss.)

The number of seals on the Pribilof Islands is decreasing. I saw positive proof of this on St. Paul Island last season. (Z. L. Tanner.)

I had an excellent opportunity to observe some of the seal rookeries during my first visit to the islands, and spent much time in studying the habits of the seals, both on the rookeries and in the adjacent waters. I was particularly impressed with the great numbers to be seen, both on land and in the water. During the summer of 1889 the *Rush* was engaged cruising in pursuit of vessels engaged in illegal sealing, so that our anchorages off the seal rookeries that season were short and infrequent, hence I did not have the opportunity to observe them as closely on land as the preceding year. During 1890 the *Rush* was not engaged in preventing sealing outside the shore limit, and we spent much time in full view of the seal rookeries and cruising about the seal islands, and I also made frequent visits to the breeding grounds. The deserted appearance of the rookeries and the absence of seals in the water was very noticeable and was a matter of general remark among the officers of the vessel who had been on former cruises. Very large tracts of the rookeries which I had formerly seen occupied by the seals were entirely deserted, and the herds were much smaller than those of 1888. My attention was also called, by those conversant with the facts, to the grass growing on the inshore side of some of the rookeries, and to the three different shades of grass to be seen, indicating the spaces that had not been occupied by the seals for several years, owing to their diminished number. The darker shade showed where the growth first commenced, and a lighter shade for each succeeding year. There were three or four differently shaded growths, reaching down to the sand of the rookeries, and on that portion of the rookeries occupied by seals they were not lying near as compact as in 1888. In our frequent passages during 1890 between the Aleutian group and the seal islands we sometimes made an entire trip without seeing a seal. This was entirely different from the experience of preceding years, indicating a great falling off of seal life. (Francis Tuttle.)

In the year 1880 I thought I began to notice a falling off from the year previous of the number of seals on Northeast Point rookery, but this decrease was so very slight that probably it would not have been observed by one less familiar with seal life and its conditions than I; but I could not discover or learn that it showed itself on any of the other rookeries. In 1884 and 1885 I noticed a decrease, and it became so marked in 1886 that everyone on the islands saw it. This marked decrease in 1886 showed itself on all the rookeries on both islands. Until 1887 or 1888, however, the decrease was not felt in obtaining skins, at which time the standard was lowered from 6 and 7 pound skins to 5 and 4½ pounds. The hauling grounds of Northeast Point kept up the standard longer than the other rookeries, because, as I believe, the latter rookeries had felt the drain of the open-sea sealing during 1885 and 1886 more than Northeast Point, the cows from the other rookeries having gone to the southward to feed, where the majority of the sealing schooners were engaged in taking seal. (Daniel Webster.)

In pursuance of Department instructions to me of May 27, 1891, I made a careful examination during the sealing season of the habits, numbers, and conditions of the seals and seal rookeries, with a view of reporting to the Department from observation and such knowledge on the subject as I might obtain whether or not, in my opinion, the seals were diminishing on the Pribilof Islands; and if so, the causes therefor. As a result of such investigation, I found, from the statements made to me by the natives, Government agents, and employees of the lessees, some of whom had been on the islands for many years, that a decrease in the number of seals had been gradually going on since 1885, and that in the last three years the decrease had been very rapid. A careful and frequent examination of the hauling grounds and breeding rookeries by myself and assistant agents during the months of June, July, and August showed that the seals had greatly diminished in number. We found large vacant spaces on all the rookeries, which in former years during these months had been covered by thousands of seals. Prior to 1888 the lessees had been able to take 100,000 skins from male seals, but I am clearly of the opinion that not more than one-third of that number of merchantable skins can be taken during the year 1891. (W. H. Williams.)

DECREASE OF SEALS.

Management of rookeries not the cause.

In studying the causes of diminution of seal life, there were found a variety of actual and possible sources of destruction which are effective in varying degrees. Fortunately, the most important of these sources were directly under my observation, and the following facts presented themselves for consideration: The restrictions upon the molestation of the breeding grounds and upon the killing of females has been imperative both on the part of the Government and lessees since the American ownership of the islands, so that in the taking of seals no injury could possibly have occurred to the females and bulls found thereon. For some years past the natives were permitted to kill in the fall a few thousand male pups for food. Such killing has been prohibited. It is not apparent how the killing of male pups could have decreased the number of females on the breeding grounds. (J. Stanley-Brown.)

If the seals were as numerous to-day on the Pribilof Islands and the manner of driving and killing conducted in the same manner as during my experience there, 100,000 male seals of from 2 to 4 years of age could be taken from the hauling grounds annually for an indefinite period without diminution of the seal herd. (Charles Bryant.)

Because of the manner of killing seals on the islands, the precautions taken to kill only males from 2 to 5 years, and the careful limitation of the numbers taken, I am fully convinced that the taking of seals on the Pribilof Islands could never affect the numbers of the seal herd or deplete the rookeries. (S. N. Buynitsky.)

I was in the employ of the Alaska Commercial Company, the former lessees of the seal islands, and their instructions were to use the utmost care in taking their quota of seals, so that there might be no diminution in number from year to year, and I personally know those instructions were rigidly enforced. (Leander Cox.)

If no other agency is at work in destroying seal life, 100,000 bachelor seals can be taken from the Pribilof Islands yearly for an indefinite period, provided the rookeries were in the same condition they were in 1871. Of this I am convinced from the fact that the seals continued to increase during all the time I was upon the islands, when 100,000 were killed every year except one, when 95,000 were taken. (Samuel Falconer.)

The management of the sealeries upon Copper Island, under Russian occupation, was left wholly to the native chiefs and ignorant laborers of the Russian-American Company. The work of killing the seals and curing the skins was done by them in a very unsystematic, careless way; but even then it was understood that as the seals are polygamous the surest way to secure an increase of the herd was to kill off surplus males and spare the females, and this was systematically practiced, resulting, as far as I am aware, most satisfactorily. After the expiration of the franchise of the Russian-American Company, in 1867 I think it was, and their abandonment of the island, and the execution of the lease to Hutchinson, Kohl & Co., in 1871, several different parties visited the island, killed seals injudiciously, and inflicted great injury upon the rookeries. They were restrained to some extent by the natives from indiscriminate slaughter, but I have no doubt they killed more males than they ought to have done, and perhaps also some females. Upon my arrival upon the island, in 1871, the native chief told me that the seals were not as plentiful as they had been formerly. I announced that we intended to secure 6,000 skins that year. They protested that it was too many, and begged that a smaller number be killed for one year at least. We, however, got the 6,000 skins as proposed, and an almost constantly increasing number in every subsequent year as long as I stayed on the islands, until in 1880 the rookeries had so developed that about 30,000 skins were taken without in the least injuring them. This is proved by the fact that the increase for the next ten years allowed still larger numbers to be killed, amounting, I think, in one of the years of the second decade of the lease to about 40,000 skins. In order to secure uniformity in the methods pursued respectively upon the Pribilof group and Commander Islands, the respective lessees of the two interests sent Capt. Daniel Webster, an expert sealer of many years' experience in the business, and who was at the time in the service of the Alaska Commercial Company at St. Paul Island, to assist and instruct me through the summer of 1874 in the best manner of handling seal droves, salting skins, and generally in the conduct of the business. In working under his direction, I found that the methods pursued by the respective parties upon the different sealeries did not differ in any essential feature. The main object in both places was to select good skins for market and spare all female seals and enough vigorous bulls to serve them. When the supply of bulls is more than enough I have no doubt the number of offspring is diminished. The bulls, when over-numerous, fight savagely for the possession of the cow seals, and unintentionally destroy many young in their conflicts. The healthiest condition of a rookery is no doubt when, under the laws of polygamous reproduction for this species, the proportion of the sexes is properly balanced. (O. F. Emil Krebŝ.)

Following the surrender of occupancy of these islands by the Russian-American Company in 1868, the sealeries were left open to all parties, and various expeditions visited them unrestricted by any governmental control. Their catches amounted in 1868 to about 15,000, in 1869 to about 20,000, and in 1870 to about 30,000 skins. In 1871 the Russian Government executed the lease to Hutchinson, Kohl & Co., and it was found necessary to restrict the killing for this year to about 6,000 skins, because the rookeries had been largely depleted by the excessive killing, unwise methods, and heedless husbandry. The result of improved methods showed themselves at once, and the rookeries steadily increased in size and number of occupants. We were thus enabled to procure an almost constantly increasing number of skins from year to year during the whole term of our lease. We were unrestricted as to the numbers to be taken, and after the first two years of the lease were urged by the Russian authorities upon the islands to take more than we wanted, in view of the condition of the seal-skin market. I revisited the islands on various occasions subsequent to 1871, and my observations confirmed the fact that we were moving in the right direction to secure an increase of the rookeries. The experience of the whole term of the lease proves conclusively that our policy in conducting the business was a wise one, and that our manner of handling, managing, and killing the seals was in every respect what it should have been. This policy was predicated upon the custom of the Russian-American Company, observed during many years and strengthened by my own actual experience in conducting the business of taking seals upon the Pribilof Islands in 1867, 1868, and 1869, and more particularly during the season of 1868, when there was unrestricted sealing done by various parties regardless of the future of the rookeries. The pernicious effects of the methods pursued by them were at once observed, and measures immediately taken by me, aided by the natives, over whom I had complete control, to correct their practices and bring them within reasonable customs already proved efficacious in preserving the rookeries from annihilation. (Gustave Niebaum.)

If the right proportion is maintained between the sexes, the greatest possible number of progeny is assured. As long as we were able to keep exclusive control, undisturbed by outside influences, we maintained the steady increase of the herd and profitable returns from the industry. When outside parties, beyond our jurisdiction, carried on their destructive work to any considerable extent, the equilibrium of the sexes was destroyed, any calculation of those in charge of the Islands was nullified or miscarried, and the speedy decrease and ultimate destruction of the seals and sealing industry made certain. (H. H. McIntyre.)

We protect and take good care of the seals, and if they were not killed in the sea we could make them increase upon the islands so that they would be as many as before. (A. Melovedoff.)

We can care for and protect the mature seals as well as the cattle on the ranges are cared for and protected, and if they could be guarded from the hunters in the sea we could by good management again make the rookeries as large as before. (S. Melovidov.)

Naturally the cause of this diminution was a matter of interest and inquiry. It was not evident that it was from causes incident to the taking of the seals upon the island. The greatest care was exercised in the driving. Under precisely similar conditions the herd had increased in former years. The number of skins originally apportioned to St. George Island was reduced at an early date, and only increased in proportion to the rookeries' expansion. No disturbance of the rookery was permitted, even the presence of dogs and use of firearms being prohibited during the presence of the seals. (T. F. Morgan.)

The management of the rookeries the first fifteen years of the Alaska Commercial Company's lease resulted in a large increase of seals. The same business management continued and the same system was pursued to the end of the term, yet in the last five years the rookeries fell off. Clearly it was through no fault of the company, and resulted from some cause beyond their control. I do not think the Alaska Commercial Company made any mistakes in managing the seal herd. They handled them in every respect as I would have done if they had been my own personal property and as I would do if they were now to come into my hands. If they erred in any particular in their management, it was in their futile attempt in 1888 and 1889 to stop the waste of the seal life at the island spigot while it was running out at the bunghole of pelagic sealing. The record shows that we did not finish the catch as early in 1885 as had been done in former years. I do not think this was from any lack of seals, but was caused by greater care in making our selection of animals to be killed. (Leon Sloss.)

I again visited St. Paul Island and remained there several days in the summer of 1885, but saw no evidence then or when formerly on the island to lead me to think that the lessees were damaging the rookeries or doing anything different from what a judicious regard for the future of the industry would dictate. In giving this evidence I am as free from prejudice as is possible when entertaining, as I do, a feeling that the late lessees treated me in some measure unjustly, nor have I any interest whatever in the seals or the products of the sealeries. (George H. Temple.)

Raids on rookeries not the cause.

It may be worth while to add that the suggestion has been made that the decrease on the number of seals is due to piratical raids upon the islands themselves during the breeding season. While it is unquestionably true that such raids have occasionally occurred during the past, and that some skins have been obtained in that way, the number of these is so trifling in comparison with the annual pelagic catch as not to affect in any way the question under consideration. It is also difficult for one familiar with the rookeries and habits of the seal to conceive of a raid being made without its becoming known to the officers in charge of the operations upon the islands. The "raid theory," therefore, may be dismissed as unworthy, in our judgment, of serious consideration. (Report of American Commissioners.)

The statistics which I have examined, as well as all the inquiries made, show that in the raids upon the rookeries themselves by marauders the loss of seal life has been too unimportant to play any part in the destruction of the breeding grounds. The inhospitable shores, the exposure of the islands to surf, the unfavorable climatic conditions, as well as the presence of the natives and white men, will always prevent raids upon the islands from ever being frequent or effective. (J. Stanley-Brown.)

During my stay upon St. George Island several attempts were made by poachers to get on shore and steal the seal, but they succeeded, as far as I am aware, only on three occasions, and in all those three I do not think they killed more than 1,200 or 1,500 seals, including pups. If any others had effected a landing we should have known it, for the rookeries are constantly watched, and the natives are very keen in this matter. (Harry N. Clark.)

We tried to make a raid on St. George, but the *Corwin* was after us and we kept out of its way. (Peter Duffy.)

During the time I was on St. George Island there never was a raid on the rookeries to my knowledge, and I never heard of any such raid ever having taken place. (Samuel Falconer.)

I have known of one or two schooners operating in Bering Sea as early as 1877 or 1878, and they were on the rookeries occasionally during the past ten years, but they can not damage the seal herd much by raiding the rookeries, because they can not take many, even were they permitted, which they are not by any means.   (John Fratis.)

Raids on the rookeries by marauders did not, while I was on the island, amount to anything, and certainly seal life there was not affected to any extent by such incursions.   I only knew of one raid upon St. Paul Island while I was there.   It was by a Japanese vessel, and they killed about 100 seals, the carcasses of which we found on board when we captured the vessel.   (H. A. Glidden.)

We sailed about January from Victoria, British Columbia; sailed along the coast until the latter part of June and went into Bering Sea, and sealed as near to St. George Island as we could.   We caught about 300 or 400 seals in the sea.   Our intention was to make a raid, but were driven away by a revenue cutter.   We left the sea about the latter part of July.   (Joseph Grymes.)

Max. Heilbronner, having been duly sworn, deposes and says: I am secretary of the Alaska Commercial Agency, and as such have in my custody all record books of the company, and among them the daily records or "log book" kept by the agents of the company on St. George Island from 1873 to 1889, inclusive, and on St. Paul Island from 1876 to 1889, inclusive.   In these books every occurrence was carefully noted from day to day by the agent in charge at the time.   They have been examined under my supervision, and show only the following raids on St. George Island during the time covered by them, to wit:

October 23, 1881: The carcasses of 15 dead pups and a cargo hook were found on a rookery.   It was supposed that the crew of a schooner seen about the island a few days previous landed in the night.

October 10, 1884: Fifteen seal carcasses were found on Zapadnie rookery.   A guard was stationed, and the following night the crew of a schooner made an unsuccessful attempt to land.   The boats were fired on by the guard and retreated.

July 20, 1885: A party landed under the cliffs in a secluded place and killed about 500 adult-female seals and took the skins away with them.   They killed about 500 pups at the same time, leaving them unskinned.

July 22, 1885: A party landed at Starry Arteel rookery and killed and skinned 120 seals, the skins of which they left in their flight, when pursued by the guard.   They killed also about 200 pups, which were left unskinned.

November 17, 1888: A crew landed and killed some seals at Zapadnie; how many is not known, but at this season of the year the number must have been small, because the seals have nearly all migrated.

September 30, 1889: Eighteen dead pups and four clubs were found on a beach near a rookery.   It is not known whether any others were killed.

An examination of St. Paul record does not show any destructive raids upon the island.   It is a fact, however, that in July, 1875, prior to the beginning of the record, the crew of the schooner *San Diego* landed on Otter Island, a small islet 6 miles from St. Paul, and killed and skinned 1,660 seals.   She was captured before leaving the island, and both the skins and vessel were condemned to forfeiture by the United States court.

The reports of the superintendent for the lessees show that it was the custom of the company's agent on the islands to frequently patrol the rookeries whenever the weather was such that a landing could be effected on them, and to keep watchmen at points distant from the villages, whose special duty it was to report every unusual or suspicious occurrence. For this purpose the northeast point of St. Paul Island was connected with the village by telephone in 1880, a distance of 12 miles, and the natives instructed in the use of the instrument. If any raids upon the islands, other than those herein mentioned, had occurred, I am sure they would have been detected and reported to this office. No such reports are on file. (Max. Heilbronner.)

H. H. McIntyre, having been duly sworn, deposes and says: I was superintendent of the seal fisheries of Alaska from 1871 to 1889, inclusive. The records above referred to were kept under my direction by my assistants on the respective islands. I was in frequent correspondence with these assistants when not personally present and am sure that anything worthy of notice would have been promptly reported to me. I believe that these records contain a true account of all destructive raids upon the islands. If there had been any others I should have heard of them. Every unusual occurrence at any point about the islands was noted by the keen-eyed natives and at once reported to the company's office, the matter was investigated, and a record of it entered in the daily journal. I am confident that the only marauding expedition that ever succeeded in killing more than a few dozen seals each were those of 1875, upon Otter Island, and of 1885 upon St. George Island, the details of which were set forth by Mr. Heilbronner in the foregoing affidavit. If there were others of which no records appear the number of seals killed was comparatively very small and had no appreciable effect upon seal life. (H. H. McIntyre.)

Sometimes they try to land on the rookeries, but we drive them off with guns, and they never get many seals that way. (Nicoli Krukoff.)

I do not mean to say that the seals were injured because a few were killed on the rookeries, when men from schooners landed on the islands in the night or when the fog was very thick, for the numbers killed in that way never amounted to much, as it is not often the raiders can land on a rookery and escape with their plunder. (Aggie Kushen.)

When on a raid we would watch for a favorable opportunity to make a landing, and then kill male and female fur seals indiscriminately. Probably for every 500 marketable skins secured, double that number of pups were destroyed. (L. M. Lenard.)

While I was on the island there were not more than three or four raids on the rookeries to my knowledge, and I think that the destruction to seal life by raiding rookeries is a small part of 1 per cent as compared with the numbers taken by killing in the water. (A. P. Loud.)

It is often difficult to entirely prevent poaching on the islands, although in my judgment it has not been of sufficient importance on the Commander Islands to have any perceptible influence in the diminution of the herd. (John Malowansky.)

I remember seeing an occasional sealing schooner in Bering Sea as long ago as 1878, but it was in 1884 they came in large numbers. At first it was supposed they intended to raid the rookeries, and we armed a number of men and kept guard every night, and we drove off any boats we found coming to a rookery. Sometimes in a dense fog or very dark night they landed and killed a few hundred seals, but the numbers taken in this manner are too small to be considered. (A. Melovedoff.)

One cause of destruction is raiding, which has been done upon the shores of the islands. A half dozen such raids are known to me personally; but while it is not possible for me to state with certainty the skins actually secured by such raids, I believe that, although such raiding is detrimental, its injurious effect as compared with the disastrous results of pelagic sealing is insignificant. (T. F. Morgan.)

There was only, as I recollect, four raids on the islands while I was there; but little or no damage was done, and seal life was not perceptibly affected by such marauding. (J. H. Moulton.)

From my personal knowledge of the number of seals killed upon the Pribilof Islands by raids upon the rookeries during my residence there, and from information gained from other sources, I conclude that the number of fur seals killed is infinitely small compared with the number killed in pelagic sealing—so small as to have no appreciable effect upon seal life upon the islands. (S. R. Nettleton.)

I am told that the diminution of seal life has been attributed to raids by poachers upon the seal islands. Very few of these have occurred, and the number of skins obtained by the poachers has been comparatively infinitesimally small. I think the whole number obtained by them in this way does not exceed 3,000 or 4,000 skins. We were accustomed always to maintain a patrol and guard upon the rookeries whenever the weather was such that poachers could land upon them, and upon the least suspicious circumstances measures were taken to forestall any attempts to steal the seals. The sea is usually rough in the fall, when poachers try to get in their work; the shores are, at most places, inaccessible from boats, and the natives are vigilant and active. If marine hunting is stopped, they can be safely trusted to defend the property upon which their very existence is dependent, as they have done repeatedly, against any single schooner's crew. (Gustave Niebaum.)

There were occasional raids made upon the islands (Commander) by poachers during our twenty years' lease, but they were generally unsuccessful in killing any considerable number of seals, and their raids had no appreciable effect upon the rookeries. (Gustave Niebaum.)

During those years the lawless occupation of seal poaching was in its infancy. Marauding vessels from time to time were seen in these waters, but the islands were so well guarded that during my term of office there never was a successful raid or landing upon either of the islands of St. Paul or St. George. The only landing upon any island of the group was made in June, 1881, upon the unoccupied island of Otter (not included in the lease), as described in my special report to the Secretary of the Treasury, dated July 4, 1881. On that occasion a predatory schooner succeeded in landing a boat's crew, who killed 40 or 50 seals, when they were driven off by a boat sent by me for that purpose from St. Paul, about 6 miles distant. (H. G. Otis.)

Until 1884 sealing schooners were seen but very seldom near the islands or in Bering Sea, and the few seals taken by the hunters who raided the rookeries occasionally are too paltry to be seriously considered, because the raids were so few, and the facilities for taking many seals off so utterly insignificant. (J. C. Redpath.)

There was but one successful raid on the rookeries while I was upon the island and but 125 seals were killed. I do not consider that raids on the rookeries have anything to do with the decrease of the number of seals. (T. F. Ryan.)

While I was on the islands there were no raids on the rookeries, and seal life was never depleted at that time by such means. (B. F. Scribner.)

There was but one raid on the rookeries while I was there, and that took place on Otter Island, about 60 skins being taken. After that raid the Government kept a man on Otter Island during the entire summer to protect it from marauders. Raids on the islands never affected seal life to any extent. (W. B. Taylor.)

I do not remember the precise date of the first successful raid upon the rookeries by sealing schooners, but I do know that for the past ten years there have been many such raids attempted and a few of them successfully carried out, and that as the number of schooners increased around the islands, the attempted raids increased in proportion, and it has been deemed necessary to keep armed guards near the rookeries to repel such attacks. Although a few of the raids were successful and a few hundred seals killed and carried off from time to time during the past ten years, the aggregate of all the seals thus destroyed is too small to be mentioned when considering the cause of the sudden decline of seal life on the Pribilof Islands. (Daniel Webster.)

DESTRUCTION OF FEMALE SEALS.

Examination of pelagic catch of 1892.

On May 7 of this year I examined 355 salted fur-seal skins, ex steamer *Umatilla* from Victoria, and found the same to be fresh skins taken off the animal within three months. They were killed in the North Pacific. On examination I found they were the skins known as the Northwest Coast seals, and belong to the herd which have their rookery on the Pribilof Islands. The lot contained 310 skins of the fur-seal cow (matured). From the shape of the skins most all of these cows must have been heavy with pup, and same cut out of them when captured. Eighteen skins of the fur-seal male (matured). Twenty-seven skins of the fur-seal gray pup, from 6 to 9 months old; sex doubtful.

On June 2 I examined 78 salted fur seal skins, ex steamer *Walla Walla* from Victoria, and found the same to be fresh skins taken off the animal within three months. They were killed in the North Pacific. On examination I found they were the skins known as the Northwest Coast seals, and belong to the herd which have their rookery on the Pribilof Islands. The lot contained 66 skins of the fur-seal cow (matured). From the shape of the skin most all of these cows must have been heavy with pup, and the same cut out of them when captured. Five skins of the fur-seal male (matured). Seven skins of the fur-seal gray pup, from 6 to 9 months old; sex doubtful.

On June 7 I examined 268 salted fur-seal skins, ex steamer *Umatilla* from Victoria, and found the same to be fresh skins taken off the animal within three months. They were killed in the North Pacific. On examination I found they were skins known as the Northwest coast seals and belong to the herd which have their rookery on the Pribilof Islands. The lot contained 212 skins of the fur-seal cow (matured). From the shape of the skin most all of these cows must have been heavy with pup, and same cut out of them when captured. Eleven skins of the fur-seal male (matured). Forty skins of the fur-seal gray pup, from 6 to 9 months old; sex doubtful.

On the same date I also examined 124 salted fur-seal skins, ex steamer *Umatilla* from Victoria, and found the same to be fresh skins taken off the animal within three months. They were killed in the North Pacific. On examination I found that they were the skins known as the North-west coast seals and belong to the herd which have their rookery on the Pribilof Islands. The lot contained 93 skins of the fur-seal cow (matured). From the shape of the skin most all of these cows must have been heavy with young, and the same cut out of them when cap-tured. Fifteen skins of the fur-seal male (matured). Sixteen skins of the fur-seal gray pup, from 6 to 9 months old; sex doubtful.

I notice on examining seals caught this spring that there is a lack of the larger size of productive animals, and the lots mostly contain the skins of the medium-sized seals, running from 2 to 3 years of age. (Charles J. Behlow.)

On the 29th instant I examined 2,170 salted fur-seal skins, ex schooner *Emma and Louise* from the North Pacific Ocean, and found same to be fresh skins taken off the animal within four months. They were killed in the North Pacific. On examination I find they were the skins known as the Northwest coast skins, and belong to the herd which have their rookery on the Pribilof Islands. The lot contained 4 skins of the fur-seal large bulls (breeding bulls); 123 skins of the fur-seal male (mostly matured); 98 skins of the fur-seal gray pup, less than 1 year old, sex doubtful; 1,112 skins of the fur-seal cow (mostly matured). From the shape of the skin most all these cows must have been heavy with pup, and same cut out of them when captured. (Charles J. Behlow.)

As a result of the work I have performed for so many years I am able to distinguish without difficulty the skin of a female seal from that of a male seal. There are generally several ways in which I can tell them apart. One of the surest ways consists in seeing whether any teats can be found. On a female skin above the age of 2 years teats can practically always be discovered; when the animal is over 3 years old even a person who is not an expert at handling skins can discover two prominent ones on each side of almost every skin. This because after the age of 3, and often even after 2, almost all females have been in pup. There are also teats on a male skin, but they are only very slightly developed. When the fur is matted, as it is in salted fur-seal skins, the male teats can not be found, but the female teats of skins more than 2 years old can be found under all circumstances.

I have been able to test all my observations as to the teats on salted fur-seal skins by following these skins through the various processes which I have described. During these processes the skins become thin-ner and thinner, and the teats more and more noticeable, and at an early stage in the dressing they must be wholly removed. There are other ways of distinguishing the skins of the two sexes. I will state a few of them.

A female has a narrower head than a male seal. By the word "head" I mean here to include that part of the body from the head down to the middle of the back. I believe all men who have handled the skins of both sexes have noticed this point. Then, again, when the whiskers have not been cut off they generally afford a safe means of distinguish-ing the sexes. Male whiskers are much more brittle and of a darker color than those of the female animal. When the male seal is over 6 years old it begins to have a mane, and for this reason it is after that age called a wig. Finally, it is generally possible for me to tell the skins of the two sexes apart by just taking a look at them or feeling them. I suppose I can do this because I have been at the business so long that I am an expert in it.

The chief classes of seal skins that I have handled are the Alaska, the Northwest coast, and the Copper Island skins. I can always distinguish the skins of these classes. The Northwest Coast skins are most easily told by the very great proportion of females contained in any given lot. Among the Alaska and Copper skins I have hardly ever seen a female skin. (John J. Phelan.)

I was sent to New York from Albany a few days ago by Mr. George H. Treadwell, with instructions to go through a certain lot of seal skins which, I understand, he had recently bought in Victoria, and to find out how many of these skins were taken from female animals. I have spent four days in doing this, working about seven hours a day.

There were several men who unpacked the skins and laid them before me, so that all of my time was spent in examining the individual skins. The lot contained 3,550 skins. I found that, with the possible exception of two dried ones, they were taken from the animal this year; they were a part of what is known as the spring catch. I know this to be the case by the fresh appearance of the blubber and of the skin as a whole. This affords a sure way of telling whether the skin has lain in salt all winter or whether it has been recently salted. I personally inspected each one of these skins by itself and kept an accurate record of the result. I divided the skins according to the three following classes: Males, females, and pups. In the class of pups I placed only the skins of animals less than 2 years of age, but without reference to sex.

I found in the lot 395 males, 2,167 females, and 988 pups. Leaving out of account the pups, the percentage of females was therefore about 82. The great majority of what I classed as male skins were taken from animals less than 3 years of age. There was not a single wig in the lot. On the other hand, nearly all the female skins were those of full-grown animals. On every skin which I classed among the females I found teats, with bare spots about them on the fur side. Such bare spots make it absolutely certain that these teats were those of female skins.

With regard to the pup skins, I will say that I did not undertake to determine whether they were males or females, because they had a thick coat of blubber which, in the case of an animal less than 2 years old, makes it very hard to tell the sex.

All of the skins that I examined were either shot or speared. I did not keep a close count, but I am of the opinion that about 75 per cent of them were shot.

The result of the examination is about what I expected it would be. The figures only confirm what I have always noticed in a general way, that nearly nine-tenths of the skins in any shipment of Northwest coast skins are those of female animals. (John J. Phelan.)

Examination of catch of vessels seized.

About seven years since I was on the revenue cutter *Corwin* when she seized the sealing schooner *San Diego* in Bering Sea. On the schooner's deck were found the bodies of some 20 seals that had recently been killed. An examination of the bodies disclosed that all of them,

with but a single exception, were females, and had their young inside
or were giving suck to their young. Out of some 500 or 600 skins on
board I only found some 5 of the number that were taken from males.
I have also been present at numerous other seizures of sealing vessels,
some 18 in number, and among the several thousand skins seized I
found on examination that they were almost invariably those of females.
There certainly was not a larger proportion of males than 5 to 100
skins. This great slaughter of mother seals certainly means a speedy
destruction of seal life. (James H. Douglass.)

While in Unalaska in September, 1891, awaiting transportation to
San Francisco, I had an opportunity to examine personally the catch
of the steam sloop *Challenge*, which had been warned out of the sea,
and was undergoing repairs at the harbor named. The catch amounted
to 172 skins, which were all taken in Bering Sea at various distances
from the seal islands, and of this number only three were those of male
seals, one of those being an old bull, and the other two being younger
males. (A. W. Lavender.)

In July, 1887, I captured the poaching schooner *Angel Dolly* while she
was hovering about the islands. I examined the seal skins she had on
board, and about 80 per cent were skins of females. In 1888 or 1889 I
examined something like 5,000 skins at Unalaska, which had been taken
from schooners engaged in pelagic sealing in Bering Sea, and at least
80 or 85 per cent were skins of females. (A. P. Loud.)

I have personally inspected skins taken upon the three schooners
Onward, Caroline, and *Thornton*, which skins, taken in Bering Sea,
were landed in Unalaska and were then personally inspected by me in
the month of May, 1887. The total number of skins so examined was
2,000, and of that number at least 80 per cent were the skins of females.
I have also examined the skins taken by the United States revenue
cutter *Rush* from one of the North Pacific islands, where they had been
deposited by what is known as a poaching schooner and taken to Unalaska, which numbered about 400 skins, and of that 400 skins at least
80 per cent were the skins of female seals. I have also examined the
skins seized from the *James Hamilton Lewis* in the year 1891, by the
Russian gunboat *Aleut*, numbering 416, of which at least 90 per cent
were the skins of female seals. From my long observation of seals and
seal skins I am able to tell the difference between the skin of a male
and the skin of a female seal. (T. F. Morgan.)

I examined over 12,000 skins from sealing vessels seized in 1887 and
1889, and of these at least two-thirds or three-fourths were the skins of
females. (L. G. Shepard.)

REASON PREGNANT FEMALES ARE TAKEN.

I think cow seals are tamer than young male seals. (Martin Benson.)

A cow seal that is heavy with pup is sluggish and sleeps more soundly
than the males, and for that reason they are more readily approached.
(Henry Brown.)

They are very tame after giving birth to their young and are easily
approached by the hunters. When the females leave the islands to
feed they go very fast to the fishing banks, and after they get their food
they will go to sleep on the waters. That is the hunter's great chance.
I think we secured more in proportion to the number killed than we did
in the North Pacific. (James L. Carthcut.)

They sleep more and are less active and more easily captured. (Simeon Chin-koo-tin.)

I think the female seal is less active and more easily approached. (Peter Church.)

I have noticed that the females when at sea are less wild and distrustful than the bachelor seals, and dive less quickly in the presence of the hunter. After feeding plentifully, or when resting after heavy weather, they appear to fall asleep upon the surface of the water. It is then they become an easy target for the hunters. (James H. Douglass.)

I think the females sleep more on the water, and are less active and more easily taken than the males. (E. Hofstad.)

When the females are with pup they sleep more, are less active, and more easily approached than the male seals. (P. Kahiktday.)

Think cows are much more plentiful on the coast, sleep more, and are more easily captured than the male seals. (John Kowineet.)

Think cows are less active and require more sleep than the young male seals. (George Lacheek.)

I am informed and believe that the reason of there being such a large proportion of females among the coast skins is because the male, which is powerful and strong, usually swims more readily and at a longer distance from the coast, and are so scattered and active and hard to catch that it does not pay to hunt them. The female heavy with young easily tires and sleeps on the water, and is easily shot while in that condition. (George Liebes.)

Mother seals heavy with young are much easier taken, for they are usually asleep on the water. (William H. Long.)

Q. Why is it, in your opinion, that more female than male seals are killed by the poachers?—A. Because, first, in the passage of the seals to the islands in the early season the females travel in groups and the males scatter; secondly, after arriving at the islands the males remain on or about the hauling grounds, while the females, having their pups to nurse, go out into the sea to obtain food.

Q. How do you tell the skin of a female from that of a male?—A. By the nipples and general appearance. (Anton Melovedoff.)

As I understand the fact to be, most of the seals killed in the open sea are females. My reasons for this conclusion are that, from my knowledge of the seal, I know that the female when heavy with young, as they are during the early part of the season when on their way to the rookeries, where they are delivered during the months of June and July, are much heavier in the water and much less able to escape, because they are capable of remaining under water to escape for a very much less period of time than when they are not heavy with young, or than the male seal would be. (T. F. Morgan.)

It is harder to take an old seal than a young one, the older ones being more on the alert and are not less active when pregnant. (W. Roberts.)

Of the seals killed, from 60 to 70 per cent are females, which, during their northerly migration, are heavy with young, slow of movement, and require an extra amount of rest and sleep, thus largely increasing their liability to successful attack. (Z. L. Tanner.)

I have been told that it is easier to catch the female seal at sea than it is to catch the male seal, but I have no personal knowledge of that point. I suppose, however, that there must be some foundation for the statement by reason of the fact that so small a proportion of male adult seals are included in what is called the northwest catch. (Emil Teichmann.)

The cows are less active, sleep more, and are more easily captured. (M. Thlkahdaynahkee.)

Cow seals sleep sounder on the water, are less active, and are easily captured. (James Unatajim.)

Cows are more easily captured because they have pups. (Rudolph Walton.)

They are less active, sleep more, and are easier captured. (Charlie Wank.)

It is my opinion that female seals are more easily captured and appear to be more tame than the male seal, and, I think, sleep more. (P. S. Weittenhiller.)

The large proportion of females killed in the North Pacific is due to the fact, as I explained before, that males pursue their way to the hauling grounds with dispatch, while the females are more leisurely in their movements and take frequent rests. (T. T. Williams.)

DECREASE OF SEALS.

Percentage lost of seals killed.

From my experience I am satisfied that 33⅓ per cent shot with a shotgun are lost, and when a rifle is used a larger per cent are lost when killed. (Peter Anderson.)

We lost three out of four we killed. (H. Andricius.)

On an average, we saved one out of three that were killed. (Bernhardt Bleidner.)

It is my honest belief that for every fur-seal skin obtained by pelagic sealers at least five other seals' lives are taken. (J. A. Bradley.)

During the trip of 1891 I don't think we got more than one seal out of six that we killed; many were wounded, and others were shot dead and sank before the boat could get to them. (Thomas Brown.)

Native hunters secure about one-third of all fur seals killed at sea, while in my belief white hunters secure even a less number in proportion to those killed. (M. Cohen.)

An average hunter will get one out of four of breaching seals and one out of three of sleepers that he kills, but a common hunter will not get so many. (Peter Collins.)

And that a vast number of the seals killed by them are lost. (Leander Cox.)

It is my experience that very few, if any, seals were lost by the hunters who use the spear, but fully 75 per cent of all those killed by the rifle were lost. (James Dalgarduo.)

From my observation of the methods employed by the open-sea hunters I believe that a very large proportion of those killed by them are lost. I have often heard sealers so express themselves. They have said to me that they get only about one out of five shot or killed; others made the loss still greater. I think the latter statement more nearly correct. (M. C. Erskine.)

Of seals killed, about four out of five are saved. (F. F. Feeny.)

An experienced hunter like myself will get two out of three that he kills, but an ordinary hunter would not get more than one out of every three or four that he kills. (Thomas Gibson.)

I lose about 50 per cent when I use the shotgun, and more are lost when rifle is used. I always shoot them in the head when possible, but if not possible, I shoot them in any part of the body that is exposed. (Gonastut.)

About 50 per cent are lost when killed with a shotgun, and a larger per cent when rifle is used. (James Gondowen.)

The hunters would get, on an average, one out of every four they killed. (James Grymes.)

On an average, I think the hunters will save about one out of three that they kill, but they wound many more that escape and die afterwards. (James Harrison.)

Formerly the seals were gentle and the approach of a vessel did not even alarm them, but when firearms came into use it so frightened them that they had to be shot at long range, entailing a loss of not less than three out of every four or five killed. (M. A. Healy.)

My experience convinces me that a large percentage of the seals now killed by shooting with rifles and shotguns are lost. My estimate would be that two out of every three killed are lost. Formerly the killing was done by spearing, and in later years it was learned that shooting them was a swifter method of killing. At the start the hunters were inexperienced and a large proportion were lost. (James Kiernan.)

I use the shotgun for taking seal, and sometimes I lose one or two out of ten that I kill. (James Klonacket.)

I have made it my business to find out what proportion of skins of seals killed are really brought into the market, and from the information which I obtained from the sealers, hunters, and those owning the skins I learned that on an average only about one out of six killed was secured, varying with the expertness of the hunter. (George Liebes.)

The number of seals actually secured to the number killed does not exceed about one in four, or about one is taken for every three destroyed, varying, of course, with the skill and experience of the hunters. (Isaac Liebes.)

From these conversations I should judge they did not secure more than one half of the seals killed; and this, I think, is a large estimate of the number secured. (A. P. Loud.)

I have frequently noticed, in the harbor of Petropaulovsky, that the natives, in killing hair seals, are only able to obtain one animal out of every four or five of those killed, and that they frequently wait about four days for the bodies to be washed ashore. (John Malowansky.)

None I lost when I used spear. About 20 per cent are lost when killed with shotgun. (Nashtau.)

An experienced A No. 1 seal hunter, in shooting sleeping seals with a shotgun, will get a large proportion of what he kills, and will get one out of four breaching seals that he kills; but an ordinary common hunter like myself will sometimes use ten cartridges and not get one seal. I can safely say that a common hunter will only get one seal out of three. (Niles Nelson.)

The white hunters who used guns in Bering Sea were banging away at the seals sometimes all day long, and they would lose a great many of those that they shot. I do not think that they brought to the schooner one-half of those that they killed, to say nothing of those that they wounded and got away. (Osly.)

But since it has become the practice to hunt seals with guns a good many are killed, wounded, and lost. Green hunters bang away and wound more than they kill, and will shoot six or seven before they get one, and sometimes more. Good hunters will do much better. I used to get most of the seals I killed, but I have killed five dead in succession and lost the whole of them. (William Parker.)

Shotgun is exclusively used by me for taking seals. Lose about 20 per cent of those killed with shotgun. (Abel Ryan.)

The captain, mate, and myself went out several times with the stern boat and we killed 15 the first time we went out. I think we went out that way three or four times, and we usually got one out of four killed. I recollect one day when we were hunting, bad weather set up and we did not get any seals.· In good weather we got more seals than we did in bad weather. (Peter Simes.)

And we got one out of five killed. (John A. Swain.)

On my first voyage I think we got two out of every five that we killed. (Adolph W. Thompson.)

When seal were struck with a spear none were lost; lose about 50 per cent when killed with shotgun. (Charlie Tlaksatan.)

I had in my employ men who are old seal hunters and who were formerly engaged in that business, and they have often told me that they lost at least two out of every three they killed. (M. L. Washburn.)

Percentage lost of seals struck.

The skill of the hunter has a great deal to do with the number of seals secured of those killed or wounded, but the most expert does not get more than half he hits, and the average for hunters in general would be about three in ten. (O. A. Abbey.)

We secure one out of about every five that we shoot at or kill. (Charles Adair.)

An experienced hunter would get one out of every three that he shot or killed, and a green hunter would get about one out of every seven or eight that he shot or killed. (Charles Adair.)

It has been my custom in the last few years to examine the logs of sealing vessels and to converse with officers and hunters of such vessels in order to obtain what information I could as to the methods employed by hunters and the loss of seals occasioned in such pursuit. From the logs I learned that in many instances 100 rounds of ammunition had been fired to each skin secured, and often more; and on an average I found that not over five seals to the hundred shots had been obtained. The logs further showed that a large number had been wounded and lost. I also ascertained from the logs and from conversation with masters of sailing schooners that not one seal out of ten killed or wounded had been caught. These inquiries I pursued at San Francisco until quite recently. The chief killing by poachers was done between the passes of Aleutian Archipelago and the Pribilof Islands. (George R. Adams.)

Have always used a shotgun and rifle in taking seal since a young man. I rarely lose any seal I shoot, as I never shoot at them unless they are very close to the boat. (Adam Ayonkee.)

Have always used a shotgun for taking seal, and lose about 40 per cent of what I shoot. (Maurice Bates.)

No seal were lost when struck with spear. About 40 per cent of seal shot with shotgun are lost, and more when the rifle is used. (Wilton C. Bennett.)

I use the shotgun for taking seal. I lose about 25 per cent of the seals shot. (Edward Benson.)

The spear and shotgun have been used by me. But few seals are lost that are struck by spear. About 66 per cent are lost when shot with shotgun, and a larger proportion are lost when rifle is used. (Martin Benson.)

On the *Pioneer* we had a couple of good hunters, who would get almost all they shot at, while some of our hunters would lose a good many that they would kill and wound. A green hunter will not get more than one out of five, and I have known one hunter on our vessel who shot eight shots and got only four seals. Indian hunters that use spears seldom lose any that are struck, and there is no wounded to go away and die. (Neils Bonde.)

This year the seals are wilder than the year before; I think it was because they were hunted so much. We did not capture as many in proportion to the number shot as we did the year previous, and did not save more than one out of six that we shot. (Thomas Brown.)

We got on an average three or five out of every twelve killed and wounded. It depends a great deal upon the weather. There were lots of seals in the water at that time. (Thomas Brown.)

The average hunter would get one out of every three that he shot; a poor hunter not nearly as many. There are 21 buckshot to a shell. * * * When they are in school sleeping we get a good many. We did not get as many as we shot at in Bering Sea as we did on the coast. If we got one out of every three we were doing pretty well. (Charles Chalall.)

I used a shotgun almost exclusively last season, and lost about one-third of all furs shot. (Julius Christiansen.)

I think about 50 per cent of the seals shot with shotgun are lost, and greater proportion are lost when shot with a rifle. (Peter Church.)

I always use the shotgun for taking seal. I think about 25 per cent are lost. (William Clark.)

Over 50 per cent are lost when shot with shotgun. (John O. Clement.)

My observation of the seal hunting by white hunters in 1888 is that they do not secure more than two or three out of every hundred shot. The number of shots fired by a hunter in an ordinary day's sealing is something enormous, and the waste of seal life in the water is dreadful to contemplate. * * * The proportion of loss of seals shot by white hunters in the Otto was quite as great in 1891 as by the hunters in the year before stated. I have never seen any black pups in the North Pacific Ocean. (Louis Culler.)

When it was rough weather we got one out of six that we killed or wounded, and in smooth weather we could get on an average one out of three and sometimes three out of five. (John Dohrn.)

On an average, all the hunters got one out of three or four seals that they killed or wounded. There were plenty of seals in the water at that time. (Richard Dolan.)

We got one out of every five or six that we killed or wounded. We wounded a great many that we did not get. (George Fairchild.)

When I was a young man the Indians used the spear for taking seals; now they have learned from the white men to use the shotgun. About three out of ten are lost that are shot. (Frank.)

The hunters used rifles and shotguns. They got about one out of every six they shot at or killed, and sometimes they got none. The great majority of them were females. We used rifles, we had experienced hunters on board, and we got one out of every three killed or wounded. (William Frazer.)

Q. What percentage of seals are taken compared to those you destroy in doing so? In other words, how many do you actually get of those you shoot?—A. About 30 per cent.

Q. Is it not a fact, when you first started in the business and was inexperienced in hunting, that you, like all other beginners, destroyed a much larger proportion than you now do?—A. Yes; a little more in proportion. (Edward W. Funcke.)

Indians lose a less number of the seals shot at and wounded or killed than white hunters. When they use spears they get nearly all they wound. When they use shotguns they do not get more than one out of eight killed or wounded. In conversation with boat steerers and boat pullers I have frequently heard them state that hunters would sometimes fire from 75 to 100 shots without bringing in a single seal. The hunters would claim they secured nearly all they fired at or killed, but it is known that this is not true. It is impossible to say what proportion of the seals fired at are killed or wounded, but taking the run of hunters, good and poor, I should say that the best get about 50 per cent of those shot at, while the poorest do not get more than one out of fifteen fired at. (E. M. Greenleaf.)

'The native hunters used spears exclusively in hunting the seals, and secured fully two-thirds of all struck. I am of the opinion that with firearms not more than one-third of the animals shot are actually secured. (A. J. Guild.)

Have always used a shotgun for taking seal, and lose about 25 per cent of the seals I shoot. (Henry Haldane.)

I use the shotgun exclusively for taking seal. About 65 per cent of the seal hit are lost. (Martin Hannon.)

Q. According to your experience, what percentage of animals that are shot are actually taken by the boats?—A. That depends a good deal on the man that shoots them. Some fellows will miss four out of five and another may miss three out of five and cripple them. I think on a general average we will get about three out of five. (H. Harmsen.)

Q. What percentage of seals are taken compared to those you destroy in doing so; in other words, how many do you actually get out of those you shoot?—A. We get about 75 per cent of them.

Q. Is it not a fact that when you first started in the business and was inexperienced in hunting, that you, like many others, destroyed a much larger proportion than you now do?—A. Yes, sir; it is. (Andrew J. Hoffman.)

The shotgun was exclusively used by our hunters. I can form no idea as to the amount of seals lost. Some hunters lost more and some less. It ranges all the way from 10 to 75 per cent, according to stories told by hunters. (O. Holm.)

We used shotguns, and secured about two seals out of five that we shot. (Alfred Irving.)

The Indian hunters with spears would not wound or lose but very few seals that they struck, but the ordinary white hunter will, on an average, lose over half that he kills and wounds. (James Jamieson.)

About 40 per cent shot with shotgun are lost. When the rifle is used a larger per cent is lost. (J. Johnson.)

Have always used shotgun and rifle for taking seal. I never lose any seal when I shoot them, because I always shoot them close to. (Johnnie Johntin.)

The spear and arrow were used to take seal when I was a boy, but now I use the shotgun and rifle. At least 50 per cent are lost when shot with shotgun. When rifle is used a larger portion of seals are lost. (P. Kahiktday.)

I always use the shotgun for killing seal. I lose about four out of ten that I shoot. (King Kashwa.)

I always use the shotgun for taking seal. Sometimes I lose two and three out of ten that I shoot. (Jim Kasooh.)

Fully one-half the seal shot with shotguns are lost, and a much larger proportion when the rifle is used. None were lost when struck with a spear. (Mike Kethusduck.)

On an average we got one or two out of every six or seven that we wounded or killed. (James Kennedy.)

Constant shooting has frightened them and made them wild, so that they have to be shot at great distances unless found asleep. Much depends for successful hunting upon the weather, as it is difficult to get accurate aim when both the hunter's boat and the seal are in motion. A poor hunter does not secure more than one out of every five shot or aimed at. Good hunters do better. (James Kiernan.)

The first sighted was August 4, longitude 136° 32′ west, latitude 52° 46′ north. During the days following August 4 canoes were lowered, but their search for seals was fruitless. On August 14, before entering Bering Sea, a seal was speared by the Indians off Marmont Island, which was bearing NW. ¼ W. 35 miles. We entered the sea at 6.30 p. m. on the 22d day of August and at 9 o'clock the following morning we got our first seal in Bering Sea. It was shot by one of the white men in a boat. We were at this time about 25 miles west by north of Northwest Cape on Unimak Pass. On the same day four other seals were shot, and three not recovered. Two sank and the other escaped badly wounded. The following day the captain shot two, losing one, and the other boat brought one seal on board. On the 25th of August we were 125 miles southteast of St. George Island. The Indian hunters were out all day and brought in three seals, the white hunters getting none. The captain informed me that day that the previous year he had taken in this locality 148 seals in one day, and that one of his hunters got 38 and lost 40, which he shot. The next day the two boats and canoes were out, and the captain brought back one, but had shot and lost six others, one of which sank. The other boat reported that they had shot seven, but all sank before they could get them, the water being so colored with blood that it was impossible to see the bodies sufficiently to recover them with the gaff. The two Indians brought back ten seals, all speared. Out of the number taken on board four were full of milk. On the 27th the Indians brought in two seals and the captain one, which were all they had seen. On the 29th seventeen seals were taken; the captain got three, having lost two, killed or wounded. The other boat brought in three, having lost two, and the cook shot one from the schooner's deck. Out of these seven were females, which covered the decks with milk while they were being skinned. I am convinced that at the very least white hunters lose 50 per cent of the seals they hit, and probably the majority of those wounded will ultimately die. (Francis R. King-Hall.)

When a seal is struck with a spear we never lose him. About 50 per cent are lost when shot with a shotgun. (Robert Kooko.)

About 60 per cent of the seals are lost when shot with a shotgun. When rifles are used a much larger proportion is lost. (James Lacheek.)

Of all the fur seals struck in the entire season by both implements more than two-thirds were actually secured, the greater proportion of losses resulting from the use of the shotgun. (James E. Lennan.)

The average hunter will fire ten times to get one seal. I think on an average he gets one seal out of every three killed. (William H. Long.)

Q. What percentage of seals are taken compared to those you destroy in doing so; in other words, how many do you actually get out of those you shoot?—A. I should say we get about 80 per cent of those we shoot.

Q. Is it not a fact that when you first started in the business and was inexperienced in hunting, you, like all other beginners, destroyed a much larger proportion than you now do?—A. There is no doubt about that. (Charles Lutjens.)

The shotgun was used exclusively. Over 60 per cent of the seals shot were lost. (George McAlpine.)

I think I lose about 66 per cent of the seals shot with shotguns. (J. D. McDonald.)

Taking the general average, we would not get more than two seals out of every ten that the hunters shot at. Out of every sixty-five seals that were brought aboard the schooner I got one, so I tried to spear as many as I could after they were shot. We caught more seals in Bering Sea than we did going along the coast, as we found more of them. * * * All the seals that we shot at in rough weather were lost. In fine weather they sleep on top of the water, and we do not lose so many of them. (William McIsaac.)

No seals are lost that are struck with spears. With a shotgun about 50 per cent are lost. (James McKeen.)

We got about one out of every five that we killed or wounded. There was any amount of them that we shot and did not get at all. It seemed as if a good many got away. * * * We had some white and Indian hunters. I do not think that we lost as many that year in proportion to those that we killed and wounded. They were better hunters. (William McLaughlin.)

Q. According to your experience, what percentage of animals that are shot at are actually taken by the boats?—A. That is according to the ammunition that we use. About one-third are taken. (Daniel McLean.)

We had Indian hunters who used shotguns. The Indian hunters are more expert than the white hunters and they do not lose so many seals as they kill. I think they would get one out of every two or three killed or wounded. (Thomas Madden.)

About 50 per cent of the seals shot with shotgun are lost. (Edward Maitland.)

There were six boats on the vessel. Some of the boats would come in without a seal after being out all day long shooting, but they would wound a great many. On an average, taking all the boats together, they got one out of every five or six that they killed or shot at. We wounded a great many that we could not get. (Patrick Maroney.)

About 50 per cent are lost that are shot with the shotgun. (Charles Martin.)

I do not think they would get more than one seal out of every six or seven they shot, and sometimes only one out of ten. (Henry Mason.)

Our hunter was a good one. His name was Joe Williams. I think he got one out of every three, on an average. He used a rifle a good deal, and was a fine shot. Some of the hunters in the other boats would shoot at the seal and not get any at all, and come in at night without any, or maybe one or two. There was one hunter from Nova Scotia that did not kill any, scarcely. (William Mason.)

I think about 33 per cent of the seals shot with a shotgun are lost. (E. Miner.)

About 20 per cent of the seals I shoot with shotgun are lost. (Amos Mill.)

Q. What percentage of seals are taken, compared to those you destroy in doing so; in other words, how many do you actually get out of those you shoot?—A. About 75 per cent. We lose about 25 per cent.

Q. Is it not a fact that when you first started in the business and were inexperienced in hunting, that you, like all other beginners, destroyed a much larger proportion than you do now?—A. Certainly; there is no doubt about that. (Frank Moreau.)

From my knowledge of the aquatic habits of the seal, and the difficulty of accurate shooting when the object is in the water, I am of the opinion that a large number of seals are killed by vessels engaged in the business of taking seals in the open seas which are not caught. I am unable to form an estimate of the number of seals shot or speared from vessels which are lost, but in the last two or three years of my residence at St. George Island, in taking 15,000 seals, I found approximately three pounds of lead, in the form of slugs, bullets, and buckshot, which I personally took from the bodies of male seals, some of which were so badly wounded that they would have died. I have personally examined the log of the schooner *Angel Dollie*, in which it was stated that the hunters from that vessel got about one seal out of every ten shot at; also that on one occasion they fired 250 rounds, and got 20 seals; on another occasion 100 cartridges, and got 6 seals, and which log also stated that the captain personally shot and killed 7 seals, of which he got only one. (T. F. Morgan.)

They lost very few of the seals they speared. They secured about all of the seals they speared. (John Morris.)

When in Bering Sea, I had an opportunity to observe the difference in the number of seals lost by killing them with shotguns and by taking them with spears. The hunters that used shotguns lost more than one-half they shot, while the hunters that used spears seldom ever lost one that they hit. (Moses.)

It is generally conceded that the Indian hunters in the use of the spear seldom lose one they kill or wound. (Morris Moss.)

When I was a boy I used a shotgun for taking seal, bought from the Hudson Bay Company at Fort Simpson, and have always used a shotgun for sealing. I think about two out of ten seal shot are lost. (Smith Natch.)

Sometimes I lose two and sometimes three seal out of ten I shoot. (Dan Nathlan.)

It depends a great deal upon the weather as to the amount of seals obtained by the hunters. After a heavy blow you see the seals lying on top of the water asleep, and you can get very close to them, and on an average you would get two or three out of every five or six you kill or wound, while in rough weather you would not get one out of five or six killed or wounded. (John O'Brien.)

Not being hunters of experience, our men lost about two-thirds of all the seal shot. Good hunters would not lose to exceed 25 per cent. (Nelson T. Oliver.)

We used shotguns, using buckshot, and I have known twenty shots to be fired at a seal before we got her. When we shot at "sleepers" we got a good many more than when we shot at "bachelors" or "rollers," and we secured on an average about one out of every three killed and wounded. The percentage of loss of those killed and wounded is fully as great as I have stated. (John Olsen.)

When the rifle is used less than one seal for five shots is secured; many shots miss, but of those seals hit about one-half are secured. (W. Roberts.)

It is very hard to estimate the number lost of those shot, but I should judge an expert hunter would lose certainly from 40 to 60 per cent, and a hunter not particularly expert would lose from 80 to 85 per cent. (L. G. Shepard.)

In some instances we ran upon schools of seal and shot five or six, all of which would be lost; in other instances we would secure about one-half of those wounded. One-half of all seals shot on the coast are lost. (William Short.)

About 25 per cent are lost when shot with a shotgun, and more are lost when shot with rifle. Shotgun and rifle are used by me for taking seal. (Jack Shucky.)

When I used a spear none were lost that were struck. When shotgun is used nearly 50 per cent are lost; when rifle is used a still larger percentage is lost. (Martin Singay.)

No seal were lost when struck with spear or arrow. Fully 50 per cent of seal shot with shotgun are lost, and a much larger per cent are lost when shot with a rifle. (Jack Sitka.)

Always use a Hudson Bay gun to take seal with. A Hudson Bay gun is a single-barreled shotgun. Sometimes I lose one and sometimes two out of ten that are shot. (Thomas Skowl.)

I think about one-third of the seal shot with shotgun are lost. (Fred Smith.)

Very few are lost when struck with a spear. About 66 per cent are lost when shotgun is used. (William H. Smith.)

An ordinary hunter will not get more than one out of four that he shoots at. (Cyrus Stephens.)

About 25 per cent of seals shot are lost. (Joshua Stickland.)

Q. What percentage of seals are taken compared to those you destroy in doing so; in other words, how many do you actually get out of those you shoot?—A. I guess we get hardly two-thirds of what we shoot.

Q. Is it not a fact that when you first started in the business and was inexperienced in hunting, that you, like all other beginners, destroyed a much larger proportion than you do now?—A. It is.

I have always understood that 33 per cent of seals shot with shotguns are lost. (W. Thomas.)

The hunters use shotguns and rifles exclusively for taking seal. I think that from what I have been able to learn about half the seal shot are lost, the hunters being unable to secure them before they sink. (John C. Tolman.)

About 60 per cent of the seal shot with shotgun are lost. A much larger per cent is lost when rifle is used. (Peter Trearsheit.)

I get most all the seals that I hit with the spear. I lose one-half of those I shoot with a gun. (John Tysum.)

When the spear was used all seal speared were secured. About 50 per cent of the seals are lost when shot with shotgun. Whenever I have used a rifle for shooting seal a much larger proportion of those killed have been lost on account of shooting them at a longer distance from the boat. (James Unatajim.)

I think I generally lose about 75 per cent of the seals shot with shotgun. (George Usher.)

I have learned from personal observation and from conversations with parties that they lose in killed and wounded at least two out of every three obtained. Other sealers have told me that their loss is much greater. (Charles T. Wagner.)

About 50 per cent are lost when shot with shotgun. When rifle is used a much larger per cent is lost. (Rudolph Walton.)

I have often conversed with the hunters relative to the percentage of the loss of seals to those taken, and some tell me they get one out of five or six. (Elkan Wasserman.)

My hunters use shotgun exclusively. They carry a rifle with them in the boat, but have not used one this season to my knowledge. I think, as near as I can estimate, about 33⅓ per cent of the seals shot are lost. (P. S. Weittenhiller.)

From my knowledge and experience in the business it is my conviction that within the last few years, since the sealers have become so numerous in the Pacific and Bering Sea, that not more than one out of three are secured. (Michael White.)

I always use the shotgun for taking seal. I think I lose about five out of every ten that I shoot. (Billy Williams.)

That for every three sleeping seals killed or wounded in the water only one is recovered. For every six traveling seals killed or wounded in the water only one is recovered. (Theo. T. Williams.)

Sometimes I lose one and sometimes two out of ten that I shoot with a shotgun. (Fred. Wilson.)

When the spear was used very few seal were lost. About 50 per cent are lost when shot with shotgun. A larger per cent are lost when killed with a rifle. I use the shotgun for taking seal, and lose about two out of ten that I shoot. (Billy Yeltachy.)

Sometimes I lose one and sometimes two out of every ten that I shoot. I always shoot the seal close to the boat, so I don't lose many. (Hastings Yethnow.)

The shotgun is used altogether for taking seal. About 33⅓ per cent of the seal shot are lost. (Alf. Yohansen.)

Always use shotgun for taking seal. I lose but very few seal, as I always shoot them very close to the boat. (Paul Young.)

Have always used the shotgun for taking seal. Think I lose about three out of ten of those I shoot. (Walter Young.)

In hunting with spears I capture nearly all that I hit. (Thomas Zolnoks.)

Wounding.

Those only wounded, whether fatally or otherwise, dive and escape capture. The less severely wounded may, and in many cases doubtless do, recover from their wounds; but, in the nature of things, many others must die of their injuries. There is a wide range of chances between an instantaneously fatal or disabling shot and a slight wound from which the victim may readily recover, with obviously a large proportion of them on the fatal side of the dividing line. (Dr. J. A. Allen.)

A good many of the seals that I have caught in the last three or four years have shot in them and some have been badly wounded. I have seen white hunters shooting seals out in the sea, and they lose a great many more than they get, and we sometimes capture some of those that they have badly wounded. (Bowa-chup.)

Have caught a great many seals that had shot in them. (Peter Brown.)

We often take seals that have been wounded with a rifle or shotgun, and in their bodies there are a large number of shot. (James Claplanhoo.)

A good many are wounded and escape, only to die afterwards. (Alfred Dardean.)

When I get seals now a great many have shot in them, a thing I never saw before until about six or seven years ago. (Frank Davis.)

Some that I shoot are wounded and get away, and probably die. I have caught a good many seals that had shot in them. (Ellabush.)

They kill and wound a great many that they do not get. I have speared a great many seals that had shot in them. (Selwish Johnson.)

I know that a great many must be lost by the white hunters, for a great many that I catch have shot in them, and some are badly wounded. (James Lighthouse.)

During the killing season on the Commander Islands we frequently find in the bodies both bullets and shot. (John Malowansky.)

I have captured a great many seals with the spear and found shot in them. (John Tysum.)

When they were wounded we had to chase them, and then sometimes would not get them. (Patrick Maroney.)

While out seal hunting last year I captured a few seals that the white hunters had wounded and lost, and found a good many shot in their bodies. I have captured a good many seals lately that had buckshot in them. (Charley White.)

At the times when the male seals are on the rookeries the large catches are made. A traveling seal is alert, cautious, quick of hearing, and easily disturbed. A sleeping seal is at the mercy of anyone. The large proportion of traveling seals shot at and lost is due to the timidity of the animal; in fact, all the hunters admit that when there is much shooting going on the seals are very difficult to get. The loss of sleeping seals, which I estimate as two lost for one saved, is due to the fact that unless the bullet or shot kills the animal instantly it will immediately dive, and it is not easy to kill a seal instantly. The head of the seal affords but a small mark. Even in the case of a sleeper, the motion of the water keeps it moving. The boat from which the hunter shoots is also moving, and while there are men who at a distance of 50 or 60 yards can shoot a small object under such circumstances, they are extremely rare. They are famous as experts and they are highly rewarded for their skill. Certainly not one in ten of all the seal hunters can truthfully assert, nor do they attempt to do so when in a confidential humor, that they kill 50 per cent of their seals dead. I was in the company of a number of them in Victoria, in 1889, and heard them talking among themselves of their prowess. Some put forward claims which the others derided. Any estimate in excess of the one I have already given called forth uncomplimentary remarks and charges of boastfulness. The disinclination of these men to state the absolute facts, and they alone know what the facts are, in relation to the number of seals shot and lost, has been intensified lately by the feeling that it is necessary for them to make a good showing to back up the claim that pelagic sealing is not absolutely of the seal herd. (Thomas T. Williams.)

Many of the seals I have speared had shot and bullets in them. This was never seen before until about eight years ago, and now it is a frequent occurrence. (Wispoo.)

A great many that I have caught in the last three or four years have shot in them, and many have been badly wounded. (Thomas Zolnoks.)

Percentage lost—General statements.

We had a row on board because some of the hunters were green hands and the men would not go out in the boats with them. They took the hunters out of our boats and put them into the other boats that made no catch, and then we kicked that they should put the green hunters into our boats, because everything they would shoot would sink on them and be lost. (Charles Adair.)

The destructiveness to seal life by pelagic hunting is very great. The majority of seals killed are pregnant females, so that two lives are often sacrificed in securing one skin. This is true whether firearms or spears are used. In addition to this, the number of skins marketed does not represent the number actually destroyed, for many are killed that are not secured, while others, though fatally wounded, still possess strength enough to escape their pursuers. (A. B. Alexander.)

Of those killed the number saved varies with the skill of the hunters. Last year we lost very few. (Charles Avery.)

A very few are lost when shot with the shotgun, as we shoot them close to the boat. (Johnny Baronovitch.)

In hunting with the spear we don't lose many that we hit. I never hunted with guns. (Peter Brown.)

Experienced hunters lose very few seal that are shot, but beginners lose a great many. (Charles Campbell.)

As to the percentage of seals lost in pelagic sealing where the use of firearms is employed, I am not able to state of my own observation and experience, but from conversation with those engaged in the business I am of the opinion that the number secured is small compared with those lost in attempts to secure them. (W. O. Coulson.)

None were lost when the spear was used. When the shotgun is used sometimes they are lost. A few more are lost when rifle is used. (Charlie Dahtlin.)

Were I engaged at present in sealing I should prefer the spear to the rifle or shotgun, and I believe its use is not near so destructive to seal life. (James Dalgarduo.)

The Indians have always hunted seal with a shotgun, and I am sorry to say that they have killed a great many more than they secured. (William Duncan.)

From the ammunition we furnished them I learned that some of the hunters on an average used from two to three rounds of shot to a seal, while others used from forty to fifty rounds. (George Fogel.)

Have always used the shotgun for killing seal, and but very few are lost. (Chief Frank.)

Have always used spear for taking seal, and but very few are lost. (Obad George.)

A very large number of shots are thrown away. In the case of the *Thistle*, in her voyage of 1891, she brought in but nine skins, while her hunters had fired away 260 pounds of shot. She had poor hunters. (E. M. Greenleaf.)

That in pelagic sealing twice as many seals are lost as are captured. (W. P. Griffith.)

Always shoot the seal close to the boat and rarely lose one; but when shot at with the rifle I lose a good many. (Hooniah Dick.)

I have always used spears in hunting seals, and seldom wounded or hit one that I did not get until in 1891, which year, and the only one, I went to Bering Sea, and used the shotgun part of the time. I found in the use of the shotgun that a great many of the seals that were killed or wounded were lost. (Alfred Irving.)

We lose but very few seals that we hit with a spear. (Selwish Johnson.)

When seals were struck with a spear none were lost; a great many are lost when the shotgun is used. (O. Klananeck.)

I have often heard them say that they only get two or three out of a school, and when they kill them, if they do not get them right away, they will sink and be lost. Further, that they lose a good many that they kill. (James Lafkin.)

Q. Do you generally shoot seals with a rifle or shotgun?—A. A shotgun. Ninety per cent are killed with a shotgun. (Frank Moreau.)

Always use the shotgun for taking seals. I lose very few, as I always shoot them close to the boat. (Matthew Norris.)

I can not say how many seals are killed and wounded, but there is no doubt that green hunters lose many, while those more experienced in business lose fewer. (Morris Moss.)

We used the spear more than the gun and secured nearly all of them that we hit with it, but lost a great many seals that we shot. We prefer to use the spear, because in so doing we do not lose so many or frighten them away. (Osly.)

The shotgun is not as fatal as the rifle, but it ruins the skins of the seals. (Adolphus Sayers.)

Breech-loading firearms (rifles and shotguns) are the instruments principally employed by pelagic fur-seal hunters, both native and white. By means of these weapons a greater number of skins are secured in a season than when spears are used; but the proportion of seals struck and lost to those actually secured is much less than when the spear is used. (John W. Smith.)

The best hunter will fire about 20 cartridges, and they get 10 or 12 seals, while a hunter of less experience will fire 100 rounds and get nothing, but will wound and disable them. (Adolph W. Thompson.)

I have always used spears in hunting the seal, and seldom lose, any I hit. (Charley White.)

In attempting to determine the sex of seals killed in the Bering Sea and the North Pacific, and of the number of seals killed in excess of those actually secured by the hunters, I had interviews with upward of fifty seal hunters, aside from interviews subsequently had with Indian hunters. I find this portion of my work by far the most difficult. Much discussion had already been had about the damaging effect of pelagic sealing, and the hunters were loath to tell how many seals were killed and not recovered, and were often averse to making truthful reports about the sex of the animals killed, but by frequenting their haunts and cultivating their company for long periods I succeeded in getting accurate statements from a number of them. (Theo. T. Williams.)

I found that at first the hunters were disposed to brag of their skill

and to overestimate their success in securing skins of seals shot at.
The reason for that was that I was about to engage in sealing enter-
prises and that I was making inquiries for the purpose of ascertaining
their skill as hunters, with a view to engaging them. The practice in
British Columbia is to pay the best hunters the highest rate per skin.
Men who could shoot fairly well, but who use a shotgun, could be
secured for a sealing voyage from $1 to $1.50 per skin, while hunters
who shot with a rifle and were of recognized skill in some instances
were paid as high as $2.50 per skin, and, generally speaking, as high
as $2 per skin. The reason for this is obvious to those who have inter-
ested themselves in the sealing business. A seal killed with buckshot
is so much punctured frequently that the pelt is of lesser value. It is
not profitable for schooners to engage as hunters men who miss their
chances of killing the seals and blaze away indiscriminately, with small
results. Even though the hunter is only paid for the skin he recovers,
the loss to the vessel by his failure to kill when an opportunity offers
is equivalent to the profit it would have made on the skin if secured.
For these reasons and on account of the general proneness of men who
consider themselves experts in the use of any weapon to brag, the seal
hunters of British Columbia, as a class, grossly exaggerate the percentage of skins they recover to the number of seals aimed at, wounded, or
killed. (Theo. T. Williams.)

In attempting to ascertain exactly the number of seals killed and lost
by the Bering Sea hunters, I found a wide divergence of statement. It
is greatly to the advantage of the seal hunter to have the reputation of
losing but few seals. He is paid by the skin, and the more he catches
the greater his remuneration; but that is not all. The hunter with the
best reputation as a sure catcher is in the greatest demand, can secure
employment in the best schooner, and the largest sum of advance
money. Besides self-interest, there comes vanity to urge the hunter
to make the biggest reputation possible for himself. To use a common
expression, the seal hunters all brag about their sureness of aim. The
best shots use a rifle and fire at a range of from 50 to 125 yards. The
poorer shots depend on a shotgun loaded with buckshot, and will fire
at a seal up to 50 yards away. The Indian hunters use spears, and
paddle noiselessly up to the sleeping seal to plunge the spear in its
shoulder. They never attempt to spear a seal that is awake. An
Indian hunter will paddle in among a lot of "sleepers" and spear them
one after the other, while a white hunter, who uses firearms, alarms
every seal in the neighborhood at the first discharge. The Indians lose
about one-third of all they spear, either from failure to kill when they
strike or because the dead seal sinks too quickly for them to secure it.
The white hunters do not get one-half of all they shoot. Some hunters
are very careful shots and will not fire unless the seal is well within
range, but the seal is likely to sink before the boat can get to it, or if
wounded, will dive like a flash to get away. A number of hunters have
boasted that they secure 95 seals for 100 shots, and some have made
affidavits of even more wonderful exploits. They presume too much
on public ignorance and credulity. (Theo. T. Williams.)

Fortunately, it is not necessary to depend on the statements of the seal
hunters. I secured access to the ship accounts of several sealers, and
found that in every case the consumption of ammunition showed more
than ten cartridges used for every seal skin captured. I spent considerable time among the Siwash Indian sealers, and, while they brag of
their individual prowess, they admitted a loss of 30 per cent at least.

On this subject I append a statement made by Captain Olsen, of the sealing bark *Bessie Ruter*, at Victoria.

Captain Olsen, of the American schooner *Bessie Ruter*, of Astoria, reached Victoria September 27, 1889. In the office of the American consul, Col. R. Stevens, he said: "I took 550 skins in Bering Sea. Of these 27 were pups, 520 females, and 3 male seals, which I killed off the island of Kadiak. Most of the female seals were with young. I had a green crew and green hunters. They used shotguns and sometimes the rifle. They got about one seal for every three they aimed at. Some they missed altogether, and some of the wounded ones got away. There is great risk of losing a traveling seal. The sleeping seals blow up an air bladder that keeps them from sinking, but the seals when awake sink easily. Hooks are used to grapple them, but if the boat is some distance from the seal when it is killed it does not often get it. For that reason rifle shooting at long range hardly pays. I will get about $7.75 for some of my skins and $8 for others. My voyage will pay, because I ran the boat on the cheap. I only had two men to the boat, and only paid my hunters $1 per skin instead of $2, which is paid to first-class hunters. Some very skillful hunters do not lose many skins. They will never fire unless a seal is at close range, and they generally kill. Of course they lose some from sinking. All the hunters brag about how few they lose, because they want the reputation of being good hunters. The better reputation they have the better chance they get.

If Bering Sea were open many new men would come into the business and the loss would be greater. Only a few men make successful hunters. It is like being a clever rifle shot. If the best hunters lose ten or fifteen in a hundred, the other kind lose ten times as many, if not more. Green hands will throw away a lot of ammunition, shooting at everything they see, whether it is in range or not. You can not stop them. They will wound more than they kill. (T. T. Williams.)

DESTRUCTION OF NURSING FEMALES.

We entered Bering Sea through the Muckawa Pass the 1st of July, and commenced hunting seals wherever we could find them, among which were a great many cows giving milk, which we killed from 30 to 150 miles from the islands. (Charles Adair.)

I have no exact information as to the proportion of male and female seals killed by pelagic hunters, but it is my firm conviction, from my knowledge of the habits of the males in not leaving the islands during the breeding season, and the well-known fact that mother seals go great distances in search of food while nursing their young, that the females are slaughtered in great numbers during their journeys to and from the islands by pelagic hunters. (George R. Adams.)

And when in Bering Sea we take seals from 10 to 120 miles from the seal islands. (William Bendt.)

And the larger proportion of those killed in Bering Sea are also cows. Have killed cow seals, with milk in them, 65 miles from the Pribilof Islands. * * * A few male seals are taken, ages ranging from 1 to 5 years. Once in a while we catch an old bull in the Pacific Ocean. (Martin Benson.)

We came out of Bering Sea the latter part of August and had caught about 1,700 seals between the Pribilof Islands and Unalaska. We caught them from 10 to 100 or more miles off St. George Island. (Niels Bonde.)

We entered Bering Sea the middle of May and captured 300 while in there. Most of these were mother seals with their breasts full of milk. (Thomas Bradley.)

·I hunted in Bering Sea in 1889 (that being the only year I ever went to that sea) and hunted seals with spears about 70 miles southwest off the islands, and our catch was nearly all cows that had given birth to their young and had milk in their teats. (Peter Brown.)

Have killed cows with milk about 60 miles off the Pribilof Islands. A few old bulls were killed by me last season. (Charles Campbell.)

At least seven out of eight seals caught in Bering Sea were mothers in milk. (Charles Challall.)

We entered the sea through the Unimak Pass, and captured therein about 40 seals, most all of which had milk in their breasts. (Louis Culler.)

The waters were full of them at that time. We caught them from 50 to 60 miles off the seal islands. (John Dalton.)

But the seals I caught in Bering Sea were most all cows in milk. (Frank Davis.)

The proportion of female seals killed in Bering Sea is equally large, but the destruction to seal life is much greater, owing to the fact that when a mother seal is killed her suckling pup left at the rookery also perishes. Impregnation having also taken place before she left the rookery in search of food, the fetus of the next year's birth is likewise destroyed. (James H. Douglass.)

We left San Francisco and fished up the coast until we entered Bering Sea, in July, and sealed about the sea until we were driven off by the revenue cutter *Corwin*. From there we went to the Copper Islands. Our whole catch amounted to 900 skins, and we killed most of them with rifles. We only got about one out of eight that we shot at, and they were most all females giving milk or in pup. When we cut the hide off you could see the milk running from the breasts of the seals. The second year we got over 1,300 skins; some of them were cows with pups in them, and most all the rest were cows giving milk, and some of the latter we killed as far from the rookeries as Unimak Pass. (Peter Duffy.)

We entered Bering Sea about April and we got 795 in there, the largest part of which were mother seals in milk. When we were skinning them the milk would run on the deck. (John Fyfe.)

I know that fully 75 per cent of those we caught in Bering Sea were cows in milk. (Thomas Gibson.)

My observation and the information obtained from seal hunters convince me that fully 90 per cent of the seals found swimming in Bering Sea during the breeding season are females in search of food, and their slaughter results in the destruction of her young by starvation. (M. A. Healy.)

While in Bering Sea we cruised around the Pribilof Islands in all directions, often coming within view of them but never landing or making any attempt to do so. The proportion of females taken to males was about 70 per cent, more than two-thirds of these being nursing cows, while the remainder were 2-year olds' and yearlings. On first entering the sea an occasional pregnant cow would be taken, but this was uncommon. Of the males taken in Bering Sea the numbers of yearlings and very young bachelors was about equal; no bulls were ever taken. (Norman Hodgson.)

Those that I secured in Bering Sea were nearly all females and had given birth to their young and were in milk. Our vessel captured about 460 seals at a distance of about 100 miles from the Pribilof Islands, most all of which were cows in milk. (Alfred Irving.)

We entered the sea and caught about a thousand there. We sealed all over on this side of Bering Sea, sometimes being over 150 miles off the seal islands, and sometimes we were closer. I did not pay any attention to the proportion of females, but I know we skinned a great many that were giving milk, because the milk would run from their breasts onto the deck when they were being skinned. We killed mother seals in milk over 100 miles from the seal islands. We generally shoot them when they are asleep on the water. * * * We caught between 300 and 400 seals on the coast and 600 in Bering Sea. We sealed on the American side of Bering Sea around the Pribilof Islands, anywhere from 10 to 150 miles off. The capture of 1890 was about the same in proportion to sex as the year before. (James Kean.)

We entered Bering Sea about the latter part of July and captured 260 seals from 20 to 100 miles off the seal islands. A large proportion of them were females nursing their young and their teats were large and full of milk. (James Kennedy.)

I have observed that those killed in the North Pacific were mostly females carrying their young and were generally caught while asleep on the water. (James Kiernan.)

The same day after a chase of an hour we were seized by the U. S. S *Mohican.* The total catch of seals at the time of seizure was 48, and at least 20 were females, the majority of which were in milk. All the seals were taken from 120 to 180 miles from St. George Island. (Francis R. King-Hall.)

When in Bering Sea we are usually from 50 to 150 miles from the Pribilof Islands. (Andrew Laing.)

I have killed females in milk in Unimak Pass, and even out in the Pacific Ocean, 200 miles from the land. (E. N. Lawson.)

In Bering Sea, where we obtained about 400 skins, males and females in about equal numbers were taken. The females were mostly nursing cows, while the males were young ones, between the ages of 2 and 5 years. (James E. Lennan.)

Another fact in connection with open-sea sealing is that the great majority of seals killed are females, and that a great part of the females are pregnant, or in milk. The milking females are most all killed while visiting the feeding grounds, which are distant 40 or 60 miles, or even farther from the islands. The female necessarily feeds so that she can supply nourishment for her young, while the males during the summer seldom leave the islands. This accounts for the large number of females killed in Bering Sea. (A. P. Loud.)

Q. Did you ever kill any seals later in the season that were giving milk?—A. Yes, sir. (Alexander McLean.)

Those we caught in Bering Sea were mostly all females with milk in their breasts. * * * The next season, 1890, we got on the way up between 100 and 200 seals, and then we entered Bering Sea about the

18th or 19th of July, and I caught 90 seals, mostly all females. * * * When we were in Bering Sea we hunted from 40 to 200 miles off the seal islands. (Patrick Maroney.)

About two-thirds of those caught in Bering Sea were females that had big teats and were giving milk. We could tell that when we were skinning them, because the milk would run out on the decks. (William Mason.)

We sealed around Unalaska, but did not go toward the Pribilof Islands. We caught 1,900 seals, all of which were captured in the sea, close to Unalaska; most all of them were cows in milk; but when we first entered the sea we killed a few cows that had pups in them. * * * That year we sealed east of the island and caught about 800 seals. I do not know how far we were from the islands, for we could not see them. The seals we caught were mostly cows with milk. (Moses.)

I was sealing in Bering Sea during July, August, and September, 1885 and 1886. I was cruising in Bering Sea around about the Pribilof Islands, and from 100 to 300 miles off. The principal portion of the cruising was between the Aleutian and Pribilof islands. One of the principal sealing grounds is off Bogslof. (Niles Nelson.)

After entering the sea we got one female with a very large pup, which I took out alive and kept it for three or four days, when it died, as it would not eat anything. All the others had given birth to their young, and their breasts were full of milk. (John Olsen.)

The seals taken in Bering Sea are nearly all grown. We get but very few young seals. I think we catch in Bering Sea more males in proportion to females than we do on the coast. We catch a good many females in Bering Sea that have given birth to their young on the islands and are in milk. I have caught plenty of cow seals in milk 100 miles or more from the islands, but seldom get any that have a pup in them in those waters. (William Parker.)

We entered Bering Sea about the 15th of August through the Unimak Pass and captured therein 1,404 seals, most of which were cows in milk. On that voyage we caught female seals in milk over 80 miles from the rookeries where they had left their young. (Charles Peterson.)

The seals captured in Bering Sea were fully 80 per cent females that had given birth to their young. A fact that I often noticed was that their teats would be full of milk when I skinned them, and I have seen them killed from 20 to 100 miles from the seal islands. (Edwin P. Porter.)

Q. How do you know that the marauders kill females principally?—
A. I know that the females, after giving birth to their young on the rookeries, frequent the open sea in search of food, whereas the males frequent the hauling grounds or waters immediately around it. At various times I have seen skins which were seized by the cutters from the poachers, and they were substantially female skins. (J. C. Redpath.)

I have been in Behring Sea but a part of one season. Of the seals taken, about one-third were males, one-third females with young, one-third barren and yearlings. (W. Roberts.)

I have taken nursing females when as much as 100 miles from Pribilof Islands. I estimate that the seals killed by pelagic hunters are at least 90 per cent females; this estimate is based on the great number of motherless pups I have observed on the rookeries, and also on statements made to me by many engaged in pelagic sealing whom I met and conversed with at Unalaska. (T. F. Ryan.)

We caught 767 seals in Bering Sea that year (1884) from 30 to 150 miles off the seal islands. The most of them were females, for the reason that they are not as cute as males. A great many of the females had their breasts full of milk, which would run out on the deck when we skinned them. * * * My third voyage was in 1889. I sailed from Yokohama on the *Arctic*, about the latter part of January. We cleared under the American flag. * * * We entered Bering Sea about the 17th of May and caught about 900 seals, the most of them around the fishing banks just north of the Aleutian Islands. The majority of them were mother seals. (James Sloan.)

The majority of seals taken in Bering Sea are cows with milk. But a very few yearlings are taken, and once in a while an old bull is taken. The male seals taken are between 2 and 4 years old. * * * I have taken female seals 80 miles off the Pribilof Islands that were full of milk. (Fred Smith.)

Have killed cow seals that were full of milk over 40 miles from the Pribilof Islands. (Joshua Stickland.)

I have never captured any cows in milk along the coast, but when in Bering Sea in 1889 I sealed off about 90 miles from the seal islands and caught cows in milk there. (John Tysum.)

The majority of seals killed in the water are females, and all the females killed in Bering Sea are mothers who have left their pups on the rookeries and gone some distance from the island in search of food. (Daniel Webster.)

Ninety-five per cent of all the seals killed in Bering Sea are females. (Theo. T. Williams.)

Thousands of the female seals were captured by the pelagic hunters in Bering Sea during the season of 1891, the most of which had to be secured quite a distance from the rookeries, owing to the presence of armed vessels patrolling the sea for miles around the islands. That the slaughter of the seals was mostly of females was confirmed by the thousands of dead pups lying on the rookeries, starved to death by the destruction of their mothers. (W. H. Williams.)

We caught a few seals in there (Bering Sea). When we first went in we did not see many, but after we were there awhile we saw plenty of them that had large breasts that were full of milk, and our catch were most all females. The average would be about one male to ten females. We killed cows in milk 150 miles from the seal islands. (John Woodruff.)

DESTRUCTION OF FEMALE SEALS.

Testimony of pelagic sealers.

My experience in seal hunting is that a much greater number of females are taken at sea than males of the fur-seal species; and of the females the majority are pregnant or milking. (Andrew Anderson.)

Q. Do you know of what sex the seals were that you have taken in the Pacific and Bering Sea?—A. Yes; I have taken both male and female seals, but I suppose the greater per cent that I have taken would be about 90 per cent, or even more.

Q. What percentage of the skins you have taken were cows?—A. About 90 per cent, for the simple reason that the bulls are not migra- tory. (George Ball.)

Most all the seals taken by me have been cows. I think cows sleep more and are more easily approached. Never killed but seven old bulls on the coast of Washington in my life, but have taken a few pups every year. (Wilton C. Bennett.)

Think the majority of the seals taken are cows. Never killed but two old bulls in my life. Have killed quite a number of yearling seals and some young males 2 or 3 years old. (Edward Benson.)

Q. Do you know of what sex the seals were that you have taken in the Pacific and Bering Sea?—A. Mostly females.
Q. What percentage of the skins you have taken were cows?—A. About 80 per cent. (Daniel Claussen.)

From my experience, observation, and conversation with seal hunters I am of the opinion that fully 75 per cent of their catch are females. (Leander Cox.)

I saw one schooner's catch examined at Unalaska in 1889, and there were found a large percentage of female seals among them. (M. C. Erskine.)

Of those taken probably four out of five are females. (F. F. Feeny.)

The seals taken by them (the *C. H. White* and the *Kate Manning*) were nearly all females. (George Fogel.)

Have never killed an old bull in my life, nor have seen one the last few years. (Luke Frank.)

Q. Do you know of what sex the seals were that you have taken in the Pacific and Bering Sea?—A. The majority of them are females. Last year I killed 72, and out of the 72 there was only 3 males.
Q. What percentage of the skins you have taken were cows?—A. About 90 to 95 per cent. (Luther T. Franklin.)

Off Cape Flattery there is hardly a dozen large males taken out of every thousand large seals whose skins are called first class; all the males taken here are small ones. (Thomas Frazer.)

Q. Do you know of what sex the seals were that you have taken in the Pacific and Bering Sea?—A. About 90 per cent of them were females.
Q. What percentage of the skins you have taken were cows?—A. About 90 per cent. (Edward W. Funcke.)

We caught about 160 seals before entering the sea. Over 100 of them were cows. (John Fyfe.)

Caught 1,500 seals on that voyage. We caught some a little ways from Victoria, and on the way up to Bering Sea, but the most of them, about 1,200, we caught in Bering Sea. I was told by the men that they were nearly all females, and I thought so, too, from the milk that I saw in their breasts when they were on the deck. I saw over a hundred little pups taken from the seals, which they threw overboard. (George Grady.)

To the best of my knowledge and belief about seven of every ten seals killed in pelagic sealing are females. (W. P. Griffith.)

Q. What sex are the seals taken by you or usually killed by hunting vessels in the North Pacific or Bering Sea?—A. Mostly females. The biggest percentage, I think, are females.

Q. What percentage of them are cows?—A. I couldn't tell you.

Q. Out of 100 seals that you would catch ordinarily what part of them would be cows?—A. I am under oath, and I could not tell you exactly. All I can say is the greater portion of them. (Charles H. Hogman.)

Think the seals taken by me have been about equally divided between females and males. Have taken a number of yearlings and some 2 and 3 year old males. Have never killed an old bull. (Henry Haldane.)

Q. Do you know of what sex the seals were that you have taken in the Pacific and Bering Sea?—A. Two-thirds of them are females.

Q. What percentage of the skins you have taken were cows?—A. Two-thirds, I should say. (William Henson.)

Q. Do you know of what sex that you have taken in the Pacific and Bering Sea?—A. The seals that I have taken were principally females.

Q. What percentage of them are females?—A. It is very seldom that you ever get hold of a male. (Gustave Isaacman.)

My experience has been that the sex of the seals usually killed by hunters employed on vessels under my command, both in the ocean and Bering Sea, were cows. I should say that not less than 80 per cent of those caught each year were of that sex. (James Kiernan.)

We caught about 400 or 500 seals before we got to Bering Sea. I don't know the precise number. They were bulls and females mixed in, but the general run of them were females. (William Isaac.)

Q. Of what sex are the seals taken by you or usually killed by hunting vessels in the North Pacific or Bering Sea?—A. Principally females.

Q. What would be your judgment as to the percentage? Out of 100 that you kill, how many of them would be females?—A. Say I would bring 2,000 seals in here, I may have probably about 100 males; that is a large average. (Alexander McLean.)

Q. What percentage of them are cows? Suppose you catch 100 seals, how many males would you have among them?—A. About 10. The seals killed by me were about half males and half females; have killed but one old bull in my life. I have killed quite a number of yearling seals, but never examined them as to sex. (Frederick Mason.)

Q. Do you know of what sex the seals were that you have taken in the Pacific and Bering Sea?—A. Mostly females.

Q. What percentage of the skins you have taken were cows?—A. I should judge about 90 per cent. (Frank Moreau.)

I can not give the exact estimate of the sex, but I know that a large portion of them are females. (Niles Nelson.)

In going up the coast to Unimak Pass we caught about 400 seals, mostly females with young, and put their skins on board the *Danube*, an English steamboat, at Alatack Bay, and after we got into Bering Sea we caught 220. We had 200 at the time the lieutenant ordered us out of the sea, the remainder we caught after. (John Olsen.)

We began sealing off Cape Flattery and captured about 300 seals along the coast, most all of which were females and yearlings. We did not capture over 50 males all told on this voyage. About 90 per cent of all the seals we captured in the water were female seals. We caught 350 seals along the coast, all of which were females excepting 20. (Charles Peterson.)

The majority of seal killed by me have been cows; have killed a few small males. (Showoosch.)

From what I have been able to learn, the majority of seals taken around Kodiak are females. (John C. Tolman.)

In my conversation with men engaged in seal hunting in the open water of the North Pacific and Bering Sea, I have not been able to get sufficient information to form a reliable estimate of the average number saved out of the total number shot nor of the percentage of females killed. As a rule, the hunters are extremely reticent about giving information on the subject to officers of the Government, but from the well-known fact that the female seal is much more easily approached than the male, and sleeps more frequently on the water, and is less active when carrying her young, I have no doubt that the female is the one that is being killed by the hunter. (Francis Tuttle.)

I believe the number they secure is small as compared with the number they destroy. Were it males only that they killed the damage would be temporary, but it is mostly females that they kill in the open waters. (Daniel Webster.)

I never paid any particular attention as to the exact number of or proportion of each sex killed in Bering Sea, but I do know that the larger portion of them were females, and were mothers giving milk. (Michael White.)

DECREASE OF SEALS.

Opinions of white sealers.

I have noticed a perceptible and gradual decrease in seal life for the past few years, and attribute it to the large number of vessels engaged in hunting them at sea. (Andrew Anderson.)

In the sea, seals are much more timid and make off as fast as possible at the approach of a vessel, while formerly they were usually quite curious and would sport and play about the vessel when come up with. I believe this decrease and timidity is due to the indiscriminate slaughter of the seals by pelagic sealers. (C. F. Anderson.)

Q. To what do you attribute the decrease?—A. I attribute the decrease to the indiscriminate slaughter of the seals. (George Ball.)

I believe that the decrease in fur-seal life, which has been constant of late years, is due principally to the number of vessels engaged in hunting them at sea. (J. A. Bradley.)

Seven or eight years ago, when seals were hunted almost wholly by Indians with spears, a vessel hunting in the vicinity of Cape Flattery was sure of getting several hundred skins in about three months, from March to the end of May, but at the present time a vessel is doing well if she gets a much smaller number, because the skins bring much higher prices. The records of "catches" in the last three or four years will confirm any person who examines them in the belief that the seals are decreasing in the Pacific Ocean on the American side. I have no reason to doubt that it is the same on the Russian side. At present they are hunted vigorously and with better methods than formerly. The hunters have had more experience and understand their habits better, but notwithstanding this, the catches are decreasing off the coast. (William Brennan.)

Seals were not as numerous in 1887 as they were in 1877, and it is my belief that the decrease in numbers is due to the hunting and killing of female seals in the water. (James L. Carthcut.)

Have noticed that seals are becoming very scarce on the coast the' last few years. The cause of the scarcity of the seals, I think, is too many schooners in the North Pacific Ocean and Bering Sea and the indiscriminate killing of females with pup in the water. (Peter Church.)

Q. Has there been any decrease in the quantity of seals as compared to the previous years?—A. I think there has. (Daniel Claussen.)

Q. If there is a decrease, to what do you attribute it?—A. To the killing and hunting of them by seal hunters. I think the indiscriminate killing of seals in Bering Sea is the cause of their scarcity along the coast. (John C. Clement.)

There were not nearly as many seals to be found in 1889 as there were in 1888. I think the decrease is caused by the great destruction of females killed in the sea by the hunters. (Peter Collins.)

I attribute this decrease [of the seals] to the terrible slaughter now going on in the sea. (Leander Cox.)

There can be but one cause for the scarcity of seals, and that is the indiscriminate killing of them in the water, and unless that is stopped the seals must soon be exterminated. The sea otter, which were plenty on this coast at one time, are now scarcely seen at all, and the indiscriminate slaughter of them in the water has almost entirely exterminated the animal. Some few remain in the far north, but they are very hard to secure. (William Duncan.)

Until hunting and killing were commenced by hunters in the open sea, I observed no appreciable decrease in the number arriving, which was about 1884. In my opinion the chasing of the seals and the shooting of them has a tendency to frighten them and disturb them, and prevents their increasing as they would if they were left undisturbed in the waters. (M. C. Erskine.)

The large decrease of seals in the waters of the ocean and sea must unquestionably be caused by the indiscriminate killing now going on by poaching schooners, and if not discontinued it will most certainly be a matter of a very few years before the seals will be exterminated. (M. C. Erskine.)

The seals have most decidedly decreased in number, caused by the continual hunting and killing in the open sea. (F. F. Feeny.)

I give them four years more, and if they keep on hunting them as they do now there will be no more seals left worth going after. I attribute the decrease in numbers to their being hunted so much. My experience is that the seal herds in the North Pacific and Bering Sea have been greatly depleted within the last few years by the constant pursuit and killing of them in the water by hunters. (George Fogel.)

In my opinion, seals and all other fur-bearing animals are decreasing, and the cause is pelagic hunting. (William Foster.)

Q. Has there been any decrease in the quantity of seals as compared with previous years?—A. I have not been on the islands in the last few years, but I should imagine there has been a great decrease.

Q. To what do you attribute the decrease?—A. To the number of vessels that are up there engaged in killing seals, nearly all of which are females. Last year there were 72 vessels fitted out from Victoria alone, to say nothing of vessels that are fitted out at other places. (Luther T. Franklin.)

The seals are not so numerous off Cape Flattery as they used to be some years ago, and it is my opinion it is owing to the constant hunting by so many schooners. (Thomas Frazer.)

Q. Has there been any decrease in the quantity of seals as compared with previous years?—A. There is a decrease of about 20 or 30 per cent.

Q. To what do you attribute the decrease?—A. I attribute it to their being overhunted. (Edward W. Funcke.)

I am decidedly of the opinion that fur-seal life has considerably decreased of late years, and believe it is due principally to pelagic sealing. (A. J. Gould.)

While at anchor off St. Paul Island, the pups playing about the vessel were very few, and while making a passage between Unalaska and Pribilof Islands, during the breeding season, did not see a dozen in the open sea during the whole trip, where formerly I met hundreds. In going from Unalaska to Atka and returning, during the last of May and the first part of June of this year (1892), I did not see a single fur seal in the water. I attribute this great decrease to the indiscriminate slaughter of the species by pelagic sealers and their wasteful methods of securing skins. (Charles J. Hague.)

Q. To what do you attribute this decrease?—A. Too many in the business, I suppose; too many after them.

Q. Would you attribute it to the killing of the females, and thereby there are not nearly as many born?—A. Certainly; it has got all to do with it.

Q. Then really you attribute the decrease to the killing of the females?—A. Yes, sir. (H. Harmsen.)

I am decidedly of the opinion that the decrease in numbers of seals in the North Pacific and Bering Sea is owing to pelagic hunting, and that unless discontinued they will soon become so nearly extinct as to be worthless for commercial purposes. (J. M. Hays.)

I think the seals are not near as plenty as a few years ago, and they are much more shy and harder to catch now than they were when I first went out sealing. I think this is caused by hunting them so much with guns. (James Hayward.)

Q. If there is a decrease, to what do you attribute it?—A. To the amount of seal hunters and hunting that is actually going on. (Andrew J. Hoffman.)

Seals have decreased very fast the last three years. The decrease is caused, I think, by the indiscriminate killing of seals in the water. (E. Hofstad.)

Q. To what do you attribute the cause?—A. Killing off the females. Whale killers and sharks kill a good many. (Gustave Isaacson.)

Q. To what do you attribute the cause of this decrease?—A. The increase of the fleet and killing off all the females. (Frank Johnson.)

My knowledge, being from long experience, is that the seals are becoming gradually scarcer in the northern waters, particularly so in later years. The cause of this decrease I believe to be the indiscriminate slaughter of the mother seals. They are hunted too much, and hence mother seals are becoming scarcer, which, if not checked, will lead to their early extermination. (James Kiernan.)

He also told me, from his own knowledge, that the Uchuckelset Indians had a few years ago caught off the coast 1,600 seals in a season, and that now they could catch hardly any; that the white men's guns were not only destroying the seals, but driving them farther from the coast. (Francis R. King-Hall.)

In my opinion, fur-seal life has not only enormously decreased in numbers since 1886, but it has become greatly scattered and grown wilder and more timid, forsaking many places where they were formerly to be found at certain seasons of the year engaged in feeding. This I attribute to the large number of vessels engaged in killing fur seals indiscriminately at sea. (James E. Lennan.)

If they keep on hunting them in Bering Sea and the North Pacific in the same way they have done in the last few years, they will exterminate them in the same way, because most all the seals killed are females. The young ones will all die, and every female seal you shoot makes the killing of two, because after the seal has given birth to her young the pup will starve to death on the land, or when you shoot them in the water they may have a pup inside. (Caleb Lindahl.)

I have observed a very great decrease in fur-seal life since 1885, and believe it is almost entirely due to the large numbers of vessels engaged in pelagic sealing. (E. W. Littlejohn.)

The seals are much less plentiful the last year I sealed than the first. I attribute this decrease to the hunting of them in the water and the increased number of boats and men engaged in the business in the last few years. (William H. Long.)

Q. Has there been any decrease in the quantity of seals as compared to previous years?—A. There has been a decrease.
Q. To what do you attribute the decrease?—A. To the hunting of the seals in Bering Sea. (Charles Lutjens.)

There can be but one reason for the decrease, and that is they are hunted too much in the open waters. (J. D. McDonald.)

There were not as many seals in 1890 as there were in 1889. I think there are so many boats and hunters out after them that they are being killed off. They are hunted too much. (William McIsaac.)

There were not as many seals as formerly. Have noticed the decrease in the last three years; caused, I think, by the indiscriminate killing of female seal. (James McKeen.)

I was also cod fishing in 1884. There were a great many more seals in the water then than there were in 1889. In 1884, when we were cod fishing, we met the steam whaler *Thrasher*, and I heard the captain remark that it was a damned shame the way they were killing the female seals in Bering Sea. (William McLaughlin.)

Q. To what do you attribute this decrease?—A. I think this is on account of killing those female seals when they have pups, and the business is getting so that so many vessels are going into it, and they are killing those pups off. A seal has not got a chance to go to work and increase.

Q. The mother seals?—A. Yes, sir. (Alexander McLean.)

Q. Have you noticed any decrease in the quantity of animals in the last few years?—A. Yes, sir.

Q. To what do you attribute the cause?—A. Killing off the females. (Daniel McLean.)

I have given up the sealing business because the slaughtering of the female seals is making them so scarce that it does not pay. (James Maloy.)

I think seals are not as plentiful as they used to be; caused, I think, by the indiscriminate killing of females with pup. (G. E. Miner.)

Q. To what do you attribute that decrease?—A. From the killing of seals, both by hunters and others. (Frank Moreau.)

Deponent further says that he thinks that the decrease in the number of seals found in the rookeries and the increase in the number of dead pups are caused directly by the open sealing in the sea, commonly called poaching. (T. F. Morgan.)

I am not able to say whether the seal herd is decreasing, but it is reasonable to suppose that where they are hunted and harassed at all times by so many hunters they are sure to be driven from their usual haunts, if not totally destroyed. (Nelson T. Oliver.)

Seals were not as plentiful in 1886 as they were in 1885. I think the principal cause of that decrease is on account of killing the females in the water, and also through their getting shy by being chased by the boats. (Niles Nelson.)

Since the use of rifles and shotguns has become common seals are much less in numbers and are more shy and timid. (William Parker.)

Seals are not near as plentiful as when I went out in 1888, and I believe the decrease is due to their being hunted so much with shotguns and rifles. (Edwin P. Porter.)

I know that the seals are rapidly decreasing, and I believe it is caused by killing females in the water. (Adolphus Sayers.)

I took very great interest in the seals, because I used to hunt them myself, and I noticed a great decrease in the number of seals from what there was formerly, when I was on sealing voyages. It was, in fact, so marked that I called the captain's attention to it, saying that we had seen very few seals. They have been getting scarcer every year since I have been going to Bering Sea, and if something is not done right away to protect them there will be no more seals in these waters. I know as a fact that they are killing them indiscriminately, and all the hunters care about it is to get a skin. I know something about it, as I have been sailing from this coast up along those waters for nineteen years, and, as I said before, I paid particular attention to them, and I firmly believe if they allow the killing in the sea to go on as they are now doing it will only be a question of a few years before there will not be enough to pay anyone to hunt them. (James Sloan.)

I think the seals are decreasing in number all the time, because there are more vessels out hunting after them and are killing off the female seals. (Cyrus Stephens.)

Q. If there is a decrease, to what do you attribute it?—A. On account of so much extermination and hunting by the seal hunters. (Gustave Sundvall.)

I have heard that seal have been decreasing the last few years; caused, I think, by pelagic sealing. (W. Thomas.)

The decrease, I think, is caused by the indiscriminate killing of female seals. (Rudolph Walton.)

From what I know seals have been decreasing very fast in recent years. Think the decrease is caused by the indiscriminate killing in the North Pacific Ocean and Bering Sea. (P. S. Weittenhiller.)

My experience is that the seals have been decreasing in numbers for the last six or seven years, and within the past two or three years very rapidly, owing to the indiscriminate killing of them by pelagic hunters and vessels engaged in that business in the waters of the North Pacific and Bering Sea. (Michael White.)

INCREASE OF SEALING FLEET.

Pelagic sealing as an industry is of recent origin and may be said to date from 1879. In 1880, according to the official report of the Canadian minister of marine and fisheries, 7 vessels and 213 men were engaged in pelagic sealing in the North Pacific, securing 13,600 skins, valued at $103,200. The same authority in 1886 20 vessels and 459 men secured 38,907 skins, valued at $389,070. In 1891 the number of United States and Canadian vessels had increased to over 100; upward of 2,000 men were engaged, and more than 62,000 skins were secured. (Report of American commissioners.)

The number of seal skins actually recorded as sold as a result of pelagic sealing is shown in the following table:

Year.	Number.	Year.	Number.	Year.	Number.	Year.	Number.
1872	1,029	1877	5,700	1882	17,700	1887	33,800
1873		1878	9,593	1883		1888	37,789
1874	4,949	1879	12,500	1884	a 14,000	1889	40,998
1875	1,646	1880	13,600	1885	13,000	1890	48,519
1876	2,042	1881	13,541	1886	38,907	1891	62,500

a Number estimated from value given.

One reason for deponent's opinion that the total number of seals in the Pacific and Bering Sea has diminished very rapidly is the fact— which deponent knows from the fact that he buys so large a portion of the poachers' catch—that there are now engaged in what is called "poaching" about 80 vessels, and that about five years ago not more than 10 vessels were engaged in poaching; that the total number of skins brought in by the whole 80 vessels is now not very much greater than the number brought in five years by 10 vessels. The poaching vessels a few years ago have been known to get as many as 3,000 or 4,000 skins, and deponent has bought 4,000 skins from one vessel, whereas no poaching vessel now gets more than a few hundred with the same size crew. One vessel last year sailing from Victoria made a catch of 1,900 skins, but this is now an altogether exceptional catch, and this vessel had a crew twice as large as poaching vessels formerly carried, and was equipped with from 12 to 15 boats instead of 5 or 6. One or two other poaching vessels also made large catches—that is, over 1,200 skins—but the average catch of the poaching vessels is not more than a few hundred each. This is true, although the poaching vessels are now equipped with much more experienced shooters, with better rifles. and with better boats than any of the vessels had five years ago. Many of the poaching vessels now have boats pointed at both ends, so that they can go backward and forward with equal ease; the old poacher only had ordinary ships' boats. Deponent knows this to be true because he has seen the boats and talked with the captains of the schooners about them. (Herman Liebes.)

I never saw many sealing schooners before 1884, but they have been coming more and more every year since, and I notice that as the schooners multiply in the sea the seals decrease on the rookeries. (Aggei Kushen.)

From 1885 to the present time the fleet of predatory vessels has constantly increased in proportion as the seal herd has decreased on the rookeries. * * * A very noticeable decrease in the herd commenced, as I have already pointed out, in 1886, and was coincident in time and proportionate in extent with the number of seals destroyed in the water. The business of pelagic sealing in Bering Sea first assumed considerable proportions in 1884, and in that year dead pup seals first became numerous enough upon the rookeries to excite remark upon the islands. As the sealing fleet increased the starved animals became more numerous. In 1887 fourteen vessels were seized for illegal sealing, and the effect was seen in the following year, when a much less number engaged in the business and the Bering Sea catch amounted, as I am informed, to about 34,000 skins against about 19,000 or 20,000 in 1888. The failure of the United States Government to vigorously pursue in 1888 and the following years the repressive policy so auspiciously begun in 1887, led to a large increase of the sealing fleet and corresponding destruction of the herd, but the prohibition of pelagic sealing nevertheless continued, and the usual proclamation was published by the Government warning all parties not to kill seals in Bering Sea or waters adjacent to the Alaskan coast. (H. H. McIntyre.)

Up to 1883 and 1884 it was only an occasional venturesome vessel that came around and secured a few hundred skins and thought itself lucky and cleared out, but since that time not even the smallest craft is satisfied unless it secures its thousands of pelts regardless of sex. (W. S. Hereford.)

While in Bering Sea during the summer of 1869 I never saw a vessel sealing about the islands or anywhere in the sea, nor did I hear any report of the presence of such sealing vessels in those waters. (J. A. Henriques.)

I do not know of any sealing schooner that went to Bering Sea until Captain McLean went there about nine years ago in the *Favorite.* (William Parker.)

Q. What effect, in your opinion, does the increase in the number of poaching vessels in Bering Sea have upon seal life?—A. Since the number of sealing vessels has increased the number of seals coming to the islands has correspondingly decreased. * * * In 1884 the sealing schooners became numerous. I believe there were about 30 in the sea that year, and they have increased very rapidly every year since, until now there are said to be about 120. (J. C. Redpath.)

I first went out in 1885, in the schooner *City of San Diego*, chartered by myself and others, and my catch for that year was between 2,300 and 2,400 seals. Of that number about 1,900 were caught in Bering Sea. There were but very few vessels sealing at that time. (Michael White.)

DECREASE OF SEALS—PELAGIC SEALING THE SOLE CAUSE.

Opinions of Indian hunters.

Fur seals were formerly much more plentiful, however, but of late years are becoming constantly scarcer. This is, we think, owing to the number of vessels engaged in hunting them at sea. (John Alexandroff.)

Fur seals were formerly observed in this neighborhood in great numbers, but of late years they have been constantly diminishing, owing to the large number of sealing vessels engaged in hunting and killing them. (Nicoli Apokchee.)

I have noticed that seal have decreased very rapidly in the last three years, owing to too many schooners engaged in sealing along the coast of Alaska and Bering Sea. (Adam Ayonkee.)

The seal are not near as plentiful as they used to be. The cause of the decrease is, I think, too many schooners hunting them off Prince of Wales Island and around Dixons Entrance. (Maurice Bates.)

Seal are not as plentiful on the coast as they used to be. They have been decreasing very fast the last few years. I think this is caused by the indiscriminate killing in the water. (Wilton C. Bennett.)

Seal are getting very scarce. I think the cause of the scarcity is too many people hunting seal. (Edward Benson.)

Seals were very plenty in the straits and around the cape until about six years ago, when the white hunters came in schooners and with shotguns and commenced to kill them all off, and now there is none in the straits and we can not get but one or two where we used to get eight or ten. They are very shy and wild, and are decreasing very rapidly. (Bowa-chup.)

White hunters came here about five or six years ago and commenced shooting the seals with guns, since which time they have been rapidly decreasing and are becoming very wild. When we hunt seals with spears we creep upon them while asleep on the water and spear them. A few years ago my people would catch from 8,000 to 10,000 each year; now we get only about 1,000 or less. * * * Seals used to be very numerous along the coast about Cape Flattery, and no decrease was ever noticed in their numbers until soon after the white hunters came around here—about seven years ago—and commenced shooting them. Since that time they have decreased fast and have become very shy. (Peter Brown.)

They were formerly much more plentiful than now, which is owing, we believe, to the number of vessels engaged in killing them at sea. (Ivan Canetak.)

Years ago seals were very plentiful from 5 to 10 miles from the shore. I could see them all around in bunches of from 10 to 20 each, but since the white man has commenced to kill them with the rifle and shotgun (in the last five or six years) they have decreased very rapidly. (Charlie.)

Fur seals have decreased very rapidly during the last five years, and we believe it is due to the large number of vessels engaged in hunting them at sea. (Vassili Chichinoff.)

Have noticed the seal are decreasing very fast the last four years; too many schooners are hunting them in the open waters of the Pacific Ocean and Bering Sea. (Chin-koo-tin.)

The last five years fur seal has been growing very scarce, and it is hard to get any now. There are too many white men with schooners hunting them off Dixons Entrance, and unless it is stopped the seal will be all gone. (William Clark.)

Seals are now very scarce and wild along the coast. I believe the cause of this is that white hunters have been hunting them so much with guns. (James Claplanhoo.)

Seals used to be very plentiful, and I never noticed any decrease in their number until white hunters commenced coming here and killing them with guns, about six or seven years ago; since that they have decreased very rapidly and have got very shy. Our tribe used to have no difficulty in catching 8,000 to 10,000 seals and now we can not get a thousand. (Circus Jim.)

I have been out sealing on the coast this spring in a schooner that carried 10 canoes, with two hunters to each canoe. We were out three days and caught 5 seals. If we had been out that long six or eight years ago with the same crew we would have taken between 60 and 100 seals. Seals are wild and shy now, and have become very scarce. I think the reason for this is that they have been hunted so much by white hunters who use firearms. (Jeff. Davis.)

Some years ago the fur seal were plenty off the islands, but since the schooners have hunted them they are nearly all gone, and it is hard for the Indians of this village to get any. (Eshon.)

Seals are not so plentiful as they were a few years ago. They began to decrease about five or six years ago. A good many years ago I used to capture seals in the Straits of San Juan de Fuca, but of late years, since so many schooners and white men have come around here shooting with guns, only a few come in here and we do not hunt in the Straits any more. I used to catch 40 or 50 seals in one day, and now if I get 6 or 7 I would have great luck. I have to go a long distance to get seals now. Seals are wild and afraid of an Indian. They have become so since the white men and the trader began to shoot them with shotguns and rifles. In a short time there will be no seals left for the Indian to kill with a spear. (Ellabush.)

Fur seals were formerly much more numerous than of late years, and are each year becoming constantly scarcer. I believe this decrease is due to the number of vessels which are engaged in hunting them at sea. (Vassili Feodor.)

When I was a young man there were lots of seals around Queen Charlotte Islands, but now they have become scarce. The last few times I was out after them I did not see a seal. They have been growing scarcer every year since the white man began hunting them in schooners. (Frank.)

Fur seal are not as plenty as they used to be, and it is hard for the Indians to catch any. I think there are too many white men in schooners hunting seals around Dixons Entrance. (Chief Frank.)

Since the white men have been hunting the seal with schooners they have become very scarce, and it is hard for the Indians to get any in their canoes. (Luke Frank.)

Seal have decreased on the coast very fast the last four years. The reason of the decrease is too much hunting and indiscriminate killing. (Chad George.)

The seal are becoming very scarce, caused, I think, by the white men hunting them too much. (Charles Gibson.)

Seal are becoming very scarce this last three or four years, and Indian hunters can hardly kill them now. Too many schooners are hunting seal, and Indian hunters have to go a long way in their canoes in order to get any, and they seldom kill one. (Gonastut.)

Have noticed that seals are decreasing the last four years, caused, I think, by too many white men hunting seal in the waters of the Pacific Ocean and Bering Sea. (James Gondowen.)

Fur seals have decreased in numbers of late years, and we believe it is due principally to the large number of vessels hunting them at sea. (Nicoli Gregoroff.)

The seal are not nearly as plentiful as they once were, and I think they are hunted too much by schooners. (Henry Haldane.)

Seals are not as plentiful now as they were before white men commenced hunting them with guns around here some six or seven years ago. They are more shy now and it is much more difficult for the hunters to creep up and spear them than it was a few years ago. (Alfred Irving.)

Years ago we could see seals all over the water. They are not so plentiful now. They have been growing less and less ever since the white man came in and began to hunt them with guns, about six or seven years ago, and so many vessels went into the business. (Ishka.)

My idea is that there are too many camp fires around on the coast of Alaska that scare the seals out to sea. The seal smell the smoke and wont come near the land; and there are a large number of people shooting seal, which scares them away also. (Jack Johnson.)

There are too many schooners hunting seal off Prince of Wales Island, and it is hard for Indians to get any in canoes. (Johnnie Johntin.)

Have noticed that seal are decreasing very fast the last few years along the coast, caused, I think, by pelagic sealing. * * * Think the seals are most all killed by the pelagic seal hunters in the waters of the North Pacific Ocean so far from the land that the Indians have no chance to get any in canoes, as they only go a short distance from the shore. (P. Kahiliday.)

Do not know why the number of the fur seal seen about these islands are now less than in former years. (Samuel Kahooroff.)

I think the seal are about as plentiful along this coast, but much more scarce farther west. The cause of this scarcity is too much pelagic hunting. (Philip Kashevaroff.)

When I was a young man the seal were very plentiful around here, but since the schooners began hunting them they have become very scarce. The white hunter destroyed the sea otter and will soon destroy the seal. I don't like to see the schooners around here hunting seal, for they kill everything they see, and unless they are stopped the seal will soon be gone. The sea otter is already gone. (King Kashwa.)

Seals have been growing scarcer the last five years, since the white man began hunting them with schooners, and if they are not stopped the seal will soon be all gone. (Jim Kasooh).

Seals have decreased very rapidly along this coast in the last three or four years. The decrease is caused, I think, by schooners using shotguns and rifles and killing mostly female seals. (Mike Kethusduck.)

The reason of the scarcity is, I think, that there are too many white hunters sealing in the open waters. (Kinkooga.)

Seal are becoming very scarce on the coast. The reason they are becoming so scarce is that hunters shoot them with guns and kill cows with pup. (O. Klananeck.)

Seal used to be plentiful, but now they are nearly all gone. They are too much hunted by the white men with schooners. (James Klonacket.)

Seal have become very scarce the last three years, and what few there are are very wild and hard to get at. I think the reason that seal have become scarce is that they are hunted too much, and too many females killed with pup. (Robert Kooko.)

Have noticed that seal are decreasing very fast the last few years. I think the cause of the decrease is that there are too many schooners hunting seal in Bering Sea and along the North Pacific Coast. (John Kowineet.)

Seal are not as plentiful as in former years; have noticed the decrease in the last three or four years. Think the cause of the decrease is the great number of schooners sealing in the North Pacific Ocean and Bering Sea. (George Lacheek.)

Seals are not nearly so plentiful now as they used to be. About seven years ago white men commenced to hunt seals in this vicinity with guns, since which time they have been decreasing in numbers, and have become wild and hard to catch. * * * Seals are not so plentiful and are more shy than they used to be, and are more difficult to catch, because they have been hunted so much for the last five or six years with guns. (James Lighthouse.)

White hunters in numbers commenced to hunt them around Cape Flattery with guns about six years ago, and since that time the seals have decreased very rapidly. (Thomas Lowe.)

Since the white man with schooners has been hunting seal they have been growing scarcer every year, and unless they are stopped the seal will soon be all gone. The Indians now have to go a long way and . suffer great hardships in order to get any. (Charles Martin.)

After careful inquiry among our oldest people and weighing my own experience and observations, I believe the decrease of the Alaskan fur seal is due altogether to pelagic hunting. (S. Melovidov.)

Since the schooners have commenced to hunt seal they are becoming very scarce, and the Indians have to go a long ways to get the few that they do. (Matthew Morris.)

Years ago seals were much more plentiful than they are now, and I could see them all around in bunches on the water, but since the white man came here and commenced to kill them with the rifle and the shotgun, within the last five or six years, they have rapidly decreased in number. (Moses.)

When I was a young man seals were very plentiful off Prince of Wales Island and Dixons Entrance, but since the schooners have begun hunting seals they have become very scarce, and Indians now are obliged to go a long ways to kill any, and sometimes they will hunt for days without getting a seal. (Nashtau.)

Since the white men with schooners began to hunt seals, the last five or six years, seals have become very scarce, and it is hard for the Indians to get any now. They have to go a long way and hunt a long time in order to get one or two seals. (Smith Natch.)

The last four or five years seals have been growing scarcer every year, owing, I think, to too many white men hunting seals in schooners off Queen Charlotte Islands and in Dixons. (Dan Nathlan.)

I think the reason of the seal becoming so scarce every year is that there are too many white men hunting seal in Bering Sea and the Pacific Ocean and it should be stopped. (Nechantake.)

Seals are not near as plenty as they used to be; too many hunters are catching them and indiscriminately killing them. (James Neishkaith.)

When I was a young man seals were much more plentiful than they are now. The last three years, since the schooners began hunting seals, they have become very scarce, and it is hard for the Indians to get any now. This year they have killed but two. (Nikla-ah.)

The Indian fur-seal hunters of my people all tell me that the fur seal are becoming very scarce; too many white men are killing them all the time, and they kill cows with pup, as well as other kinds. I am the chief of my people, and they all tell me what they know. (Peter Olsen.)

Seal are getting very scarce along the coast; cause of the scarcity is, I think, too many schooners hunting them off Prince of Wales Island. (Abel Ryan.)

Since the schooners have hunted seal off the Prince of Wales Island the seals have become scarce, and it is hard for the Indians to get any in canoes. In former times they used to get plenty. (Jack Shucky.)

The disappearance of the fur seal is due to the killing by pelagic seal hunters, who appear in large numbers off this part of the coast; and the scarcity of the fur seals is in proportion to the number of vessels engaged in seal hunting. (Alex. Shyha.)

Seal have become very scarce the last few years. Too many white men are engaged in killing seal. (Martin Singay.)

Have noticed a large decrease in seal in the last three years, caused, I think, by pelagic sealing in Bering Sea and the North Pacific Ocean. (Jack Sitka.)

Since the white men have been hunting with schooners they have become very scarce, and Indians are obliged to go a long way and stop away from home a long time in order to get any, and after being away there four or five days they frequently return without killing one seal, they have become so scarce. (Thomas Skowl.)

There are no seal left now; they are most all killed off. The last ten years the seal have been decreasing very fast—ever since the white men with schooners began to hunt them. (George Skultka.)

Seal have been growing scarce along the coast the last four years. Think there are too many schooners engaged in sealing in the North Pacific Ocean and Bering Sea. (M. Thlkahdaynahkee.)

Have noticed a large decrease the last four years. I think that pelagic sealing in Bering Sea is the cause of the seal becoming scarce along the coast. (Charlie Tlaksatan.)

Have heard all the Indians of different tribes say that seal are becoming very scarce in the last three or four years. They also say that unless the schooners are stopped from sealing in Bering Sea and the North Pacific Ocean the seal will all be gone, and none will be left for the Indians or anyone else. The seal have become so scarce of late years that I don't know much about them. (Twongkwak.)

During the last five or six years seals have decreased in numbers very rapidly. A great many of the white men are poor hunters and lose a great many of the seals that they shoot. They shoot, and shoot, and shoot, and don't get any seals, and that makes them wild, so that an Indian can't get near them with a spear. (John Tysum.)

Have noticed the seal have been decreasing along the coast the last four years. Think the cause of the decrease is that there are too many schooners engaged in pelagic sealing in Bering Sea. (James Unatajim.)

Last year was a very bad season. The Indians think scarcity of seals is due to the method of hunting them adopted by the whites, by which the seals are scared away. (Francis Verbeke.)

Have noticed the seal are decreasing very fast, particularly the last four years, caused by the indiscriminate killing of seal in the waters of the North Pacific Ocean and Bering Sea. (Charlie Wank.)

So many schooners and white men are hunting them with guns all along the coast that they are getting all killed off. (Watkins.)

Formerly the Indians hunted them for food, but nowadays white men and Indians hunt them for their fur, and they are rapidly diminishing in number. (Weckenunesch.)

Seals were always plenty in the Straits of San Juan de Fuca and along the coast until the white hunter came here and commenced shooting them some six or eight years ago; since that time they have decreased very rapidly. (Charley White.)

Seals are becoming very scarce since the white men began hunting them in schooners. (Billy Williams.)

Seals have become scarce the last three or four years, and the cause of it is, I think, the indiscriminate killing of seals in the water. (Fred Wilson.)

Seals are not near so plenty as they were seven or eight years ago. I think the cause of this is that they have been hunted so much by white hunters, who use shotguns and rifles. (Wispoo.)

Have noticed the seal are decreasing very fast, owing to so many schooners hunting seals in the waters of the North Pacific Ocean and Bering Sea. (Michael Wooskort.)

The seal, like the sea otter, are becoming very scarce. I think if the schooners were prohibited from taking seal in Bering Sea and along the coast of Alaska the seal would become plentiful and the Indians could kill them once more in canoes. (Yahkah.)

Since the white men with schooners began to hunt seal off Prince of Wales Island the seal have become very scarce, and unless they are stopped from hunting seal they will soon be all gone. If the white men are permitted to hunt seal much longer the fur seal will become as scarce as the sea otter, which were quite plenty around Dixons Entrance when I was a boy. The Indians are obliged to go a long way for seal now and often return after two or three days' hunt without taking any. (Hastings Yethnow.)

Seal have been decreasing very rapidly the last few years, and it is hard for our people to get them. There are too many white men hunting them with schooners off Prince of Wales Island. (Paul Young.)

Since the white man began to hunt seal they are becoming very scarce. (Walter Young.)

Within the last five or six years seals have decreased in number very fast and are becoming very shy, and it is difficult to creep upon them and hit them with the spear. Years ago the heads of seals along the coast would stick up out of the water almost as thick as the stars in the heavens, but since the white men with so many schooners have come and began to shoot and kill them with the guns they have become very scarce. (Hish Yulla.)

If so many white hunters keep hunting the seal with shotguns as they do now it will be but a short time before they will be all gone. (Thomas Zolnoks.)

DECREASE OF SEALS—RESULTS OF INDISCRIMINATE SLAUGHTER.

It is impossible to distinguish the sex of a seal in the water, unless it is an old bull. I am unable to state anything as to the proportion of females taken, but the seal hunter shoots every kind of seal he sees. (C. A. Abbey.)

I can not tell the difference between the male and female seal while in the water, except it be an old bull. (Peter Brown.)

I shoot all seal that come near the canoe and use no discrimination, as I can not distinguish a young bull from a cow in the water. All hunters shoot everything that comes near their boats. (Akatoo.)

No discrimination is or can be used; everything is game that comes within range of the hunter's weapon. (A. B. Alexander.)

It is impossible to distinguish the male from the female at a distance in the water. (H. Andricius.)

It is impossible to distinguish sex when the seals are swimming, and killing is indiscriminate. (Charles Avery.)

The sex of seal can not be told in the water. I shoot everything that comes near the boat. (Adam Ayonkee.)

I used no discrimination, but killed everything that came near the boat in shape of a seal. Never stopped to ask if it is female or not. A few old bulls have been taken by me. (Johnny Baronovitch.)

Everything that comes near the boat in the shape of a seal is shot, regardless of sex. (Maurice Bates.)

The sex of the seal can not be told in the water; I shoot everything that comes near the boat. (Wilton C. Bennett.)

We kill everything that comes near the boat, and use no discrimination, but shoot them regardless of sex. (Edward Benson.)

We kill everything regardless of sex; the sex of the seal can not be told in the waters. (Martin Benson.)

It is almost impossible to distinguish the female seals from the male in the water unless it is an old bull. (Bernhardt Bleidner.)

It is not possible to make any distinction between males (other than large bulls) and females of the fur-seal species at sea and there is none attempted. Full-powered bulls are, however, readily recognized at sea by their much larger size and darker fur; they are seldom taken, their pelts being comparatively valueless. The slaughter is therefore indiscriminate, the object being to secure all the skins possible. (J. A. Bradley.)

We used to shoot at anything we ran across, and got about a third of what we killed or wounded. I do not know how many miles off the seal islands we were when we caught them, as I did not know the distances. (Thomas Bradley.)

It is not easy to tell a bull from a cow or either from a year-old pup when they are in the water, and the hunters must shoot at all the seals they see. If they get them they are fortunate, for at the best many are lost. Some hunters rarely miss a seal they fire at, but many are wounded, and a seal with a charge of bullets and buckshot in him must be in very vigorous health to recover. Some hunters never miss a seal during the season, but if others get one out of four they wound they are doing well. (William Brennan.)

It is practically impossible to distinguish the age or sex of seals in the water while approaching them while at a reasonable gunshot distance from them, excepting in the case of old bulls. (Henry Brown.)

Use no discrimination, but kill all seal that come near the boat. The best way to shoot seal to secure them is to shoot them in the back of the head when they are asleep with their noses in the water. (Peter Brown.)

I can not distinguish male seals from female at a distance in the water, unless it be an old bull with a long wig. (Landis Callapa.)

I can not distinguish male seals from female in the water except in the case of an old bull, which is told by its size. Use no discrimination, but kill everything in the shape of a seal that comes near the boat. (Charles Campbell.)

There is no way of distinguishing the sex of fur seals (except large bulls) in the water at sea, nor do hunters ever make any effort to do so, but, on the contrary, kill all seals they can indiscriminately. (Vassili Chichinoff.)

Sex of the seal can not be told in the water unless it be an old bull. All seal are shot that come near the boat, regardless of sex. (Simeon Chin-koo-tin.)

It is impossible to distinguish the sex of the fur seal in the water at sea, and no effort was made to do so. We killed all fur seals indiscriminately. (Julius Christiansen.)

The sex of the seal can not be distinguished in the water. I shoot everything that comes near enough. (Peter Church.)

I am unable to tell a male seal from a female while in the water, unless it be an old bull with a long wig. (James Claplanhoo.)

The sex can not be told in the water, and all are shot that come near the boat. No discrimination is used; hunters kill everything they see. (John O. Clement.)

In pelagic sealing no distinction is made by hunters as to the sex of the seals, the killing being done indiscriminately. It is not possible to distinguish between the male and female seals at sea, even if a hunter so desired, and this is the reason why pelagic sealing will soon result in the total extermination of the species. (M. Cohen.)

The hunters will kill any seals that come along, it being impossible to tell the sex in the water. (Peter Collins.)

All seal are killed that come near the canoe, whether it is male or female; I make no difference. In former years there were lots of seal, but now there are very few. Too many schooners hunting them all the time in the water, killing the mother seals as well as others. (Charlie Dahtlin.)

We tried to shoot them while asleep, but shot all that came in our way. (Alfred Dardean.)

Use no discrimination in killing seal, but shoot everything that comes near the boat in the shape of a seal. Hunters shoot seal in the most exposed part of the body. (George Dishow.)

I can not tell the sex of the seal in the water. (Peter Duffy.)

I never examine them to know whether they are male or female seal. I can not tell the difference in the water, and shoot everything without knowing whether they are male or female. (Echon.)

While there is some difference in the appearance of the female and old male seals, I do not think it would be possible for the hunters to tell that difference in the sea at any great distance. (M. O. Erskine.)

Everything in shape of seal that comes near the boat is killed. (Chief Frank.)

I can not tell the sex of a seal in the water; use no discrimination, but kill everything that comes near the boat. (Luke Frank.)

There is no way by which hunters can distinguish sex while the seals are in the water, nor do we aim to do so; the killing is always done in an indiscriminate way. (Thomas Frazer.)

I could not tell whether a seal was a male or female while it was in the water, unless it was an old bull. (William Frazer.)

There is no way that I know of to distinguish the sex of a seal when it is in the water. No attempt is made to discriminate the sex so as to kill only males. (F. F. Feeny.)

Can not distinguish the sex of seal in the water, but spear everything that comes near the boat, regardless of sex. (Chad. George.)

I have never examined the seal as to sex. I shoot everything that comes near the boat, and use no discrimination whatever. (Charles Gibson.)

I kill everything that comes near the boat, and use no discrimination, as the sex can not be told in the water, except it be an old bull, which is told by its size. (Gonastut.)

Can not distinguish sex of seal in the water. Hunters use no discrimination, and kill everything that comes near the boat. (James Gondowen.)

We have no way of distinguishing fur seals in the water at sea as to whether males or females, and do not try to do so, but kill all we can indiscriminately. (Nicoli Gregoroff.)

Every seal is shot that comes near the boat, regardless of sex; hunters use no discrimination. (James Griffin.)

Among all other fur seals at sea no distinction is possible, and none is attempted. The killing is indiscriminate, the object being to secure all the pelts possible. Bulls are, however, readily recognized at sea by their larger size and darker fur. (A. J. Gould.)

I always shoot everything that comes near the boat; can not tell the sex in the water. (Henry Haldane.)

I use no discrimination in sealing, but shoot everything that comes near the boat, regardless of sex. (Martin Hannon.)

I can't tell a male from a female while in the water at a distance. (James Harrison.)

My experience has been that the vessels employed in hunting seals shoot, indiscriminately, pups, male and female seals, regardless of age or sex; and even should sealers wish to discriminate in the killing it would not be possible for them to do so. My study of them in a long experience has not enabled me to positively distinguish the sex of a seal while in the water. It is the custom to pay seal hunters per skins taken; hence it is the object of the hunters to secure as many as possible, without reference to sex, age, or condition. While hunting they use small rowboats, with two or three men in each boat armed with shotgun and rifle, chiefly the former, and it would be simply impossible for the master or owners, even should they desire it, to supervise ten or a dozen hunters as to the killing of any particular sex or kind. (M. A. Healy.)

It is difficult to tell the sex of a seal which you shoot at in the water; but you can tell an old seal from a young seal. (William Hermann.)

It is impossible to distinguish positively between females and males (other than large bulls) in the water at sea, and no effort is made to do so. Full-powered bulls are readily recognized by their great bulk and darker fur. The killing of the fur seals is therefore absolutely indiscriminate, as the object is to secure all the skins possible, irrespective of sex, age, or condition. (Norman Hodgson.)

Hunters use no discrimination in shooting seal, but kill everything that comes near the boat. They could not discriminate if they wanted to, as the sex can not be told in the water. (O. Holm.)

Everything in the shape of a seal that comes near the boat is killed. (Jack Johnson.)

I am unable to distinguish a male seal from a female seal at a distance in the water. (Selwish Johnson.)

I shoot everything in the shape of a seal that comes near the boat, and use no discrimination. (Johnnie Johntin.)

The sex of the seal can not be told in the water unless in the case of an old bull, which is told by its size. We use no discrimination in shooting seal. Everything is killed that comes near the boat, regardless of sex. (Philip Kashevaroff.)

We can not tell the difference between a male and a female in the water, but kill everything that comes near the boat. (King Kashwa.)

All killing of seals in the water must of necessity be indiscriminate slaughter, as it is impossible to tell the sex or the exact age of a seal until it has been taken into the boat, whereas on land careful discrimination can be made. (Francis R. King-Hall.)

Hunters use no discrimination in hunting seal, but shoot everything that comes near the boat. (Kinkooga.)

Hunters always kill all seal that come near the boat, regardless of sex. (O. Klananeck.)

I kill everything that comes near the canoe, regardless of sex. (Robert Kooko.)

I always kill every seal that comes near the boat; hunters use no discrimination. (John Kowineet.)

Have never killed but few old bulls in my life. The only seal that can be distinguished in the water is the old bull, which can be told by its size. Everything in shape of seal that comes near the boat is killed if possible, regardless of sex. (George Lacheek.)

We can not distinguish between the sexes of fur seals in the water at sea, nor do we try to. On the contrary, everything in sight is taken if possible, except large bulls, whose skins are worthless. (E. L. Lawson.)

It is impossible to distinguish between males and females of the fur-seal species in the water at sea, excepting large bulls, and no effort is made to do so. The object is to get all the marketable skins possible, and the killing is consequently indiscriminate. The pelts of large bulls, whose fur is coarse and of little value, and of yearlings of both sexes, whose skins are too small, not being strictly marketable skins, they were not taken. (James E. Lennan.)

Of late years most of the catches of Northwest skins are sold at a certain price per skin without particular examination. The dealers, knowing the location from which the skins are obtained, make an average price, and owners and hunters are, therefore, less particular than they were in former years as to the class of animals they capture. They kill everything they see without regard to age or sex, their only object being to swell the total number of the catch to the highest possible figure. (Isaac Liebes.)

But of course you could not tell when you shot a seal lying asleep whether it was a male or female. We shoot at all the seals when we get a chance, but it is only the ones that we find asleep that we catch. (Caleb Lindahl.)

It is impossible to distinguish the sex of fur seals at sea (excepting large bulls) and no effort is made to do so, the object being to secure all the skins possible; hence the killing is indiscriminate. (E. W. Littlejohn.)

It is impossible to tell the sex of a seal in the water. (William H. Long.)

Everything was killed that came near the boat; we did not use any discrimination. (George McAlpine.)

The sex can not be distinguished in the water unless it be the case of an old bull, which is distinguished by its size. Everything is killed in the shape of a seal that comes near the boat. (J. D. McDonald.)

When we find weather we are out in the boats killing all the seals we can get. We can not hunt in rough weather. (William McIsaac.)

Sex of seal can not be told in the water. We use no discrimination and kill all seal that come near the boat. Seal are not shot in any particular place; shoot them in the head if possible; if not, in the body. (James McKeen.)

It makes no difference if a seal is a male or female; we shoot everything that comes near enough. (Edward Maitland.)

I know it to be the custom of seal hunters to shoot seals at sea when they are at rest upon the surface of the water, and that those generally obtained are females, and constitute but a very small portion of those killed and lost. (John Malowansky.)

Everything that comes near the boat in shape of a seal is shot. I can not tell the sex of a seal till after it is dead. (Frederick Mason.)

We hunted with shotguns and shot them mostly when they were asleep on the water, or any chance we could get. I was a boat puller, and the hunters shot at everything in sight. (Henry Mason.)

We generally tried to kill them while asleep in the water, but fired at everything that came around us. (Thorwal Mathasan.)

I use no discrimination in shooting seals; shoot everything that comes near the boat, and all other hunters do the same. (G. E. Miner.)

Q. If awake, do you shoot them while breaching?—A. Yes, sir; we shoot at them anywhere, either while they are breaching or heads up, or any way. (Frank Moreau.)

We shot at everything in sight. We killed more females than males, and we lost a good many that we killed. (Eddie Morehead.)

Shoot everything that comes near the boat in shape of a seal, and use no discrimination. (Matthew Morris.)

The sex of the seal can not be told in the water. Hunters use no discrimination, but kill everything they can. (Nashtau.)

We shoot everything that comes near the canoe in shape of a seal, regardless of sex. The sex can not be told in the water unless it be an old bull. (Dan Nathlan.)

Everything is killed that comes near the canoe in shape of a seal. We can not tell a male from a female in the water. (Joseph Neishkaith.)

I can not tell the age or sex of a seal in the water. (Niles Nelson.)

I can not tell the difference between a male and female seal in the water, and I shoot every seal that comes near the canoe. (Nikla-ah.)

Sex can not be distinguished while the seals are in the water, nor do the hunters try to do so, for they kill everything they can shoot. (Nelson T. Oliver.)

I am unable to tell the sex of the seal while it is in the water, unless it be an old bull with a long wig. (Osly.)

It is impossible to distinguish the male seal from the female when they are in the water at a reasonable gunshot distance. (Charles Peterson.)

Yearlings are rarely taken in North Pacific. The age or sex of a seal in the water can not be distinguished, except that when close the apparent size is an indication of age. (W. Roberts.)

I use a shotgun to hunt for seal. Have lost very few seal, as I always shoot them near the boat. Everything in shape of a seal that comes near the boat is killed. I use no discrimination. (Rondtus.)

Everything in the shape of a seal that comes near the boat is shot. Hunters use no discrimination, but kill everything that puts its head above the water. (Abel Ryan.)

It is impossible to distinguish a male from a female seal in the water, except in the case of a very old bull, when his size distinguishes him. Therefore open-sea sealing is entirely indiscriminate as to sex or age. (L. G. Shepard.)

All seal are killed that come near the boat. I never stop to consider whether it is a male or female, but kill it off if I can. (Jack Shucky.)

Hunters use no discrimination, but shoot everything that comes near them. Their sex can not be told unless in the case of an old bull, which is distinguishable by its size. (Jack Sitka.)

The sex of the seal can not be told in the water. I kill everything that comes near my canoe in shape of a seal, and all other hunters do the same. (Thomas Skowl.)

Always shoot everything that comes near the boat in shape of a seal, regardless of sex. (George Skultka.)

Hunters use no discrimination, but shoot everything that come near the boat. (Fred Smith.)

It is impossible to distinguish between male and female seals at sea, even if the hunters so desired, except in the case of full-powered bulls, when they are readily recognized by their greatly superior size. Large bulls are rarely taken. No distinction is thought of by pelagic sealers, and the killing is done indiscriminately, the object being to secure as many skins as possible. (John W. Smith.)

I can not tell the sex of the seal in the water, unless he is an old bull. A hunter will blaze away at anything he sees in the water. (E. W. Soron.)

Hunters use no discrimination, but shoot everything in the shape of a seal that comes near the boat. (Joshua Stickland.)

All seals are killed that come near the boat, regardless of their sex. I never look to see whether I have killed a male or female seal until I have the seal dead in the boat. (M. Thlkahdaynahkee.)

Hunters use no discrimination in killing seal, but kill everything that comes near the boat, regardless of sex. (W. Thomas.)

The sex of the seal can not be told in the water when hunting. We use no discrimination, but kill everything in the shape of a seal that comes near the boat. (Charlie Tlaksatan.)

Hunters use no discrimination in taking seal, but kill everything that pokes its head out of the water near the boat. (John C. Tolman.)

The sex of the seal can not be told in the water. Hunters use no discrimination and everything in the shape of a seal that comes near the boat is killed. (Peter Trearsheit.)

Sex of seal can not be distinguished in the water, except in the case of an old bull, which can be told by its size. No discrimination is used in taking seal; everything that comes near the boat is shot at. (James Unatajim.)

I always shoot everything that comes near the boat, regardless of sex. We use no discrimination. (George Usher.)

Sex of seal can not be distinguished in the water. No discrimination is used in seal hunting; all are killed that come near. (Rudolph Walton.)

The sex of seal of same age can not be distinguished in the water. The only seal that can be distinguished is an old bull. We use no discrimination in seal hunting; everything is killed that comes near the boat. Pelagic hunters have become so plentiful and seals have become so wild that we are obliged to take long shots at them. (Charlie Wank.)

Our purpose and practice was to take all the seals we could get, regardless of their age or sex, without any discrimination whatever. (Michael White.)

Everything in the shape of seal that comes near the boat is shot. I can't tell the difference between a young cow and a male seal. (Fred Wilson.)

The seals are getting wild and hard to catch. There are a great many green hands in the business. We shot at everything that came along. We were getting 50 cents for every skin obtained. Our boats went 30 and 40 miles from the schooner. Sometimes they would leave in the morning at 5 and not return until the next day at 4 or 5 in the evening. (John Woodruff.)

The sex of seal can not be told in the water. No discrimination is used in seal hunting. All seal are killed that come near the boat. The only seal that can be distinguished in the water is an old bull. (Michael Wooskort.)

I can not distinguish the sex of a seal in the water, but kill every seal that comes near the canoe, if possible. (Billy Yeltachy.)

I can not tell the sex of a seal in the water, and use no discrimination, but kill everything that comes near my canoe in the shape of a seal. (Hastings Yethnow.)

We use no discrimination in killing seal, but shoot everything that comes near the boat. What seals we have seen this year are very wild and hard to get at. The cause of their being wild is the indiscriminate shooting of them in the water. (Alf. Yohansen.)

I use no discrimination, and kill everything that comes near the boat in the shape of a seal. (Paul Young.)

I can not tell the difference between a male and a female in the water; use no discrimination, but kill everything that comes near the boat. (Walker Young.)

We fired at all the seals we could, regardless of their sex. We got one out of every six or seven we shot at or killed. (George Zammett.)

DESTRUCTION OF PREGNANT FEMALES.

We caught about 185 seals, mostly females in young, and we killed them while they were asleep on the water. (Charles Adair.)

Most of the seals killed by me have been females with pup. (Akatoo.)

We sealed along the coast and captured 154. Most all of them were pregnant females. (Charles Avery.)

Most all seals that I have killed were pregnant cows. Have taken a few male seals from 1 to 4 years old, I think. Have never killed an old bull. (Adam Ayonkee.)

Q. What percentage of the cows you have taken were with pup?— A. About 99 per cent of the cows taken were with pup. There may be one in a hundred that is either without pup or has had one. (George Ball.)

Most all the seals taken are females with pup. (Johnnie Baronovitch.)

Seventy-five per cent of the seal taken on the coast are cows with pup. (Martin Benson.)

We left Port Townsend in May and sealed south to Cape Flattery, and then went north along the coast until we came to Unimak Pass, and captured from 300 to 400 seals. Most all were females and had pups in them. I think fully two-thirds of all we caught were females, and a few were bulls. * * * We secured 500 skins along the coast, most all of which were pregnant females. (Bernhardt Bleidner.)

I have never killed any full-grown cows on the coast that did not have pups in them, and I have hunted all the way from the Columbia River to Barclay Sound. (Bowa-chup.)

We left Victoria about May, going north, and sealed all the way to Bering Sea. We had about 60 before entering Bering Sea, nearly all of which were females with young pups in them. (Thomas Bradley.)

Our last catch of seals on the coast were almost exclusively gravid females. (Henry Brown.)

We had 250 seals before entering the sea, the largest percentage of which were females, most of them having young pups in them. I saw some of the young pups taken out of them. (Thomas Brown.)

On my last sealing cruise this spring we caught five seals; two of them were females and had pups in them; three of them were young and smaller seals and had black whiskers. None but full-grown cows have white whiskers, but young cows and young bulls have black whiskers. About half of all the seals captured along the coast have white whiskers and are cows with pups in them. Most all full-grown cows that are caught have pups in them. Once, late in the season, I caught a full-grown barren cow with white whiskers. (Landis Callapa.)

Seventy-five per cent of seals shot in the North Pacific Ocean are females heavy with young. (John C. Cantwell.)

Most of the seals we killed going up the coast were females heavy with pup. I think nine out of every ten were females. (Charles Challall.)

Not quite half of all seals caught along the coast are cows with pups in them. About half are young seals, both male and female, and the rest (a small number) are medium-sized males. We never get any old bulls worth speaking of, and we do not catch as many gray pups now as formerly. Have not caught any gray pups this year. Do not know what has become of them. Have never caught any full-grown cows without pups in them, and have never caught any cows in milk along the coast. (Charlie.)

Of those secured, the larger part by far were females, and the majority of these were pregnant cows. (Julius Christiansen.)

Most of the seals taken by me have been females with young. A few male seals have been taken by me, their ages ranging from 1 to 5 years old. Killed three large bulls during my life. (Peter Church.)

A great many years ago we used to catch about one-half cows and one-half young seals. I never caught any seals along the coast that had given birth to their young and that had milk in their breasts. I never captured any barren cows. * * * We secured ten seals in all, five of which had pups in them. I know this because I saw the pups when we cut the carcasses open. * * * The other five seals were smaller and probably male and female. (Circus Jim.)

About half the seals killed by me have been cows with pup. I never shot but two old bulls in my life. Have shot a few yearling seals. The young males I have killed were between 2 and 3 years old, I think. (William Clark.)

The seals we catch along the coast are nearly all pregnant females. It is seldom we capture an old bull, and what males we get are usually young ones. I have frequently seen cow seals cut open and the unborn pups cut out of them and they would live for several days. This is a frequent occurrence. (Christ Clausen.)

Q. What percentage of the cows you have taken were with pup?—A. About 70 per cent. (Peter Collins.)

The majority of seals taken are cows with pup; once in a while we take an old bull. A few yearlings are taken also. (Charlie Dahtlin.)

From 75 to 80 per cent of all the seals taken were mothers in young, and when cut open on deck we found the young within them. (James Dalgarduo.)

We had between 100 and 300 seals before entering the sea. Most all of them were females with pups in them. (John Dalton.)

Of the seals that were caught off the coast fully 90 out of every 100 had young pups in them. The boats would bring the seals killed on board the vessel and we would take the young pups out and skin them. If the pup is a good, nice one we would skin it and keep it for ourselves. I had eight such skins myself. Four out of five, if caught in May or June, would be alive when we cut them out of the mothers. One of them we kept for pretty nearly three weeks alive on deck by feeding it on condensed milk. One of the men finally killed it because it cried so pitifully. (Alfred Dardean.)

In all my experience in sealing on this coast I have killed but one cow seal that had milk in her breast, and that had given birth to her pup. I have killed a very few barren cows along the coast. Nearly all of the full-grown cows along the coast have pups in them. (Frank Davis.)

We sailed from San Francisco to Queen Charlotte Island, and caught between 500 and 600 seals, nearly all females heavy with young. I have seen a young live pup taken out of its mother and kept alive for three or four days. We sealed from 10 to 120 miles off the coast. (Joseph Dennis.)

A large proportion of all seals taken are females with pup. A very few yearlings are taken. Never examine them as to sex. But very few old bulls are taken, but five being taken out of a total of 900 seals by my schooner. (George Dishow.)

We left Victoria the latter end of January, and went South to Cape Blanco, sealing around there two or three months, when we started north to Bering Sea, sealing all the way up. We had between 200 and 300 seals before entering the sea, a great many of them being females with pups in them. (Richard Dolan.)

The Indians left their homes in March and remained away until May. Their hunting lodges were on some small islands outside of Dundas Island. From what they tell me the majority of seals taken by them have been females with young. (William Duncan.)

We went north to Bering Sea, sealing all the way up, and got 110 seals before entering the sea. Most of them were cows, nearly all of which had pups in them. We took some of the pups alive out of the bodies of the females. (George Fairchild.)

Most all of the females taken are with young, or mothers. (F. F. Feeny.)

There were cow seals with pup among the seals that I have taken, but don't know how many. I have never taken an old bull in my life. (Chief Frank.)

I think the seals taken by me are about half females with pup, and the rest are 1 and 2 year old males and yearlings; never examined the yearlings as to sex. (Luke Frank.)

Q. What percentage of the cows you have taken were with pup?— A. All that are killed in the Pacific are with pup, and those that are killed in Bering Sea have been delivered of pups on the islands and are with milk. (Luther T. Franklin.)

Q. What percentage of the cows you have taken were with pup?— A. About 60 per cent were with pup. (Edward W. Funcke.)

Most all the seals taken by me were females with pup. Most of the seals killed in Bering Sea have been cows with milk. Have never taken a bull seal off the coast of Washington, but have taken a few farther north. A few young males are taken off the coast of Washington. (Chad George.)

I did not pay much attention to the sex of seals we killed in the North Pacific, but know that a great number of them were cows that had pups in them, and we killed most of them while they were asleep on the water. (Thomas Gibson.)

Most of the seals killed are cows with pup. A few males are killed, averaging from 1 to 4 years old. Have killed but one old bull in my life. A few yearlings are taken, the majority of which are females. (James Gondowen.)

We captured 63 seals, all of which were females, and all were pregnant. With regard to pregnancy, I may note that the seals taken off the coast of Vancouver Island were not so far advanced as those taken farther north. * * * I am acquainted with the hunters and masters who sail from this port, and board all incoming and outgoing vessels of that class. These men all acknowledge that nearly all the seals taken off the Pacific Coast are females, and that they are nearly all with young. (E. M. Greenleaf.)

We began sealing off the northern coast of California and followed the sealing herd northward, capturing about 700 seals in the North Pacific Ocean, two-thirds of which were females with pup; the balance were young seals, both male and female. We captured between 900 and 1,000 on the coast, most all of which were females with pups. (Arthur Griffin.)

The catch was mostly females. Those we got in the North Pacific were females in pup, and those taken in Bering Sea were cows giving milk. (Joseph Grymes.)

Of the skins taken in this region fully nine-tenths are pregnant and milking females, but I never saw a young pup in the water. Large bulls were never taken, their skins being practically valueless. (A. J. Guild.)

Q. What percentage of the cows are taken with pups?—A. All the large ones have—all the grown ones have. Very seldom you find a barren one. (Charles H. Hagman.)

A large majority of seals taken are females with young. Only two old bulls were taken by me last year out of the 100 seals taken. But very few yearlings are taken. Paid no attention to sex. A few male seals are taken between 2 and 4 years old, I think. (Martin Hannon.)

Q. What percentage of the cows taken are with pups?—A. You can safely say about four-fifths of them. You get about 800 out of 1,000 seals. (H. Harmsen.)

I am told the white hunter kills mostly cow seals with pup. (Sam Hayikahtla.)

I have often conversed with masters, seamen, and hunters engaged in hunting the fur seals, and their statements to me have always been that the capture of a male seal was a rarity; that nearly all of their catch were cow seals heavy with young, or those who had given birth to their young on the islands and gone out to the fishing bank to feed, and that they lose a large proportion of those killed and wounded. (J. M. Hays.)

Q. What percentage of the cows you have taken were with pup?—A. At least 60 per cent were with pup. (William Henson.)

Of the seals secured in a season fully 70 per cent are females, and of these more than 60 per cent are pregnant and milking cows. The males taken are about equally divided in numbers between yearlings and bachelors from the ages of 2 to 5 years; bulls are seldom shot. (Norman Hodgson.)

Q. What percentage of the cows you have taken were with pup?—A. About the same amount (about 95 per cent) were with pup. (And. J. Hoffman.)

Most all seals taken are females with young. * * * A few male seal are taken. I would say they are generally 3 or 4 years old. A few yearlings are killed, mostly females. About five bull seal are killed out of every hundred taken. (E. Hofstad.)

About one-half of those caught along the coast were full-grown cows with pups in them; a few were medium-sized males, and the rest were younger seals of both sexes. I have never caught a full-grown cow in the straits or along the coast that did not have a pup in her. (Alfred Irving.)

Q. What percentage of the cows taken are with pup?—A. In the early part of the season, up to June, all the full-grown cows are with pup.

Q. Did you ever kill any cows whose young were born, and were giving milk?—A. That I don't remember taking notice of. I can not answer that question. (Gustave Isaacson.)

The female seals go through the passes from the Pacific Ocean into Bering Sea between June 25 and July 15. Females killed previous to this time I found with pups, but none with pups after that latter date. (Victor Jacobson.)

We began to seal when about 20 miles off Cape Flattery. We worked toward the northwest, and captured between 60 and 100 seals on the coast, about two-thirds of which were females with pup; the balance were yearlings consisting of male and female; after which we ran into Barclay Sound for supplies, from which place we worked to the northward toward Bering Sea. We captured about 80 seals while en route to the sea; about two-thirds of these were females with pup, the balance being yearlings about one-half male and one-half female. (James Jamieson.)

We began sealing off Barclay Sound, and caught three skins only, all of which were females with pup. * * * In hunting along the coast, I think about 80 per cent of those we caught were females, and most of them were carrying their young. We seldom caught any bulls, but caught a few of the younger males. I have seen the unborn young cut out of the mother seal and live for a week without food. We used to skin some, but threw most of them overboard. (James Jamieson.)

A majority of the seal taken on the coast are cows with pup. A few young males are taken, the ages ranging from 1 to 5 years. Once in awhile an old bull is taken in the North Pacific Ocean. (J. Johnson.)

Most of the seals taken are females with pup. Once in awhile an old bull is killed. (Jack Johnson.)

A large proportion of seals killed by me were cows with pup. Have killed a very few old bulls and some yearlings. (Johnnie Johntin.)

Most of the seal I have taken have been pregnant cows. But a very few young male seal are taken by me along the coast. (P. Kahiktday.)

The majority of seal are cows with pup. A few males are taken, about 4 or 5 years old. (Philip Kashevaroff.)

About half of the seals killed are females with pup. Have killed some yearling seals, but never killed any old bull. The young males I killed were between 2 and 3 years old. (Jim Kasooh.)

We caught somewhere about 500 seals before entering the sea, of all kinds. There were a good many females among them; there was a good many more of them than males, but the exact number I do not know. The old females had young pups in them. I saw them taken out and a good many of them skinned. (James Kean.)

We sailed from Victoria, British Columbia, and bore due north to Bering Sea. When we arrived there we had some 75 to 80 seals, the greater part of which were females, some of which had pups in them. (James Kennedy.)

Most of the seals taken by me were females with pup; have taken a few male seals from 1 to 4 years old. A very few yearlings have been killed by me, mostly females. (Mike Kethusduck.)

Those taken in Bering Sea were nearly all mother seals, in milk, that had left their young and were in search of food. (James Kiernan.)

Most all seals killed by me have been cows. * * * Have not killed a bull seal for three years. I have taken a few yearlings, mostly females. (John Kowineet.)

All the seals which I have seen killed were females, and the majority of these were pregnant cows. (Olaf Kvam.)

Most all seals that I have taken were cows with pup. A few male seals have been taken by me from 1 to 2 years old. (George Lacheek.)

A good many have pups in them, and when the boats come aboard loaded with seals, after they got through skinning them they would have a big pile of pups on deck. (James Laflin.)

We·had a good catch, having taken 1,400 skins, more than 1,000 of which we secured on the coast. Of the latter more than 75 per cent were female pelts, and of these about 60 per cent were taken from pregnant cows. (James E. Lennan.)

I have often cut a seal open and found a live young one inside. (Caleb Lindahl.)

Of all the seals captured by me about one-half of them, I think, were cows with pups in them, and it is very seldom that I have ever caught a full-grown cow that was barren or did not have a pup in her; nor have I, in my long experience, caught a cow that was in milk, or that had recently given·birth to her young. I seldom ever kill an old bull, for there are but very few of them that mingle with the herd along the coast. (James Lighthouse.)

In the year 1885 600 fur seals were caught during the month of March off the Farallon Islands (California). In subsequent years we have had to go farther north each year in order to secure a good spring catch. My experience has been that fully 90 per cent of all seals taken were females, and of these two-thirds were mothers in milk. (E. W. Littlejohn.)

I know that a large proportion of the seals taken were mothers in pup or mothers giving milk, but I paid no particular attention to the percentage. (William H. Long.)

On my last trip this year, when hunting seals off the cape, I caught 10 seals, 5 of which had pups in them; the rest of them were from 1 to 2 years old, part male and part female. I think that fully one-half of the seals caught along the coast are full-grown females with pups in them. We sometimes catch a few medium-sized males, the rest being younger ones, both male and female. (Thomas Lowe.)

Q. What percentage of the cows you have taken were with pup?— A. About 70 per cent, I should say. (Charles Lutjens.)

Most of the seals taken were females with pup. A few male seals were killed, ages ranging from 1 to 5 years. One old bull was taken. (George McAlpine.)

Most of the seals taken by me have been females with pup. The female seals are easier killed than the male, and we aim to get them. A few yearlings have been killed by me, mostly females. (J. D. McDonald.)

Several of the females that we caught in the ocean were in pup, but the pup taken out of the belly was of no use for anything, and we would throw it overboard. (William McIsaac.)

We had 300 or 400 seals altogether before entering Bering Sea. They were most all females, which had young pups in them. (William McLaughlin.)

Q. What percentage of the cows taken are with pup?—A. The females are mostly all with pup—that is, up until the 1st of July. (Daniel McLean.)

We came down each year to the coast of Oregon, then went along up the coast to Bering Sea. I do not recollect the exact number of seals we caught in 1888, 1889, 1890, but last year we caught about 150 along the coast. I did not pay much attention to the sex of the seals, but I seen lots of little pups taken out of them. (Thomas Madden.)

We sailed up the coast and caught a few seals until we got to Bering Sea. We caught 1,100 seals, nearly all of which were caught in Bering Sea. We caught them around St. George Island. I think out of the 1,100 we caught there were 600 females. Out of that 600 there were over 400 that had pups inside of them, and we threw them all overboard. (James Maloy.)

About half of the seals killed by me, I think, were cows with pup. Have never killed an old bull, but have killed a few yearlings in my life. Never examined the latter as to sex. (Charles Martin.)

The biggest part of my year's catch off the coast were females with pups in them. (Patrick Maroney.)

In 1890 I went sealing in the schooner *Argonaut*. She sailed from Victoria about the 8th of April, and sealed along the coast up to the pass in Bering Sea. We caught about 250 seals that year. Most of the seals we caught in the North Pacific were females. A good many of them also had pups inside. (Henry Mason.)

I noticed in the seals that we caught along the coast that a great many of them were females and had pups. I think most of them were females. I know that in my boat the catch was most all females and they had pups in them. They were usually shot when sleeping on the water. (William Mason.)

We caught over 1,000 seals off the coast, almost all females, and a great number of them had pups in them. * * * Entered Bering Sea in July and was chased out by the cutters. Did not catch any seals in the American waters in Bering Sea, but went over across on the Russian side and sealed there. The whole catch for that year was about 1,500 seals. Those that we killed on the Russian side were about in the same proportion as to females as those killed on this side. (Thorwal Mathasan.)

Q. What percentage of the cows you have taken were with pup?— A. About 75 per cent were with pup. (Frank Moreau.)

Most all the seals killed by me have been females with pup. (Amos Mill.)

We began sealing off Cape Flattery; sailed and sealed to the north-ward, and captured about 800 seals along the coast. There were not over ten males in the whole lot. The females had pups in them, and we cut them out of their mothers and threw them overboard into the ocean. (John Morris.)

About half of the seals caught along the coast are cows with pups in them. A few medium-sized males are also taken, and the rest are young seals of both sexes. We scarcely ever see an old bull seal, nor can we tell the sex of the seals in the water. I have never caught any full-grown cows along the coast that did not have pups in them. (Moses.)

About half the seals taken by me are cows with pup. I have taken a few old bulls in my life, but not many. Have taken quite a number of yearlings. The male seals taken are between 2 and 3 years. (Nashtau.)

About one-half of the seals I have taken were females with pup. Have taken a very few yearlings. Once in a while I take an old bull, but not often. The male seals that I have killed are 2 and 3 years old, I think. (Dan Nathlan.)

Think about half of the seals taken by me have been cows with pup. The rest are yearlings and young males 2 and 3 years old. Have never seen an old bull in my life. (Joseph Neishkaith.)

Almost every female that has arrived at the age of maturity is preg-nant. We follow them on from there into Bering Sea, and almost all the females taken are pregnant. (Niles Nelson.)

We sailed south as far as Blanco, sealing around there for two or three months, when we headed north into Bering Sea, having caught 250 or 300 seals before entering the sea, of which 60 per cent of them were females, mostly all of them having pups in them. (John O'Brien.)

In the beginning of the season we killed mostly yearling seals, but as the season advanced we got almost all mothers in young in the vicinity of Cape Flattery or from the Columbia River to Vancouver. (Nelson T. Oliver.)

The catch along the coast for the last six or seven years, since the rifle and shotgun have come into use, is principally females, and the grown ones have pups in them. The catch of young seals is much less in proportion to the number caught than they were when Indians used to take them by spearing. (William Parker.)

We began sealing off Cape Flattery and sealed right up toward Be-ring Sea, capturing 16 seals along the coast, all of which were females with pup. We captured 250 females with pup on the coast and then returned to Victoria, after which we sailed again in a short time on the same vessel with the same crew for the North Pacific Ocean and Be-ring Sea, capturing about 250 female seals while en route to Bering Sea, also a few male yearlings. (Charles Peterson.)

My experience in four years' sealing is that nearly all the seals taken along the coast are pregnant females, and it is seldom that one of them is caught that has not a young pup in her. (Edwin P. Porter.)

I have been out sealing this year and caught 16 seals; 5 of them were full-grown cows that had pups in them. The rest were young seals about 2 years old, both male and female, excepting one, and that was a gray pup. (Wilson Porter.)

Most of the seals taken by me have been cows with pup. (Rondtus.)

The majority of seals taken by me have been females with pup. Once in a great while I catch an old bull. A few yearlings have been taken and the majority of males are 2 and 3 year olds. (Abel Ryan.)

While cruising along the coast our principal catch was female seals with pup, the balance being principally yearlings, about half male and female. (William Short.)

We had 315 skins when we arrived here. Mostly all of them were females heavy with pup asleep on the water, and we killed them with shotguns. (Peter Simes.)

Most of the seal taken by me were cows with pup. (Aaron Sim son.)

Most of the seals taken are cows with young. (Jack Sitka.)

Most of the seals taken by me are females with pup. Never killed but one old bull in my life. Have killed but a few yearlings and never looked to see if they were male or female. The young males killed by me were between 1 and 3 years old. (Thomas Skowl.)

I think 3 females with pup out of every 10 killed. I kill lots of yearlings, but never examined them as to sex. Never shoot any old bulls, although I have seen a good many. (George Skultka.)

We sailed from here on the *Flying Mist* on the 17th day of April, 1871, and caught altogether on that voyage about 875 seals, of which a large majority were either females with pups or with their breasts full of milk. I saw it flowing on the deck when we were skinning them. * * * Went to Okhotsk Sea and sealed there about two months. We got there some 500 seals, of which more than one-half were females, and the most of them had pups in them. (James Sloan.)

I am informed by our London sales agent, and believe, that nearly or quite nine-tenths of the Victoria catch is composed of females. (Leon Sloss.)

A very large majority of the seal taken in the North Pacific are cows with pup. (Fred Smith.)

We left San Francisco in February, and fished all the way up to Kadiak Island. We caught about 475 seals and about 40 otters. To the best of my judgment, the greatest portion of these were cows heavy with young. We could see the milk running out of their teats when they were skinned. I saw pups inside of the seals that we cut, and we saved some of them and fed them." (E. W. Soron.)

We left here with the *City of San Diego* in February, 1888, and arrived in Bering Sea in June, 1888. As soon as we got into the ocean we commenced shooting seals and continued shooting all the way up to the Aleutian Islands. The seals became more plentiful as we were going north. We caught about 650 during that voyage. We killed a portion in Bering Sea. We killed 1 large bull that I recollect, and the rest were nearly all females with pup, or mothers giving milk. (Cyrus Stephens.)

Most of the seals taken are females with pup. Out of 111 seals last year I killed but 3 bulls. A very few yearlings have been taken by me.

A few male seal have been taken by me from 2 to 4 years old. (Joshua Stickland.)

We commenced sealing as soon as we got outside of the cape, and captured about 270 seals along up the coast. Most of the seals caught were pregnant females, and when we would skin them the milk would run out of them on the deck. We began sealing off the Columbia River and then sealed northward up the coast to Bering Sea, and captured about 320 seals in the North Pacific Ocean, mostly all females, and nearly all had young pups in them. (John A. Swain.)

Most of the seals taken by me were cows with pup. * * * A few male seal have been taken from 1 to 4 years old. But very few old bulls have ever been taken by me. Have killed a few yearlings every year. (M. Thlkahdaynahkee.)

Most of the seals taken on this coast are cows with young. Quite a large number of yearlings are taken, most of which are females. (Charlie Tlaksatan.)

Most of the seal taken by me have been females with pup. A few male seals have been taken by me, ages ranging from 1 to 4 years, I should think. Some yearlings have been taken, a majority of which were females also. Very few old bulls have been killed by me. (James Unatajim.)

Most of the seals taken have been cows with pup. I have taken but a very few old bulls. I have killed plenty of young males, and have taken quite a number of yearlings, but never examined them as to sex. (George Usher.)

The majority of seal taken are cows. A few yearlings are killed, mostly females. (Rudolph Walton.)

In purchasing fur seals from hunters I have noticed that not less than 75 per cent of the catch taken previous to May 25 are female seals, and from the development of the teat on the skin were evidently females with pup. After that the catch is mostly young seals, and I paid no attention to the sex. (M. L. Washburn.)

Most of the seals captured along the coast are cows with pups in them. I have never captured any cows in milk or that had given birth to their young that year on the coast, and I do not recollect of ever having caught an old bull. (Watkins.)

Out of 50 seals taken so far this season 46 are females with pup and 4 are males. Only one yearling seal has been taken this season among the males. I should think the male seals taken this year were between 2 and 3 years old. (P. S. Weittenhiller.)

While out hunting this year we caught 16 seals; one-half of them were cows with pup, the remainder were yearlings and 2-year-olds of both sexes. (Charley White.)

In my captures off the coast between here and Sitka 90 per cent of my catch were females, but off the coast of Unimak Pass there was a somewhat smaller percentage of females, and nearly all the females were cows heavy with pup, and in some instances the time of delivery was so near at hand that I have frequently taken the live pup from the mother's womb. (Michael White.)

I think about one-half the seals killed by me have been females with pup, and the balance were divided up between yearlings and 1 and 2 year old males. Never examined the yearlings as to sex; have never killed an old bull in my life. (Billy Williams.)

Think that most of the seals I have taken were females with pup. Have also taken some 2 and 3 year old males and some yearlings. Never killed but one old bull in my life. (Fred Wilson.)

Most all the seals caught by me along the coast were cows that had pups in them. I never killed a barren cow or one that was in milk. (Wispoo.)

About half the seals I have killed were females with pup, and the balance were yearling seals and 2 and 3 year old males. Never killed an old bull in my life, nor have I ever seen one. (Billy Yeltachy.)

Some years ago there were more male seals taken than are taken now, but now about one-half are females with pup. The rest are yearling seals and 1 and 2 year old males. I have never examined the yearling seal to ascertain their sex. Have not killed any old bull seal for a number of years, but used to kill them. (Yethnow.)

Most of the seals I have killed were females with pup. Once in a while an old bull is taken. (Paul Young.)

I have been out on the Pacific Ocean this year seal hunting and caught three seals. They were large cow seals, and had pups in them. One and 2 year old seals are about equally male and female. (Hish Yulla.)

Almost half the seals I now catch are cow seals, and have little pups in them. (Hish Yulla.)

About one-third of all the cows I caught along the coast were cows with pups in them; never caught any old bulls, and used to catch more gray pups than I do now. Most all the rest of the seals I caught have been 1 and 2 years old, and are about equally male and female. (Thos. Zolnoks.)

OFFICE SPECIAL AGENT TREASURY DEPARTMENT,
Washington, D. C., December 30, 1892.

SIR: I have the honor to hand you herewith a series of tables setting forth the number of fur seals killed on the Pribilof Islands, for all causes whatsoever, during the term of the lease of the Alaska Commercial Company—that is, from 1870 to 1889, both inclusive.

These tables have been compiled by me with great care from the official records of the Pribilof Islands, and are correct, careful comparisons having been made. They include every seal killed from any cause, intentional or accidental, incident to the taking of seal skins on the islands of St. Paul and St. George.

JOSEPH MURRAY,
First Assistant Special Agent.

Hon. CHARLES FOSTER,
Secretary of the Treasury.

H. Doc. 92, pt. 2——17

Tables showing in detail all killing of fur seals, for whatsoever purpose, on the Pribilof Islands during the term of the lease of the Alaska Commercial Company—that is, from 1870 to 1889, both inclusive.

NOTE.—There is a misapprehension in regard to the names of the several rookeries and hauling grounds, and they are often confounded by people who are not thoroughly acquainted with them. Zoltoi and Garbotch are local subdivisions of the Reef Rookery and are treated as rookeries by some of the Treasury agents, while others ignore them altogether. Zapadnie and Southwest Bay are one. Polivina and Halfway Point are one; some men using the Russian while others use the English names. English Bay and Middle Hill are separate and distinct rookeries, and yet they are often spoken of as though they were one. Near is a local subdivision of North Rookery. Little East is a subdivision of East Rookery. Sea Lion Rock, Southwest Point, and Rocky Point are neither rookeries nor hauling grounds in the strict sense of the term; the seals come and go at will, for it is only under the most favorable conditions of wind and water they can be reached, and it is but seldom there are many of them. By keeping these facts in mind it will be seen that seals were driven from all of the hauling grounds on both islands from 1870 to date.

ST. PAUL ISLAND.

[No record of daily killings for 1870.]

Date.	Rookery.	Total killed for all purposes.	Date.	Rookery.	Total killed for all purposes.
1870.....	Not indicated in the records of this year	15,314	**1871.** Oct. 30	Tolstoi	2,992
1871.			31	English Bay	30
May 16	Tolstoi	186	Nov. 7	Reef	729
24	Reef	246	Dec. 19	.....do	647
June 1	Tolstoi	579	19	.....do	3,877
2	Reef	222			
6	.....do	917		Total	81,803
8	English Bay	1,682	**1872.**		
10	Southwest Bay	2,701	May 11	Northeast Point	5
14	Zoltoi	874	14	Reef	227
15	Zoltoi and English Bay	1,167	24	.....do	455
16	Zoltoi	1,309	June 1	.....do	759
20	Southwest Bay	1,971	3	Tolstoi	278
22	Lukannon	1,283	5	Reef	293
23	Zoltoi	518	10	Tolstoi and English Bay	209
24	Reef	796	11	Southwest Bay	1,616
24	Northeast Point	2,654	12	Zoltoi and Reef	662
29	Zoltoi and Reef	1,014	13	Tolstoi and English Bay	1,057
28	Tolstoi and English Bay	2,401	14	.....do	1,730
July 1	Lukannon	1,133	15	Reef and Northeast Point......	4,714
3	Northeast Point	2,038	17	Zoltoi	395
6	Lukannon and Zoltoi..	3,623	19	English Bay	2,828
7	Reef	1,189	20	Reef	1,169
8	Lukannon	756	21	Lukannon	1,705
18	Zoltoi	1,040	22	Northeast Point and Ketova....	5,547
21	English Bay	1,940	24	Reef and Zoltoi	910
22	Ketova	801	27	English Bay	4,618
22	Northeast Point	3,404	28	Tolstoi and Northeast Point....	6,427
24	Zoltoi	1,179	29	Ketova and Lukannon........	1,151
26	Zoltoi and Lukannon...........	1,807	July 1	Zoltoi	1,841
28	Tolstoi	1,418	5	English Bay	3,265
28	Northeast Point	2,845	6	Zoltoi and Northeast Point....	6,765
31	Lukannon	657	9	English Bay	3,139
Aug. 11	Zoltoi	205	12	Zoltoi and Lukannon.........	2,071
18	.....do	150	13	English Bay	2,329
24	.....do	118	16	Zoltoi	1,116
29	Ketova	60	17	Halfway Point	1,664
Sept. 4	Lukannon	193	19	English Bay, Lukannon, and Northeast Point..............	7,388
11	Zoltoi	178	22	Zoltoi	1,384
18	Ketova	105	24	English Bay	2,080
26	.....do	77	25	Zoltoi	638
29	Tolstoi	130	30	.....do	34
Oct. 2	Ketova	1,250	Aug. 1	Northeast Point	10
9	Halfway Point	1,308	6	Zoltoi	119
10	Ketova	5,083	9	Northeast Point	7
13	Tolstoi	896	13	Tolstoi	90
14	Reef	506	17	Northeast Point	3
16	Northeast Point	632	20	Zoltoi	114
17	Ketova	683	29	Lukannon	161
17	Reef	1,158	Sept. 7	Zoltoi	99
19	Tolstoi	3,150	12	.....do	122
21	English Bay	3,666	20	.....do	118
21	Northeast Point	2,181	Oct. 3	English Bay	93
25	.....do	2,142	3	Zoltoi	490
27	.....do	3,042	10	.....do	127
28	.....do	679	21	Ketova	91
28	.....do	495			

Tables showing in detail all killing of fur seals, for whatsoever purpose, on the Pribilof Islands during the term of the lease of the Alaska Commercial Company—that is, from 1870 to 1889, both inclusive—Continued.

ST. PAUL ISLAND—Continued.

Date.	Rookery.	Total killed for all purposes.	Date.	Rookery.	Total killed for all purposes.
1872.			**1874.**		
Oct. 29	Zoltoi, Reef, and Lukannon....	1,284	June 3	Southwest Bay	2,395
Nov. 29	Tolstoi	753	3	Reef	538
30	Northeast Point	724	4	Tolstoi and English Bay	536
30	....do	1,286	6	Northeast Point	4,062
Dec. 5	Reef	112	8	Reef and Zoltoi	639
6	Zoltoi	426	9	Southwest Bay and English Bay	1,898
Nov. —		5,121	10	Tolstoi	634
			11	Reef and Garbotch	540
	Total	81,819	13	Southwest Bay and English Bay	1,982
			13	Tolstoi	622
1873.			13	Northeast Point	4,737
May 23	Reef	193	15	Reef and Zoltoi	891
23	Southwest Bay	104	17	English Bay and Tolstoi	2,689
June 3	Reef and Tolstoi	803	17	Zoltoi	474
4	Southwest Bay	703	19	Southwest and English bays	3,419
6	Reef and Tolstoi	920	20	Zoltoi and Lukannon	3,033
10	Southwest Bay and English Bay	2,597	20	Northeast Point	7,217
11	Reef and Zoltoi	1,666	23	Zoltoi and Lukannon	3,982
13	Tolstoi and English Bay	2,029	25	English Bay	3,270
13	Northeast Point	3,243	26	Reef and Zoltoi	1,921
16	Tolstoi	1,770	27	Ketova and Lukannon	1,321
16	Lukannon	677	27	Northeast Point	8,172
16	Reef	465	30	Zoltoi and Tolstoi	1,212
18	Southwest Bay and English Bay	3,946	July 1	English Bay and Tolstoi	2,209
21	Zoltoi	652	3	Reef, Tolstoi, and Lukannon	2,621
21	Northeast Point	3,412	4	Tolstoi	1,538
23	Tolstoi and English Bay	1,803	4	Lukannon	538
24	English Bay	3,159	6	Zoltoi	3,014
26	Reef and Tolstoi	2,210	6	English Bay	1,564
27	Zoltoi and Lukannon	1,147	8	English Bay and Tolstoi	2,702
27	Northeast Point	5,020	9	Zoltoi	1,987
30	Tolstoi	1,848	9	Tolstoi and Lukannon	1,580
July 1	Lukannon and Zoltoi	2,337	10	Zoltoi	432
2	Tolstoi	1,938	10	Northeast Point	3,367
3	Zoltoi	2,212	13	Tolstoi and Lukannon	1,664
5	....do	710	14	English Bay	2,169
8	Zoltoi and Tolstoi	1,510	15	Zoltoi	475
9	....do	2,494	16	Lukannon and Ketova	1,099
9	Northeast Point	6,278	16	Zoltoi	668
14	Tolstoi	925	17	Lukannon	533
15	Zoltoi	1,248	17	Northeast Point	4,004
16	....do	1,547	23	Tolstoi	130
17	English Bay	1,501	28	Zoltoi	167
18	Zoltoi	929	Aug. 3	....do	112
19	Lukannon, Ketova, and Zoltoi	1,047	10	....do	107
19	Northeast Point	5,696	17	....do	131
21	English Bay	754	26	....do	89
22	Lukannon and Ketova	1,079	31	....do	201
22	....do	446	Sept. 7	....do	197
24	Northeast Point	2,727	16	....do	163
Aug. 4	Zoltoi	179	25	....do	174
13	....do	168	Oct. 1	....do	179
20	....do	95	19	Reef	175
20	Lukannon	155	29	....do	236
Sept. 1	Zoltoi	110	Nov. —	Pups for natives	4,897
9	....do	109	Dec. 17	Reef	1,541
29		122			
30	Southwest Bay	10		Total	98,139
Oct. 8	Ketova	104			
16	....do	80	**1875.**		
21	Zoltoi	154	Jan. 1	Northeast Point	25
Nov. —	For natives' food	5,489	Feb. 10	....do	6
Dec. 9	Reef	231	16	....do	9
30	Garbotch	267	17	....do	16
			May 7	Southwest Bay	496
	Total	81,987	10	Northeast Point	9
			14	....do	20
1874.			18	Reef	143
Apr. 27	Northeast Point	10	24	....do	657
May 6	Southwest Bay	404	31	....do	492
19	Reef	340	June 1	English Bay and Southwest Bay	1,201
25	....do	301	1	Tolstoi	203
30	....do	217	5	Zoltoi and Tolstoi	692

*Tables showing in detail all killing of fur seals, for whatsoever purpose, on the Pribilof Islands during the term of the lease of the Alaska Commercial Company—that is, from 1870 to 1889, both inclusive—*Continued.

ST. PAUL ISLAND—Continued.

Date.	Rookery.	Total killed for all purposes.
1875.		
June 7	Zoltoi and Reef	711
7	Southwest Bay	1,560
7	Northeast Point	27
10	English Bay, Southwest Bay, Tolstoi	1,456
12	Reef and Zoltoi	631
12	Northeast Point	4,065
14	Tolstoi and English Bay	739
16	Halfway Point	2,115
16	Tolstoi and reef	707
17	Lukannon	452
18	Southwest Bay, English Bay	3,303
19	Zoltoi	1,363
19	Northeast Point	5,252
22	Tolstoi	1,830
22	Zoltoi	1,150
24	English Bay	3,009
25	Lukannon	262
26	Northeast Point	7,349
26	Southwest Bay	4,047
28	Reef	1,527
30	Zoltoi and English Bay	3,927
July 2	Lukannon, Ketova, Zoltoi	2,534
3	Northeast Point	5,024
6	Zoltoi and Lukannon	1,248
8	English Bay	3,370
9	Tolstoi and English Bay	2,093
10	Lukannon and Ketova	1,125
10	Northeast Point	5,937
13	Zoltoi	1,565
14	Tolstoi and English Bay	1,810
15	Lukannon and Ketova	748
16	English Bay	2,700
16	Zoltoi	1,205
17	Northeast Point	7,439
22	Zoltoi	557
28	...do	159
Aug. 4	...do	235
14	Ketova	192
21	Zoltoi	159
Sept. 2	...do	210
12	...do	143
21	...do	146
30	...do	153
Oct. 12	...do	115
Nov. 5	...do	172
17	Tolstoi	1,990
29	Southwest Bay	24
	Pups killed for food	3,745
Dec. 2	Northeast Point	15
4	Tolstoi	694
	Total	**94,960**
1876.		
Jan. 12	Tolstoi	914
May 23	Southwest Bay	223
31	Reef	189
June 3	Southwest Bay	830
6	Reef and Zoltoi	673
7	Tolstoi	468
8	Southwest Bay	566
10	Zoltoi	184
11	Northeast Point	1,585
13	Zoltoi	868
14	Halfway Point	811
15	Tolstoi, Zoltoi, Reef	1,509
17	Southwest Bay	2,641
17	Northeast Point	3,120
20	Zoltoi	2,942
21	Tolstoi	3,161
22	Zoltoi	480
24	English Bay and Northeast Point	10,696

Date.	Rookery.	Total killed for all purposes.
1876.		
June 26	Zoltoi	862
28	English Hill	3,017
29	Zoltoi	1,442
July 1	Tolstoi and Northeast Point	11,495
4	Tolstoi and Ketova	2,844
5	...do	2,848
7	English Bay	2,267
8	Lukannon	2,126
8	Northeast Point	2,116
10	Zoltoi	2,039
13	English Hill	1,074
22	Zoltoi	53
29	...do	1,040
Aug. 1	Lukannon and Zoltoi	8,677
10	Zoltoi	120
17	Ketova	134
23	...do	215
Sept. 1	Zoltoi	179
11	Ketova	130
19	Zoltoi	133
28	...do	146
Oct. 6	...do	133
14	...do	136
18	...do	120
31	Tolstoi	163
Nov. 24	Southwest Bay	636
24	...do	665
	Pups killed for food	3,958
Dec. 15	Tolstoi	825
	. **Total**	**83,157**
1877.		
May 22	Reef	342
June 4	...do	548
5	Southwest and English Bay	799
9	...do	1,705
12	Reef	449
13	Halfway Point	1,095
14	English Bay and Southwest Bay	1,647
15	Tolstoi	1,507
16	Zoltoi	1,094
18	...do	1,013
19	Tolstoi	1,458
20	Southwest Bay	1,631
21	Lukannon	1,172
22	Halfway Point	1,224
23	Northeast Point	5,965
23	Tolstoi	1,050
25	Zoltoi	1,250
26	Halfway Point	430
27	Tolstoi	2,029
28	Zoltoi and Lukannon	1,401
29	English Bay	2,166
30	Tolstoi	1,917
30	Northeast Point	6,449
July 1	Zoltoi	1,849
3	Tolstoi and Lukannon	1,534
4	Tolstoi and English Bay	2,522
6	Lukannon and Ketova	2,275
7	Zoltoi	1,113
7	Northeast Point	5,660
9	Zoltoi	495
10	Tolstoi	2,086
10	Northeast Point	2,172
14	Zoltoi	1,066
22	...do	75
Aug. 6	...do	165
11	...do	172
20	...do	190
30	...do	200
Sept. 12	...do	196
21	...do	171
29	...do	163

Tables showing in detail all killing of fur seals, for whatsoever purpose, on the Pribilof Islands during the term of the lease of the Alaska Commercial Company—that is, from 1870 to 1889, both inclusive—Continued.

ST. PAUL ISLAND—Continued.

Date.	Rookery.	Total killed for all purposes.
1877.		
Oct. 6	Zoltoi	171
16	...do	157
24	...do	146
Nov. 7	Zoltoi and Ketova	2,715
9	Zoltoi and Lukannon	1,535
12	Zoltoi and Reef	757
19	Tolstoi	222
27	...do	1,259
29	...do	383
Dec. 20	Northeast Point	20
	Total	67,810
1878.		
May 19	Sea Lion Rock	206
30	Southwest Bay	703
June 8	Reef	857
11	Tolstoi and English Bay	2,409
12	Reef and Zoltoi	556
13	Southwest Bay and English Bay	1,099
14	Tolstoi	887
15	Ketova, Reef, Zoltoi	1,283
17	Tolstoi	1,501
18	Southwest Bay and English Bay	2,278
19	Lukannon, Ketova, Zoltoi	998
20	Tolstoi	1,731
21	Southwest Bay and English Bay	1,457
22	Zoltoi, Lukannon, Ketova	1,309
22	Northeast Point	5,900
24	Halfway Point	1,473
25	Tolstoi	1,552
26	Zoltoi, Ketova, Lukannon	1,896
27	English Bay	2,672
28	Zoltoi and Ketova	1,661
29	Tolstoi	1,131
29	Northeast Point	6,375
July 1	Halfway Point	2,237
2	Zoltoi and Ketova	3,903
3	Lukannon	791
4	Zoltoi	2,010
5	Tolstoi	2,622
6	Zoltoi and Ketova	1,036
6	Northeast Point	7,231
8	Halfway Point, Lukannon, Ketova	1,369
9	Zoltoi	916
10	...do	2,288
10	Northeast Point	3,322
12	Tolstoi and Middle Hill	3,600
13	Zoltoi	2,101
16	Middle Hill	1,986
17	Zoltoi	2,337
18	...do	1,549
18	Lukannon	272
30	Zoltoi	404
Aug. 10	...do	294
22	...do	173
Sept. 2	...do	211
14	...do	156
24	Lukannon	144
Oct. 2	...do	148
10	...do	149
Nov. 1	...do	1,380
4	...do	2,000
6	Reef	1,255
8	...do	571
26	Tolstoi and Reef	1,144
28	...do	853
Aug. to Dec.	Northeast Point (watchmen)	133
	Total	88,519

Date.	Rookery.	Total killed for all purposes.
1879.		
May 19	Southwest and English bays	278
26	Reef	525
June 2	...do	162
7	English and Southwest bays, Tolstoi	1,627
9	Reef	434
10	Halfway Point	1,188
11	Southwest and English bays	1,462
12	Tolstoi	498
13	Reef, Zoltoi, Ketova	730
14	Southwest Bay and Middle Hill	997
16	Halfway Point	522
17	Southwest Bay and Middle Hill	1,331
18	Reef, Ketova, Zoltoi	914
19	Southwest Bay and English Bay	1,110
20	Tolstoi and Middle Hill	1,176
21	Reef, Ketova, Lukannon	1,053
16	Northeast Point	1,528
17	...do	966
18	...do	1,860
19	...do	1,745
20	...do	1,289
23	Tolstoi and Middle Hill	2,300
24	Southwest and English bays	1,822
25	Reef, Zoltoi, Ketova	1,995
26	Tolstoi, Middle Hill, English Bay, Zoltoi	1,542
27	Halfway Point	1,900
28	Ketova, Reef, Lukannon, Zoltoi	1,206
23	Northeast Point	1,550
24	...do	1,414
25	...do	1,339
26	...do	1,074
27	...do	1,665
30	Tolstoi and Middle Hill	2,617
July 1	English Bay	2,148
2	Lukannon and Zoltoi	1,885
3	Zoltoi and Middle Hill	1,932
4	English Bay	2,106
5	Lukannon and Ketova	1,168
June 30	Northeast Point	2,524
1	...do	1,628
2	...do	866
3	...do	1,988
July 4	...do	2,077
7	Zoltoi and Middle Hill	1,528
8	Zoltoi, Ketova, Lukannon	1,920
9	Zoltoi and Ketova	983
10	...do	948
7	Northeast Point	2,418
8	...do	1,264
9	...do	1,519
10	...do	308
14	Zoltoi	2,652
15	Zoltoi and Ketova	1,233
16	Middle Hill, Lukannon, Tolstoi	2,882
16	...do	157
25	Zoltoi	278
Aug. 2	Northeast Point	13
2	Zoltoi	273
11	...do	195
20	...do	206
27	Specimens	18
28	Zoltoi	203
Sept. 5	Southwest Bay	1
8	Zoltoi	184
9	Northeast Point	16
18	Zoltoi	174
18	Halfway Point	2
18	Southwest Bay	5
18	Northeast Point	4
29	Zoltoi	198
29	Southwest Bay	6
29	Northeast Point	7
Oct. 7	Ketova	109

Tables showing in detail all killing of fur seals, for whatsoever purpose, on the Pribilof Islands during the term of the lease of the Alaska Commercial Company—that is, from 1870 to 1889, both inclusive—Continued.

ST. PAUL ISLAND—Continued.

Date.	Rookery.	Total killed for all purposes.	Date.	Rookery.	Total killed for all purposes.
1879.			**1880.**		
Oct. 20	Lukannon	263	Oct. 12	Lukannon	260
29	...do	544	25	English Bay	193
30	...do	335	Nov. 2,? 3,5,6	Lukannon, Reef, Ketova	4,410
31	Ketova	999	Dec. 6	Reef	1,176
31	...do	107	9	Southwest Bay	13
Nov. 3	Lukannon	1,153	24	Northeast Point	82
3	...do	31	Nov. 11	Middle Hill	270
5	Garbotch	1,466	29	English Bay	270
5	...do	79			
10	...do	356		Total	84,779
13	Reef	260			
27	...do	172	**1881.**		
Dec. 6	Garbotch	1,206	Jan. 1–3	Tolstoi and reef	1,058
10	Reef	1,308	May 4	Additional skins found on recount	8
26	Northeast Point	62	14	Additional skins found in salt	3
26	Southwest Bay	5	29	Reef	165
			June 6	Reef and Zoltoi	423
	Total	88,221	7	Southwest and English bays, Tolstoi	1,250
1880.			9	Zoltoi	127
May 14	Southwest Bay	209	10	Halfway Point	474
22	Reef	225	14	Zoltoi	196
22	Northeast Point	19	15	Southwest and English bays, Tolstoi	2,387
June 1	Reef	216	16	Zoltoi, Reef, Lukannon	724
8	Southwest Bay	1,497	17	Halfway Point	539
9	Reef	926	18	Tolstoi	1,229
11	English Bay, Tolstoi	889	20	Zoltoi and Lukannon	1,614
12	Southwest and English bays	763	21	Tolstoi, Middle Hill, Northeast Point	4,103
14	Halfway Point	1,204	22	English Bay, Middle Hill, Tolstoi, Northeast Point	3,049
15	Reef and Zoltoi	765	23	Reef and Tolstoi	1,164
16	Zapadnie and English Bay	990	24	Halfway Point, Middle Hill, English Bay, Tolstoi	3,082
17	Ketova, Zoltoi, Reef	18	25	Middle Hill, Lukannon, Zoltoi, Reef	1,275
18	Tolstoi and English Bay	1,619	27	Middle Hill, English Bay, Tolstoi, Northeast Point	2,043
19	English Bay and Middle Hill	802	28	English Bay, Southwest Bay, Northeast Point	3,318
14/19	Northeast Point	5,279	29	Lukannon, Zoltoi, Northeast Point	2,967
21	Halfway Point	1,459	30	English Bay, Middle Hill, Northeast Point	4,596
22	Ketova, Zoltoi, Reef	1,035	July 1	Tolstoi and Lukannon	3,358
23	Tolstoi and Middle Hill	1,702	2	Halfway Point	943
24	Reef, Zoltoi, Ketova	1,437	4	English Bay, Middle Hill, Tolstoi, Northeast Point	3,758
25	English Bay and Middle Hill	2,582	5	Zoltoi, Ketova, Northeast Point	1,949
26	Ketova, Zoltoi, Reef	1,062	6	Southwest Bay, English Bay, Northeast Point	4,853
21/26	Northeast Point	6,202	7	Zoltoi, Tolstoi, Middle Hill, Northeast Point	3,421
28	Halfway Point	1,516	8	Halfway Point, Northeast Point	2,269
29	Ketova, Lukannon, Zoltoi	1,743	9	English Bay, Middle Hill, Tolstoi	2,631
30	Tolstoi and Middle Hill	2,297	12	Zoltoi, Tolstoi, Lukannon	3,075
July 1	...do	1,622	13	English Bay, Middle Hill	1,782
2	English Bay and Middle Hill	2,374	14	English and Southwest bays	1,473
3	Ketova, Zoltoi, Reef	1,386	15	English Bay, Middle Hill, Tolstoi, Zoltoi Ketova, Lukannon	3,561
3	Northeast Point	7,167	16	Zoltoi, Ketova (711), Northeast Point, food (16)	727
5	Halfway Point	789	18	Middle Hill, Tolstoi, Zoltoi, Ketova, Lukannon	2,455
5	Lukannon and Ketova	651	19	English Bay, Middle Hill, Tolstoi	2,301
6	Tolstoi	1,577	20	Tolstoi, Zoltoi, Ketova, Lukannon	2,536
7	Tolstoi and Lukannon	1,654			
8	Zoltoi, Reef, Lukannon, Ketova	2,221			
9	Tolstoi, Lukannon, Middle Hill	1,428			
10	Zoltoi, Ketova, Lukannon	1,221			
5–10	Northeast Point	7,073			
12	Zoltoi, Ketova, Lukannon	817			
13	Tolstoi and English Bay	1,763			
14	Reef, Zoltoi, Lukannon, Ketova	2,640			
15	English Bay	1,834			
16	Middle Hill and Lukannon	2,461			
17	Zoltoi	534			
28	Northeast Point	43			
30	Halfway Point	228			
31	Zoltoi	218			
Aug. 11	...do	253			
19	...do	160			
28	...do	189			
Sept. 8	...do	195			
18	...do	239			
30	...do	227			

Tables showing in detail all killing of fur seals, for whatsoever purpose, on the Pribilof Islands during the term of the lease of the Alaska Commercial Company—that is, from 1870 to 1889, both inclusive—Continued.

ST. PAUL ISLAND—Continued.

Date.	Rookery.	Total killed for all purposes.	Date.	Rookery.	Total killed for all purposes.
1881.			**1882.**		
July 27	Zoltoi	202	July 8	Halfway Point and Northeast Point	2,454
18-29	Northeast Point	38	10	Reef, Ketova, Lukannon, Northeast Point	3,291
Aug. 5	Zoltoi	224	12	...do	2,833
10	...do	276	13	Southwest Bay and Northeast Point	3,420
17	Northeast Point	10	14	English Bay, Middle Hill, Tolstoi, Northeast Point	3,087
20	...do	6	17	English Bay, Middle Hill, Northeast Point	2,593
26	Zoltoi	240	18	Zoltoi	1,012
29	Northeast Point	7	19	Zoltoi, Lukannon, Ketova	1,276
Sept. 6	Zoltoi	249	20	Southwest Bay	727
17	Northeast Point	11	25	Zoltoi	204
19	Zoltoi	205	27	Northeast Point	19
28	Zoltoi (208), Northeast Point (5)	213	Aug. 4	Zoltoi	252
Oct. 10	Zoltoi (211), Northeast Point (5)	216	14	...do	226
25	Zoltoi, Northeast Point	227	26	...do	234
Nov. 5	Tolstoi	209	Sept. 4	...do	191
16	Middle Hill	683	15	...do	237
17	Tolstoi	494	27	...do	227
21	Middle Hill	462	30	Northeast Point	25
Dec. 7	Southwest Bay	1,022	Oct. 6	Zoltoi	201
9	Reef	1,294	17	...do	261
	Total	83,774	30	Reef	166
1882.			30	Northeast Point	23
Jan. 12	Reef	80	Nov. 17	Middle Hill and Tolstoi	248
Feb. 8	...do	103	28	Reef	383
10	...do	8	30	Tolstoi	429
May 22	Tolstoi	126	Dec. 4	Reef	498
31	Reef	246	6	...do	388
June 2	Southwest Bay, Middle Hill, Tolstoi	400	12	...do	400
7	Southwest Bay	849		Total	79,834
8	Zoltoi and Reef	428	**1883.**		
10	Tolstoi and Reef	488	May 26	Southwest Bay	230
12	Southwest Bay and Northeast Point	2,223	June 4	Southwest and English bays, Tolstoi	592
13	Halfway Point	217	6	Halfway Point	354
13	Northeast Point	306	9	Reef	177
14	Southwest Bay and Tolstoi	803	11	Southwest Bay, Middle Hill, Tolstoi	405
16	Halfway Point, Reef, Tolstoi	1,458	12	Reef and Zoltoi	352
17	Southwest Bay and Tolstoi	1,070	13	Halfway Point	252
19	Ketova, Reef, Zoltoi, Northeast Point	1,829	14	Southwest and English bays	490
20	Southwest Bay and Northeast Point	3,069	15	English Bay and Tolstoi	440
21	English Bay, Tolstoi, Northeast Point	1,617	16	Reef and Zoltoi	341
22	Ketova, Reef, Zoltoi, Northeast Point	2,811	18	Southwest Bay	417
23	Halfway Point and Northeast Point	2,528	19	English Bay, Middle Hill, Tolstoi	735
24	Middle Hill, Tolstoi, Northeast Point	1,560	20	Halfway Point and Lukannon	908
26	Southwest Bay and Northeast Point	3,164	27	Reef and Tolstoi	972
27	English Bay, Middle Hill, Northeast Point	3,226	22	English Bay, Middle Hill, Tolstoi	1,401
28	Reef, Tolstoi, Northeast Point	4,270	23	Reef, Zoltoi, Lukannon	1,078
29	Middle Hill, Tolstoi, Northeast Point	2,239	18-23	Northeast Point	3,279
30	Halfway Point, Lukannon, Northeast Point	2,469	25	English Bay, Middle Hill, Tolstoi	1,428
July 1	Northeast Point and Reef	1,601	26	Zoltoi and Lukannon	838
3	Southwest Bay and Northeast Point	3,805	27	Southwest and English bays	1,640
4	English Bay, Middle Hill, Northeast Point	2,575	28	Zoltoi, Tolstoi, Reef, Lukannon	1,612
5	Zoltoi and Northeast Point	3,034	29	English Bay, Middle Hill, Zoltoi	1,519
6	Middle Hill, Zoltoi, Northeast Point	1,540	30	Lukannon, Reef, Zoltoi	1,191
7	English Bay, Middle Hill, Northeast Point	2,673	25-30	Northeast Point	5,012
			July 2	Halfway Point	1,700
			3	Southwest Bay	2,151
			4	Middle Hill and Tolstoi	1,494
			5	Zoltoi and Lukannon	2,346
			6	Zoltoi, Middle Hill, Tolstoi	1,755

Tables showing in detail all killing of fur seals, for whatsoever purpose, on the Pribilof Islands during the term of the lease of the Alaska Commercial Company—that is, from 1870 to 1889, both inclusive—Continued.

ST. PAUL ISLAND—Continued.

Date.	Rookery.	Total killed for all purposes.	Date.	Rookery.	Total killed for all purposes.
1883.			**1884.**		
July 7	Zoltoi, Reef, Lukannon	939	July 1-4	Northeast Point	5,799
8	Middle Hill	1,164	5	Zoltoi, Reef, Ketova	1,251
2-9	Northeast Point	5,066	7	Halfway Point	1,935
10	English Bay, Middle Hill, Tolstoi, Zoltoi	1,923	8	Reef, Zoltoi, Lukannon	2,071
			9	Southwest Bay	1,966
12	Halfway Point, Lukannon, Zoltoi	1,659	10	English Bay, Middle Hill, Tolstoi	1,920
13	Southwest Bay	2,444	7-10	Northeast Point	3,003
14	English Bay, Middle Hill, Zoltoi	2,136	12	Middle Hill, Tolstoi, Lukannon, Zoltoi, Reef	3,067
16	....do	2,060	14	Halfway Point and Zoltoi	2,515
17	Halfway Point and Zoltoi	1,116	15	Southwest Bay	2,052
18	Southwest Bay	1,876	16	English Bay, Middle Hill, Ketova	1,526
19	Middle Hill, Lukannon, Zoltoi	8,183	17	Zoltoi and Lukannon	1,782
Aug. 1	Zoltoi	191	18	English Bay, Tolstoi, Middle Hill	1,872
10	....do	250	14-18	Northeast Point	5,089
20	....do	102	19	Middle Hill, Tolstoi, Lukannon, Ketova, Zoltoi, Reef	2,529
Sept. 1	....do	278	21	Middle Hill, Zoltoi, Tolstoi, Ketova	1,911
12	....do	123	Aug. 1	Zoltoi	198
24	....do	286	5	....do	92
Oct. 6	....do	200	12	....do	80
20	....do	200	19	....do	90
29	Reef	1,562	26	....do	78
Nov. 1	Ketova	336	Sept. 2	....do	160
2	Reef	884	12	....do	147
5	....do	119	19	....do	131
15	....do	134	27	....do	150
26	....do	155	Oct. 5	....do	142
27	Tolstoi	84	14	Lukannon	144
27	Northeast Point (to date)	66	22	Reef	115
Dec. 12	Reef	420	30	English Bay	179
19	....do	421	Nov. 3	Reef	1,956
	Total	63,295	5	....do	785
			10	Tolstoi	182
1884			24	Reef	153
Jan. 2	{Reef	177	Dec. 5	....do	495
	{Northeast Point (to date)	36	6	English Bay	265
May 15	....do	20	24	Reef	244
21	Halfway Point and Reef	187			
24	Northeast Point (to date)	15		Total	88,861
27	Southwest Bay and Reef	427			
June 3	Reef	318			
5	Southwest and English bays, Zoltoi	767	**1885.**		
9	Reef and Halfway Point	1,239	May 19	Sea Lion Rock	181
10	Ketova and Tolstoi	426	27	Reef	141
11	Southwest Bay and Halfway Point	1,364	June 3	Zoltoi	49
12	Zoltoi, Ketova, English Bay	865	6	Zoltoi and Reef	73
13	Zoltoi and Reef	771	11	....do	125
14	Halfway Point	838	13	Tolstoi and English Bay	587
16	Southwest Bay	1,424	15	Halfway Point	741
17	English Bay, Tolstoi, Reef, Zoltoi	1,206	16	Zoltoi and Reef	973
18	Halfway Point	912	17	Southwest Bay	1,700
19	English Bay and Tolstoi	487	18	English Bay, Middle Hill, Zoltoi	617
20	Southwest Bay	1,793	19	Lukannon and Halfway Point	1,309
21	Zoltoi and Reef	1,117	20	Zoltoi and Reef	986
21	Northeast Point	3,902	22	Zoltoi, Reef, Lukannon	789
23	Halfway Point and Zoltoi	2,163	22	Northeast Point	1,532
24	Zoltoi, Reef, Lukannon, Ketova	1,729	23	Halfway Point and Zoltoi	1,143
25	Southwest Bay	1,197	23	Northeast Point	847
26	English Bay, Middle Hill, Tolstoi	2,546	24	English Bay, Southwest Bay, Middle Hill	1,733
27	Zoltoi, Reef, Ketova	1,830	25	Reef and Lukannon	1,681
27	Northeast Point	5,134	25	Northeast Point	1,051
28	Tolstoi, Middle Hill, Lukannon	1,500	26	Halfway Point	1,373
30	Zoltoi and Halfway Point	1,662	26	Northeast Point	667
July 1	Zoltoi, Reef, Tolstoi, Lukannon	1,826	27	Lukannon, Zoltoi, Reef	1,328
2	English Bay and Middle Hill	1,888	27	Northeast Point	539
3	Southwest Bay, Zoltoi, Middle Hill, Lukannon	1,340	29	Southwest Bay	1,602
			29	Northeast Point	553
4	English Bay and Tolstoi	1,522	30	English Bay and Middle Hill	2,681

Tables showing in detail all killing of fur seals, for whatsoever purpose, on the Pribilof Islands during the term of the lease of the Alaska Commercial Company—that is, from 1870 to 1889, both inclusive—Continued.

ST. PAUL ISLAND—Continued.

Date.	Rookery.	Total killed for all purposes.	Date.	Rookery.	Total killed for all purposes.
1885.			**1886.**		
June 30	Northeast Point	1,053	May 29	Reef	153
July 1	Reef, Lukannon, Zoltoi, Ketova	1,777	June 4	....do	562
1	Northeast Point	1,096	8	Tolstoi, English Bay, Southwest Bay	1,323
2	Tolstoi, Middle Hill, English Bay	1,465	9	Halfway Point	299
2	Northeast Point	631	10	Reef and Zoltoi	634
3	Halfway Point and Zoltoi	2,134	11	English Bay	214
3	Northeast Point	658	14	Lukannon Reef	427
4	Lukannon and Zoltoi	976	14	Northeast Point	1,343
4	Northeast Point	161	15	Southwest Bay	1,106
6	Southwest Bay	1,271	15	Northeast Point	1,116
6	Northeast Point	522	16	English Bay, Middle Hill, Tolstoi	850
7	Middle Hill, Southwest Bay, Tolstoi	2,664	16	Northeast Point	585
7	Northeast Point	1,184	17	Halfway Point	833
8	Zoltoi, Lukannon, Ketova	2,550	17	Northeast Point	761
8	Northeast Point	746	18	Reef and Zoltoi	651
9	Middle Hill and Tolstoi	1,294	18	Northeast Point	376
9	Northeast Point	793	19	English Bay, Middle Hill, Tolstoi	1,064
10	Halfway Point	2,304	19	Northeast Point	371
10	Northeast Point	671	21	Southwest Bay	1,891
13	Southwest Bay	2,134	21	Northeast Point	1,161
13	Northeast Point	822	22	English Bay and Tolstoi	1,007
14	English Bay and Tolstoi	2,692	22	Northeast Point	659
14	Northeast Point	955	23	Halfway Point	1,770
15	Zoltoi	2,139	23	Northeast Point	955
15	Northeast Point	363	24	Zoltoi	1,555
16	Halfway Point and Lukannon	2,137	24	Northeast Point	408
16	Northeast Point	757	25	Reef, Zoltoi, English Bay, Middle Hill, Tolstoi	2,158
17	Zoltoi	2,203	25	Northeast Point	581
17	Northeast Point	647	26	....do	441
18	Reef, Tolstoi, Middle Hill	1,552	28	Southwest Bay	1,070
18	Northeast Point	616	28	Northeast Point	926
20	Southwest Bay	1,591	29	English Bay, Tolstoi, Zoltoi	1,503
20	Northeast Point	828	29	Northeast Point	794
21	Middle Hill and English Bay	2,723	30	Halfway Point	490
21	Northeast Point	474	30	Northeast Point	1,056
22	Zoltoi and Lukannon	2,743	July 1	English Bay and Tolstoi	1,319
22	Northeast Point	687	1	Northeast Point	1,202
23	Middle Hill, English Bay, Zoltoi	1,603	2	Southwest Bay	856
23	Northeast Point	631	2	Northeast Point	566
24	Halfway Point, Middle Hill, Ketova	2,498	3	Reef and Zoltoi	1,263
25	English Bay, Zoltoi, Ketova, Middle Hill, Lukannon, Reef	2,215	5	English Bay and Tolstoi	1,163
27	Zoltoi, Reef, Middle Hill	983	5	Northeast Point	1,180
Aug. 3	Zoltoi	147	6	Halfway Point	942
12	....do	179	6	Northeast Point	806
21	Zoltoi and Reef	185	7	Zoltoi, Reef, Lukannon	1,969
Sept. 5	Zoltoi	135	7	Northeast Point	1,187
12	....do	155	8	Southwest Bay	1,460
25	....do	152	8	Northeast Point	952
Oct. 7	....do	78	9	English Bay, Middle Hill, Tolstoi	1,563
14	....do	122	9	Northeast Point	636
20	....do	85	10	Reef, Zoltoi, Ketova, Lukannon	1,133
Nov. 2	Reef	1,524	12	Halfway Point	1,014
4	....do	904	13	Southwest Bay and Southwest Point	1,442
7	....do	300	12	Northeast Point	1,501
9	Middle Hill	332	14	English Bay and Middle Hill	1,074
21	Reef	148	14	Northeast Point	602
Dec. 4	....do	1,096	15	Zoltoi, Reef, Ketova, Lukannon	1,957
17			15	Northeast Point	899
31	Northeast Point	48	16	Halfway Point	937
			16	Northeast Point	1,013
	Total	88,880	17	Southwest Bay and Southwest Point	2,057
1886.			17	Northeast Point	407
Jan. 21	Sea Lion Rock	84	19	Reef and Zoltoi	2,312
29	Southwest Bay	49	19	Northeast Point	753
May 5	Northeast Point	7	20	English Bay, Middle Hill, Tolstoi	3,140
8	....do	5	20	Northeast Point	801
17	Southwest Bay and Reef	300			
10–17	Northeast Point	49			

Tables showing in detail all killing of fur seals, for whatsoever purpose, on the Pribilof Islands during the term of the lease of the Alaska Commercial Company—that is, from 1870 to 1889, both inclusive—Continued.

ST. PAUL ISLAND—Continued.

Date.	Rookery.	Total killed for all purposes.	Date.	Rookery.	Total killed for all poses.
1886.			**1887.**		
July 21	Halfway Point...................	1,476	July 8	Northeast Point..............	795
21	Northeast Point..............	312	9	Southwest Bay................	2,065
22	Southwest Bay and Southwest		9	Northeast Point..............	429
	Point.......................	2,015	12	English Bay, Tolstoi, Lukannon	2,595
22	Northeast Point..............	923	12	Northeast Point..............	2,219
23	Reef, Zoltoi, Lukannon, Ketova.	3,147	13	Reef, Zoltoi, Ketova...........	3,029
23	Northeast Point..............	739	13	Northeast Point..............	1,930
24	English Bay and Middle Hill...	1,625	14	Halfway Point................	1,201
24	Northeast Point..............	658	14	Northeast Point..............	826
26	Halfway Point, Southwest Bay,		15	Tolstoi and Lukannon..........	1,298
	Lukannon, Zoltoi.............	1,993	15	Northeast Point..............	803
Aug. 3	Zoltoi.......................	75	16	Reef and Zoltoi..............	986
9	....do.......................	152	16	Northeast Point..............	546
19	....do.......................	134	17	West Point...................	617
30	....do.......................	96	18	Southwest Bay................	2,108
Sept. 6	....do.......................	148	18	Northeast Point..............	1,671
17	....do.......................	146	19	English Bay, Middle Hill, Tol-	
29	Reef.........................	148		stoi	2,038
Oct. 11	....do.......................	144	19	Northeast Point..............	922
28	....do.......................	152	20	Reef, Zoltoi, Lukannon. Ketova.	3,209
Nov. 5	....do.......................	768	20	Northeast Point..............	11,092
6	Reef and Lukannon...........	445	21	Halfway Point and Lagoon.....	1,397
8	Reef.........................	900	21	Northeast Point..............	798
10	....do.......................	711	22	English Bay, Tolstoi, Middle	
22	....do.......................	379		Hill	1,877
23	Tolstoi......................	289	22	Northeast Point..............	1,082
Dec. 1	Reef.........................	380	23	Southwest Bay, Reef, Zoltoi,	
21	Tolstoi......................	191		Lukannon...................	2,226
			24	Tolstoi......................	232
	Total..................	88,085	31	Northeast Point..............	39
			Aug. 1	Zoltoi.......................	137
1887.			8	....do.......................	113
May 24	Reef and Southwest Bay......	275	16	Reef and Lukannon...........	209
June 1	Northeast Point..............	138	24	English Bay..................	521
6	Tolstoi......................	419	Sept. 5	Middle Hill..................	403
9	Reef and Zoltoi..............	315	6	English Bay..................	356
11	Tolstoi......................	501	15	Zoltoi.......................	192
13	Southwest Bay................	407	28	....do.......................	100
15	Reef and Zoltoi..............	526	Oct. 6	....do.......................	116
16	Halfway Point................	750	17	....do.......................	108
17	English Bay and Tolstoi.......	765	26	....do.......................	76
20	Southwest Bay................	523	Nov. 1	Reef.........................	1,013
20	Northeast Point..............	1,899	3	....do.......................	1,132
21	Reef.........................	1,641	4	....do.......................	32
21	Northeast Point..............	452	6	Zoltoi.......................	65
22	English Bay, Tolstoi, Lukannon	1,004	7	Middle Hill..................	611
22	Northeast Point..............	1,172	25	Reef.........................	82
23	Halfway Point	1,314	26	Middle Hill and Tolstoi.......	185
	Northeast Point..............	521	Dec. 9	....do.......................	450
24	Reef and Zoltoi..............	1,165	15	Sea Lion Rock and Southwest	
24	Northeast Point..............	709		Bay	169
23	Tolstoi, Middle Hill, English				
25	Bay	1,961		Total..................	89,092
27	West Point and Southwest Bay	1,180			
27	Northeast Point	1,205	**1888.**		
28	Reef, Zoltoi, Ketova, Lukannon.	2,964	Jan. 25	Northeast Point.............	545
28	Northeast Point..............	691	May 19	Tolstoi, Reef, Sea Lion Rock...	131
29	Middle Hill and Tolstoi.......	1,895	24	Reef.........................	113
29	Northeast Point..............	1,144	28	....do........................	82
30	Halfway Point................	1,604	31	Northeast Point..............	82
30	Northeast Point..............	1,203	June 2	Reef.........................	121
July 1	English Bay..................	1,162	7	Reef and Zoltoi..............	175
1	Northeast Point..............	1,201	9	Tolstoi......................	342
2	Reef and Zoltoi	1,616	11	Southwest Bay................	543
2	Northeast Point..............	624	12	English Bay..................	587
4	Tolstoi and Middle Hill.......	1,703	15	Halfway Point................	428
4	Northeast Point..............	1,196	16	Reef and Zoltoi..............	789
5	Reef, Zoltoi, Lukannon.........	2,023	18	Southwest Bay................	764
5	Northeast Point..............	1,056	18	Northeast Point..............	1,490
6	Halfway Point................	......	19	Tolstoi......................	490
6	Northeast Point..............	1,247	19	Northeast Point..............	930
7	English Bay and Tolstoi.......	1,622	21	Reef and Zoltoi..............	1,400
7	Northeast Point..............	994	21	Northeast Point..............	1,604
8	Reef and Zoltoi	1,125	22	Halfway Point	801

Tables showing in detail all killing of fur seals, for whatsoever purpose, on the Pribilof Islands during the term of the lease of the Alaska Commercial Company—that is, from 1870 to 1889, both inclusive—Continued.

ST. PAUL ISLAND—Continued.

Date.	Rookery.	Total killed for all purposes.	Date.	Rookery.	Total killed for all purposes.
1888.			**1888.**		
June 22	English Bay, Tolstoi, Middle Hill	702	Sept. 6	Zoltoi	114
23	Northeast Point	565	15	....do	100
22	....do	973	22	....do	98
25	Southwest Bay	1,440	29	....do	98
25	Northeast Point	870	Oct. 10	Lukannon	83
26	English Bay, Middle Hill, Tolstoi	1,158	18	Middle Hill	98
26	Northeast Point	1,509	27	....do	111
27	Reef, Zoltoi, Ketova, Lukannon	2,005	Nov. 3	Middle Hill and Zoltoi	126
27	Northeast Point	850	5	Reef	761
28	Halfway Point	911	7	....do	547
28	Northeast Point	1,180	8	....do	716
29	Southwest Bay	1,098	9	....do	154
29	Northeast Point	625	15	Middle Hill	277
30	Middle Hill, English Hill, Tolstoi	1,625	26	Reef	111
30	Northeast Point	964	30	....do	129
July 2	Reef and Zoltoi	2,071	Dec. 17	Tolstoi	206
2	Northeast Point	1,413	26	Sea Lion Rock	78
3	Halfway Point	1,188			
3	Northeast Point	1,439		Total	86,270
4	Southwest Bay	822			
4	Northeast Point	1,241	**1889.**		
5	English Bay, Tolstoi, Lukannon	1,942	May 22	Sea Lion Rock	124
5	Northeast Point	446	25	Reef	41
6	Reef and Zoltoi	1,491	28	....do	234
6	Northeast Point	1,609	31	Northeast Point	133
7	Halfway Point	490	June 5	Reef	201
7	Northeast Point	906	10	....do	120
9	English Bay, Tolstoi, Lukannon, Middle Hill	2,398	12	Tolstoi	947
9	Northeast Point	1,740	14	Zoltoi and Reef	762
10	Reef and Zoltoi	1,083	15	Southwest Bay	340
10	Northeast Point	959	17	Halfway Point	895
12	English Bay, Middle Hill, Tolstoi	1,557	17	Northeast Point	1,054
12	Northeast Point	927	18	English Bay, Tolstoi, Middle Hill	1,161
13	Southwest Bay	1,337	18	Northeast Point	1,270
13	Northeast Point	912	19	....do	494
14	Halfway Point	773	19	Reef, Zoltoi, Lukannon	1,561
15	Northeast Point	550	20	Southwest Bay	253
15	West Point	481	21	Northeast Point	1,205
16	Reef and Zoltoi	2,004	22	English Bay, Tolstoi, Middle Hill	1,355
16	Northeast Point	1,038	24	Northeast Point	754
17	English Bay and Tolstoi	2,055	24	Reef and Zoltoi	2,578
17	Northeast Point	1,328	25	Halfway Point and Lukannon	979
18	Southwest Bay	2,216	25	Northeast Point	1,407
18	Northeast Point	1,004	26	English Bay and Middle Hill	1,314
19	Halfway Point	1,410	26	Northeast Point	441
19	Northeast Point	705	27	Southwest Bay	311
20	Reef and Zoltoi	2,018	27	Northeast Point	844
20	Northeast Point	646	28	Reef, Zoltoi, Ketova	1,349
21	English Bay and Tolstoi	1,157	28	Northeast Point	470
21	Lagoon	190	29	....do	335
21	Northeast Point	742	29	English Bay and Tolstoi	1,038
23	Reef, Lukannon, Zoltoi	1,209	July 1	Northeast Point	1,200
23	Northeast Point	917	1	Reef, Zoltoi, Lukannon	1,023
24	Halfway Point	347	2	Halfway Point	834
24	Northeast Point	970	2	Northeast Point	968
25	English Bay, Tolstoi, Middle Hill	1,619	3	English Bay, Tolstoi, Middle Hill	1,841
25	Northeast Point	1,028	4	Reef, Zoltoi, Lukannon	1,706
26	Reef, Zoltoi, Lukannon	1,353	4	Northeast Point	1,559
26	Northeast Point	650	5	Southwest Bay	1,255
27	Southwest Bay and Zoltoi	950	5	Northeast Point	1,524
Aug. 2	Zoltoi	177	6	English Bay, Tolstoi, Middle Hill	1,302
8	....do	140	6	Northeast Point	376
16	....do	159	8	Reef, Zoltoi, Lukannon	814
23	Middle Hill and Lukannon	364	8	Northeast Point	914
24	Zoltoi	321	9	English Bay and Tolstoi	1,314
24	....do	19	9	Northeast Point	641
			10	Halfway Point	654
			10	Northeast Point	800
			12	Reef and Zoltoi	2,004

Tables showing in detail all killing of fur seals, for whatsoever purpose, on the Pribilof Islands during the term of the lease of the Alaska Commercial Company—that is, from 1870 to 1889, both inclusive—Continued.

ST. PAUL ISLAND—Continued.

Date.	Rookery.	Total killed for all purposes.	Date.	Rookery.	Total killed for all purposes.
1889.			**1889.**		
July 13	Southwest Bay	1,006	July 27	Zoltoi and Lukannon	1,105
13	Northeast Point	793	29	English Bay and Middle Hill	1,643
15	English Bay and Middle Hill	3,085	29	Northeast Point	1,624
15	Northeast Point	1,838	30	Halfway Point	973
16	Reef, Zoltoi, Lukannon	1,911	30	Southwest Bay	615
16	Northeast Point	1,156	31	Northeast Point	538
17	Halfway Point and Lukannon	1,931	31	Zoltoi	160
17	Northeast Point	948	Aug. 6	Lukannon	163
18	Lagoon, English Bay, Middle Hill	2,046	14	Zoltoi	131
18	Northeast Point	1,282	22	....do	141
19	Southwest Bay	2,017	22	Tolstoi	179
19	Northeast Point	834	Sept. 9	Zoltoi	141
20	Reef and Zoltoi	1,913	18	....do	110
20	Northeast Point	243	25	....do	107
22	English Bay, Middle Hill, Lukannon	1,943	Oct. 5	....do	120
22	Northeast Point	350	15	....do	103
23	Reef, Zoltoi, Ketova	1,122	26	Lukannon	132
23	Northeast Point	740	Nov. 4	Zoltoi	1,169
24	Halfway Point	1,384	19	Tolstoi	1,460
24	Northeast Point	616	21	Reef	347
25	English Bay and Middle Hill	1,756	27	....do	192
25	Northeast Point	1	27	Zapadnie	10
26	Southwest Bay	680	30	Reef	240
26	Northeast Point	1,483	Dec. 11	Zapadnie	243
				Total	73,982

ST. GEORGE ISLAND.

Date.	Rookery.	Total killed for all purposes.	Date.	Rookery.	Total killed for all purposes.
1870.			**1872.**		
	Not indicated in the records of this year.		June 3	Southwest Bay	140
			5	North	26
	Alaska Commercial Co., Hutchinson, Kohl & Co ⎰	1,200	8	East	49
		473	10	Southwest Bay	162
	⎱	6,786	11	North	81
	Total	8,459	11	Starry Arteel	175
1871.			12	Southwest Bay	98
June 4	Near	123	12	East	61
6	Northeast	98	15	Starry Arteel	140
8	Near	69	15	North	188
9	Southwest Bay	277	17	East	405
13	Starry Arteel	322	19	North	300
15	Southwest Bay	301	19	Starry Arteel	212
17	Northern	434	19	Southwest Bay	261
20	Southwest Bay	172	22	East	860
22	Northeast	518	22	North	349
24	Starry Arteel	594	22	Starry Arteel	701
26	Southwest Bay	298	24	Southwest Bay	261
27	Northeast	462	24	East	629
28	Starry Arteel	571	25	Starry Arteel	500
July 1	Northern	875	25	North	237
3	Southwest Bay	303	27	Starry Arteel	805
5	Starry Arteel	518	28	North	400
8	Southwest Bay	612	29	Starry Arteel	560
10	Northern	1,769	29	Southwest Bay	643
12	....do	1,021	July 1	East	981
14	Southwest Bay	481	2	North	454
15	Northern	1,038	2	Starry Arteel	431
18	....do	1,264	3	Southwest Bay	245
20	Southwest Bay	484	4	East	641
21	Northern	945	5	Starry Arteel	300
23	Southwest Bay	542	6	Southwest Bay	574
25	Northern	792	7	East	718
27	....do	1,054	9	Starry Arteel	367
28	Starry Arteel	730	9	North	300
30	Southwest Bay	1,270	10	East	610
31	Northern	893	11	Southwest Bay	1,412
	Taken in October	237	12	North	482
	Pups for food	2,090	14	East	1,332
			15	Starry Arteel	600
	Total	21,157	15	North	583
			17	East	770
			18	Starry Arteel	575

Tables showing in detail all killing of fur seals, for whatsoever purpose, on the Pribilof Islands during the term of the lease of the Alaska Commercial Company—that is, from 1870 to 1889, both inclusive—Continued.

ST. GEORGE ISLAND—Continued.

Date.	Rookery.	Total killed for all purposes.	Date.	Rookery.	Total killed for all purposes.
1872.			**1874.**		
July 19	Southwest Bay	1,171	June 29	Starry Arteel and North	686
20	East	400	July 1	East	800
22	Starry Arteel	600	3	Starry Arteel and North	792
22	North	320	8	East	641
23	East	703	9	....do	548
25	Starry Arteel	300	14	East and North	263
25	East	400	15	East	534
25	North	252	16	Starry Arteel	568
27	East	350	18	Southwest Bay	411
27	North	85	20	East	871
27	Starry Arteel	200	22	East and North	778
27	Southwest Bay	227	24	East	640
	Killed for natives' food	2,000	24	North	156
				Pups killed for food	2,446
	Total	27,000		Total	12,446
1873.					
June 4	North	198	**1875.**		
5	Starry Arteel	240	June 1	Starry Arteel and North	302
6	Southwest Bay	285	9	....do	256
9	East and Starry Arteel	190	11	East	177
10	Southwest Bay	275	14	Starry Arteel and North	307
12	North	300	16	East	358
13	Southwest Bay	521	18	Starry Arteel and North	334
16	North and Starry Arteel	378	19	Southwest Bay	1,294
17	Southwest Bay	174	23	East	666
19	East	313	24	Starry Arteel and North	540
21	Starry Arteel and North	506	28	East	692
21	Southwest Bay	870	30	Starry Arteel and North	1,412
23	East	180	July 5	East	717
24	Southwest Bay	499	7	Starry Arteel and North	1,019
25	Starry Arteel and North	195	12	East	1,073
26	East	241	14	North	676
27	Southwest Bay	301	17	....do	177
28	Starry Arteel and North	493		Killed for food	1,500
30	Southwest Bay	310			
30	East	168		Total	11,500
July 2	Starry Arteel	332			
3	Southwest Bay	564	**1876.**		
4	East	592	June 1	North	415
5	Starry Arteel	517	8	Starry Arteel and North	372
8	Southwest Bay	743	12	East	388
8	East	616	14	Southwest Bay	599
9	Starry Arteel and North	690	15	Starry Arteel and North	784
11	East	974	22	East	581
11	Southwest Bay	602	25	Starry Arteel and North	2,067
12	Starry Arteel and North	474	27	East	1,168
13	East	345	29	Starry Arteel and North	1,023
14	Southwest Bay	337	July 3	East	1,259
16	Starry Arteel and North	480	6	Starry Arteel	1,027
17	East	1,097	7	East	317
18	Southwest Bay	913		Pups for food	1,500
20	Starry Arteel and North	1,359			
21	East	1,810		Total	11,500
23	Southwest Bay	513			
23	Starry Arteel	889	**1877.**		
25	East	1,710	June 1	North	198
26	Southwest Bay	600	12	....do	702
28	Starry Arteel	588	13	East	578
28	East	1,528	14	Southwest Bay	1,389
	Pups killed for food	2,190	18	North and Starry Arteel	1,154
			20	East	838
	Total	27,190	22	North	871
1874.			23	East	552
June 1	North	56	26	North and Starry Arteel	1,360
8	....do	81	29	East	1,589
11	East	116	July 3	North and Starry Arteel	1,669
12	Starry Arteel and North	154	6	East	2,164
14	Southwest Bay	250	9	North	300
16	East	170	10	East	880
18	Starry Arteel and North	354		Killed for food	256
22	East	178		....do	1,500
23	Starry Arteel and North	378			
27	Southwest Bay	575		Total	16,500

Tables showing in detail all killing of fur seals, for whatsoever purpose, on the Pribilof Islands during the term of the lease of the Alaska Commercial Company—that is, from 1870 to 1889, both inclusive—Continued.

ST. GEORGE ISLAND—Continued.

Date.	Rookery.	Total killed for all purposes.	Date.	Rookery.	Total killed for all purposes.
1878.			**1880.**		
June 10	North	325	June 14	East	352
14	Southwest Bay	1,074	15	Southwest Bay	738
17	North, Starry Arteel, East	858	17	....do	254
19	Southwest Bay	717	17	North and Starry Arteel	559
22	North and Starry Arteel	570	19	East	599
25	East	324	19	Southwest Bay	223
27	Southwest Bay	851	21	North and Starry Arteel	1,183
28	....do	517	22	Southwest Bay	518
	Killed for food to date	405	23	East	814
July 1	East	644	25	Southwest Bay	839
2	North and Starry Arteel	930	25	North and Starry Arteel	1,322
4	Southwest Bay	1,433	28	East	1,770
8	East	793	29	Southwest Bay	846
10	North and Starry Arteel	1,333	30	Starry Arteel	808
12	Southwest Bay	328	July 1	North	392
13	....do	1,025	2	East	956
15	East	1,892	2	Southwest Bay	961
17	East and North	1,290	5	North and Starry Arteel	515
19	North and Starry Arteel	1,577	6	East	1,483
21	East	1,291	7	Southwest Bay	1,814
	Killed for food to May 19, 1879..	2,627	9	East	949
				During season perished on the drives	28
	Total	20,804	16	North	72
			17	Zapadnie	7
1879.	North	69	20	....do	8
June 3	East	450	28	North	60
10	Southwest Bay	105	Aug. 6	....do	51
11	Starry Arteel	413	11	North and East	226
12	Southwest Bay	372	Sept. 1	North	40
13	East	445	2	Zapadnie	35
16	Southwest Bay	502	27	North	47
17	Starry Arteel and North	755	Oct. 5	....do	62
19	East	473	28	....do	501
20	Southwest Bay	434	Nov. 1	....do	765
20	North and Starry Arteel	515	9	Zapadnie	30
23	Southwest Bay	576	18	....do	10
23	East	888	Dec. 1	East	05
25	Southwest Bay	524	3	Zapadnie	16
25	....do	278			
26	North and Starry Arteel	1,179		Total	20,939
27	East	1,595			
30	North and Starry Arteel	1,414	**1881.**		
July 3	Southwest Bay	849	May 21	North	32
3	....do	351	31	....do	55
4	North	535	June 9	Starry Arteel, East, North	612
5	East	1,775	13	....do	920
6	North and Starry Arteel	1,263	15	Zapadnie	408
8	East	1,840	16	Starry Arteel, North, East	622
14	North	863	20	....do	445
16	Southwest Bay	1,360	21	East and Zapadnie	1,030
16	....do	8	23	Starry Arteel, North, Zapadnie	518
24	North	63	24	East	553
28	....do	48	27	Starry Arteel and North	815
Aug. 6	....do	08	28	East and Zapadnie	1,119
10	North	54	30	Zapadnie, Starry Arteel, North	1,034
Sept. 1	....do	47	July 1	East	1,378
23	....do	58	4	Starry Arteel and North	1,182
Oct. 2	....do	48	6	Zapadnie	476
15	....do	18	7	East	1,356
Nov. 3	....do	318	8	Starry Arteel and North	363
8	Killed for food	1,506	11	East	1,310
8	Southwest Bay and Starry Arteel		12	Starry Arteel	498
Dec. 6	teel	113	12	Zapadnie	771
			14, 15	East	1,715
	Total	22,190	14	Zapadnie	592
			16	East, North, Starry Arteel	1,639
1880.			26	North	43
May 18	North	14	30	....do	45
26	Starry Arteel	23	Aug. 9	....do	57
June 3	North	82	15	....do	92
9	East	338	26	....do	52
12	North and Starry Arteel	564	Sept. 3	....do	52

Tables showing in detail all killing of fur seals, for whatsoever purpose, on the Pribilof Islands during the term of the lease of the Alaska Commercial Company—that is, from 1870 to 1889, both inclusive—Continued.

ST. GEORGE ISLAND—Continued.

Date.	Rookery.	Total killed for all purposes.	Date.	Rookery.	Total killed for all purposes.
1881.			**1883.**		
Sept. 13	North	65	July 31	Starry Arteel, North, East	473
24	East	88		During theseason, perished on drives	38
Oct. 3	North	68	Aug. 13	East	100
22	East	50	21	Near	50
Nov. 2	North	559	31	Little East	60
10	....do	472	Sept. 14	East	54
30	Starry Arteel	113	Oct. 19	Near	120
	Total	21,289	29	Near and North	540
			Nov. 2	North	340
1882.			19	North and Starry Arteel	171
May 22	North	12			
29	....do	48		Total	16,214
June 6	....do	26			
12	Starry Arteel, North, East	509	**1884**		
16	....do	890	May 21	East	15
19	....do	927	26	North	52
22	....do	847	June 4	East	119
24	....do	1,195	10	Zapadnie	1,222
27	....do	1,044	12	Starry Arteel, North, East	604
29	....do	1,273	16	Zapadnie	585
July 1	....do	1,065	18	Starry Arteel, North, East	572
3	....do	911	21	Zapadnie	592
5	....do	1,384	23	Starry Arteel, North, East	603
8	....do	1,955	26	Zapadnie	578
10	East	1,371	28	Starry Arteel, North, East	489
11	Starry Arteel and North	1,115	July 1	Zapadnie	289
12	East	1,070	1	....do	11
14	Starry Arteel and North	527	3	Starry Arteel, North, East	71
15	East	640	7	Zapadnie	90
17	Starry Arteel and North	1,022	10	Starry Arteel, North, East	1,269
18	East	1,086	12	Zapadnie	973
19	Starry Arteel and North	655	14	East and Little East	302
20	North	50	15	Starry Arteel and North	465
Aug. 5	East	40	16	Zapadnie	726
11	Starry Arteel	45	18	East and Little East	996
17	East	34	19	Starry Arteel and North	506
25	....do	44	23	Zapadnie	797
Sept. 15	....do	46	24	Starry Arteel and North	744
Nov. 22	Starry Arteel	119	25	East and Little East	597
28	....do	19	26	Zapadnie	573
	Total	19,978	30	Starry Arteel, North, East	640
			Aug. 4	East	225
1883.			4	....do	19
May 26	North	40	14	North	48
June 4	....do	78	21	East	64
12	Starry Arteel and East	136	Sept. 2	Near	60
15	....do	287	15	North	60
19	Starry Arteel	61	Nov. 3	....do	515
22	East and Starry Arteel	380	5	....do	482
25	East, North, Starry Arteel	684	12	....do	503
28	....do	443	26	Starry Arteel	27
30	....do	611			
July 2	....do	340		Total	16,573
4	....do	200			
7	....do	647	**1885.**		
9	Zapadnie	1,336	May 18	North	40
10	Little East and East	307	June 1	....do	38
10	Zapadnie	507	11	Starry Arteel, North, East	750
11	North and Starry Arteel	263	15	Zapadnie	77
12	Little East and East	546	18	....do	698
13	North and Starry Arteel	321	17	Starry Arteel and North	802
16	Little East and East	775	15	East	825
17	Zapadnie	1,017	22	Zapadnie	414
17	Starry Arteel and North	130	27	Starry Arteel, North, East	1,775
18	Little East and East	467	29	Zapadnie	401
19	Zapadnie	1,216	July 3	Starry Arteel, North, East	2,287
20	Little East and East	281	7	Zapadnie	789
21	Zapadnie	1,150	10	Starry Arteel, North, East	2,158
23	North, East, Starry Arteel	766	13	Zapadnie	1,011
25	East	78	17	Starry Arteel, North, East	2,222
27	Starry Arteel, North, East	606	20	Zapadnie	483
30	....do	505	25	North	35
			Aug. 3	....do	23

272 · ALASKA INDUSTRIES.

Tables showing in detail all killing of fur seals, for whatsoever purpose, on the Pribilof Islands during the term of the lease of the Alaska Commercial Company—that is, from 1870 to 1889, both inclusive—Continued.

ST. GEORGE ISLAND—Continued.

Date.	Rookery.	Total killed for all purposes.	Date.	Rookery.	Total killed for all purposes.
1885.			**1887.**		
Aug. 6	North	50	Aug. 3	Near	52
21	East	65	6	East	41
25	North	37	12	....do	87
Nov. 4	Near	250	20	....do	23
6	North	700	22	....do	61
18	....do	120	Sept. 5	Near	44
21	Starry Arteel	41	10	Zapadnie	24
	Perished on drives during year.	53	19	....do	45
			Oct. 24	East	126
	Total	16,144	Nov. 1	Near	766
			7	....do	614
1886.					
May 18	East	39		Total	16,668
28	North	102			
June 8	Captured skins	4	**1888.**		
14	Zapadnie food skins	81	June 6	North	121
10	Starry Arteel, North, East	1,430	11	Zapadnie	272
14	Zapadnie	779	12	Starry Arteel, North, East	455
17	Starry Arteel, North, East	1,438	16	....do	227
21	Zapadnie	843	18	Zapadnie	427
22	Starry Arteel and North	742	19	Starry Arteel, North, East	324
23	East	343	22	....do	764
24	Zapadnie	306	25	Zapadnie	911
28	....do	288	26	Starry Arteel and North	895
29	Starry Arteel and North	632	27	East	438
July 1	East	482	29	Starry Arteel and North	343
5	Zapadnie	620	July 2	Zapadnie	343
6	Starry Arteel and North	503	3	East	532
7	East	650	4	Starry Arteel and North	503
10	Starry Arteel and North	867	6	East	650
12	Zapadnie	745	9	Zapadnie	389
14	East	888	10	Starry Arteel and North	1,170
15	Starry Arteel and North	712	12	East	820
19	Zapadnie	663	13	Starry Arteel and North	518
21	Starry Arteel and North	853	16	Zapadnie	705
22	East	527	17	North	410
23	East and North	295	18	Starry Arteel and East	692
Aug. 2	Near North	14	19	Zapadnie	366
2	....do	11	20	Starry Arteel, North, East	554
9	East	66	23	Zapadnie	179
17	North	42	24	Starry Arteel, North, East	405
23	East	70	25	Zapadnie	159
Sept. 6	North	76	26	Starry Arteel, North, East	521
Oct. 26	....do	759	27	Zapadnie	144
28	East	24	27	Starry Arteel, North, East	410
Nov. 8	North	527	Aug. 9	Near	39
13	East	3	9	North	39
	Perished on drive	12	15	...do	37
			20	East	191
	Total	16,436	28	Starry Arteel and North	494
			Sept. 5	East and Starry Arteel	428
1887.			27	North	40
May 20	North	25	Oct. 20	North and East	73
28	...do	32	Nov. 1	North	610
June 9	North and East	390	5	...do	368
14	North, East, Starry Arteel	465	Aug. 20	Zapadnie (for watchmen)	68
15	Zapadnie	427			
20	....do	261		Total	17,034
21	Starry Arteel, North, East	974			
25	East	533	**1889.**		
27	Zapadnie	599	May 22	North and East	60
28	Starry Arteel and North	847	June 4	East	156
30	East	410	10	Zapadnie	207
July 1	Starry Arteel and North	100	17	...do	244
4	Zapadnie	883	18	Starry Arteel, North, East	773
6	East, North, Starry Arteel	1,321	21	East	176
8	East	421	22	North and Starry Arteel	284
11	Zapadnie	701	24	Zapadnie	596
12	Starry Arteel, Near, North	1,296	25	East and North	496
14	East	1,509	27	Zapadnie	223
18	Zapadnie	1,077	29	Starry Arteel and East	429
19	Near, North, Starry Arteel	894	July 1	Zapadnie	167
21	East	1,130	2	North, East, Starry Arteel	275
22	Starry Arteel and North	489	5	....do	418

Tables showing in detail all killing of fur seals, for whatsoever purpose, on the Priblof Islands during the term of the lease of the Alaska Commercial Company—that is, from 1870 to 1889, both inclusive—Continued.

ST. GEORGE ISLAND—Continued.

Date.	Rookery.	Total killed for all purposes.	Date.	Rookery.	Total killed for all purposes.
1889.			**1889.**		
July 8	Zapadnie	229	Aug. 10	North	55
10	North, East, Starry Arteel......	270	19	....do...........................	56
12	Zapadnie	192	30	.....do..........................	48
13	North, East, Starry Arteel	667	Sept. 7	East.............................	64
15	Zapadnie	371	21	.....do..........................	50
16	North, East, Starry Arteel.....	1,028	30	North	33
18	Zapadnie	439	Oct. 11	....do...........................	37
19	North, East, Starry Arteel	1,140	21	Starry Arteel...................	32
22	Zapadnie	500	31	North	4
23	North, East, Starry Arteel.....	628	Nov. 6	....do...........................	606
24	Zapadnie	279	12	....do...........................	477
25	North, East, Starry Arteel.....	1,430	25	Starry Arteel...................	61
27	....do	942			
28	Zapadnie	568		Total.....................	15,225
29	North, East, Starry Arteel.....	515			

Fur seals killed on the island of St. Paul, for all purposes, from 1870 to 1889, both inclusive.

[Compiled from tables on file in the Treasury Department.]

Year.	Seals killed for natives' food.				Seals killed for skins for lessees.			Total of bachelors killed, accepted and rejected.			Grand total of seals killed for all purposes.
	Pups.	Bachelors.	Skins accepted.	Skins rejected.	Bachelors.	Skins accepted.	Skins rejected.	Bachelors.	Skins accepted.	Skins rejected.	
1870	2,800	6,449	......	6,449	6,065	6,017	48	12,514	6,017	6,497	15,314
1871	2,877	2,341	2,290	51	75,585	74,628	957	77,926	76,918	1,008	81,803
1872	5,121	6,916	5,365	1,551	69,782	69,576	206	76,698	74,941	1,757	81,819
1873	5,489	2,090	1,198	892	74,408	73,884	524	76,498	75,082	1,416	81,987
1874	4,897	4,874	4,225	649	88,368	88,258	110	93,242	92,483	759	98,139
1875	3,745	6,282	5,784	498	84,933	84,800	73	91,215	90,644	571	94,960
1876	3,958	5,061	3,064	1,997	74,138	71,137	1	79,199	77,201	1,998	83,157
1877	5,007	4,041	2,853	1,186	58,762	58,732	30	62,803	61,585	1,218	67,810
1878	5,206	4,718	3,632	1,086	78,595	78,570	25	83,313	82,202	1,111	88,519
1879	5,071	5,070	3,898	2,072	77,280	77,280	......	83,250	81,178	2,072	88,321
1880	4,413	4,466	3,408	1,418	75,900	75,872	28	80,306	78,920	1,446	84,779
1881	......	7,538	6,068	1,470	76,236	76,169	67	83,774	82,226	1,537	83,774
1882	......	5,175	3,362	1,813	74,659	74,581	78	79,834	77,943	1,891	79,834
1883	2,982	3,168	2,194	974	57,145	57,070	75	60,313	59,264	1,049	63,295
1884	2,741	3,907	2,582	1,325	82,213	82,086	127	86,120	84,668	1,452	88,861
1885	2,788	3,184	2,508	676	82,908	82,866	42	86,092	85,374	718	88,880
1886	2,824	3,081	2,450	601	82,180	82,150	30	85,261	84,630	631	88,065
1887	2,177	4,207	3,975	232	82,708	82,679	29	86,915	86,654	261	89,092
1888	2,178	3,762	3,700	62	80,330	80,314	16	84,092	84,014	78	86,270
1889	2,280	3,400	2,570	830	81,712	81,698	14	85,112	84,268	844	87,392
Total .	67,554	90,630	64,796	25,834	1,463,907	1,461,427	2,480	1,554,537	1,526,212	28,314	1,622,091

NOTE.—The above statement includes all seals killed from all causes, either intentional or accidental, incident to the taking of seal skins on the island of St. Paul.

Fur seals killed on the island of St. George, for all purposes, from 1870 to 1889, both inclusive.

[Compiled from tables on file in the Treasury Department.]

Year.	Seals killed for natives' food.				Seals killed for skins for lessees.			Total of bachelors killed, accepted and rejected.			Grand total of seals killed for all purposes.
	Pups.	Bachelors.	Skins accepted.	Skins rejected.	Bachelors.	Skins accepted.	Skins rejected.	Bachelors.	Skins accepted.	Skins rejected.	
1870	1,200				7,259	7,259		7,259	7,259		8,459
1871	2,090	237	237		18,830	18,830		19,067	19,067		21,157
1872	2,000				25,000	25,000		25,000	25,000		27,000
1873	2,190				25,000	25,000		25,000	25,000		27,190
1874	2,446				10,000	10,000		10,000	10,000		12,446
1875	1,500				10,000	10,000		10,000	10,000		11,500
1876	1,500				10,000	10,000		10,000	10,000		11,500
1877	1,500	256	256		14,744	14,744		15,000	15,000		16,500
1878	1,500	1,532	1,216	316	17,772	17,772		19,304	18,988	316	20,804
1879	1,506	843	564	279	19,841	19,758	83	20,684	20,322	362	22,190
1880	1,330	702	565	137	18,907	18,830	77	12,609	19,395	214	20,939
1881	1,031	812	509	303	19,446	19,360	86	20,258	19,869	389	21,289
1882		483	371	112	19,495	19,440	35	19,978	19,811	167	19,978
1883	1,000	475	468	7	14,739	14,675	64	15,214	15,143	71	16,214
1884	1,500	345	223	122	14,728	14,620	108	15,073	14,843	230	16,573
1885	1,080	319	304	15	14,745	14,686	59	15,064	14,990	74	16,144
1886	1,286	544	413	131	14,606	14,578	28	15,150	14,991	159	16,436
1887	1,356	585	471	114	14,727	14,725	2	15,312	15,196	116	16,668
1888	978	1,409	1,321	88	14,647	14,582	65	16,056	15,903	153	17,034
1889	1,071	512	280	232	13,642	13,641	1	14,154	13,921	233	15,225
Total	28,064	9,054	7,198	1,856	318,128	317,500	628	327,182	324,698	2,484	355,246

NOTE.—The above statement includes all seals killed from all causes, either intentional or accidental, incident to the taking of seal skins on the island of St. George.

Fur seals killed on the islands of St. Paul and St. George, for all purposes, from 1870 to 1889, both inclusive.

	Seals killed for natives' food.				Seals killed for skins for lessees.			Total of bachelors killed, accepted and rejected.			Grand total of seals killed for all purposes.
	Pups.	Bachelors.	Skins accepted.	Skins rejected.	Bachelors.	Skins accepted.	Skins rejected.	Bachelors.	Skins accepted.	Skins rejected.	
St. Paul Island	67,554	90,630	64,796	25,834	1,463,907	1,461,427	2,480	1,554,537	1,526,212	28,314	1,622,091
St. George Island	28,064	9,054	7,198	1,856	318,128	317,500	628	327,182	324,698	2,484	355,246
Total	95,628	99,684	71,994	27,790	1,782,035	1,778,927	3,108	1,881,719	1,850,910	30,798	1,977,337

Seals taken on St. Paul Island in 1890.

Date.	Rookery.	Total.	Date.	Rookery.	Total.
May 28	Southwest Bay..................	119	July 4	Tolstoi, English Bay, Middle Hill	494
June 6	Reef	116	4	Northeast Point................	321
11	.....do..........................	574	5	Reef............................	526
13	Tolstoi..........................	182	5	Northeast Point................	74
16	Reef	317	7	English Bay, Middle Hill, Tol-	
17	Northeast Point................	16		stoi, Lukannon, Ketova.......	411
17	Halfway Point	167	7	Northeast Point................	336
18	Tolstoi and Middle Hill........	274	8	Halfway Point.................	261
18	Northeast Point................	78	8	Northeast Point................	379
20	Reef and Lukannon............	339	9	Southwest Bay.................	163
20	Northeast Point................	438	9	Northeast Point................	271
21	Southwest Bay.................	292	10	Reef............................	378
21	Northeast Point................	96	10	Northeast Point................	112
23	English Bay and Lukannon....	521	12	English Bay, Middle Hill, Tol-	
23	Northeast Point................	179		stoi, Lukannon, Ketova......	633
24	Reef and Zoltoi................	426	13	Halfway Point	211
24	Northeast Point................	205	13	Northeast Point................	658
25	Halfway Point.................	266	14	Reef............................	104
25	Northeast Point................	166	15	English Bay, Middle Hill, Tol-	
26	Southwest Bay.................	117		stoi, Lukannon, Ketova......	315
27	English Bay and Middle Hill...	396	15	Northeast Point................	245
27	Northeast Point................	230	16	.....do..........................	312
28	Reef	206	17	Polavina, Lukannon, Ketova...	372
28	Northeast Point................	79	17	Northeast Point................	485
30	Tolstoi, Middle Hill, English		18	.....do..........................	405
	Bay, Ketova..................	209	18	Zapadnie.......................	236
30	Northeast Point................	98	19	Reef and Zoltoi	556
July 1	Reef............................	246	19	Northeast Point................	446
1	Northeast Point................	131	20	English Bay, Middle Hill, Tol-	
2	Halfway Point	242		stoi, Ketova, Rocky Point....	780
2	Northeast Point................	96	20	Northeast Point................	556
3	Southwest Bay.................	183			
3	Northeast Point................	180		Total....................	17,124

Seals taken on St. George Island in 1890.

Date.	Rookery.	Total.	Date.	Rookery.	Total.
June 2	North	71	July 8	East and Little East............	24
16	East.............................	218	9	Starry Arteel and North........	193
18	North...........................	118	11	East.............................	60
19	East and Little East............	181	12	Starry Arteel and North........	103
20	Zapadnie........................	304	14	Zapadnie........................	53
23	Starry Arteel and North........	164	15	East.............................	132
25	East and Little East............	184	16	Starry Arteel and North........	119
28	Starry Arteel and North........	189	18	East.............................	71
30	Zapadnie........................	180	20	Starry Arteel and North........	641
July 1	East and Little East............	149	20	Zapadnie........................	527
3	Starry Arteel and North........	238			
5	East and Little East............	57		Total...................	4,133
7	Zapadnie........................	58			

PAST AND FUTURE OF THE FUR SEAL.[1]

BY JOSEPH STANLEY-BROWN.

There are but two groups of fur seals to furnish to the world its supply of seal skins, the fur seal of the north and the fur seal of the south.

When Sir Francis Drake circumnavigated the globe in 1577–1580 the *Arctocephalus,* or southern fur seal, was to be found at not less than thirty localities, and their numbers aggregated millions. To-day the contributions of these southern waters are from three resorts, and do not usually reach 15,000 skins annually.

When Vitus Bering, in 1741, was wrecked upon the Commander Islands, off the coast of Kamchatka, and Pribilof searched out, in 1786–87, the group of islands in Bering Sea that bears his name, there were discovered, not only the chief breeding grounds of the northern fur seal, *Callorhinus ursinus,* but some of the most superb seal rookeries the world has ever known. It is questionable if mortal vision ever rested upon more magnificent displays of amphibian life than were to be seen on the island of St. Paul at the time of its discovery. To-day these subarctic resorts are prostrate; their glory also has departed, and they furnish a home for but a mere remnant of the seals that formerly swarmed in myriads along their rocky shores.

For two years the hopes of thoughtful persons were high, that through the medium of international negotiations and the deliberations of wise and able men the safety of the fur seal would be at last secured. To-day, when the decision of the Paris Tribunal is common property, we find public opinion divided on the question as to whether the practical application of the decision will preserve the fur seal as a commercial commodity.

CHARACTERISTICS OF THE SEAL.

The condition of affairs thus briefly outlined is all the more deplorable when we consider the characteristics of the animal with which we are dealing. It is a creature peculiarly adapted by its habits to man's management. It occupies no territory needed, as were the buffalo's feeding grounds, for the subsistence of more valuable domestic animals; no herders are required to prevent its being lost in the wastes of the ocean, and no expense is incurred either to protect it from the inclemency of the weather or to provide a winter food supply; yet with more certainty than the ranchman's flocks and herds seek the home range do the seals annually return to their breeding grounds where, under proper management, they can without injury to the parent stock be made to yield a profit equal to if not greater than that derived from the cattle of the plains or the sheep of the mountains.

THE SOUTHERN FUR SEAL AND ITS DESTRUCTION.

Despite these characteristics, which must have been apparent to the most ignorant and unobservant, what has been the course of events? Turning first to the fur seal of the south we find that as early as 1690 some little interest was manifested in its capture, but it was not until the close of the last century that the pursuit was begun in earnest. Hardy mariners, stimulated by the hope of sharing in the profits of the

---

[1] From Bulletin United States Fish Commision, 1893, pp. 361–370.

fur trade which the Russians had developed with the Chinese, searched out the resorts of the southern fur seal; ravaged them year after year, in season and out of season; slaughtered the helpless creatures with clubs on land regardless of age or sex; gathered a harvest of 16,000,000 or 17,000,000 skins, and by 1830 had practically destroyed, in the southern seas, this valuable fur-bearing animal. If all these resorts were in their original condition and under wise and prudent direction, they could easily supply to the fur trade annually something like a half a million skins, with corresponding advantage to an army of skilled artisans. As it is, indiscriminate butchery has left only the Lobos Islands rookeries at the mouth of the La Plata River and a few insignificant resorts at Cape Horn and the Cape of Good Hope, the total yearly yield of which is, as before stated, less than 15,000 skins. Such destruction is left absolutely without justification in the face of man's entire ability to maintain the fur seal rookeries at the highest possible limits permitted by the operation of nature's restrictions, or when depleted to develop them again. This is not idle speculation, but rests upon a firm foundation of fact furnished by the history of the fur seal of the north.

THE NORTHERN FUR SEAL AND ITS RELATION TO THE SEAL-SKIN INDUSTRY.

The two great resorts of the northern fur seal are the Pribilof and Commander islands in Bering Sea. Robbens Reef, a rocky islet in the Okhotsk Sea, has a small rookery, and a few localities of minor importance are found along the Kurile Islands. While the Russians who first discovered these resorts prohibited all interference from outsiders, their own treatment of the seals was similar to that practiced by the sailors in the south. No attention was paid to sex, season, or period of procreation, and it was not long before the end came there just as it had done in the south. The Russians were taught by this severe lesson that the only way in which the rookeries could be restored and perpetuated was to protect the females from death and the breeding grounds from molestation. This course, accompanied by practically a suspension of killing during certain years, was rigidly adhered to, with the result that when the rookeries of the Pribilof Islands were turned over to the United States in 1867 their condition, instead of being one of exhaustion, approximated that which existed when they were first discovered. The truth of this will be more apparent when it is stated that in 1868, before the United States could assume and exercise control over its newly acquired possessions, nearly a quarter of a million skins were improperly taken from the islands of St. Paul and St. George by unauthorized persons without apparently producing any diminution of the numbers which came the following year.

Although there are but four of these northern localities, and Russian mismanagement from time to time played such havoc with them that the catch was an uncertain quantity, still they have contributed since their discovery between 5,000,000 and 6,000,000 skins to the fur trade, or about one-third as many as have been furnished by the southern resorts. From the time that the fur seal of the south ceased to be of commercial importance trade has relied upon these rookeries. Thanks to the more enlightened policy employed by the Russians, and adopted and improved upon by the United States, these rookeries of Bering Sea contributed to commerce for the twenty years ending with 1889 a uniform yearly quota of nearly 150,000 pelts, which formed the basis of and made possible the systematized seal-skin business of modern

times. As a raw commodity they sold for an average of $2,500,000 at
the annual London trade sales, and the Pribilof quota yielded the Government of the United States in revenue more than the $7,200,000
originally paid for the entire Territory of Alaska. The value of raw
seal skins is now represented by about $15 for skins taken at sea and
$30 for Pribilof skins. At the present revenue rate, if it were now
possible to take from the Pribilof Islands the former yearly quota, the
Government income would be nearly $1,000,000 annually.

IMPORTANCE OF THE SEAL-SKIN INDUSTRY.

The seal-skin industry is of no slight importance, and its proportions
are but roughly indicated by the first profit on the raw skins. These
peltries must be gathered in remote regions; they form part of the
transportation business of railroad and steamship lines; coopers must
make casks for their shipment; they must pass through the hands of
many laborers before they reach the 40 buyers in London who purchase
them, and the 2,000 skilled artisans who convert them into fabrics
suited to the use of trade; and when all this is done there must still be
stores maintained and clerks employed in order that they may find
their way to the wealthy consumers. The labor incident to the taking,
transporting, manipulating, and disposing of these peltries demands
the employment of thousands of persons each year, and when we recall
the prices paid for these skins when converted into the garments dic-
tated by fashion, it will readily be seen that it is an industry the ultimate value of which is represented by millions of dollars annually.
Above all it is a peculiarly worthy industry, in that it gives occupation
to many, while the profits come from the purses of those best able to
pay them.

CAUSE OF THE DESTRUCTION OF THE NORTHERN FUR SEAL.

Some ten years ago there was put in operation on the American side
of the Pacific Ocean an agency of destruction, the growth of which, if
uninterrupted, promised to prove as effective as did the sailors' clubs
upon the southern resorts. Its promise has been generously kept, and
from its deadly though partially controlled effects the rookeries are
now suffering. That agency was pelagic sealing, or the taking of seals
at sea by means of weapons. The source of the injury is the indiscriminate killing. Whether this is practiced on land, as in the south,
or at sea, as in the north, the outcome is the same. No animal which
produces but a single offspring each year can long survive an attack
which involves the death of the producing class, the females. I am
aware that there is another side to this question, and that two great
nations point each a finger at the other and say: "You did it." The
subject-matter of that contention is only germane to such a paper as
this in so far as it touches upon the career of the seal, and only to that
extent will it be referred to.

England and Canada hold the theory (which, in justice to them,
should be stated) that the decline of the northern rookeries was due to
excessive killing on the islands, pelagic sealing being a factor of only
secondary importance. If this theory meant that after pelagic sealing
had made serious inroads upon the seal herds it was excessive killing
to continue taking the annual quota of 100,000 skins, it would be a
sound one, and the United States would be culpable to that extent,
but England and Canada would not accept this limitation; they want
it to account for much more. They fail, however, to sustain their

theory until they show by clearest proof that the decline of the rookeries began prior to the development of pelagic sealing, and also get rid of the awkward fact that for the first twelve or fifteen years there was no difficulty in securing the annual quota allowed by law. Why did this alleged decadence through excessive killing on land take so long to manifest itself? Certainly the evils of indiscrimination is not inherent in land killing; on the contrary, selection can be exercised at the rookeries as readily as it can be at the abattoir, and there is no more necessity for molesting the females than there would be for a farmer to ship all of his herd to Kansas City and have the selection of the killable males made at the stock yards. The briefest recital of the facts of seal life will make this plain.

THE FACTS OF SEAL LIFE.

The northern fur seals, unlike their southern relatives, are forced each year by Arctic cold and the necessity for food to leave their homes on the approach of winter and to seek the southern waters and the abundant fish supply along the continental shores. The migration routes of the Alaskan and Asiatic herds do not coalesce, nor do the seals intermingle. Late in April or early in May, depending upon the character of the season, the breeding males, bulls, or "seecatchie," first return to their resorts from this migration. About a month later the mature females or "matkie" begin to seek the breeding grounds, and between the time of arrival of these two classes the young males or "hollustchikie" are swimming in the water near the rookery fronts or hauling out upon the hauling grounds some distance away from the areas occupied by the mature seals. The young males are not permitted to gather upon the breeding grounds until, by reason of age and strength, they are able to maintain a position there.

Each old bull when he arrives in the spring selects and maintains, often by desperate combat, a little area upon which he hopes to establish his household. The male weighs four or five times as much as his consort, and, as is usually the case where the male preponderates in size, they are extremely polygamous. Their vitality and virility is almost beyond belief. For eighty or ninety days, while they are making secure their position, and while guarding and presiding over their families or "harems," they are debarred from both food and water. When the season of propagation is past they again betake themselves to the sea, and the breeding grounds are given up to the intermingling of young males, females, and pups, but during that eighty or ninety days the immature males from 1 to 5 years of age have been compelled to consort together upon the hauling grounds, and thus there is given an opportunity without in any way interfering with the course of events upon the breeding grounds, to drive away, select, and slaughter such of these young males as will furnish desirable pelts. These are the only skins shipped from the islands.

Can anyone successfully maintain that in the case of polygamous animals the taking of the surplus male life and reserving the females can destroy the herd? If this can be demonstrated, then our stock-raisers are at fault, and the evidence derived from Russian management goes for naught.

THE FACTS OF PELAGIC SEALING.

Before the breath of life can be breathed into this theory of decadence through excessive killing on the islands there must be removed from the record books certain well-established facts concerning pelagic sealing. It will be necessary to dispose of the fact that while in 1878 there was but 1 vessel engaged in pelagic sealing, the number steadily increased until in 1892 there were 122 to follow on the migration tracks of the herds, to harry them eight months out of the twelve, and, if permitted, to accompany them to and even upon their chosen resorts. There must also be a successful refutation of the fact that there is a loss of at least 10 per cent inherent in the methods of taking seals at sea; that pelagic sealing strikes at the very life of the rookeries, by killing 75 or 80 per cent of the females, more than half of which are mothers whose death involves that of their unborn offspring; and that the period of gestation being nearly twelve months, a mother killed in Bering Sea means that three seal lives may pay the penalty.

It is equally important to the maintenance of this theory that there be an elimination of the fact that during the four seasons, ending with the past one of 1893, there were taken on the Pribilof Islands only a total of 50,000 skins of young males, while during that same period there were actually marketed by the sealers over 200,000 skins, which represented only about half the injury done the seal herds, an injury falling heaviest upon the producing class, the females. For four years there has been practically a closed time on these islands, and pelagic sealing has had full swing in the North Pacific. The rookeries have not improved under these conditions, and until the records of the real cause of destruction stand impeached it is idle to offer obscure and improbable explanations for the present condition of seal life.

It has only been profitable to follow this question of the cause of the decadence to indicate what might be expected from pelagic sealing. Whenever and to whatever extent carried on, its deadly effects are certain and continuous, the amount of injury being limited only by the magnitude of the enterprise. Improprieties on land can be guarded against, but the disastrous consequences of pelagic sealing are inherent to the business and are beyond man's control. They can be lessened, but only through the curtailment of the number of seals taken. The injurious effect upon the herd, while proportionately less, remains a constant factor.

In following the career of an animal possessing such capacity for self-perpetuation and ready adaptability to the uses of man, the student of natural history or of economics is struck by the wanton and needless destruction which pursues it wherever found. As to its future he turns, for what comfort he may be able to extract, to the decision of that court of recent if not last resort—the Paris Tribunal of Arbitration.

THE PARIS TRIBUNAL OF ARBITRATION.

The causes which led to the arbitration are known to all. For some years the Alaskan fur seal, when on its migration route, had been the eagerly sought quarry of the pelagic hunters. This route, which by reason of its vast extent and proximity to inhabited shores makes this herd especially vulnerable to attack, extends from the Pribilof Islands southward through the passes of the Aleutian chain, expands in the broad Pacific, but ultimately brings the seals in more compact masses to the North American Coast, and thence along its shores, back through the passes, to the Pribilof Islands again. Realizing the peril of the rookeries, the Government of the United States attempted to partially protect them by seizing sealing schooners in Bering Sea. Each year it was thought that at least so far as these waters were concerned the danger would cease, but each year it increased as the vessels multiplied and the skill and knowledge of the sealers became greater and was ultimately extended to the Asiatic herd which frequents the Russian or Commander Islands. The continued seizing of schooners by the United States met with remonstrances on the part of Canada and England, and finally, after much irritation and heat, became the subject of diplomatic negotiations, the peaceful outcome of which was the Paris Tribunal of Arbitration.

Three duties were intrusted to the Tribunal of Arbitration: It was to settle certain jurisdictional questions, to decide the question of property rights, and in the event of the matter being left in such shape that the concurrence of Great Britain was necessary to establish regulations for the purpose of protecting and preserving the fur seal, it was to frame such regulations as would be applicable outside of the jurisdiction of the respective Governments and to indicate the nonterritorial waters over which these regulations should extend. As it is not important in this connection to consider the jurisdictional phases of the case there will be taken up at once the property question and the regulations—the two points that immediately concern us; the former from the standpoint of general interest, and the latter by reason of their intimate relation to the future of the seals.

THE AMERICAN POSITION.

The able representatives of the United States took the position that the tribunal was bound by no precedents, and possessed, by virtue of its very origin, a creative as well as a judicial function. They urged upon the tribunal the taking of high ground and the settlement of the question upon broad and comprehensive principles. They pointed out that man, by means of invention, was rapidly extending his dominion over the water, as he had over the land, and, by employing methods which were not even dreamed of when many existing municipal and international laws were enacted, threatened the very existence of many creatures useful to man. Turning from the citations of voluminous authorities vindicating the justness of their claim of property right in the seals and in the industry, they pleaded with sturdy argument and great eloquence that the tribunal would fail of its high duty did it not lend its aid to such an extension of the world's idea of property right as was needed to meet the demands of the advancing age. They asked that the narrow ground be not taken that this great tribunal was called into existence solely for the purpose of settling a dispute between two nations, but that it was given an opportunity and was vested with the power to make a substantial contribution to international law, and that its verdict, while disposing of the immediate matter in dispute, should be such a formulation, upon broader lines, of our conception of rights of property and of protection as would be of value to all mankind, irrespective of nations. They pointed out that the material progress of the world was based upon the fundamental principle of ownership, and that the most effective way of preventing the commercial annihilation of certain great groups of creatures was by lodging in the nation best qualified by its geographic position to protect them a custodianship, to be exercised over them for the benefit of all. It was shown that the adoption of this principle would dispose of the question of the relation of other governments to the subject; would make possible the rehabilitation of many of the seal rookeries of the south; that it would protect such industries as the coral and pearl fisheries, and that it would be useful in controlling the rapid inroads man's ingenuity is now making on the denizens of the sea. In short, that it would be a direct, useful, and common-sense way of settling the whole matter.

THE BRITISH POSITION.

With equal skill of argument and eloquence of address the advocates of Great Britain and Canada held that the tribunal possessed but one function—that its duty was to declare the law and not to make it; but that, whatever its function might be as an international body, it was not vested with the power to make international law, but must keep to the straight and narrow way of settling a contention between two nations and adjusting two conflicting methods of catching seals. They asked that the tribunal provide for the continuation of pelagic sealing under the most favorable conditions consistent with carrying out the terms of the treaty. True, nothing was said in the treaty about preserving the business of pelagic sealing, but before so patient and generous a court it was not difficult to confuse the issue of preserving the seals and continuing pelagic sealing and to take up a large share of the proceedings with pleadings in behalf of the latter. They demanded that the question of property right be settled from the standpoint that the seals were wild animals, which man could only reduce to possession by destroying. They insisted that the law relating to wild animals, regardless of its origin, had been accepted by nations as the years ran on; it was very old law and very good law; but, whether good or bad, it was *the* law, and from its teachings the tribunal must not allow itself to be enticed away by the seductive citations and insidious arguments of learned counsel on the other side. There must be no making of laws to suit new conditions; the old stand-bys must be adhered to, whether applicable or not. They urged that the seals being wild animals, the United States had done nothing to encourage or develop in them the animum revertendi—the inclination to return to their homes, as in the case of bees and similar creatures—and thus had lost their claim to a property in them, and if the world or a part of it desired to turn out in boats and to destroy the industry by shooting the seals in the water they had a perfect right to do so, for a wild animal was free to all. No matter if seal mothers roaming the sea for food did fall before the gun or spear of the pelagic hunter and their helpless pups starve on the rookeries, the hand of destruction must not be stayed, for the United States had no rights anyone was bound legally to respect when the seals were 3 miles off shore, and humanitarian considerations had no place in the controversy. They insisted that the tribunal had no authority in law to declare a property right in the seals or in the industry, but if the tribunal contemplated disregarding the law and settling this question on lines of their own choosing they must refrain from doing so, because it would interfere with that wonderful invention, the immemorial right on the high seas, an interference nations not only would not brook, but which they would actively resent.

THE TRIBUNAL'S DECISION.

The tribunal, true to the conservatism of the Old World, accepted this interpretation of their powers, recognized the potency of venerable legal relics, assented to the arguments of the counsel for Great Britain and Canada based thereon, and contented itself with deciding that the United States had no right of protection or property in the fur seals.

THE REGULATIONS.

The next task to which the tribunal addressed itself was the framing of regulations. These regulations furnish the last hope for the preservation of the fur seal as a commercial commodity. It is not probable that any other nations having seal interests will be content with less than the United States secured, nor is it likely they will obtain more, and thus they represent the measure of protection all seals are likely to receive in the future.

After listening to an enormous mass of testimony—some good, some bad, and some very indifferent—concerning seal life, the tribunal proposes to preserve the Alaskan branch of the northern fur seal by prohibiting sealing within a zone of 60 miles around the Pribilof Islands by establishing a closed time, or time of no killing at sea, from May 1 to July 31; by permitting only sailing vessels to engage in the business of seal hunting, and requiring them to carry a distinctive flag, to take out a special license, and to keep a daily record of the catch and the sex of the seals taken, these records to be communicated to each of the two Governments at the close of the sealing season; by limiting the weapons of capture to shotguns in the North Pacific and spears in Bering Sea, and by requiring the two Governments to take such measures as will determine whether the hunters are fit to handle with sufficient skill the weapons by means of which the seals are to be captured. These regulations, which are to remain in force until they have been in whole or in part abolished or modified by common agreement between the Governments of the United States and Great Britain, are to be submitted every five years to a new examination, so as to enable both Governments to consider whether, in the light of past experience, there is occasion for any modification of them.

The three prime points in the regulations are: The zone around the islands; the closed time of three months injected into the middle of the sealing season, thus breaking it up, and the restriction of the use of firearms to the North Pacific.

First as to the zone: If there was any one fact clearly established by the testimony of the pelagic sealers themselves and official experts it was that in the summer season great numbers of seals, and especially females, are found at long distances from the islands of Bering Sea, distances two or three times greater than that of the protecting zone provided by the regulations. Now, as the object was to preserve the fur seals, it is proper to assume that the tribunal, prompted by a desire to protect them, and acting in good faith, established such a zone as they believed would practically prohibit the attack of the pelagic sealer; but if this was so, then mere amount of distance was immaterial, and in view of the fact that incessant fogs brood over the waters of Bering Sea during the summer season, rendering it difficult to tell when a vessel is within or without a zone, the limit of which can not be marked, why not at once adopt that natural and well-defined boundary line, the Aleutian chain? Just here arises the question: When vessels are seized, whose word shall be accepted as to the locality of seizure—the pelagic sealer's or the seizing officer's? Does not this uncertainty, having as it does an important bearing on the question of conviction, weaken the regulations restraining influence on pelagic sealing? Aside from questions of protection it seems to me that this part of the decision will tend to increase dispute and bitterness rather than to diminish it.

The adoption of the closed time means the recognition on the part of

the tribunal that the destruction by the pelagic sealer has been excessive, and the cutting off of one month of the sealing season in Bering
Sea clearly shows that it realized the danger to the herd from allowing
sealing there. Why, then, was sealing not prohibited altogether in
those waters? Is the danger less in August and a portion of September? The seals are still going long distances from the islands and the
sealer can continue his work until stopped by the September gales.
Bering Sea is the focal point, the great massing ground of seal life, and
the seals are more readily taken there than anywhere else. In 1891 the
catch of the Canadian fleet in the North Pacific was a little over 21,000
seals, and before the modus vivendi could be enforced a portion of the
fleet sealed from three to five weeks on the American side of Bering Sea,
and with fewer vessels and with fewer small boats they took in that time
as many seals as they had previously secured in the Pacific. During
the three years ending with and including 1891 the Canadian fleet (and
I only quote from Canadian records, because they are so reliable) took,
in five months, in the North Pacific, an average of 567 skins per vessel;
with ten vessels less, they took in Bering Sea 727 skins per vessel in
about two and one-half months.

The proposed regulations still allow at least five weeks' sealing in
Bering Sea; but, say the regulations, the hunters can only use spears
in Bering Sea, thereby intimating that spears are less effective than the
shotguns allowed in the North Pacific, and that an additional safeguard
has therefore been provided in Bering Sea. Just why the shotgun is
pernicious in Bering Sea and is not in the North Pacific is not indicated;
but if we turn to the testimony of the Northwest Coast Indians, who
ship on the schooners and accompany them to Bering Sea, we find that
they claim that they can do better work with the spear than with the
shotgun. The latter makes the game wild, while the former does not.
The spear makes no noise, and they are thus able to take seal after seal
as they sleep on the water, and get all in sight, while at the sound of a
gun's discharge the comrades of the captured or wounded seal swim
away.

It is evident from an inspection of these regulations as a whole that
the tribunal, taking into account the interests of both nations, endeavored
to frame measures which, while protecting the seals, would permit the
continuation of pelagic sealing. This seems to me a task the accomplishment of which is an impossibility. The evils of pelagic sealing
appear to have been clearly recognized by the tribunal, but instead of
adopting prohibitive measures it took the middle course of throwing
some protection around the seals, and while at the same time appearing
to concede something to the pelagic sealers, made the conditions just
sufficiently hard as to prevent them from engaging successfully in the
business. It is admitted that these regulations possess value in limiting and discouraging pelagic sealing, but their inherent weakness is
that while they now seem to possess some deterring power, changed
conditions may at any time arise which will negative their influence
and offer inducements sufficient to enable the sealers to again engage
in this business on a large and injurious scale. This contingency is
not so remote as may appear at first sight. In 1889 the average price
paid in Victoria for skins taken at sea was $6.83; in 1890 it had risen
to $10.70; in 1891 it was $15. In 1889 the cost of each skin in wages
was from $2 to $3; in 1890 and 1891 it was $3.50; in 1892 it was $4;
in other words, an advancing price for both master and hunter.

Now, it is evident that it will be some time before the Pribilof Islands
can very greatly increase their annual output of skins. The maximum

output of the Commander Islands has been reached, and probably will have to be decreased in the future. There must be through these regulations some curtailment of the contribution of the scaling schooners, and the result of all this will be that seal skins will demand a higher price. Should that price reach a figure which will compensate for the obstacles which the regulations place in the way of the pelagic sealer, then we will have the changed conditions referred to, and pelagic sealing with its attendant evils will go on as before. If there is doubt in the minds of anyone upon this point it is only necessary to turn to the history of the sea otter, which, though nearly exterminated, is as eagerly sought after to-day as it ever was, simply because the ever-increasing price the trade is willing to pay for its skin still compensates for the small numbers now taken. There is no reason to believe that the career of the fur seal will be different from that of the sea otter.

Another possible source of changed conditions lies in the regulations themselves, for they provide, as we have seen, for their own modification every five years, and the pressure will come heaviest from the pelagic sealers' side of the case. Indeed, the regulations require that each pelagic sealer—an interested party—shall keep records which are to be made available when the question of modifications of the regulations arises. Now, while there never was a more fearless and courageous set of men than these pelagic sealers, it will be something entirely new in their history if their records do not appeal in the strongest possible terms for a modification of the regulations in their favor.

The final question that arises in regard to these regulations is, will they, as they now stand, ever be put in operation? The interested powers have yet to agree upon measures for giving effect to them. Is it likely that, when a neutral tribunal found the making of regulations so tedious and difficult, the interested powers will be able without interminable delay and possibly irreconcilable conflict to agree upon "concurrent measures" putting them in force? England has won on the great law points of the case, but these regulations are objectionable to Canada, for they bear somewhat heavily upon pelagic sealing; and these "concurrent measures" offer tempting fighting ground for securing their modification in favor of the Dominion.

Under the circumstances it is only to be expected that the arts of diplomacy will be vigorously exercised in that direction. There is but one course, however, for the United States to pursue—permit no modifications, stand squarely for the prompt carrying out of these regulations. and let time reveal how much value they possess for protecting the seal herd. England will champion no plan of greater protection; she has all to gain and nothing to lose from delay, and it will require all the energy and firmness of the Executive to put effectively in force the regulations as adopted by the tribunal.

CONCLUSIONS.

After more than two years of close study of this question it is my conviction that the only way in which the world can secure the largest benefit commercially from the fur seal wherever found is by taking the surplus immature males upon land under the most favorable conditions suggested by experience; that securing seals by any other methods introduces the fatal element of indiscrimination; that the life of the herd is jeopardized in proportion to the number of females killed; that the injury inflicted on the northern herds by pelagic sealing increases from January to August, grows greater as Bering Sea is approached, and culminates in those waters; that the shotgun and spear are both deadly, the latter by reason of its noiseless efficiency, the former by reason of its ready use by all classes, and that the disposition of this question on the basis of adjusting two conflicting interests is futile and illogical, but material issues are not alone involved; it presents biologic features as well and has to do with forces of nature beyond man's control.

Regulations can not be framed by human ingenuity which will preserve the seal herds in their greatest possible proportions and permit the continuation of successful pelagic sealing. It would be reconciling the irreconcilable. It would be accomplishing a feat equal to that of making two bodies occupy the same space at the same time. Either the regulations will be prohibitive in their operation—in which case it would be more straightforward to make them so in the first instance—or, if allowing successful pelagic sealing, they will be valueless in preventing the extermination of the seal. In general it may be said that no pelagic sealing can be carried on which is not inherently and uncontrollably injurious to the life of the seal herd—the amount of injury being proportionate to the magnitude of the attack.

AWARD OF THE TRIBUNAL OF ARBITRATION CONSTITUTED UNDER THE TREATY CONCLUDED AT WASHINGTON THE 29TH OF FEBRUARY, 1892, BETWEEN THE UNITED STATES OF AMERICA AND HER MAJESTY THE QUEEN OF THE UNITED KINGDOM OF GREAT BRITAIN AND IRELAND.

[English version.]

Whereas, by a treaty between the United States of America and Great Britain, signed at Washington, February 29, 1892, the ratifications of which by the Governments of the two countries were exchanged at London on May 7, 1892, it was, amongst other things, agreed and concluded that the questions which had arisen between the Government of the United States of America and the Government of Her Britannic Majesty, concerning the jurisdictional rights of the United States in the waters of Bering Sea, and concerning also the preservation of the fur seal in or habitually resorting to the said sea, and the rights of the citizens and subjects of either country as regards the taking of fur seals in or habitually resorting to the said waters, should be submitted to a tribunal of arbitration to be composed of seven arbitrators, who should be appointed in the following manner, that is to say: Two should be named by the President of the United States; two should be named by Her Britannic Majesty; His Excellency the President of the French Republic should be jointly requested by the high contracting parties to name one; His Majesty the King of Italy should be so requested to name one; His Majesty the King of Sweden and Norway should be so requested to name one; the seven arbitrators to be so named should be jurists of distinguished reputation in their respective countries, and the selecting powers should be requested to choose, if possible, jurists who are acquainted with the English language;

And whereas it was further agreed by Article II of the said treaty that the arbitrators should meet at Paris within twenty days after the delivery of the counter cases mentioned in Article IV, and should proceed impartially and carefully to examine and decide the questions which had been or should be laid before them as in the said treaty provided on the part of the Governments of the United States and of Her Britannic Majesty, respectively, and that all questions considered by the tribunal, including the final decision, should be determined by a majority of all the arbitrators;

And whereas by Article VI of the said treaty, it was further provided as follows:

In deciding the matters submitted to the said arbitrators, it is agreed that the following five points shall be submitted to them in order that their award shall embrace a distinct decision upon each of said five points, to wit:

1. What exclusive jurisdiction in the sea now known as Bering Sea, and what exclusive rights in the seal fisheries therein, did Russia assert and exercise prior and up to the time of the cession of Alaska to the United States?

2. How far were these claims of jurisdiction as to the seal fisheries recognized and conceded by Great Britain?

3. Was the body of water now known as Bering Sea included in the phrase Pacific Ocean, as used in the treaty of 1825 between Great Britain and Russia; and what rights, if any, in Bering Sea were held and exclusively exercised by Russia after said treaty?

4. Did not all the rights of Russia, as to jurisdiction and as to the seal fisheries in Bering Sea east of the water boundary, in the treaty between the United States and Russia of the 30th of March, 1867, pass unimpaired to the United States under that treaty?

5. Has the United States any right, and if so, what right, of protection or property in the fur seals frequenting the islands of the United States in Bering Sea when such seals are found outside the ordinary 3-mile limit?

And whereas, by Article VII of the said treaty, it was further agreed as follows:

If the determination of the foregoing questions as to the exclusive jurisdiction of the United States shall leave the subject in such position that the concurrence of Great Britain is necessary to the establishment of regulations for the proper protection and preservation of the fur seal in, or habitually resorting to, Bering Sea, the arbitrators shall then determine what concurrent regulations, outside the jurisdiction limits of the respective Governments, are necessary, and over what waters such regulations should extend;

The high contracting parties furthermore agree to cooperate in securing the adhesion of other powers to such regulations;

And whereas, by Article VIII of the said treaty, after reciting that the high contracting parties had found themselves unable to agree upon a reference which should include the question of the liability of each for the injuries alleged to have been sustained by the other, or by its citizens, in connection with the claims presented and urged by it, and that "they were solicitious that this subordinate question should not interrupt or longer delay the submission and determination of the main questions," the high contracting parties agreed that "either of them might submit to the arbitrators any question of fact involved in said claims and ask for a finding thereon, the question of the liability of either Government upon the facts found, to be the subject of further negotiation;

And whereas the President of the United States of America named the Hon. John M. Harlan, Justice of the Supreme Court of the United States, and the Hon. John T. Morgan, Senator of the United States, to be two of the said arbitrators, and Her Britannic Majesty named the Right Hon. Lord Hannen and the Hon. Sir John Thompson, minister of justice and attorney-general for Canada, to be two of the said arbitrators, and His Excellency the President of the French Republic named the Baron de Courcel, senator, ambassador of France, to be one of the said arbitrators; and His Majesty the King of Italy named the Marquis Emilio Visconti Venosta, former minister of foreign affairs and senator of the Kingdom of Italy, to be one of the said arbitrators; and His Majesty the King of Sweden and Norway named Mr. Gregers Gram, minister of state, to be one of the said arbitrators;

And whereas we, the said arbitrators so named and appointed, having taken upon ourselves the burden of the said arbitration, and having duly met at Paris, proceeded impartially and carefully to examine and decide all the questions submitted to us, the said arbitrators, under the said treaty, or laid before us as provided in the said treaty on the part of the Governments of Her Britannic Majesty and the United States, respectively;

Now we, the said arbitrators, having impartially and carefully examined the said questions, do in like manner by this our award decide and determine the said questions in the manner following; that is to say, we decide and determine as to the five points mentioned in Article VI as to which our award is to embrace a distinct decision upon each of them:

As to the first of the said five points, we, the said Baron de Courcel, Mr. Justice Harlan, Lord Hannen, Sir John Thompson, Marquis Visconti Venosta, and Mr. Gregers Gram, being a majority of the said arbitrators, do decide and determine as follows:

By the ukase of 1821 Russia claimed jurisdiction in the sea now known as Bering Sea to the extent of 100 Italian miles from the coast and islands belonging to her; but, in the course of the negotiations which led to the conclusion of the treaties of 1824 with the United States and of 1825 with Great Britain, Russia admitted that her jurisdiction in the said sea should be restricted to the reach of cannon shot from shore, and it appears that from that time up to the time of the cession of Alaska to the United States Russia never asserted in fact or exercised any exclusive jurisdiction in Bering Sea or any exclusive rights in the seal fisheries therein beyond the ordinary limit of territorial waters.

As to the second of the said five points, we, the said Baron de Courcel, Mr. Justice Harlan, Lord Hannen, Sir John Thompson, Marquis Visconti Venosta, and Mr. Gregers Gram, being a majority of the said arbitrators, do decide and determine that Great Britain did not recognize or concede any claim upon the part of Russia to exclusive jurisdiction as to the seal fisheries in Bering Sea outside of ordinary territorial waters.

As to the third of the said five points, as to so much thereof as requires us to decide whether the body of water now known as Bering Sea was included in the phrase "Pacific Ocean" as used in the treaty of 1825 between Great Britain and Russia, we, the said arbitrators, do unanimously decide and determine that the body of water now known as Bering Sea was included in the phrase "Pacific Ocean" as used in the said treaty.

And as to so much of the said third point as requires us to decide what rights, if any, in Bering Sea were held and exclusively exercised by Russia after the said treaty of 1825, we, the said Baron de Courcel, Mr. Justice Harlan, Lord Hannen, Sir John Thompson, Marquis Visconti Venosta, and Mr. Gregers Gram, being a majority of the said arbitrators, do decide and determine that no exclusive rights of jurisdiction in Bering Sea and no exclusive rights as to the seal fisheries therein were held or exercised by Russia outside of ordinary territorial waters after the treaty of 1825.

As to the fourth of the said five points, we, the said arbitrators, do unanimously decide and determine that all the rights of Russia as to jurisdiction and as to the seal fisheries in Bering Sea east of the water boundary, in the treaty between the United States and Russia of the 30th March, 1867, did pass unimpaired to the United States under the said treaty.

As to the fifth of the said five points, we, the said Baron de Courcel, Lord Hannen, Sir John Thompson, Marquis Visconti Venosta, and Mr. Gregers Gram, being a majority of the said arbitrators, do decide and determine that the United States has not any right of protection or property in the fur seals frequenting the islands of the United States in Bering Sea when such seals are found outside the ordinary 3-mile limit.

And whereas the aforesaid determination of the foregoing questions as to the exclusive jurisdiction of the United States, mentioned in Article VI, leaves the subject in such a position that the concurrence of Great Britain is necessary to the establishment of regulations for the proper protection and preservation of the fur seals in or habitually resorting to Bering Sea, the tribunal having decided by a majority as to each article of the following regulations, we, the said Baron de Courcel, Lord Hannen, Marquis Visconti Venosta, and Mr. Gregers Gram, assenting to the whole of the nine articles of the following regulations, and being a majority of the said arbitrators, do decide and determine in the mode provided by the treaty that the following concurrent regulations outside the jurisdictional limits of the respective Governments are necessary, and that they should extend over the waters hereinafter mentioned, that is to say:

REGULATIONS.

ARTICLE 1.

The Governments of the United States and of Great Britain shall forbid their citizens and subjects, respectively, to kill, capture, or pursue at any time and in any manner whatever the animals commonly called fur seals within a zone of 60 miles around the Pribilof Islands, inclusive of the territorial waters.

The miles mentioned in the preceding paragraph are geographical miles, of 60 to a degree of latitude.

ARTICLE 2.

The two Governments shall forbid their citizens and subjects, respectively, to kill, capture, or pursue, in any manner whatever, during the season extending each year from the 1st of May to the 1st of July, both inclusive, the fur seals on the high sea, in the part of the Pacific Ocean, inclusive of Bering Sea, which is situated to the north of the thirty-fifth degree of north latitude and eastward of the one hundred and eightieth degree of longitude from Greenwich, till it strikes the water boundary described in Article I of the treaty of 1867 between the United States and Russia, and following that line up to Bering Straits.

ARTICLE 3.

During the period of time and in the waters in which the fur-seal fishing is allowed, only sailing vessels shall be permitted to carry on or take part in fur-seal fishing operations. They will, however, be at liberty to avail themselves of the use of such canoes or undecked boats, propelled by paddles, oars, or sails as are in common use as fishing boats.

ARTICLE 4.

Each sailing vessel authorized to fish for fur seals must be provided with a special license issued for that purpose by its Government, and shall be required to carry a distinguishing flag, to be prescribed by its Government.

H. Doc. 92, pt. 2——19

ARTICLE 5.

The masters of the vessels engaged in fur-seal fishing shall enter accurately in their official log book the date and place of each fur-seal fishing operation, and also the number and sex of the seals captured upon each day. These entries shall be communicated by each of the two Governments to the other at the end of each fishing season.

ARTICLE 6.

The use of nets, firearms, and explosives shall be forbidden in the fur-seal fishing. This restriction shall not apply to shotguns when such fishing takes place outside of Bering Sea during the season when it may be lawfully carried on.

ARTICLE 7.

The two Governments shall take measures to control the fitness of the men authorized to engage in fur-seal fishing. These men shall have been proved fit to handle with sufficient skill the weapons by means of which this fishing may be carried on.

ARTICLE 8.

The regulations contained in the preceding articles shall not apply to Indians dwelling on the coasts of the territory of the United States or of Great Britain, and carrying on fur-seal fishing in canoes or undecked boats not transported by paddles, oars, or sails, and manned by not more than five persons each in the way hitherto practiced by the Indians, provided such Indians are not in the employment of other persons, and provided that, when so hunting in canoes or undecked boats, they shall not hunt fur seals outside of territorial waters under contract for the delivery of the skins to any person.

This exemption shall not be construed to affect the municipal law of either country, nor shall it extend to the waters of Bering Sea or the waters of the Aleutian Passes.

Nothing herein contained is intended to interfere with the employment of Indians as hunters or otherwise in connection with fur-sealing vessels, as heretofore.

ARTICLE 9.

The concurrent regulations hereby determined with a view to the protection and preservation of the fur seals shall remain in force until they have been in whole or in part abolished or modified by common agreement between the Governments of the United States and of Great Britain.

The said concurrent regulations shall be submitted every five years to a new examination, so as to enable both interested Governments to consider whether in the light of past experience there is occasion for any modification thereof.

And whereas the Government of Her Britannic Majesty did submit to the Tribunal of Arbitration by Article VIII of the said treaty certain questions of fact involved in the claims referred to in said Article VIII, and did also submit to us, the said tribunal, a statement of the said facts, as follows, that is to say:

FINDINGS OF FACT PROPOSED BY THE AGENT OF GREAT BRITAIN AND AGREED TO AS PROVED BY THE AGENT FOR THE UNITED STATES AND SUBMITTED TO THE TRIBUNAL OF ARBITRATION FOR ITS CONSIDERATION.

1. That the several searches and seizures, whether of ships or goods, and the several arrests of masters and crews, respectively, mentioned in the schedule to the

British case, pages 1 to 60, inclusive, were made by the authority of the United States Government. The questions as to the value of the said vessels or their contents, or either of them, and the question as to whether the vessels mentioned in the schedule to the British case, or any of them, were wholly or in part the actual property of the citizens of the United States have been withdrawn from and have not been considered by the tribunal, it being understood that it is open to the United States to raise these questions or any of them, if they think fit, in any future negotiations as to the liability of the United States Government to pay the amounts mentioned in the schedule of the British case.

2. That the seizures aforesaid, with the exception of the *Pathfinder*, seized at Neah Bay, were made in Bering Sea at the distances from shore mentioned in the schedule annexed hereto, marked C.

3. That the said several searches and seizures of vessels were made by public armed vessels of the United States, the commanders of which had, at the several times when they were made, from the Executive Department of the Government of the United States, instructions, a copy of one of which is annexed hereto, marked A, and that the others were, in all substantial respects, the same; that in all the instances in which proceedings were had in the district courts of the United States resulting in condemnation, such proceedings were begun by the filing of libels, a copy of one of which is annexed hereto, marked B, and that the libels in the other proceedings were in all substantial respects the same; that the alleged acts or offenses for which said several searches and seizures were made in each case were done or committed in Bering Sea at the distances from shore aforesaid; and that in each case in which sentence of condemnation was passed, except in those cases when the vessels were released after condemnation, the seizure was adopted by the Government of the United States; and in those cases in which the vessels were released the seizure was made by the authority of the United States; that the said fines and imprisonments were for alleged breaches of the municipal laws of the United States, which alleged breaches were wholly committed in Bering Sea at the distances from the shore aforesaid.

4. That the several orders mentioned in the schedule annexed hereto and marked G, warning vessels to leave or not to enter Bering Sea, were made by public armed vessels of the United States, the commanders of which had, at the several times when they were given, like instructions as mentioned in finding 3, and that the vessels so warned were engaged in sealing or prosecuting voyages for that purpose, and that such action was adopted by the Government of the United States.

5. That the district courts of the United States in which any proceedings were had or taken for the purpose of condemning any vessel seized, as mentioned in the schedule to the case of Great Britain, pages 1 to 60, inclusive, had all the jurisdiction and powers of courts of admiralty, including the prize jurisdiction, but that in each case the sentence pronounced by the court was based upon the grounds set forth in the libel.

ANNEX A.

TREASURY DEPARTMENT, OFFICE OF THE SECRETARY,
Washington, April 21, 1886.

SIR: Referring to Department letter of this date, directing you to proceed with the revenue steamer *Bear*, under your command, to the seal islands, etc., you are hereby clothed with full power to enforce the law contained in the provisions of section 1956 of the United States Revised Statutes, and directed to seize all vessels and arrest and deliver to the proper authorities any or all persons whom you may detect violating the law referred to, after due notice shall have been given.

You will also seize any liquors or firearms attempted to be introduced into the country without proper permit, under the provisions of section 1955 of the Revised Statutes, and the proclamation of the President dated February 4, 1870.

Respectfully, yours,

C. S. FAIRCHILD, *Acting Secretary.*

Capt. M. A. HEALY,
Commanding Revenue Steamer Bear, San Francisco, Cal.

ANNEX B.

In the district court of the United States for the District of Alaska—August special term, 1886.

To the Hon. LAFAYETTE DAWSON, *Judge of said District Court:*

The libel of information of M. D. Ball, attorney for the United States for the District of Alaska, who prosecutes on behalf of said United States, and being present here in court in his proper person, in the name and on behalf of said United States, against the schooner *Thornton*, her tackle, apparel, boats, cargo, and furniture, and against all persons intervening for their interest therein, in a cause of forfeiture, alleges and informs as follows:

That Charles A. Abbey, an officer in the Revenue-Marine Service of the United States, and on special duty in the waters of the District of Alaska, heretofore, to wit, on the 1st day of August, 1886, within the limits of Alaska Territory, and in the waters thereof, and within the civil and judicial District of Alaska, to wit, within the waters of that portion of Bering Sea belonging to the said district, on waters navigable from the sea by vessels of 10 or more tons burden, seized the ship or vessel commonly called a schooner, the *Thornton*, her tackle, apparel, boats, cargo, and furniture, being the property of some person or persons to the said attorney unknown, as forfeited to the United States, for the following causes:

That the said vessel or schooner was found engaged in killing fur seal within the limits of Alaska Territory, and in the waters thereof, in violation of section 1956 of the Revised Statutes of the United States.

And the said attorney saith that all and singular the premises are and were true, and within the admiralty and maritime jurisdiction of this court, and that by reason thereof and by force of the statutes of the United States in such cases made and provided, the aforementioned and described schooner or vessel, being a vessel of over 20 tons burden, her tackle and apparel, boats, cargo, and furniture, became forfeited to the use of the said United States, and that said schooner is now within the district aforesaid.

Wherefore the said attorney prays the usual process and monition of this honorable court issue in this behalf, and that all persons interested in the before-mentioned and described schooner or vessel may be cited in general and special to answer the premises, and all due proceedings being had, that the said schooner or vessel, her tackle, apparel, boats, cargo, and furniture may, for the cause aforesaid, and others appearing, be condemned by the definite sentence and decree of this honorable court, as forfeited to the use of the said United States, according to the form of the statute of the said United States in such cases made and provided.

M. D. BALL,
United States District Attorney for the District of Alaska.

ANNEX C.

The following table shows the names of the British sealing vessels seized or warned by the United States revenue cruisers, 1886–1890, and the approximate distance from land when seized. The distances assigned in the cases of the *Carolena*, *Thornton*, and *Onward* are on the authority of the United States Naval Commander Abbey (see Senate Ex. Doc. No. 106, pp. 20, 30, 40, Fiftieth Congress, second session). The distances assigned in the cases of the *Anna Beck*, *W. P. Sayward*, *Dolphin*, and *Grace* are on the authority of Captain Shepard, United States Revenue Marine (Blue Book, United States, No. 2, 1890, pp. 80–82. See Appendix, Vol. III).

Name of vessel.	Date of seizure.	Approximate distance from land when seized.	United States vessel making seizures.
Carolena	Aug. 1, 1886	75 miles	Corwin.
Thornton	...do	70 miles	Do.
Onward	Aug. 2, 1886	115 miles	Do.
Favourite	...do	Warned by Corwin in about same position as Onward.	
Anna Beck	July 2, 1887	66 miles	Rush.
W. P. Sayward	July 9, 1887	59 miles	Do.
Dolphin	July 12, 1887	40 miles	Do.
Grace	July 17, 1887	96 miles	Do.
Alfred Adams	Aug. 10, 1887	62 miles	Do.
Ada	Aug. 25, 1887	15 miles	Bear.
Triumph	Aug. 4, 1887	Warned by Rush not to enter Bering Sea	
Juanita	July 31, 1889	66 miles	Rush.
Pathfinder	July 29, 1889	50 miles	Do.
Triumph	July 11, 1889	Ordered out of Bering Sea by Rush. (?) As to position when warned.	
Black Diamond	...do	35 miles	Do.
Lily	Aug. 6, 1889	66 miles	Do.
Ariel	July 30, 1889	Ordered out of Bering Sea by Rush	
Kate	Aug. 13, 1889	...do	
Minnie	July 15, 1889	65 miles	Do
Pathfinder	Mar. 27, 1890	Seized in Neah Bay. (?)	Corwin

And whereas the Government of Her Britannic Majesty did ask the said arbitrators to find the said facts as set forth in the said statement, and whereas the agent and counsel for the United States Government thereupon in our presence informed us that the said statement of facts was sustained by the evidence, and that they had agreed with the agent and counsel for Her Britannic Majesty that we, the arbitrators, if we should think fit so to do, might find the said statement of facts to be true:

Now we, the said arbitrators, do unanimously find the facts as set forth in the said statement to be true.

And whereas each and every question which has been considered by the tribunal has been determined by a majority of all the arbitrators:

Now we, Baron de Courcel, Lord Hannen, Mr. Justice Harlan, Sir John Thompson, Senator Morgan, the Marquis Visconti Venosta, and Mr. Gregers Gram, the respective minorities not withdrawing their votes, do declare this to be the final decision and award in writing of this tribunal in accordance with the treaty.

Made in duplicate at Paris and signed by us the 15th day of August, in the year 1893.

And we do certify this English version thereof to be true and accurate. (Bering Sea Arbitration. Indexes to the British case, p. 19.)

DECLARATIONS MADE BY THE TRIBUNAL OF ARBITRATION AND REFERRED TO THE GOVERNMENTS OF THE UNITED STATES AND GREAT BRITAIN FOR THEIR CONSIDERATION.

[English version.]

I.

The arbitrators declare that the concurrent regulations, as determined upon by the Tribunal of Arbitration, by virtue of Article VII of the treaty of the 29th of February, 1892, being applicable to the high sea only, should, in their opinion, be supplemented by other regulations applicable within the limits of the sovereignty of each of the two powers interested and to be settled by their common agreement.

II.

In view of the critical condition to which it appears certain that the race of fur seals is now reduced in consequence of circumstances not fully known, the arbitrators think fit to recommend both Governments to come to an understanding in order to prohibit any killing of fur seals, either on land or at sea, for a period of two or three years, or at least one year, subject to such exceptions as the two Governments might think proper to admit of.

Such a measure might be recurred to at occasional intervals, if found beneficial.

III.

The arbitrators declare moreover that, in their opinion, the carrying out of the regulations determined upon by the Tribunal of Arbitration should be assured by a system of stipulations and measures to be enacted by the two powers; and that the tribunal must, in consequence, leave it to the two powers to decide upon the means for giving effect to the regulations determined upon by it.

We do certify this English version to be true and accurate, and have signed the same at Paris this 15th day of August, 1893.

(Bering Sea Arbitration. Indexes to the British case, p. 3.)

Vessels composing the Canadian sealing fleet for 1894.

1. Arctic.	21. Geneva.	41. Pioneer.
2. Ainak.	22. Henrietta.	42. Rosie Olsen.
3. Aurora.	23. Kate.	43. Shelby.
4. Annie C. Moore.	24. Kilmeny.	44. San Jose.
5. Agnes McDonald.	25. Katherine.	45. Sapphire.
6. Arietas.	26. C. D. Rand.	46. Saucy Lass.
7. Annie E. Paint.	27. Libbie.	47. Sadie Turpel.
8. Brenda.	28. Labrador.	48. Theresa.
9. Beatrice.	29. Louis Adair.	49. Triumph.
10. Borealis.	30. Minnie.	50. Umbrina.
11. G. G. Cox.	31. May Bell.	51. Viva.
12. Cosco.	32. Maud S.	52. Vera.
13. Charlotte.	33. Mary Taylor.	53. Venture.
14. City of San Diego.	34. Mascot.	54. W. B. Hall.
15. Dora Sieward.	35. Mary Ellen.	55. W. P. Sayward.
16. Diana.	36. Mermaid.	56. Wanderer.
17. E. B. Maroin.	37. Otto.	57. Walter L. Rich.
18. Enterprise.	38. Ocean Bell.	58. W. A. Earle.
19. Fawn.	39. Osca and Hattie.	59. Favorite.
20. Florence M. Smith.	40. Penelope.	

Vessels composing the American sealing fleet for 1894.

1. Alton.	18. Henry Dennis.	35. Rosie Sparks.
2. Alexander.	19. Herman.	36. St. Paul.
3. Anaconda.	20. Ada Etta.	37. Sophia Sutherland
4. Anna Matilda.	21. Jane Grey.	38. San Diego.
5. Allie I. Alger.	22. Kate and Ann.	39. Stella Erland.
6. Bonanza.	23. Louis D.	40. Teresa.
7. Bowhead.	24. Louis Olsen.	41. Volunteer.
8. G. G. White.	25. Lillie L.	42. Willard Ainsworth.
9. Emma and Louisa.	26. Josephine.	43. Winchester.
10. Emma.	27. Mary H. Thomas.	44. Amature.
11. Eppinger.	28. Mascot.	45. Columbia.
12. Edward E. Webster.	29. Mattie T. Dyer.	46. C. C. Perkins.
13. Ella Johnson.	30. Mathew Turner.	47. Deeahks.
14. Ethel.	31. Penelope.	48. Dart.
15. Geo. Peabody.	32. Prescott.	49. Felitz.
16. Geo. R. White.	33. Retriever.	50. James G. Swan.
17. H. C. Wahlberg.	34. Rattler.	51. Puritan.

MANAGEMENT OF ROOKERIES—DECREASE OF SEALS.

SEAL ISLANDS, ALASKA, *July 16, 1889.*

GENTLEMEN: I regret to report that the season's seal catch is progressing very unfavorably, and that the condition of the breeding rookeries, already past the date of fullest occupation for the year, indicates a large falling off in productiveness—much greater, in fact, than I have heretofore reported.

During the period from 1873 to 1883, as my reports from year to year will show, we experienced no difficulty in obtaining the full catch of seals early in the season, and the skins were all of the best marketable size and quality, for we had at that time a large surplus of killable animals from which to make our selection. It was customary during that period to secure in the month of June nearly one-half. of our catch, all of the primest and best, and at the same time turn back to the rookeries for breeding animals, or as being undesirable for market, a very large percentage, averaging for the ten years in question perhaps 30 per cent of the whole number driven. In July in each of those years the percentage of rejected animals was still larger, amounting from 50 to 80 per cent of the number driven; but of those a large majority were under size for killing and required the additional year's growth that we were enabled by the very abundant supply of seals to give them. We could confidently count on their return the next season in prime condition.

The season's work for a catch of 100,000 skins was then finished from the 14th to the 20th of July, determined by our ability to do the work and not by the condition of the hauling grounds, for we had always seals enough in sight after the 10th of June, and sometimes even earlier, to keep our force fully occupied.

The breeding rookeries, from the beginning of the lease till 1882 or 1883, were, I believe, constantly increasing in area and population, and my observations in this direction are in accordance with those of Mr. Morgan, Mr. Webster, and others who have been for many years with me in your service, and of the late Special Treasury Agent J. M. Morton, who was on the islands from 1870 to 1880. Even as late as 1885 Special Treasury Agent Tingle reported a further increase of breeding seals, but his estimates were made in comparison with those of Prof. H. W. Elliott in 1872–73, and he was probably not fully aware of the fact that the increase had occurred prior to 1883, and that in 1885 there was already perhaps a slight diminution of breeders.

The contrast between the present condition of seal life and that of the first decade of the lease is so marked that the most inexpert can not fail to notice it. Just when the change commenced I am unable from personal observation to say, for, as you will remember, I was in ill health and unable to visit the islands in 1883, 1884, and 1885. I left the rookeries in 1882 in their fullest and best condition, and found them in 1886 already showing a slight falling off, and experienced that year for the first time some difficulty in securing just the class of animals in every case that we desired. We, however, obtained the full catch in that and the two following years, finishing the work from the 24th of June to the 7th of July, but were obliged, particularly in 1888, to content ourselves with much smaller seals than we had heretofore taken. This was in part due to the necessity for turning back to the rookeries many half-grown bulls, owing to the scarcity of breeding males. I should have been glad to have ordered them killed instead, but, under your instructions to see that the best interests of the rookeries were conserved, thought best to reject them. The result of killing from year to year a large and increasing number of small animals is very apparent. We are simply drawing in advance on the stock that should be kept over for another year's growth, reserving as far as possible, of course, all desirable half-grown bulls for breeders, but at the same time killing closer, I believe, than a wise policy would indicate.

The deduction need hardly be drawn, as it is only too apparent that the lessees, for the next two or three years at least, must, in any event, if the rookeries are to be stocked up to their best condition, be content with very small catches. I estimate that not more than 15,000 or 20,000 desirable skins can be obtained next year, and it is possible that taking even a much smaller number would sooner restore the rookeries to their former vitality.

The change in the breeding rookeries, though not so immediately alarming as that observed in the hauling grounds, owing to the large number of seals still in sight, is sufficiently marked to excite curious inquiry as to its cause. Large patches of ground on the outskirts of every rookery, which were covered with breeding seals and their young a few years ago, are now bare; the lanes and paths across the rookeries, along which the nonbreeding seals pass to and from their grounds, are growing wider, and what is still more disturbing to the experienced seal hunter there is a very noticeable sparseness of population, both male and female, on every rookery. I should certainly be within bounds in stating that at least one-third less seals landed on the islands this year than ten years ago.

You will remark that I have, at the beginning of this statement, referred back for comparison only to 1873. Prior to that time we were suffering from the excessive killing of 1868, when, in the absence of any restriction, more than 200,000 seals were killed in a single year. The deficiency of male breeding seals, caused by this excess, continued for four or five years, and is referred to by Special Treasury Agent Bryant in his report to the Department under date of September 5, 1872. I allude to this only for the purpose of calling attention to the fact that any improper handling of the seal industry is immediately followed by marked results.

For the cause of the present diminution of seal life we have not far to look. It is directly traceable to the illicit killing of seals of every age and sex during the last few years in the waters of the North Pacific and Bering Sea. We are in no way responsible for it. During the first thirteen years of the lease comparatively few seals were killed by marauders, and we were then able not only to make good the deficiency caused by the slaughter of 1868, but, under our careful management, to produce a decided expansion of the breeding rookeries.

The history of fur-seal killing on the British Columbia coast would, no doubt, carry us back to an earlier date than the transfer of Alaska to the United States, but it was done mostly up to 1875 by the use of rude appliances, and the hunters were unable to pursue their vocation, to any great extent, on the open sea. About 1875–76, under the stimulus of better prices for skins, induced by the improved methods applied by us to the fur markets of the world, it was found profitable to fit out more expensive ventures from Victoria, and the seals were followed along the British Columbia and United States coasts as far north as Sitka; but prior to 1882–83 it had not come to the knowledge of the hunters that their work could be profitably pursued farther to the northwestward. The catch was too small up to this time to seriously affect seal life. An occasional predatory schooner came into Bering Sea before 1882, and the *San Diego*, with her cargo, was seized in 1876 and condemned to forfeiture under section 1956 of the United States laws. In one or two other cases certificates of probable cause of seizure were issued by the courts to the revenue officers, thus affirming the illegality of killing seals in Alaskan waters.

About 1882–83 the British seal hunters discovered that profitable voyages could be made to Bering Sea, and the few vessels engaged in those years were soon joined by others, until, in 1885, a fleet of twelve or fifteen schooners, some of them propelled by steam, were engaged in the business, and the catch sent into Victoria amounted to about 25,000 skins. The fleet sent more than 40,000 skins to market in the following year. More stringent orders were, however, issued to our revenue vessels, and three of the twenty or more engaged in that year were seized and still lie rotting on the beach in Unalaska Harbor. In 1887 a still larger fleet appeared, but was badly demoralized before the end of the season by the capture of fourteen of the vessels and the confiscation and sale of a large part of them, together with a large number of skins; in all, some 12,000, I think. Had this repressive policy of the Government been firmly adhered to from that time we should probably be little troubled with marauders this year; but pending negotiations with for eign powers sealed orders to be opened in Bering Sea were given to the revenue officers, directing them not to make seizures, and while these orders were withheld from American hunters they appear to have been published to the British fleet, for the usual number of British vessels made a profitable season's cruise, sending into market more than 19,000 skins; at the same time our American vessels were deterred by the tone of the published regulations of our Government from undertaking their usual voyages.

The operations of the marauders in the North Pacific and Bering Sea beyond the jurisdiction of British Columbia, and exclusive of what is known as the "Victoria catch" proper, may be summarized, not with absolute accuracy, but correctly enough for all practical purposes, about as follows:

1883, 1,000, and 1884, 5,000 skins, estimated without reliable data at hand; 1885, 12,000; 1886, 27,500; 1887, 25,000; and 1888, 19,000 skins reported by Messrs. C. M. Lampson & Co., of London; 1889, 10,761 skins to August 1, landed at Victoria, British Columbia.

Add to this the Victoria catch for the same seven years, which has averaged about 12,000 skins per annum—84,000—and we have a total of 184,261 skins sent to market in less than seven years. To represent the destruction of seal life, this number should be nearly doubled to include the loss of one young seal in embryo or left to starve upon the islands for nearly every adult killed; and again doubled, perhaps, to compensate for the unknown factor of waste in killing. Just what proportion of seals killed are actually secured we do not know, but we are confident that the loss of dead seals in the rough water of the open sea, and the wounding and subsequent death of many more, is a large percentage of those taken. Beyond this, we must also take into the account the demoralization of the herd, the infraction of their steady migratory habits and their possible deviation from their accustomed haunts, and the consequent destruction of the industry within our borders if indiscriminate slaughter is continued. I append a list of vessels reported engaged in sealing the present season.

I have at hand data from this year only on which to base an estimate of the respective numbers of seals killed in the waters of the North Pacific and Bering Sea. It appears that during the present season at least 5,201 skins, exclusive of the catch of the American vessels, were taken after the sealers left the Straits of Fuca and before they passed the Alaskan peninsula, for that number were transshipped to the British schooner *Wanderer* at Sand Point and sent back to Victoria to avoid possible capture by our revenue vessels. The British schooners *Pathfinder*, *Viva*, and *Sapphire* landed in Victoria their spring catches, amounting to 1,719 skins, early in June, and again sent down by the *Wanderer* 2,039 skins about the middle of July. This latter number must have been captured in the Pacific in less than six weeks, and many of them among the Shumagin Islands and along the coast to the westward of Kodiak, clearly within American waters.

Attention should also be directed to the fact that by preconcerted action all the British vessels rendezvoused at Sand Point, Ounga Island, Alaska, where there is neither port of entry nor customs officer stationed, and there, in utter disregard of customs law or international right, transshipped cargo, received supplies brought from a foreign port, and landed and sold whisky to the Alaska natives.

Until the present season we have been under the impression that the catching of seals in the waters of the North Pacific would be difficult and unprofitable, and that seal life could be preserved by maintaining the closure of Bering Sea. Now, however, it seems doubtful whether it will not be necessary to extend protection over the waters of the North Pacific.

Of course it will be asked if this can be legally effected; I see no obstacle in the way of doing it. We would have no difficulty whatever in proving to the satisfaction of any fair-minded nation that all the seals in the eastern part of the North Pacific and Bering Sea are born and reared on the Pribilof Islands, and those in the western part of the same waters have their habitat on the Commander Islands; nor do they resort for breeding to any other than these two places in the Northern Hemisphere, excepting only the very small number found on Robben Island in the Okhotsk Sea. They can be positively identified as our property. The seals found in these respective places differ so much that expert skin assorters can distinguish between them in handling the skins; and, in any event, this matter concerns only the United States and Russia. When the seals on which the British are now poaching are found in the Pacific they are simply astray; but are, nevertheless, either our property or that of Russia, and should be respected and protected as such.

After twenty-one years of careful study of the subject, I am entirely satisfied that the usual migratory course of the seals leads them to the southward from the Pribilof Islands, mostly through the passes into the Pacific, to the eastward of and including the pass of longitude 172° west; thence they turn to the eastward along the Aleutian Archipelago, through the Shumagin group, and past Kodiak, to appear in February and March down about Vancouver Island and in the straits and channels to the northward and eastward of Vancouver, where large numbers are annually killed in the early spring months. The more notable proofs of this are:

(1) That many young seals are killed in November, December, and January by the Alaskans among the Aleutian Islands, and more could and would be taken if the natives were not restrained by our agents from hunting them.

(2) Fur seals are fish eaters and naturally keep upon such banks and shoals, within easy soundings, as furnish them an abundant food supply.

(3) They are rarely seen in the waters of the North Pacific at any considerable distance from soundings, but are plentiful along the Alaskan coast during all the winter months.

(4) A large proportion of the several thousand seals killed annually about the British Columbia coast in March and April are pregnant females in just that stage of gestation that would be expected in animals whose period of eleven months terminates in June.

(5) Almost simultaneously with their disappearance from the British Columbia coast in April they are again found in increasing numbers in the Aleutian Archipelago and, a little later, in Bering Sea.

(6) The most careful search for other breeding grounds than those at the Pribilof Islands has been fruitless. It can be positively asserted that none exist.

The best season for marauding in Bering Sea is the latter part of July and August, for the female seals, having left their young on the islands, are then off on the feeding grounds to the southward, and the destruction of the mother at this time is followed by the loss of the pup, which dies for want of nourishment. This was vividly illustrated in the heavy storms of last fall, when several thousand pups, too weak and feeble to withstand their violence, were thrown upon the beaches and killed. In the earlier years of the lease no such destruction of the young was observed during the autumn storms as we have lately witnessed.

The greater part of the illegal killing is done with firearms, but some of the vessels manned by Indians, and notably the *Black Diamond*, seized this summer, are fitted only with spears, and these, skillfully handled, are even more effective than guns, as they can be used without alarming the neighboring seals. The crew of the *Black Diamond* secured 143 skins in less than two days preceding her capture. It is also regarded by old hunters as quite feasible to catch them still more rapidly and surely by the use of seines and purse nets, though I am not aware the attempt has as yet been made.

The effect of this extensive and indiscriminate slaughter I have already pointed out. If unrestricted killing is to be continued we have no occasion to inquire in regard to any further franchise. The renewal of the lease would hardly be worth taking as a gift, and with the assurance of fullest protection against marauders and poachers, the fact should not be lost sight of that under the most intelligent management some years must elapse before the rookeries can be restored to their former productiveness. The protection, too, must extend beyond Bering Sea and over the North Pacific to insure perpetuation of the industry; and ought, indeed, in order to make it complete, to include all the waters along the British Columbia coast, for even the comparatively small number killed there is no inconsiderable item to the lessees in the present status of the rookeries.

Different plans for the preservation of the seals are suggested:

1. It is certainly in the interest of the whole world, excepting a few Canadian seal hunters, that the seals should be propagated and killed under proper restrictions. This is particularly true for the English, for they have more capital invested in the business and more people dependent upon the seal industry than any other nation. If, therefore, a territorial limit can be defined beyond which no seals shall be killed in the water, such limit being agreed upon by convention with England and Russia, and acquiesced in by the powers that have nothing at stake in the matter, protection will be afforded to such an extent as the limit proves restrictive. My own idea is that it should cover all the aquatic resorts of the seals, but if it be decided that British Columbia hunters are right in killing seals in British Columbia waters, then the limits might be defined, say, by restricting their operations to the eastward of longitude 153° west from Greenwich, to the southward of latitude 54° north, and to the northward of Cape Flattery. If at the same time restrictions are needed for the protection of Russian interests in the Northwestern Pacific, similar limitations, as the facts may indicate, may be marked out and seal life respected at all points beyond such limits.

2. If restriction by territorial limitation is likely to be difficult to enforce, or if for any other reason it appears objectionable, a close season could be agreed upon by convention within which no seals should be killed in the water. Such season should begin, if it be determined to allow seals to be killed in British Columbia waters, at about the time when the seals leave the vicinity of Vancouver Island in the spring and continue until the next winter, say about the middle of May until about the 1st of February.

3. To facilitate the enforcement of the regulation, both the territorial limitation and close season might be adopted. The vast extent of water to be patrolled, and the eagerness with which the seals are pursued, make it necessary to throw every possible safeguard around them if they are to be preserved.

It would unquestionably be unwise, from a financial point of view, on the part of any of the nations interested, to allow pregnant female seals (and scarcely any other are taken there) to be killed on the British Columbia coast; but, if something must be conceded to the poachers, perhaps the opportunity to pursue their destructive occupation along this coast is the least that will reasonably content them.

In the present status of the seal fishery the value of a renewal of the franchise for another term of twenty years is very difficult to estimate. The outlook for the next three or four years is decidedly bad. The chief elements of uncertainty are:

(1) Doubt as to the intention of our Government in regard to protecting the fisheries against poachers.

(2) Question whether, in case a restrictive policy is decided upon, the Government will be able to successfully patrol the broad extent over which the seals are scattered. A failure to protect them without further delay will be fatal to any considerable catch on the islands.

(3) The fact that the rookeries are already badly depleted, and that all the best seals, for the next two or three years at least, must be reserved for breeders.

(4) The positive knowledge that the seals that would otherwise come forward for killing during the next two years have already been slaughtered, and that the catch must for several years to come be much smaller than heretofore.

I was of opinion two years ago that the next twenty years' lease could be more valuable than the present term, but am now greatly in doubt whether we can afford to pay as much as the present rental, even with a guaranty from the Government of entire protection outside of British Columbia waters. Without such guaranty there is "nothing in it" beyond a small prospective catch from such animals as may escape the toils of the hunter this summer. At the present rate of diminution the rookeries will soon do little more than support the natives dependent on them and pay the expenses of the necessary outfit to follow the business.

Very truly, yours,

H. H. McINTYRE,
Superintendent.

The ALASKA COMMERCIAL COMPANY,
San Francisco, Cal.

CAUSES OF DECREASE IN SEAL HERD.

WEST RANDOLPH, VT., *December 15, 1890.*

MY DEAR SIR: I have at hand extracts from the report of Prof. H. W. Elliott to the Secretary of the Treasury in the fall of 1890, relative to the decrease of the seal rookeries of St. Paul and St. George islands, Alaska; and knowing that you take a deep interest in the matter, beg to call your attention to a few conspicuous errors into which he has fallen.

He asks, "Why is it that we find now only a scant tenth of the number of young male seals which I saw there in 1872? When did this work of decrease and destruction so marked on the breeding grounds begin? And how?" He answers these questions as follows:

(1) From overdriving, without heeding its warning, first begun in 1879, dropped then until 1882, then suddenly renewed again with increased energy from year to year, until the end is abruptly reached, this season of 1890.

(2) From the shooting of fur seals (chiefly females) in the open waters of the North Pacific Ocean and Bering Sea, begun as a business in 1886, and continued to date.

Professor Elliott is a naturalist, and a very good one. He is thoroughly familiar with the size, form, color, comparative anatomy, domestic habits, and whatever goes to make up the natural history of the seal. He is tolerably familiar with the seal as viewed from the hunter or business man's standpoint. He is also fairly capable of deducing from given facts a theory in regard to the increase or decrease of the seal. Given correct premises, he would perhaps come as near the truth in his deduction as the average observer. But when his premises are wrong, his deductions are more mischievous than those of the average man, because he asseverates his findings with such positiveness, and such an air of knowing all about it, as to carry the investigator along with him to the pitfalls digged by theory from wrong hypotheses.

He says, in brief, that there was overdriving in 1879, none in the two following years, beginning again in 1882 and continuing "until the end is abruptly reached in the season of 1890." As he writes forcibly in the same connection against the practice of driving the long distance from Southwest Bay (Zapadnie) to the village killing ground—about 4 miles—pointing out most disastrous effects from this practice, I suppose he means by "overdriving" the driving too great distance. If this is it I quite agree with him, and always have, and for that reason, except on very rare occasions, did not allow seals to be driven the long distances he describes, and it has never been habitually done. Boats were almost invariably sent to Southwest Bay and carts to Halfway Point to bring in the skins, and the animals were as invariably killed, during the last ten years of the Alaska Commercial Company's lease, as near the rookeries as seemed prudent. The windmill he fights through several paragraphs of alleged "reasons" is less worthy of attack than Don Quixote's. It exists only in his imagination.

Then, the end was not "abruptly reached." I repeatedly pointed out to our company and to the special Treasury agents, during the seasons of 1887, 1888, and 1889, that the seals were rapidly diminishing, and that in order to get the full quota allowed by law we were obliged to kill, in increasing numbers in each of those years, animals that should have been allowed to attain greater size; and, finally, the catch of 1889 was mostly of this class. If they had been contented with the same class in 1890 a much larger catch could have been obtained.

Again he is in error in saying that marauding in Bering Sea began in 1886. It commenced in 1884 with a catch of 4,000 skins, and was followed with a take of almost 10,000 in 1885.

This brings us to the second reason given by him for the decrease, to wit, "the shooting of seals (mostly females) in the open waters of the North Pacific Ocean and Bering Sea." And here he strikes the key note of what should have been his warning, but he strikes it so flat as to throw his chorus quite out of tune; but he was not there present on the islands during any of those six years of active poaching prior to the season just past, nor, in fact, for several years previous to those six years, and does not know what he is talking about. His second "reason" should have been his first, and I assert most positively, with knowledge drawn from an accurate personal cognizance of the facts, that the diminution of the seal was exactly coincident in the time of the decrease, and in its ratio from year to year, with the time and extent of the piratical marauding of the Canadian and American vessels in the waters of Bering Sea, and prior to the beginning of such marauding was not perceptible and did not exist.

I regret that Professor Elliott did not urge this one true reason with all the strong force of which he is capable, because it is fully time that the serious import to the seal fisheries of further poaching was understood by our Government, premising always that it is really in earnest about what it is doing to preserve the fisheries, which, indeed, I find myself already doubting.

    I am, very truly, yours,

                                     H. H. McINTYRE.

Gen. N. L. JEFFRIES,
    Washington, D. C.

LETTER FROM SECRETARY OF TREASURY TRANSMITTING ELLIOTT'S REPORT TO STATE DEPARTMENT.

TREASURY DEPARTMENT,
OFFICE OF THE SECRETARY,
Washington, D. C., February 23, 1893.

    SIR: In compliance with the request contained in your communication of February 11, I have the honor to transmit herewith a certified copy (together with the photographic reproductions of the illustrations and maps) of the report of H. W. Elliott on the Pribilof Islands for 1890. The original of this report will be placed in the custody of Special Agent W. H. Williams for such use as you may find necessary.

    In handing you this report I deem it my duty to acquaint you with certain facts in connection with my refusal to permit its publication.

    Upon its examination I became convinced that it was pervaded by a spirit of aggressive criticism instead of being a dispassionate statement of facts; that Mr. Elliott's views had been unduly influenced by his relations toward certain individuals; that the report contained much matter, and particularly that referring to the fur seal, which had already been published by the Government in two forms at least; that the illustrations being made from sketches possessed inherent defects which rendered them valueless as records of the diminution or growth of the rookeries, while the scale of the rookery charts was too small to accurately indicate the condition of seal life at the time these observations were made.

    I therefore declined to permit Mr. Elliott's return to the islands, and deemed it expedient to withhold publication of his report pending the sending of other officers to the islands for the verification of his statements and for the procuring of data on a systematic plan, aided by photography.

    On further examining Mr. Elliott's report in the light of this additional information and comparing his statements with the island records now on file in this Department, I find that not only do the objections against the report cited above still hold good, but that Mr. Elliott has so used extracts from the records of the islands as to make them appear to substantiate his assertions that mismanagement on the part of the United States has played an important part in the diminution of seal life, which assertions are unsupported by the unabridged records. In view of its inaccuracies, its misleading character, and its disagreement with the information brought to me independently by at least three other officers whom I sent to the islands, and the further grave fact of the misuse of official data by Mr. Elliott, I do not believe that the Government would be justified in publishing this report.

    Very respectfully,

                                    CHARLES FOSTER, *Secretary.*

Hon. JOHN W. FOSTER,
    Secretary of State.

MR. PHELP'S ARGUMENT BEFORE PARIS TRIBUNAL.

I.

On April 4, 1893, Mr. Phelps made to the Tribunal the following
. statement in regard to Mr. Elliott's report of 1890:

This paper was produced and furnished to the British commissioners during their
session at Washington and remained in their possession as long as they cared to keep
it. It will thus be seen that there has been no disposition on the part of the United
States Government to withhold or to conceal this document.

II.

The report is of little value as an authority and quite as likely to
mislead as to guide. The author is utterly untrustworthy as an
observer.

(1) His field notes show this on their face: A field note should be a
bare and clear and uncolored record of facts observed. These are a
record not only of facts, but of conjectures, opinions, predictions,
reflections, emotions, etc.

An observer should be severely objective. Elliott is always sub-
jective. It is his own conjectures and reasonings which he is most
concerned with. A perusal of pages 236 and 237 (entry of July 10) will
afford amusing proof of this.

(2) It is the misfortune of Mr. Elliott and of those who rely upon
him that he has written at different times on the subject of fur seals,
and his representations of the facts at these different times vary
in some cases according to the theories which he was interested to
establish.

Thus, in 1872–1874, he observed that a certain detached rock or islet
was then covered with the forms of fur seals; but in 1890, writing with
the purpose of showing that injurious redriving was practiced, he represents that the presence of seals at this place was a wholly recent
phenomenon, occasioned by a too severe working of the neighboring
sealing grounds.

(3) His assertions of important matters of fact are shown to be erroneous by evidence far better than his. For instance, in his report for
1890 he represents certain places which on his earlier visits he found
abounding in young seals to be absolutely destitute of them, whereas
it is proved by the records of the islands that at those very times young
seals were driven and killed from those same places.

Thus he writes July 19, 1890: "Not a single holluschak on Zoltoi
Sands this morning and not one had hauled there this season." The
official records for 1890 (British case, Appendix, Vol. III, United States,
No. 2, 1890, pp. 16, 23) show: (a) That on that very day 3,956 seals
were driven from Zoltoi in connection with Reef rookeries, of which
number 556 were killed; (b) that a drive had already been made from
those same places June 24, on which occasion 426 seals were killed.

(4) Mr. Elliott appears to be guilty of great inaccuracy in quoting
statements which have been made to him. Thus he attributes to Daniel
Webster the following:

He says that over since 1876–77 he has observed a steady shrinking of the hauling
grounds at Northeast Point.

In the United States case (Appendix, Vol. II, p. 181), Daniel Webster
makes, however, a sworn statement which is wholly at variance with the
above:

My observation has been that there was an expansion of the rookeries from 1870 up
to at least 1879. In the year 1880 I thought I began to notice a falling off from the

year previous of the number of seals on Northeast Point rookery, but this decrease was so very slight that probably it would not have been observed by one less familiar with seal life and its conditions than I.

(5) As a reasoner he is equally bad. He is dominated by a favorite theory, and when this comes in collision with facts he can not yield the former, and is consequently constrained to accommodate the latter to it. ·

(6) The counsel for Great Britain, in order to establish credit for Mr. Elliott as an authority, point to the circumstance that Mr. Blaine referred to him with respect in his letter of March 1, 1890. Mr. Blaine was, undoubtedly, as many others were upon the first appearance of Mr. Elliott as a writer upon seals, under the impression that he was a trust-worthy witness. But such was not, at that time, the view of those representing the British Government.

In order to discredit Mr. Elliott as a theorist and reasoner, Mr. Tupper cites, in a letter dated March 8, 1890 (British case, Appendix, Vol. III, United States, No. 2, 1890, p. 441), the following criticism made upon Mr. Elliott by Mr. W. L. Morris in 1879:

This man seems to be the natural foe of Alaska, prosecuting and persecuting her with the brush and the pen of an expert whenever and wherever he can get an audience, and I attribute the present forlorn condition of the Territory more to his ignorance and misrepresentation than to all other causes combined.

Mr. Tupper then goes on to say:

His evidence in 1888 is open advocacy of the United States contention. His writings and reports prior to the dispute will be referred to, and it will be submitted that his statements and experiences before 1888 hardly support his later theories.

(7) Dr. Dawson's (one of the British commissioners) estimate of Professor Elliott in the fall of 1891 is thus told by Judge Swan (United States counter case, p. 414), who quotes Dr. Dawson as follows:

Elliott's work on seals is amusing. I have no hesitation in saying that there is no important point that he takes up in his book that he does not contradict somewhere else in the same covers. * * * His work is superficial in the extreme.

III.

The avowed purpose of Mr. Elliott in this report of 1890 is to show that the Alaskan herd has been generally diminished in numbers and to point out the causes of the diminution.

The only true cause of this decrease which can be gathered from any facts stated by him is pelagic sealing; but he has a theory that there is another cause, namely, overdriving and redriving, which he assumes, not only without proof, but against the proof, to have been practiced to a considerable extent for a long period of time prior to 1890.

It is important to understand just what he means by overdriving and redriving. He does not mean careless handling or undue urging of the seals during any given drive, for he specially states that the drives were and are very carefully made (infra, under fourth, 3).

What he does mean by his charges concerning overdriving is this: That in the face of a diminishing number of seals it was still endeavored to take 100,000 skins per annum, which necessitated, at a date as early as 1884 to 1885, the following:

Driving from the rookery margins, where alone the young males were found in these later years, with consequent disturbance to the breeding seals.

The turning away from the killing grounds of an increasing number of unkillable seals, which seals ran the risk of being several times redriven in the same season.

(b) When did this scraping of the rookery margins and overdriving begin? There is no evidence that either began prior to 1890. Elliott failed to observe or record either between 1872 and 1876. He advances in his report of 1890 no evidence whatever on the subject, though he alleges at a single place that the natives assured him "that they had been driving seals in this method ever since 1885; had been obliged to or go without seals."

This statement attributed to the natives is wholly uncorroborated, nor does it appear in their examinations, which are given at pages 300 to 304; that it is in conflict with the evidence of Mr. Goff will be shown hereafter.

(c) Mr. Elliott thinks that the necessity which leads to overdriving, namely, a scarcity of killable males, began to exhibit itself as early as 1879, and in proof of this he alleges that a hitherto untouched reserve was then resorted to, namely, Zapadnie. Furthermore, he supposes that this scarcity of killable seals, making redriving essentially necessary, was decisively manifested in 1882 and continually thereafter by the fact that a constant resort was from that time made to theretofore "untouched sources of supply" (VI, VIII, IX). In this particular also he is totally in error. No such supposed "untouched sources of supply" then, or ever, existed. Zapadnie and Polavina are intended by him. They had been systematically drawn upon from the first. (Vol. II, Appendix to United States case, pp. 117–127; United States counter case, pp. 78, 79.)

Mr. Elliott's error in this respect is the more inexcusable, since the official island records were at his disposal and appear to have been examined by him. The following tables showing the drives that were actually made from Zapadnie and Polavina are taken from the British argument, page 103:

Year.	Southwest Bay (including Zapadnie).	Halfway Point (Polavina).		Year.	Southwest Bay (including Zapadnie).	Halfway Point (Polavina).
1871	4	1		1881	5	4
1872	1	1		1882	10	5
1873	3	0		1883	9	5
1874	6	0		1884	9	9
1875	7	1		1885	6	8
1876	8	1		1886	12	9
1877	6	3		1887	8	6
1878	6	3		1888	8	8
1879	7	3		1889	8	7
1880	5	4				

(In examining tables in the United States case, Appendix Vol. II, pp. 117–127, it should be remembered that "Zapadnie" and "Southwest Bay" are two names for the same place, and so also are "Polavina" and "Halfway Point," the latter term being the English for Polavina.)

(d) Upon this basis of utter misapprehension Elliott proceeds forthwith to construct a theory, and his theory as far outruns his supposed facts as those supposed facts do the truth. For he proceeds to assume that the driving and redriving of seals have been gradually increasing from year to year and very rapidly since 1884–85, that the process of driving in any form renders those seals which are turned back from the killing grounds worthless for rookery service, and that the work of destruction thereby produced "set in from the beginning, twenty years before 1890" (pp. 7 to 10).

(e) He introduces no proof that driving, overdriving, or redriving of any sort ever injured the generative organs of a seal which was allowed eventually to return to the water, save the following (see pp. 150, 203, 271), which he has mistranslated from Veniaminof. But here, too, he has led himself into error:

Elliott's translation (p. 203).	*Correct translation.*
Nearly all the old men think and assert that the seals which are spared every year, i. e., those which have not been killed for several years, are truly of little use for breeding, lying about as if they were outcasts or disfranchised.	Nearly all the old travelers think and assert that sparing the seals for some years, i. e., not killing them for some years, does not contribute in the least to their increase and only amounts to losing them forever.

Veniaminof thus makes no reference whatever to driving, still less does he refer to any supposed effects of driving upon the reproductive powers of the seal.

It should be added that both the British commissioners and the British Government have been misled by Elliott's erroneous translation. (See British Commissioners' Report, sec. 712, and British counter case, p. 263.)

(*f*) The notion that the mere driving of a seal even over rough ground renders it impotent is in itself sufficiently absurd, but it becomes still more so when considered in connection with the following extract from Mr. Elliott's field notes (p. 244):

I have sat for hours at a time watching the seals come up and go down in ceaseless files of hundreds and thousands, actually climbing up in places so steep that it was all an agile man could do to follow them safely.

(*g*) It follows from the above that so far as Elliott's report is relied on to show considerable cause of injury to the herd, it fails entirely. His belief upon this point was founded upon an utter mistake, assuming that he did not wish to be misled. He never saw any redriving or overdriving until 1890 (when it did exist); nor had any other witness ever seen any worthy of notice previous to 1890.

The counsel for Great Britain, seeking for another evidence to prove redriving, have recourse to the report of Mr. Goff for 1890. But he disproves the assertion by distinctly contrasting the large numbers of young seals turned back in 1890, with the small number theretofore turned back. (British counter case, p. 265.)

(*h*) Eliminating this clear and manifest error from Elliott's report, the latter proves, and alone proves the following: That in 1872–1874 the herd was in a condition of full and abounding prosperity; that when he next observed it in 1876, its condition was not perceptibly changed; that in 1890, when he last observed it, it had become greatly diminished in numbers, so as to make it difficult to obtain the quota of 100,000 without redriving.

And this is just what the United States have from the first maintained.

IV.

His melange of observations, reasonings, conjectures, predictions, and criticisms, when scrutinized, will be found to support the positions of the United States in nearly every particular, certainly in each of the following:

(1) That it is in the power of the United States and its lessees under normal conditions to gather the whole annual increase of the seals without diminishing the normal numbers of the herd.

Page 69: The polygamous habit of this animal is such that, by its own volition, I do not think that more than one male annually out of fifteen born is needed on the breeding grounds in the future.

Page 118: In this admirably perfect method of nature are those seals which can be properly killed without injury to the rookeries, selected and held aside by their own volition, so that the natives can visit and take them without disturbing in the least degree the entire quiet of the breeding grounds where the stock is perpetuated.

Page 129: When the "holluschickie" are up on land they can be readily separated into their several classes as to age by the color of their coats and size, when noted; namely, the yearlings, the 2, 3, 4, and 5 year old males.

(2) That the methods adopted on the Pribilof Islands were from the first, according to his observations of 1872–1874, admirably adapted to accomplish the gathering of the annual increase (pp. 71, 74).

(See also description of drive in the parts quoted from his report of 1874, pp. 122–128.)

(3) That the methods pursued in 1890 (with the single exception of alleged redriving and overdriving, already noticed) were in all respects as good as, and in some better than, those pursued in 1872 to 1874.

Page 269: I should remark that the driving of the seals has been very carefully done; no extra rushing and smothering of the herd, as it was frequently done in 1872. Mr. Goff began with a sharp admonition, and it has been scrupulously observed thus far by the natives.

Page 283: Yesterday afternoon I went back to Tolstoi over the seal road on which the drive above tallied was made in the night and morning of the 7th instant; the number of road "faints" or road skins was not large, which shows that the natives had taken great care in driving these seals. This they have uniformly done thus far (see also p. 129).

Mr. Elliott draws a bill (p. 217) for the restoration of the herd, but it contains no designed improvement in the methods.

Elsewhere, however, he suggests the following: That no culling of the herds be allowed, i. e., that every seal driven up be killed (p. 73), and that no driving be allowed after July 20 (p. 179).

These are the sole improvements which even he has to suggest.

(4) That according to his observations of 1872–1874 and 1876 the herd could safely support a draft far larger than 100,000, probably as large as 180,000 annually (p. 69).

(He was first on the islands during the three years 1872 to 1874. This report, written in 1890, represents the herd in 1874 as being in a flourishing condition. He was again on the islands in 1876. He does not intimate anywhere in this report of 1890 that the condition of 1876 was not in all respects as good as that of 1872, 1873, and 1874.)

(5) That female seals should never be killed.

Page 74: We do not touch or disturb these females as they grow up and live, and we never will if the law and present management is continued.

Page 213: In 1835, for the first time in the history of this industry on these islands, was the vital principal of not killing female seals recognized.

(6) That pelagic sealing is essentially destructive in its nature, and that at least 85 per cent of the pelagic catch is composed of females.

Page IX: I could figure out from the known number of skins which these hunters had placed on the market a statement of the loss and damage to the rookeries, to the females and young born and unborn, for that is the class from which the poacher secures at least 85 of the 100 of his catch.

Page 13: The young male seals have been directly between the drive, club, and poacher since 1882, while the females have had but one direct attack outside of the natural causes. They have been, however, the chief quarry of the pelagic sealer during the last five years.

(7) That the loss through the wounding and sinking of seals is enormous.

Page 214: Five thousand female seals, heavy with their young, are killed in order to secure every 1,000 skins taken. (See also p. 85, footnote.)

(8) That it is an absolute necessity that pelagic sealing should be suppressed on the ground that it is an immoral pursuit, and one which

is "repugnant to the sense of decency and simplest instincts of true manhood." He makes the following recommendation (see p. 214):.

That pelagic sealing in the waters of Bering Sea be prohibited and suppressed throughout the breeding season, no matter how, so that it is done, and done quickly. This step is equally imperative. The immorality of that demand made by the open-water sealer to ruin within a few short years and destroy forever these fur-bearing interests on the Pribilof Islands—the immorality of this demand can not be glossed over by any sophistry. The idea of permitting such a chase to continue where 5,000 females, heavy with their unborn young, are killed in order to secure every 1,000 skins taken, is repugnant to the sense of decency and the simplest instincts of true manhood.

I can not refrain from expressing my firm belief that if the truth is known—made plain to responsible heads of the civilized powers of the world—that not one of these Governments will hesitate to unite with ours in closing Bering Sea and its passes of the Aleutian chain to any and all pelagic fur sealing during the breeding season of that animal.

(9) That cows suckle no pups other than their own. Referring to the driving of nursing cows, he says:

Page 297: * * * That means death or permanent disability, even if the cows are driven but once—death to both cow and her pup left behind, since that pup will not be permitted to suckle any other.

(10) That pups learn to swim; that in the beginning of August a large majority of them are wholly unused to water (p. 255), and that a number of them do not "get into the water" before September 1 (p. 260).

(11) That the seals are of a gentle disposition, are not frightened by the presence of man, and should not be regarded as wild animals.

Page 123: *Docility of fur seals when driven.*—I was also impressed by the singular docility and amiability of these animals when driven along the road; they never show fight any more than a flock of sheep would do.

Page 98: *Gentleness of the seals.*—Descend with me from this sand-dune elevation of Tolstoi and walk into the drove of holluschickie below us; we can do it; you do not notice much confusion or dismay as we go in among them; they simply open out before us and close in behind our tracks, stirring, crowding to the right and left as we go, 12 to 20 feet away from us on each side. Look at this small flock of yearlings—some 1, others 2, and even 3 years old, which are coughing and spitting around us now, staring up at our faces in amazement as we walk ahead; they struggle a few rods out of our reach and then come together again behind us, showing no further notice of ourselves. You could not walk into a drove of hogs at Chicago without exciting as much confusion and arousing an infinitely more disagreeable tumult; and as for sheep on the plains they would stampede far quicker. Wild indeed! you can now readily understand how easy it is for two or three men, early in the morning, to come where we are, turn aside from this vast herd in front of us and around us 2,000 or 3,000 of the best examples, and drive them back, up and over to the village.

(12) That virgin females go to the islands when 2 years old and are there impregnated.

Page 18: It must be borne in mind that perhaps 10 or 12 per cent of the entire number were yearlings last season and came up onto these breeding grounds as virgins for the first time during this season; as 2-year-old cows they of course bear no young. (Ibid.) This surplus area of the males is also more than balanced and equalized by the 15,000 to 20,000 virgin females which come onto the rookery for the first time to meet the males. They come, rest a few days or a week, and retire, leaving no young to show their presence on the ground.

Page 139: Next year these yearlings which are now trooping out with the youthful males on the hauling grounds will repair to the rookeries, while their male companions will be obliged to come again to this same spot.

V.

Again this report not only supports all positions taken by the United States on the main points, but as clearly condemns all of the special assertions made on the part of Great Britain for the purpose of weakening those positions:

(1) Mr. Elliott holds that coition is never effected in the sea (p. 83).

(2) He repudiates the notion that the seals have in any respect changed their habits, either in dates of arrival at the islands or otherwise (pp. 80, 104, 105, 108, 240, 242, 291).

(3) His observations are to the effect that in the years 1872 to 1876 the herd was in a condition of abounding prosperity. The British Government cites Bryant to prove that during this period a decrease in certain classes of the seals had been observed (pp. 69, 77, 78, 79, 124, 151).

(4) The British assertion that the effect of raids upon the island has been considerable is contradicted by him (pp. 57, 58).

(5) He states, contrary to the contention of Great Britain, that there has been a gradual improvement in the methods of driving and that the actual driving to-day is carried on with the greatest of care (pp. 269, 283).

(6) He states that the condition of the natives has improved since the Americans took possession of the islands, and that they are to-day in every respect well off (pp. 163, 185).

(7) He states that no reduction took place in the standard weight of skins until 1887 (p. 143).

(8) He states that the seals have great powers of locomotion on land (pp. 53, 244, 258).

Elsewhere Mr. Elliott says (Fur-seal Fisheries of Alaska, p. 136):

Its forefeet or flippers are exceedingly broad and powerful, and when it comes out of the water it moves forward, stepping with considerable rapidity and much grace.

(9) That the latest date for properly observing the rookeries is July 20 or thereabouts, for after that date disintegration sets in (pp. iii, 16, 21, 83, 236, 249).

It follows that the rookery observations of the British commissioners, who did not reach the islands in 1891 until July 27, are worthless. It follows, also, that Mr. Macoun (whose observations in 1891 were even less extensive than those of the British commissioners) is not in a position to institute any comparison between the appearance of the rookeries in 1891 and 1892, respectively.

(10) Mr. Elliott gives no countenance to the idea that there exist independent pelagic schools of young seals which do not visit the islands. His report is replete with instances where he has observed large numbers of yearlings and 2-year olds of both sexes on the islands.

Page 105: By the 14th to the 20th of June, they (the holluschickie) then appear in their finest form and number for the season, being joined now by the great bulk of the 2-year olds, and quite a number of yearlings. By the 10th of July their numbers are beginning to largely increase, owing to the influx at this time of that great body of the last year's pups or yearlings. By the 20th of July the yearlings have put in their appearance for the season in full force. Very few yearlings make their appearance until the 15th of July, but by the 20th they literally swarmed out, in 1872-1874, and mixed up completely with the young and older males and females as the rookeries relax their discipline and "pod" or scatter out.

Page 243: I took notice of a large proportion of small or 2-year-old females, and the unusual slowness of hauling, compared with 1872, which was now at its greatest activity July 7. (Tolstoi, July 1, 1890.)

Page 253: The holluschickie are chiefly 1-year olds; nine-tenths of the several pods hauled out here to-day are yearlings. A great many yearling females are hauling down at landings in and among the scattered harems, aimlessly paddling about; their slight forms and bright backs, white throats and abdomens, are shining out very brightly. (North Rookery, July 30, 1870.)

Page 298: I observed a very large proportion of yearling cows scattered all over the breeding ground from end to end near the sea margin, while the yearlings of both sexes are completely mixed upon the outskirts of the rookery, here and everywhere else commingled with the adult cows and their young pups. (St. George, July 30, 1890.)

References to the report showing that yearlings and 2-year olds come to the islands might be multiplied almost indefinitely. (See pp. 98, 139, 140, 143, 147, 253, 255, 256, 277, 280, 291.)

(11) Mr. Elliott scorns the notion upon which the framers of the British case have sought to base the moral title of Canada to a special benefit from the herd, namely, that the seals consume food which would otherwise support fisheries valuable to Canada, for he shows that the true enemy of these fisheries is the dogfish, of which the seal is, in its turn, the greatest destroyer.

Page 307: Suppose, for argument, that we could and did kill all the seals, we would at once give the deadly dogfish (*Squalno ancarthias*), which fairly swarms in these waters, an immense impetus to its present extensive work of destruction of untold millions of young food fishes, such as herring, cod, and salmon.
A dogfish can and does destroy every day of its existence hundreds and thousands of young cod, salmon, and other food fishes—destroys at least double and quadruple as much as a seal. What is the most potent factor to the destruction of the dogfish? Why the seal himself, and unless man can and will destroy the dogfish first, he will be doing positive injury to the very cause he pretends to champion if he is permitted to disturb this equilibrium of nature and destroy the seal.

VI.

If Mr. Elliott's views, as an observer of facts, as a discoverer of causes, as a reasoner, or as an authority in any particular upon seal life upon the Pribilof Islands, are of any value whatever, it should be to show that in the years from 1884 to 1890 the male seals had, in consequence of overdriving, become so few in number and so destitute of virile power that they were not competent to the task of impregnating even the diminished number of female seals which the herd then contained.

Do those who represent the Government of Great Britain really wish to persuade the Tribunal that this is true? Such would seem to be the only conceivable purpose for which such a struggle was made to introduce this report into the evidence. Unless it tends to prove this, it has no tendency except to overthrow every position taken on the part of Great Britain.

But yet the same learned counsel have produced more than one hundred witnesses who swear that in the years 1891 and 1892 the seals were found upon the seas in unprecedented numbers, and some twenty-five of them note specifically having taken young, small, or 2-year-old seals, some of the catches consisting exclusively of this class, which must have owed their existence to the impotent bull observed by Mr. Elliott.

What are we to believe—that Elliott's statements are worthless, or that these witnesses are testifying to what is false, or that these few supposed impotent bulls were endowed with procreative powers wholly unprecedented even in the case of the Alaskan bull seal? Let these contradictions be reconciled as best they may. In the view of the Government of the United States, both these conflicting statements are alike untrustworthy and should be disregarded.

(1) The names of the one hundred witnesses and upward are given in the British counter case (Appendix, Vol. II, pp. 29–33).

(2) The names of the twenty-five witnesses above mentioned, who caught young, small, or 2-year olds in 1892, are as follows (see British counter case, Appendix, Vol. II, pp. 14–22):

Capt. Abel Douglas, George Roberts, William G. Goudie, James Shields, George F. French, Andrew Mathison, Capt. Ernest Lorenz, Capt. Charles Campbell, Capt. James W. Todd, Henry Paxton, George

Heater, Capt. James D. Warren, Capt. Micajah Pickney, Capt. Michael Keefe, William F. Roland, P. Garlson, Kasado, Schoultwick, Clahapisum, Hanaisum, Clahouto, W. Watt, Clat-ka-koi, Kickiana, and Ehenchesut.

And Mr. Elliott himself seems to have observed the usual number of yearlings in 1890. And yet the impotency which he imagines to have been brought about as the result of redriving through a series of years must, if it existed at all, have been nearly as marked in 1889, when these yearlings were begotten.

PELAGIC SEALING IN BERING SEA.

CORRESPONDENCE OF THE TREASURY WITH OTHER DEPARTMENTS UPON THE SUBJECT.

PELAGIC SEALING IN BERING SEA.

CORRESPONDENCE OF THE TREASURY WITH OTHER DEPARTMENTS UPON THE SUBJECT.

TREASURY DEPARTMENT,
Washington, D. C., January 19, 1895.

SIR: I have the honor to inclose herewith our official statement of the American pelagic fur-seal catch of 1894, taken from the record of the custom-houses at the ports of San Francisco, Port Townsend, and Astoria, that the same may be transmitted to the British Government in compliance with article 5 of the Bering Sea arbitration award.

It appears in said statement that the total number of seal skins taken by American vessels and entered at American ports was 26,095. The catch was distributed between the Asiatic Coast, the British Columbia and Northwestern Coast, and Bering Sea, the two latter constituting the total pelagic catch taken from the American herd of so-called Alaska fur seals.

It will be observed by reference to said statement that in many instances the latitude and longitude have been omitted. The explanation of this offered by the collector at San Francisco is that the several masters of the said vessels deposed under oath that they cleared without notice of the pending award, and consequently were ignorant of its requirement.

An examination of the London sales of North Pacific pelagic fur skins, which have recently been held, discloses that 125,269 skins were sold and about 10,000 reserved for future sale; to this should be added the skins retained in the United States, estimated at 10,000, the total being 145,269. The unofficial returns of the British catch, transmitted to the State Department by our consul at Victoria, added to our official returns, make a total of 121,143, or about 24,126 skins less than the estimated catch of 145,269 based on trade sales and estimate of skins retained in the United States. It is possible that this number may have been transshipped by American or British vessels at Yokohama. We have no record of any transshipments except as regards 6,760 skins, which arrived in the port of San Francisco, and appear in our official returns, and which were undoubtedly taken on the Japan and Russian coasts. It is possible that said balance of 24,126 skins may have been entered at Victoria or shipped via Suez Canal.

Should the identity of these skins be ascertained, this Department will submit a supplemental report thereon.

All of the skins, of which the sex is indicated in the accompanying statement, were carefully examined by an expert inspector at the time of their entry.

I have the honor to request that you call upon the British Government for its official returns of the pelagic catch for 1893 and 1894, and that you specifically request information on the following points as to the catches of 1893 and 1894:

(1) As to the total number of seals taken by British vessels.

(2) The total number of skins landed at British ports by said vessels.

(3) The total number transshipped in Japan and Russian ports, and landed ultimately at Victoria.

(4) The number of skins landed at Victoria by American vessels.

(5) A report as to the sex of seals taken in Bering Sea and the North Pacific Ocean.

(6) Location of the place of catch by latitude and longitude.

(7) The number of seal hunters employed, indicating whites and Indians, and also the number of the crew and the tonnage of each vessel.

I have further to request that you inquire whether or not pelagic skins were specially examined as to the sex by the British Government, as was done in the case of the skins entered in United States ports.

I have the honor, etc.,

J. G. CARLISLE, *Secretary.*

The SECRETARY OF STATE.

TREASURY DEPARTMENT,
Washington, D. C., February 12, 1895.

SIR: I have the honor to inform you that the statistics heretofore furnished to you by this Department, containing a statement of the pelagic catch of seals taken by American vessels in the north Pacific Ocean and Bering Sea during the season of 1894, were not extended so as to show the operations by latitude and longitude for each day. I herewith transmit two copies of a detailed statement of such operations, which statements include four vessels additional to those named in the papers heretofore sent to you, viz, the *Louis Olsen*, which entered at Victoria; *Rose Sparks, Therese*, and *Jane Grey*, which entered at San Francisco. These statements have been compiled under the direction of the United States Fish Commission, from the records of the custom-houses and personal observations of Prof. C. H. Townsend, who is connected with the Commission. It will be noted that, as you have been advised heretofore, the remaining vessels cleared without notice of the regulations of the Paris award, and therefore kept no record of latitude and longitude. The collector of customs at San Francisco reports that he required the masters to make oath to this fact on entry.

Respectfully, yours,

C. S. HAMLIN, *Acting Secretary.*

The SECRETARY OF STATE.

˙ (For inclosure, see Reports on Fur Seals by Fish Commission, Sen. Doc. 137, part 2, 54th Cong., 1st sess., pp. 59–60.)

TREASURY DEPARTMENT,
Washington, D. C., February 12, 1895.

SIR: As requested by Mr. Bax-Ironsides, I take pleasure in transmitting for your information copy of a letter dated the 11th instant, addressed by me to the President, in relation to the resolution recently introduced in the Senate, calling for reports, documents, and other papers, including logs of vessels, pertaining to the enforcement of the regulations respecting fur seals, adopted by the Governments of the United States and Great Britain, in accordance with the decision of the Tribunal of Arbitration convened at Paris, and the regulations under which such reports are required to be made, etc.

I desire to state, also, that I have to-day transmitted to the honorable the Secretary of State two copies of a statement prepared by Prof. C. II. Townsend, of the Fish Commission, wherein is given further information as to the latitude and longitude in which seals were taken by pelagic sealers in American vessels during the season of 1894, one of which copies, it was suggested, should be transmitted for the information of the British Government in compliance with article 5 of the Bering Sea arbitration award.

Respectfully, yours,

J. G. CARLISLE, *Secretary.*

Sir JULIAN PAUNCEFOTE,
British Ambassador, Washington, D. C.

[Inclosure.]

TREASURY DEPARTMENT,
Washington, D. C., February 11, 1895.

To the PRESIDENT:

Referring to Senate resolution of January 8, 1895, calling for all reports, documents, and other papers, including logs of vessels, relating to the enforcement of the regulations respecting fur seals adopted by the Governments of the United States and Great Britain, in accordance with the decision of the Tribunal of Arbitration convened at Paris, and the resolutions (regulations?) under which said reports are required to be made, as well as relating to the number of seal taken during the season of 1894 by pelagic hunters and by the lessees of the Pribilof and Commander islands, I have the honor to transmit herewith a statement taken from the logs of vessels showing the latitude and longitude in which fur seals were taken in Bering Sea. It will be noticed that only 10 of the 32 American vessels engaged in fur-seal fishing have made returns as to said latitude and longitude. The collector of customs at San Francisco reports that the captains of the other vessels deposed under oath that they cleared without notice of the regulations and therefore made no record of the places of capture.

I have further to state that during the season of 1894 about 142,000 seals were killed by pelagic hunters in the North Pacific Ocean, including Bering Sea. Of this number about 60,000 were taken either in Bering Sea or on the American side of the North Pacific Ocean; 15,033 seals were taken on the Pribilof Islands by the North American Commercial Company, lessees under the contract with this Government dated March 12, 1890. This Department has no official statistics of the catch on the Commander Islands, but is unofficially informed that it amounted to 27,285.

I also transmit herewith extracts from the logs of the United States Revenue-Marine steamers *Rush*, *Corwin*, and *Bear*, with relation to the enforcement of the regulations respecting fur seals as determined by the Paris Tribunal of Arbitration.

The papers above referred to constitute all the reports and documents in this Department which it is deemed compatible with the public interests to transmit to Congress at this time.

I have the honor, etc.,

J. G. CARLISLE, *Secretary.*

TREASURY DEPARTMENT,
Washington, D. C., May 6, 1895.

SIR: The experience of the past sealing season—the first under the provisions of the Paris award of August 15, 1893—has disclosed certain defects both in the form and scope of the award and of the legislation, especially that enacted by the British Government, for carrying out its provisions. I deem it my duty to call these defects to your attention with the request that you communicate with the British Government and endeavor to secure by mutual arrangement appropriate legislation in order that the object of the award—the preservation of the fur-seal fisheries—may be more effectually accomplished.

The contention of the British Government that regulations framed for the purpose of carrying out the award should be coextensive with and limited by the terms of the award would seem to be sound. It therefore only remains to consider certain aspects in which the award fails to provide for contingencies which, in the opinion of this Department, should be promptly guarded against by concurrent legislation not attainable with the assent of Great Britain in the form of regulations because of the limitations hereinafter referred to.

The most radical defect in the award is Article VI, which prohibits the use of nets, firearms, and explosives in fur-seal fishing, the only exception being that of guns when used outside of Bering Sea. The prohibition is directed simply to the use of these weapons for one particular purpose—seal fishing—leaving both the possession and use lawful for all other purposes, such as killing whales, walrus, sea otter, hair seal, and other animals found within said sea.

Experience has shown it to be almost a practicable impossibility to detect a sealing vessel in the act of using such firearms for this one prohibited purpose. Although the searching officer may be morally certain that firearms have been used and may properly consider the mere presence of firearms on the vessel, if accompanied with bodies of seals, seal skins, or other suspicious evidences, sufficient justification (even apart from the provisions of section 10 of the act of Congress of April 6, 1894, which is applicable only to American vessels) for the seizure of such a vessel, it must be apparent that in proceedings for condemnation brought in a court thousands of miles away from the place of seizure it will be almost an impossibility to secure conviction and forfeiture on the ground of illegal use of weapons. Furthermore, under the procedure necessary following the seizure of a British vessel the United States seizing officer delivers the vessel, with such witnesses and proof as he can procure, to the senior British naval officer at Unalaska. At the trial no representative of our Government is present, and the British Government must conduct the prosecution and must trust to such proofs and witnesses as the American officer could collect and furnish at the time. Under such circumstances forfeiture of the vessel could not be secured except in the clearest cases of guilt.

The prohibition of the use of firearms in seal fishing in Bering Sea was effectually accomplished only by prohibiting the possession of firearms in said sea for any purpose whatever.

The provisions of section 10 of the act of Congress of April 6, 1894, raising a presumption of illegal use from the possession of implements forbidden then and there to be used, is of great value in enforcing the award, but the act is limited to American vessels. It is to be regretted that there is no equivalent provision in the British act of Parliament enacted April 18, 1894, for carrying out said award.

In this connection it is significant that in the act passed by Parliament for carrying out the modus vivendi of June 15, 1891, prohibiting all sealing in Bering Sea (54 and 55 Victoria, Chap. XIX), a provision similar to that contained in the act of Congress above cited was inserted, as follows:

If a British ship is found within Bering Sea having on board thereof fishing or sealing implements or seal skins or bodies of seals it shall lie on the owner or master of such ship to prove that the ship was not used or employed in contravention of this act.

It is to be regretted that the late British act for carrying out the award contained no similar provision, modified, of course, to suit the terms of the award. Although an amendment bringing it into harmony with the American law would render the task of enforcing the award much easier and the result more efficacious, yet, as stated above, the most satisfactory amendment would consist in common legislation rendering a vessel subject to forfeiture if found with firearms in its possession in Bering Sea.

The above suggestions are prompted by certain reports just received from the United States Fish Commission containing statements of officers of the Commission employed last season in Bering Sea that firearms were used by sealers fishing in said sea. Although little or no direct evidence was submitted by these officers as a basis of their opinion, yet the opportunity they had of observing the operations of the sealing fleet and of boarding vessels and inspecting catches renders their opinion of the greatest value and prompts the Department to suggest the necessity of an immediate change in the law.

It should further be provided by legislation that sealing vessels having implements or seal skins on board, desiring to traverse the area covered by the award during the close season if licensed, and during any season if unlicensed, should have such implements duly sealed and their catch noted on the log book (a privilege now accorded at the option of the master, under the regulations of 1895, Article IV), under penalty of forfeiture for violation of this provision. This privilege, however, as above stated, should not be accorded vessels having firearms in Bering Sea.

It is further to be noted that under the British act of Parliament the provisions of the merchant shipping act of 1854 with respect to official logs (including the penal provision) are made applicable to sealing vessels. Said penal provisions, however, do not appear in the schedules attached to the copy of the act in the possession of the Department. I have therefore to request that you ascertain of the British Government whether such penalties include forfeiture of the vessel and cargo. The United States act, section 8, expressly provides that any violations of the award or regulations will render the vessel and cargo liable to forfeiture. It is feared that because of the specific reference to the penal provisions of the merchants' act as to official logs the failure of a vessel to keep log entries might not bring her within the general liability to forfeiture contained in the British act, unless said merchants' act, made a part thereof, contains similar provisions. During the past season log-book entries were duly made by the United States sealing vessels in Bering Sea and were transmitted to Congress.

The Department is also informed that similar entries were made by British vessels in Bering Sea, which entries have been duly transmitted by the British Government. Many vessels, however, had cleared for the coasts of Japan and Russia as early as January, long before the passage of either the act of Congress of April 6, 1894, or the act of Parliament of April 18, 1894.

Inasmuch as the award was not self-operative and contained no penalties for its violation the Department considered that the penalties provided in the subsequent legislation were not retroactive and could not properly be applied to acts or omissions before the passage of said legislation. Entry was therefore permitted of the catch of seals on receipt of the master's oath that he cleared in ignorance of the provisions as to log-book entries. During the coming season collectors have been instructed to rigidly enforce the law as to log-book entries. The exact status, therefore, of the British law becomes important.

I have further to call your attention to the fact that, by acts of Congress making appropriations for sundry civil expenses of the Government for the fiscal years ending 1895 and 1896, provision was made for the appointment of seal experts to examine all seal skins landed in the United States as to the number and sex, with the purpose of verifying the log-book entries. All skins entered during the past season at United States ports, except Port Townsend, were duly examined by said inspectors as to number and sex; by an error, however, the skins entered at Port Townsend, although duly counted, were not examined as to sex.

I would respectfully request that you ascertain whether during the past season the British Government employed similar inspectors to verify the log-book entries of British vessels as to number and sex of seal skins landed.

I have also further to suggest that you request of the British Government that its consent be given to stationing United States inspectors at British Columbian ports, for the purpose of verifying said log entries of British vessels and examining the skins as to sex, freely according to the British Government a like privilege in United States ports.

I would also call your attention to the fact that under the British act it is nowhere made the duty of the British naval officers to seize ships when found in violation of the law. Section 11 of the United States act imposes said duty on United States officers duly designated by the President. This latter point, I believe, has already been called by you to the attention of the British Government.

Should these suggestions as to new legislation meet with your approval and be enacted by the respective Governments, I am confident that the award can be enforced so as to better subserve the purpose for which it was intended—the preservation of the fur-seal fisheries.

I have the honor to request that you communicate these suggestions, if approved by you, to the British Government.

Respectfully, yours,

J. G. CARLISLE, Secretary.

The SECRETARY OF STATE.

TREASURY DEPARTMENT,
Washington, D. C., May 6, 1895.

SIR: In my letter of even date herewith I had the honor to suggest that you endeavor to secure the cooperation of the British Government to the end that certain necessary legislation be enacted to render the Paris award more effective for the purpose of preserving the fur-seal herd. I have now the honor to transmit certain further suggestions as to widening, by mutual agreement, the scope of the award, which I believe to be warranted by the information now in possession of the Department.

The sealing season just closed was the first during which the provisions of the Paris award were applicable, and I regret to have to state that the pelagic catch of seals both without and within the award area was the largest ever known in the history of pelagic sealing.

In my communication to Congress, January 21, 1895 (Ex. Doc. No. 243, Fifty-third Congress, third session),[1] I was able to state the catch

----

[1] A copy of this communication is appended hereto.

as reported in the United States and British Columbia custom-houses as 121,143. I stated, however, that it was known that a large number of skins were transshipped in Japan ports and sent to London by way of the Suez Canal. Reliable information as to the sale of fur-seal skins in London for the season of 1894 discloses that 138,323 skins, taken at sea in the North Pacific Ocean from the American and Russian or Japanese herds during the season of 1894, were sold in London. Careful estimates show that about 3,000 were retained in the United States for dressing and dyeing, making a total of 141,323. To this should be added about 800 which were known to have been on a vessel believed to have been lost, making the total about 142,000. Of this amount 55,686 were taken within the area covered by the Paris award.

The following table gives the number of skins taken within said Paris award area during the years 1890 and 1894, inclusive:

```
1890............................................................... 40,809
1891............................................................... 45,941
1892............................................................... 46,642
1893............................................................... 28,613
1894............................................................... 55,686
```

It would be within moderate bounds to state that these figures of seals killed and recovered represent only about one-third of all killed, but whose bodies were not recovered.

A perusal of these figures must satisfy the most skeptical that the fur-seal herd will be speedily exterminated unless an immediate change is made in the scope as well as the form of the award.

So far as the articles of the award relating to the North Pacific Ocean, exclusive of Bering Sea, are concerned, forbidding all seal fishing from May to August, it must be admitted much good has been accomplished, and favorable results were apparent on the islands early in the season. The fatal defect in the scope of the award, however, was in opening Bering Sea during August and September to sealers, prohibiting only therein the use of firearms. It has been claimed, and with some evidence in its favor, that the spear is as destructive in Bering Sea as the shotgun. Some experts believe that even greater destruction is accomplished in Bering Sea by the use of the spear than by guns, for the reason that the noise of the shotguns frightens away many seals who might be easily killed sleeping on the water by spearsmen. While the herd is traveling in the North Pacific Ocean away from the islands it is very difficult to kill seals with spears, as they are constantly traveling and rarely found asleep on the water. In Bering Sea, however, the females leave their pups on the islands and go out for a distance of 100 or 200 miles from the islands, far beyond the prohibited zone of 60 miles, to feed. They are there found in large numbers asleep on the water, and can easily be killed by the silent, skillful spearsmen. The large number of pups found dead from starvation on the islands during the latter part of September and October, 1894—12,000 by actual count on the accessible parts of the rookeries, and 20,000 by careful estimates—shows the destructive effect of permitting sealing at all in Bering Sea. Should Bering Sea be forever closed to pelagic sealers, and should the closed season now provided by the award in the North Pacific Ocean be maintained, the Department believes that the seals would receive fair protection, and that fur-seal fishing might continue to be a profitable one, both on land and on sea. Unless this limitation in the scope of the award be made, within a very few years at the most the seals will be exterminated commercially and the industry destroyed.

H. Doc. 92, pt. 2——21

The Department understands that you have already suggested to the British Government the appointment of a commission, to consist of members appointed by the Governments of Great Britain, Russia, Japan, and the United States, to examine further into the sealing question, and that pending their examination and report a modus vivendi be agreed upon, one clause of which provides for closing Bering Sea to pelagic sealing absolutely. This communication is sent you to further inform you that the official figures of last season's catch, now definitely known, fully bear out the wisdom and necessity of such a change in the provisions of the award.

Trusting that some arrangement as above suggested may be agreed upon, I have the honor to be,

Respectfully, yours,

J. G. CARLISLE, *Secretary.*

The SECRETARY OF STATE.

[Inclosure.]

TREASURY DEPARTMENT,
Washington, D. C., January 21, 1895.

SIR: I have the honor to acknowledge the receipt of the following resolution, dated the 11th of December, 1894, of the House of Representatives:

"*Resolved,* That the Secretary of the Treasury be requested to furnish the House with information—

"1. As to whether the articles of the Bering Sea Tribunal, convened at Paris in 1893, for the regulation of the fur-seal industry of Alaska, have resulted during the last season in saving the fur-seal herds from that destruction which these articles were intended to prevent.

"2. Whether the Secretary has authentic information of the probable number and sex of Alaskan fur seals killed in the last season by pelagic sealers; and if so, what such information is; and in view of such facts, what, in the judgment of the Secretary, will be the practical result of these articles if carried out in good faith for the future.

"3. What is the present condition of the fur-seal herds on the Pribilof Islands?

"4. What has been the revenue derived by the Government from the fur-seal herds of Alaska during the past season, and also what has been the expenditure during the same period in executing the regulations of the Paris award?"

In reply to the first inquiry I have to state that the number of seals taken by pelagic sealers in the North Pacific Ocean for the season just expired and entered at United States and British Columbia ports, as contained in the accompanying table, compiled from official reports of collectors of customs in the United States and reports transmitted by the United States consul at Victoria, British Columbia, compiled by him from the official statements of the collector of customs at that port, aggregate 121,143. Of this number 55,686 were taken from the so-called Alaska seal herd in the North Pacific Ocean and in Bering Sea; 58,621 were taken off the coasts of Japan and Russia, leaving 6,836 undetermined. Ninety-five vessels were employed, 60 belonging to Great Britain and 35 to the United States.

As appears in said table, the actual number of seals killed in 1894 exceeds the amount of skins entered as above by about 20,000, making the total catch about 142,000. This balance of 20,000 skins was probably shipped to London via Suez Canal from the Asiatic Coast.

From these figures it becomes evident that during the present season there has been an unprecedented increase over preceding years in the number of seals killed by pelagic sealers, both in American and Asiatic waters. This increase has caused an alarming decrease in the number of seals on the islands, as hereinafter explained. A significant fact in this connection is the unprecedented number of dead pups found on the islands this season which presumably died of starvation, their mothers being killed at sea. Our agent counted over 12,000 on the accessible portions of the rookeries alone. He estimates, upon said count, a total of nearly 20,000. It should be remembered that at the close of the season of 1893, when pelagic sealing was prohibited in Bering Sea, less than 1,000 were found on St. Paul Island, no count having been made on the Island of St. George.

The alarming increase in the number of seals killed by pelagic sealers, and the further fact that in four or five weeks the vessels in Bering Sea, only about one-third of the total number, killed more seals than were taken in the four months sealing on the American side of the North Pacific, emphasize the conclusion expressed in my annual report to Congress, that long before the expiration of the five years, when the regulations enacted by the Tribunal of Arbitration are to be submitted to the respective Governments for reexamination, the fur seal will have been practically exterminated.

My answer to the first inquiry is, therefore, that the operation of the articles of the Bering Sea Tribunal for the regulation of the fur-seal industry of Alaska has not resulted in saving the fur-seal herd from that destruction which those articles were intended to prevent.

As to the number and sex of Alaskan fur seals killed during the past season by pelagic sealers, I have to state that of the American catch of 26,095 seals, 3,099 were males, 15,976 females, and 7,020 pups and seals the sex of which was unknown. Each seal landed in the United States was carefully examined as to sex by experts appointed by the Department.

Of the catch of the British Columbia schooners of 95,048 seals, as reported by our consul, only those killed in Bering Sea—26,425—were classified as to sex. Of these, 11,723 were reported as males and 14,702 as females. •

With reference to the present condition of the fur-seal herds on the Pribilof Islands, I have to report a dangerous decrease. Information on file in the Department indicates a falling off of at least one-half during the past four seasons. It thus appears that the condition of the Alaskan fur-seal herd is most critical. All facts point to its speedy extermination unless the present regulations, enacted in the award of the Paris Tribunal, are changed at an early date, so as to afford a greater measure of protection to the seal herd.

In reply to the inquiry concerning the revenue derived by the Government from the fur-seal herds during the past season, and the expenditures during the same period in executing the requirements of the Paris award, I have to state that 15,000 seals were taken on the Pribilof Islands in the year last past, and 1,031 remained on hand from last year. The amount to be paid by the lessees of the islands, according to the provisions of their contract, on or before April 1 next, will be $214,298.37, the items being as follows:

Rental	$60,000.00
Tax of $2 per skin on 16,031 skins	32,062.00
Bonus of $7.625 per skin on 16,031 skins	122,236.37

As to expenses, I have to state that the honorable the Secretary of the Navy reports that the expenditure incident to the presence of the United States naval vessels in Bering Sea during the past year was $158,188.25. The expenses attending the presence of the revenue steamers *Bear*, *Corwin*, and *Rush* aggregate $40,116.24. The amounts named do not include the pay of officers or men or the rations supplied to them. Of the $1,500 appropriated to enable the Secretary of the Treasury to pay the necessary expenses of enforcing the provisions of section 4 of the act approved April 6, 1894, under which two experts were employed to examine and classify pelagic seal skins, the sum of $250 has been expended. The salaries and expenses of the agents of the Seal Islands, whose duties would require them to be present on said islands without regard to the Bering Sea controversy, have not been included in preparing this answer to the resolution. The aggregate expenses would, therefore, seem to be $198,554.49.

In this connection I have to state that suit has been instituted against the North American Commercial Company for the recovery, under the terms of its lease of the Seal Islands, of the sum of $132,187.50, covering the season of 1893. The company named, under its lease, is required to pay the sum of $60,000 per annum rental, $2 tax on each seal taken, and, in addition, $7.625 for each seal skin accepted. It is claimed by said company that, as it was denied the right to take the number of seals contemplated at the time the lease was executed, by reason of the operation of the modus vivendi, a reduction in the rental and in the item of $7.625 per skin should be made. This claim, under advice of the Attorney-General, has not been admitted by the Treasury Department, and, as hereinbefore stated, suit has been instituted. I find that the following balances for the years specified are due from said company under its lease, notwithstanding the fact that my predecessors have accepted payments in less amounts than those mentioned in the contract hereinbefore referred to: 1890, $47,403; 1891, $133,628.64; and 1892, $108,686.52.

Action by this Department on the above-mentioned unpaid amounts will be determined by the result of the suit pending for the amount due for the year 1893.

Respectfully, yours,

J. G. CARLISLE, *Secretary.*

Hon. CHARLES F. CRISP,
    Speaker of the House of Representatives.

Summary of pelagic seal catches for 1893 and 1894, based on the official returns from ports of entry.

Year.	Nationality.	British Columbia and Northwest coasts.	Bering Sea.	Japan Coast.	Russian Coast.	Locality undeter-mined.	Total.	Grand total.
1893...	American...		(Modus viven-di in opera-tion.			8,342	8,342	} * 78,083
	Canadian ...	28,613		29,173	11,955		69,741	
1894...	American...	12,308	5,160	1,500	201	6,836	26,095	} † 121,143
	Canadian ...	11,703	26,425	49,483	7,437		95,048	
	Total ..	24,101	31,585	50,983	7,638	6,836		

* *Notes concerning catch for 1893.*—The United States consul at Victoria states (Consular Reports No. 161, p. 279) that American schooners in 1893 transshipped, at Yokohama and Hakodadi, between 17,000 and 18,000 skins. These skins, added to those which in all probability were transshipped by British Columbia vessels on the Asiatic Coast, and including the estimated number retained in America for treatment, would swell the total catch to about 109,000. The accuracy of these figures is corroborated by the fact that the trade sales of London (all seal skins are sold there) account for the disposition of 109,669 skins in 1893.

† *Notes concerning catch for 1894.*—The catch of 6,836 noted in the column headed "Locality undetermined" were skins, 76 of which were landed at Astoria without statement as to place of capture; 641 were transshipped at Unalaska, and the remaining 6,119 were transshipped from Yokohama. All were entered and recorded in American ports of entry and they are quite certainly a mixture of Northwest Coast and Japan skins. It has been ascertained from the sales of seal skins in London that about 125,000 skins were actually sold, and about 14,000 withheld for future sale in 1894. In addition thereto it is estimated that about 3,000 skins were retained in this country and elsewhere for treatment. It thus appears that about 142,000 is a figure much more closely representing the number of skins taken in 1894 than the official returns of 121,143. The balance, about 20,000 skins, was probably shipped to London, via Suez Canal, from the Asiatic Coast.

Number of schooners reported as having taken skins.

Year.	American.	Canadian.	Total.
1893	28	* 56	84
1894	35	* 60	95

* Indian canoe catch counted as one (1) vessel. In destructive effects the canoe catch is about equal to three average schooner catches.

Number of schooners reported as having made catches in Bering Sea.

Year.	American.	Canadian.	Total.
1893	Modus vivendi in operation.		
1894	10	27	37

TREASURY DEPARTMENT,
Washington, D. C., May 15, 1895.

SIR: I beg to return herewith the letter of the British ambassador, dated the 11th instant,[1] handed me by you, transmitting the declination of his Government to agree upon concurrent regulations for carrying out the provisions of the Paris award during the present season. The reason given for such declination is that the provisions of the award relating to the special license and distinguishing flag are already provided for in the British order in council of February 2, 1895; that concurrent regulations similar to those agreed upon for last season by the representative Governments as to sealing up the outfit and arms of sealing vessels are not considered necessary for the present season, inasmuch as the possession by vessels within the award area and during the closed season of said outfit and arms is nowhere forbidden by

---

[1] Not furnished.

the terms of the award. On the subject of the regulations of last season it is stated that "the arrangement in question has not in practice been worked for the protection of British sealers from interference as Her Majesty's Government had hoped would have been the case," reference being specially made to the seizure by United States officers of the British schooners *Wanderer* and *Favorite.*

Attention is further called to the fact that in making said seizures the United States officers were under the erroneous impression that the act of Congress of April 6, 1894, was applicable to British vessels, and, in fact, cited said act as justification for the seizure, whereas its provisions are applicable only to American vessels, the right to seize British vessels being limited to offenses under the British act of Parliament only to be exercised by virtue of the power given in the order in council of April 30, 1894. Request is also made that United States officers engaged in patrolling the award area during the present season be instructed accordingly.

In reply, I have the honor to state that on December 15, 1894, a draft of proposed concurrent regulations for the season of 1895 was transmitted by you to the British ambassador for the approval of his Government. This draft had been prepared by me and sent to you for this purpose. Subsequently the British ambassador obtained your consent to confer directly with me upon the subject, and a number of interviews were accorded him by myself and Assistant Secretary Hamlin upon this matter. He submitted a draft, now in our possession, of proposed concurrent regulations containing certain suggested improvements over the draft submitted by myself; after preliminary negotiations covering considerable period of time a draft was finally agreed upon satisfactory to each of us, the understanding being that I should submit a copy of the same to the President for his approval and promulgation and that he, on his part, should forward a copy for the approval of his Government and for insertion in an order in council shortly to be passed. He stated that it would be necessary to insert the regulations in a new order in council for the reason that the last order bearing upon the subject was limited in its operations to the sealing season of 1894.

I accordingly presented a copy of the proposed regulations to the President, who signed the same, understanding that they received the approval of the British ambassador and would be forwarded by him to his Government, as above stated.

While I did not for a moment understand or believe that the British ambassador had authority or had undertaken definitely to bind his Government without a formal transmission of the proposed regulations, yet I had every reason to believe that the draft agreed upon by us would be promptly accepted by the British Government, or its declination as promptly communicated. The order in council alluded to by the British ambassador was enacted February 2, 1895. On that date the copy of said proposed regulations must have been in the possession of the home Government, as it was given to Sir Julian on January 17 for transmission. I would further call to your attention the fact that in said order in council a reference is made to arrangements which have been entered into by the respective Governments, which can only refer to these proposed regulations and which by necessary implication recognizes the same as valid and subsisting. The preamble of this order recites that—

Whereas arrangements have been made between Her Majesty's Government and the Government of the United States for giving effect to articles 4 and 7 of the scheduled provisions, and it is expedient that effect should be given to those arrangements by an order in council. * * *

The word "arrangements" in said preamble, as I have stated, can only refer to these proposed regulations for the season of 1895, for no other agreement or arrangement than that contained in said regulations has been entered into this year between the respective Governments as to any of the award provisions, the arrangements for last season having been by their terms limited to the sealing season of 1894.

It may be pointed out that said order in council related only to the special license and distinguishing flag; these were, however, the only matters embraced in said regulations which depended upon the order in council for their binding force, the remaining provisions being optional with the masters of vessels.

That this word "arrangements" can only refer to the agreement of understanding upon which said regulations were based on our part is made clear by the use of the same word in the previous orders in council of April 30 and June 27, 1894, respectively. In the order in council of April 30, 1894, it is recited—

Until arrangements for giving further effect to articles 4 and 7 of the said scheduled provisions shall have been made between Her Majesty and the Government of the United States, the following provisions should have effect. * * *

Following this order, to wit, on May 4, 1894, the President of the United States signed and approved regulations for the season of 1894 based upon an agreement made by Sir Julian and myself for the respective Governments, articles 7 and 8 of which provided for a special license and distinguishing flag.

The order in council following on June 27, 1894, contains this significant language:

And whereas arrangements have been made for giving further effect to the said articles and for regulating during the present year the fishing for fur seals in accordance with the scheduled provisions. * * *

It thus would seem that the word "arrangements" as contained in the orders in council of April 30 and June 27, 1894, respectively, could only mean the preliminary agreement upon which were based the regulations of 1894; this agreement was expressly limited by its terms to the sealing season of 1894. It would therefore seem to follow that the reference to "arrangements" in the order in council of February 2, 1895, could only relate to the agreement as contained in the proposed regulations approved by the President and transmitted to the British Government as aforesaid.

While, therefore, it would seem that the British Government by necessary implication has ratified and recognized as subsisting the proposed regulations, submitted as above, by the passage of the order in council of February 2, 1895, I nevertheless recognize that this notification and concurrence can at any time be withdrawn upon notice to our Government. I feel it, however, to be my duty to express deep regret that the British Government should have communicated its declination at this late period of the season after our consuls have been instructed and our patrolling fleet has sailed under orders based on the assumption that the privileges afforded by said regulations should be accorded during the present season as during last season to British as well as to American vessels. It is to be regretted, also, that the chief reason assigned for such declination—the seizure of the schooners *Wanderer* and *Favorite* should not have prompted an early refusal to enter upon preliminary negotiations for new regulations, thus saving much trouble and uncertainty now apparently unavoidable. The British fleet engaged in sealing last season numbered sixty vessels; of those the *Wanderer* and *Favorite* were the only ones seized. These seizures were made because of a direct infraction of the regulations of 1894, agreed upon, as above stated, by both Governments. The *Wanderer* was seized June 9, 1894, and the *Favorite* on August 27, 1894. The master of the *Wanderer* before the seizure stated to the boarding officer that all his arms were sealed up, which upon examination was found to be true.

The Department is unaware that any objection has ever been made by the British Government because of these seizures until the present time, and it must express its regret that these facts, in possession of said Government during all of the preliminary negotiations above set forth as to the regulations for this season, should only now be brought forward as a ground for its refusal to adopt concurrent regulations.

In view of this communication from the British Government, it is presumed by the Department that no British sealing vessel now at sea has applied, or will hereafter apply, for the privilege of having its outfit and arms sealed up. The officers of the patrol fleet will, however, be instructed not to regard the fact that the outfit and arms are not sealed as evidence in considering whether or not a British vessel has violated the act of Parliament. They will also be instructed to refuse to grant this privilege in the future to British vessels. I have to request that you at once give similar instructions to our consuls in Japanese and British Columbia ports, and, further, that you request that the British Government shall notify its officers engaged in patrolling the award area to seal up the outfit and arms of American vessels applying for this privilege, in accordance with sections 4 and 7 of the regulations promulgated by the President January 18, 1895.

In closing I would further inform you that the instructions already given our officers as to patrolling the award area during the present season will not admit of any error or doubt as to the proper scope and limitation of the act of Congress approved April 6, 1894.

I have the honor, etc.,

J. G. CARLISLE, *Secretary.*

The SECRETARY OF STATE.

TREASURY DEPARTMENT,
Washington, D. C., June 11, 1895.

SIR: The Department is in receipt of a communication dated Sitka, May 15, in which the seizure of the British sealing schooner *Shelby* on May 11 by Captain Munger, of the United States revenue cutter *Corwin*, is reported. The declaration of seizure states that the boarding officer, Lieut. W. E. W. Hall, "found the following evidence that she was engaged in sealing unlawfully: She was found within the area of prohibited waters, latitude 52° 50' 10" north, longitude 134° 10' 58" west, with 124 seal skins on board, and all the implements and outfits for sealing, including 3 tons of salt, 3 boats, and 7 shotguns and ammunition for me."

The declaration of seizure prepared by Captain Munger and delivered to the commanding officer of H. M. S. *Pheasant* states that the vessel was seized for disregarding the proclamation of the President of the United States and the act of Congress, April 6, 1894. From an examination of the report of Captain Munger it would appear that the seizure was made on the ground that there was cause to believe that said vessel had killed fur seals within the award area during the closed season, the reason of such belief being found in the possession by the vessel of seal skins, implements, and outfits, together with salt, shotguns, and ammunition.

On receipt of said report, Captain Hooper, commanding officer of the patrolling fleet, was reminded that the act of Congress of April 6, 1894, was applicable only to American vessels. He was also directed, if on investigation he found that said vessel was seized on the charge of illegal killing during the closed season, to instruct Captain Munger to deliver to the commanding officer of H. M. S. *Pheasant* an amended declaration of seizure, assigning as the cause the violation of the second article of the regulations of the Paris award, as set forth in the schedules annexed to the British act of Parliament known as the Bering Sea award act of 1894.

In this connection the receipt signed by the commander of H. M. S. *Pheasant* is called to your attention:

SITKA, *May 13, 1895.*

In accordance with the provisions of section 12, article 9, of the Bering Sea fisheries award, I have this day received from C. L. Hooper, captain, U. S. R. C. S., commanding Bering Sea fleet, the British schooner *Shelby,* of Victoria, British Columbia, C. Classen, master, with her tackle, furniture, cargo, and documents, seized by the United States revenue steamer *Corwin,* Capt. F. M. Munger, commanding, for violation of the acts of Congress and of the British Parliament regulating the fur-seal fisheries.

FRANK A. GARFORTH,
Lieutenant, Royal Navy, Commanding H. B. M. S. Pheasant.

I have the honor to suggest for your consideration the expediency of obtaining the consent of the British Government to the appointment of counsel to represent the United States Government in condemnation proceedings against the *Shelby* and such other British vessels as may be seized this season by the United States officers for violation of the regulations of the Paris award. I believe that such action would greatly assist in the proper enforcement of the award provisions. In this connection I would suggest the advisability of notifying at once the British Government that the declaration of seizure will be amended to the end that the libel in admiralty may set forth the breach of the British act of Parliament known as the Bering Sea award act of 1894.

I have the honor, etc.,

J. G. CARLISLE, *Secretary.*

The SECRETARY OF STATE.

DEPARTMENT OF STATE,
Washington, May 23, 1895.

SIR: Referring to your letter of the 30th ultimo, I have the honor to inclose for your information a copy of a dispatch from the consul at Victoria reporting that the commander in chief of that station ordered the release of the schooner *Wanderer,* having reached the conclusion that no case could be made out against the vessel.

The collector of the port informed the consul that the Government at Ottawa instructed him to take no official action in the matter.

I have the honor, etc.,

EDWIN F. UHL,
Acting Secretary.

The SECRETARY OF THE TREASURY.

[Inclosure.]

CONSULATE OF THE UNITED STATES,
Victoria, British Columbia, May 13, 1895.

SIR: In reply to your dispatch, No. 19, of the 4th instant, received to-day, inclosing copy of letter from the Treasury Department requesting information with respect to the final disposition of the sealing schooner *Wanderer*, seized during the sealing season, 1894, I beg leave to inform the Department that the said schooner was released by the commander in chief of the Pacific station, Admiral Stephenson, of H. M. S. *Royal Arthur*.

This schooner was seized by the U. S. S. *Concord* by reason of the fact that one unsealed gun was found in one of her berths.

She was turned over to H. M. S. *Pheasant* and brought to this port. Upon investigation it was found that all her other guns and her entire armament were sealed, and that her master was not aware and had no knowledge that there was a gun aboard unsealed in violation of the provisions of the Paris award.

Collector A. R. Milne, of this port, and from whom I get this information, advises me that his Government at Ottawa instructed him to take no official action whatever in the matter, and that the commander in chief of this station, after careful investigation, and acting under legal advice, ordered the release of the schooner, the conclusion having been reached that no case could be made out against her.

I am, sir, etc.,

W. P. ROBERTS, *United States Consul.*

Hon. EDWIN F. UHL,
Assistant Secretary of State, Washington, D. C.

TREASURY DEPARTMENT,
Washington, D. C., June 12, 1895.

SIR: I have the honor to acknowledge the letter dated May 23, from the Acting Secretary of State, inclosing for my consideration a communication from the United States consul at Victoria, British Columbia, to the effect that the British sealing schooner *Wanderer*, seized June 9, 1894, in the North Pacific Ocean by the commander of the United States cruiser *Concord* and formally delivered to the commander of H. M. S. *Pheasant*, was subsequently taken to Victoria and released by Admiral Stephenson, of H. M. S. *Royal Arthur*.

It is further stated in said communication that upon investigation it was found that all the guns of the *Wanderer* except one were secured under seal; that her master had no knowledge that there was a gun on board unsealed, in violation of the provisions of the Paris award, and further, that the "commander in chief of this station, after careful investigation, and acting under legal advice, ordered the release of the schooner, the conclusion having been reached that no case could be made out against her."

The Department also understands that the sealing schooner *Favorite*, seized in Bering Sea August 24, 1894, by the commanding officer of U. S. S. *Mohican*, was similarly released on being turned over to the British naval authorities. I deem it to be my duty to call to your attention this action of the naval authorities of Great Britain, with the suggestion that it is not in accord with evident intent and spirit of the legislation enacted by the respective Governments for carrying out the provisions of the Paris award.

These vessels were seized under authority of the order in council of the British Government dated April 30, 1894, authorizing United States officers duly commissioned and instructed by the President to seize any British vessel which has violated the Paris award regulations, as contained in the act of Parliament known as the Bering Sea award act, 1894, and bring her for adjudication before any British court of admiralty, or in lieu thereof to deliver her to any British officer for adjudication before the court.

The plain spirit and intent of the law would seem to require proceedings in admiralty for condemnation and forfeiture of every vessel seized by the United States officers and delivered to the British authorities as aforesaid. In the cases in question, however, it would appear that Admiral Stephenson, in discharging said vessels, took upon himself to decide a question which, under the law, could properly be decided only by a British court of admiralty.

The evidence in the case of the *Wanderer* clearly would seem to justify the suspicion and belief that some, at least, of the 400 seal skins found on board had been taken during the prohibited season by means of shotguns, in violation of the award regulations and of the British and American law.

In the case of the *Favorite* 1,230 seal skins were found on board, together with a shotgun whose barrel was partly cut off, but leaving about 12 inches. It was found that it would shoot accurately for a distance of 50 yards.

The causes particularly assigned for these seizures, the carrying of firearms unsealed, taken in connection with the fact that such weapons were forbidden then and there to be used, and that there were also found seal skins on board, would plainly justify the belief that said firearms had been used in violation of article 6 of the award as contained in the Bering Sea award act of 1894 (British) and the act of Congress, April 6, 1894 (American). That the notices of seizure, as prepared by the United States seizing officers, do not, with particularity, specify the illegal use of these weapons, but rely chiefly upon their presence on board unsealed, clearly would not prevent such use being proved in subsequent proceedings in court of admiralty for condemnation and forfeiture, said notices being merely to acquaint the authorities to whom the ships are turned over of the seizure and of the particular offenses relied upon for maintaining a libel in condemnation proceedings. It would seem perfectly clear that additional breaches of the law could be assigned and made the subject of condemnation proceedings at any time before the trial took place.

The instructions issued by the British Government to the commanders of its cruising vessels for the season of 1894 would, it is submitted, have imposed upon such officers under similar circumstances the duty of seizing these vessels. Said instructions, in part, were as follows:

If you are satisfied that a vessel has hunted contrary to the act you will seize her. * * * Whether the vessel has been engaged in hunting you must judge from the presence of seal skins or bodies of seals on board and other circumstances and indications. (See Senate Ex. Doc. 67, p. 116, Fifty-third Congress, third session.)

In the case of the *Wanderer*, the master gave his guns and ammunition to the commander of the United States cruiser *Yorktown*, to be secured under seal. Later in the same day he was boarded by the cruiser *Concord*, and stated that the guns and ammunition sealed up by the *Yorktown* was all he had on board. After a search, however, a breech-loading shotgun and a bag of loaded shell were found concealed in the extreme forward part of the vessel under a pile of iron cans between decks. While the officer was making an entry in the log book as to this weapon the master of the vessel was heard to say to the mate, "God damn it, I told you you ought to have had that put in with the others," or words to that effect. This deception of the master, together with the concealed weapons, the presence on board of seal skins, and other suspicious evidence revealed on search, clearly should have been submitted to a court of admiralty in condemnation proceedings. I respectfully call these facts to your attention, with the suggestion that a formal protest against said action of the British naval authorities be communicated to the British Government, with the request that in future every vessel seized by United States officers shall be proceeded against for condemnation in the admiralty court having jurisdiction in the premises.

I have the honor, etc.,

J. G. CARLISLE, Secretary.

The SECRETARY OF STATE.

TREASURY DEPARTMENT,
Washington, D. C., June 12, 1895.

SIR: I have received a copy of the communication of the British Foreign Office to the British ambassador of the 17th ultimo, in answer to his dispatch of January 24 last, conveying the proposition of this Government for the appointment of an international commission by the Governments of the United States, Great Britain, Russia, and Japan, respectively, for investigating the fur-seal fisheries of the North Pacific Ocean, and, pending a report of said commission, for a modus vivendi prohibiting sealing in Bering Sea and extending the regulations of the Paris award along the thirty-fifth degree of north latitude to the shores of Asia.

The communication opens with the proposition that our Government, because of its contention before the Paris tribunal that the Asiatic and American fur-seal herds are distinct and do not commingle, can not now with propriety draw any inference as to the other effects of pelagic sealing on the American fur-seal herd from figures indicating increased catches over previous seasons in the total of seals killed on the Asiatic and American sides of the Northern Pacific Ocean. The claim is further advanced that although the catch of fur seals during last season on the Asiatic side was greater than in any previous year, yet the catch taken from the American herd—that is, within the Paris award area—while admittedly larger than in most previous seasons, was, in fact, not as large as that of the season of 1891. Our Government is further reminded therein that the success or failure of the regulations established by the Paris tribunal must be judged " solely by their effect on the herd which they were intended to protect."

I have the honor to reply that during the hearings before the Tribunal of Arbitration at Paris it was earnestly contended by counsel representing Great Britain that the Asiatic and American herds did commingle. That fact was disputed by the American counsel in the light of the evidence before them. The tribunal, however, was not called upon to make any definite finding upon this important question. While I do not wish to be understood as expressing any opinion upon the subject, yet, in view of the admission contained in said communication, in which I cordially join, that "our knowledge of seal life is still far from complete," I feel that the question as to whether said herds intermingled requires most careful consideration and study. It has been suggested that the American herd seal, even if not naturally commingling with the Asiatic herd, may have been driven over to Asiatic shores by incessant slaughter during the past seasons. If such be found to be the fact on careful investigation—which investigation is unfortunately refused by the British Government—it might appear that the total slaughter of fur seals on both sides of the North Pacific Ocean has a more intimate connection with the present condition of the American fur-seal herd than is now admitted.

However this may be, the British Foreign Office seem to have fallen into the serious error of assuming that the proposition of the United States Government, contained in your letter of January 23 last to the British ambassador, was purely selfish in its character, having application only to the material interests of the United States Government in the American fur-seal herd. Nothing could be further from the truth. It is to be presumed that the President in said letter was actuated by the desire to protect the fur-seal fisheries on both sides of the North Pacific Ocean, Asiatic as well as American, for the benefit of mankind. Incidentally this would result in benefit to the interests of the United States; but the proposition was founded on broad humanitarian principles, no especial benefit being sought save what would accrue to all mankind from the proper regulation of these valuable fisheries. A proposition of similar nature, but limited to Bering Sea, was made by Secretary Bayard to our ministers in England, Japan, Russia, Sweden, and Norway for formal transmission to the respective Governments as far back as 1887. Subsequently, at the request of Lord Salisbury, the British minister of foreign affairs, its scope was broadened so as to embrace the whole Northern Pacific Ocean from the Asiatic to the American shores north of the forty-seventh degree of north latitude. Unfortunately the British Government withdrew its approval of this arrangement. (See letter of White to Bayard, June 20, 1888.)

The closed season established by the Paris award has induced many sealing vessels to go over to Japan and Russian waters, thus causing a startling increase in the pelagic slaughter off these shores. The figures given in said communication include only the slaughter in Japan waters; adding the seals killed in Russian waters we have a total of over 73,000 in 1893 and over 79,000 in 1894. It was to regulate the killing in these waters as well as in American waters that the proposition contained in said letter of January 23 was made.

But even if we assume that the American and Asiatic herds are distinct and have never commingled, the fact still remains that the slaughter of the American herd during the past season has been greater than any season in the history of pelagic sealing. The communication of the Foreign Office states that about 12,500 fewer seals were killed in the award area in 1894 than in 1891. It is suggested, however, that their computation of seals killed in Bering Sea in 1891 (29,146) consisted partly of seals taken on the western side of the sea in the vicinity of the Russian seal islands, while the figures for the catch in said sea in 1894 (31,585) included only seals killed on the eastern side of the sea, embraced within the area of the Paris award.

It is a matter of history that after the promulgation of the modus vivendi of June 15, 1891, between the dates of June 29 and August 15, 41 British vessels were warned out of the American side of Bering Sea by American cruisers (see report of British commissioners in report of Paris tribunal). Of these vessels so warned it is believed that many went over to the Russian side of Bering Sea and made catches there. From statistics in the possession of this Department it would appear that about 8,432 seals were taken; of these 6,616 by British vessels and 1,816 by American vessels. There should be deducted, therefore, from the British figures 6,616, leaving about 23,000 as the catch of said vessels in the award area in Bering Sea during the season of 1891. (A similar result (23,041) was reached by deducting from the catch stated in Consul Myers's report, 28,605 (United States counter case), the number of 5,847, estimated to have been killed off the Russian Coast. This estimate was reached after a careful examination by an expert of all the catches of 1891 and by affidavits scattered through the case and counter case of the United States and Great Britain.)

The number of seals stated by this Government to have been killed in Bering Sea during 1891 (23,041) does not include any caught by American vessels during that season, for the reason that the expert who prepared the figures could not obtain exact information on this question. Other statistics now in the possession of the Department indicated that 6,736 seals were killed in Bering Sea, from the Asiatic to the American shores, by American vessels in 1891; subtracting from this total 1,816 seals supposed to have been killed in Russian waters and we have as a result 4,920; adding this to 23,041, the total number of seals killed within the American area in Bering Sea for the season of 1891 falls below 28,000.

The communication of the Foreign Office states the total catch of American and British vessels within the award area in 1891 as 68,000. This Department is in the possession of a careful computation, prepared by an expert of the catch for 1891, based on a careful study of all the evidence as disclosed in the case and counter case of both Governments. This estimate places the number of seals known to have been killed within the award area at 45,000, leaving about 16,000 undetermined. Taking, however, the figures as given by the Foreign Office, 68,000, and subtracting the number supposed to have been killed in Russian waters, 8,432, we have left 59,568 as the maximum catch within the award area for that season.

The official statement of the catch for 1892, contained in the report of the Canadian department of marine and fisheries, credits 14,805 out of a total of 53,912 to the Asiatic shores; the report for 1891 gives only the total of 52,995, crediting none to Russian waters; nor does the report of the British commissioners of the catch of 1891 give any number as killed in said waters. It is respectfully suggested that to this extent these reports are in error.

In computing the catch of 1894 the Foreign Office states that 55,602 seals were killed within the award area, including 17,558 as the catch of American vessels. It should be remembered, however, that in my communication to Congress, from which the totals contained in the letter of January 23 to the British ambassador were taken, 6,836 skins taken by American vessels were stated as undetermined as to location. Assuming that they were divided as between the American and Asiatic herds in the same proportion as the other skins landed during the season of 1894 at American ports by United States vessels, we should have the total catch of American vessels within the award area 55,686+ 6,152, making a total of 61,838. This total justifies my repeating the statement contained in previous letters to you that the pelagic catch within the award area last season was the largest in the history of pelagic sealing, the nearest approximation being the season of 1891, in which, even on the theory of the British figures, not exceeding 59,568 seals were taken. The significance of this catch of 1894 will be better appreciated when it is understood that only 95 vessels were employed, as against 115 in 1891.

It is further contended in said communication that the increased catch, with proportionately fewer vessels, indicates an increased number of seals in 1894 as compared with 1891, and consequently a better condition of the fur-seal herd. When, however, the startling decrease of seals on the Pribilof Islands, pronounced by experts to be at least one-half since 1890, taken in connection with the great destruction of pups from starvation on the islands last season caused by the slaughter of their mothers at sea is considered, it will appear conclusively demonstrated that the increased catch is but a measure of the increased inefficiency of the crews employed as hunters on the sealing vessels and that the seal herd is rapidly diminishing in numbers and is in danger of speedy extermination unless changes are made in the regulations established by the Paris award.

It is correctly stated by the Foreign Office that the catch in the award area of last season outside of Bering Sea was less than during the season of 1893. It should be remembered, however, that it falls only a little short of the catch of 1893, and that it was taken during the four months, January to April, while the catch of 1893 was taken during the seven months, January to July. Unquestionably, however, the prohibition in the award regulations of pelagic sealing during the months of May, June, and July was calculated to do much good to the herd, and some favorable results might naturally have been expected on the islands. After, however, the sealing fleet had entered Bering Sea the startling increase in dead pups (by accurate estimate about 20,000) found on the islands revealed unmistakably the fatal defect in the award regulations in opening Bering Sea to pelagic sealing.

The marvelously increased efficiency of the pelagic seal hunters in the use of the shotgun and spear, as shown by the enormous catches of late years, especially of last season, under said award regulations, will, in my judgment, speedily deplete the fur-seal herd. The pelagic catches must soon decrease in like degree with that necessitated in the land catches on the Pribilof Islands. Reports at hand of the coast catch of the season of 1895 would seem to indicate that this condition has already been reached. It is to be presumed, however, that for some few years the pelagic slaughter of Bering Sea—the great nursery of the fur-seal herd—can be maintained at figures approximating or even exceeding those of last year. That such slaughter as has taken place within the last year—largely of nursing females—gives conclusive evidence that the regulations as established by the Paris award area are not giving that measure of protection that the arbitrators intended, can not, it is respectfully submitted, be longer seriously denied. Commercial extermination of the fur-seal herd, Asiatic as well as American, is imminent. It is to be deeply regretted, therefore, that the British Government has declined our proposition for the appointment of an international commission and for the suggested modus vivendi.

The suggestion, however, is made by the Foreign Office that resident agents be appointed by the United States and Great Britain, to be stationed on the Pribilof Islands, to investigate jointly during the next four years and to report from time to time as to the condition of the fur-seal fisheries. The appointment of similar agents on the Commander Islands is also suggested.

While I believe that this suggestion of the British Government is inadequate and will not satisfactorily supply the need of an international commission of scientists, yet I am disposed to favor a new proposition to said Government, based largely on their suggestion, as follows: That three agents be appointed by Great Britain, Russia, Japan, and the United States, respectively, twelve in all, who shall be stationed on the Kurile, Commander, and Pribilof islands, respectively; that they be instructed to examine carefully into the fur-seal fishery and to recommend from time to time any suggestions as to needed changes in the regulations of the Paris award, and desirable limitations of the land catches on each of said islands; that within four years they shall present a final report to their respective Governments, and that, pending said report, a modus vivendi be entered into extending the award regulations along the line of the thirty-fifth degree of north latitude from the American to the Asiatic shores.

I have the honor, etc.,

J. G. CARLISLE, *Secretary.*

The SECRETARY OF STATE.

DEPARTMENT OF STATE,
Washington, June 25, 1895.

SIR: I have the honor to inclose for your information and consideration a copy of a dispatch, No. 450, of the 14th instant, from the United States ambassador at London, in regard to British legislation in reference to sealing in the North Pacific Ocean.

You will observe that Mr. Bayard, for the purpose of better comparing the pending bill with the act of 1893, which it is intended to replace, incloses a copy of the act last mentioned and also of the merchant shipping act of 1894, which is recited and referred to in the act now proposed.

I have the honor, etc.,

RICHARD OLNEY.

The SECRETARY OF THE TREASURY.

[Inclosure.]

EMBASSY OF THE UNITED STATES,
London, June 14, 1895.

SIR: I have to-day obtained and have now the honor to inclose herewith copy of the proposed bill regulating sealing in Bering Sea and other parts of the Pacific Ocean adjacent to Bering Sea. For the purpose of better comparing this bill with the act of 1893 (which it is intended to replace), I also inclose herewith a copy of the act last mentioned and also of the merchant shipping act of 1894, which is recited and referred to in the act now proposed.

Since obtaining these copies but little time is left before the mail closes for criticism of the proposed modifications in the phraseology of the act of 1893, and I defer comments thereon at this writing.

I have the honor, etc.,

T. F. BAYARD.

Hon. EDWIN F. UHL,
Acting Secretary of State.

JULY 3, 1895.

SIR: I have the honor to acknowledge the receipt of your communication of June 25, inclosing a copy of dispatch No. 450, of the 14th ultimo, from the United States ambassador at London, in regard to proposed British legislation with reference to sealing in the North Pacific Ocean. I have further to acknowledge the inclosures therein, namely, a copy of the said proposed act, seal fishery (North Pacific) act, and of the merchants' shipping act of 1894.

I have carefully read the same, and desire to call to your attention the significant omission in said pending bill of the sixth clause of the

first section of the seal-fishery (North Pacific) act of 1893, which expired
by limitation on the 1st instant, and for which the present pending bill
is to be substituted. This clause is as follows:

> If during the period and within the seas specified by the order a British ship is
> found having on board thereof fishing or shooting implements, or seal skins, or bodies
> of seals, it shall be on the owner or master of such ship to prove that the ship was
> not used or employed in contravention of this act.

A similar provision was contained in the seal-fisheries (Bering Sea)
act of 1891, upon which the modus vivendi of 1891 and 1892 was founded,
clause 5 of section 1 of which provided as follows:

> If a British ship is found within Bering Sea having on board thereof fishing or
> shooting implements, or seal skins, or bodies of seals, it shall be on the owner or
> master of such ship to prove that the ship was not used or employed in contraven-
> tion of this act.

Inasmuch as the pending bill expressly states that its provisions shall
not be in derogation of the provisions of the Bering Sea award act of
1894, but in addition thereto, this omission is significant and becomes
of the utmost importance.

Under the Bering Sea award act of 1894, enacted to carry out the
provisions of the Paris award, the subject-matter of orders in council is
strictly limited to provisions for carrying into effect the schedule provisions (that is, the Paris award and the merchants' shipping act), and
for giving the necessary authority to United States officers to seize
British vessels which have violated the award provisions.

The scope of such orders in council as may be issued under said act
is also limited to the area designated in said award.

The seal-fishery (North Pacific) act of 1893, however, extends the
scope of the orders in council to all of the Pacific Ocean and Bering
Sea north of the forty-second parallel of latitude, and, further, gives
the widest latitude to said orders as to limitations, conditions, qualifications and exceptions, which appear to Her Majesty in council expedient for carrying into effect the object of this act as expressed in the
title, "For prohibiting the catching of seals at certain periods in Bering
Sea and other parts of the Pacific Ocean adjacent to Bering Sea."

If, therefore, the pending bill should reenact the clause above quoted,
in which the presumption of illegality is drawn from the presence of
implements or seal skins on board, it would be possible by subsequent
orders in council to bring the British law into harmony with that
enacted by Congress upon this question, to which I have had the honor
in previous communications to call to your attention. Should, however,
the pending bill become law with said clause omitted, I fear that it may
prove a source of embarrassment in the effort to properly enforce the
provisions of the Paris award in the future. I have the honor to request,
if such course be approved by you, that our ambassador at London be
instructed to present these views to the British Government.

Very respectfully,

O. S. HAMLIN, *Acting Secretary.*

The SECRETARY OF STATE.

JULY 18, 1895.

SIR: Referring to my letter to you of the 8th instant, wherein it was
stated that on the 27th ultimo the collector of customs at San Francisco had reported to the United States attorney at said port the action
of the master of the sealing schooner *Sophia Sutherland* in taking seals

during the closed season within the limits of the area of the award of the tribunal of Paris, and without the special license provided for in article 4 of said Paris award and section 3 of the act of Congress dated April 6, 1894, and requesting that instructions be given to the district attorney to proceed promptly in said case and to report the result, I have the honor to inform you that I have received a letter, dated the 8th instant from the United States attorney with reference to said case and recommending that no action be commenced against either the master or the vessel, and that the skins be released to the proper owner. The basis of the United States attorney's recommendation is that the skins in question are those of seals secured within the prohibited waters by members of the crew of the vessel, who during the voyage were unruly and defiant, and that the action of said members of the crew was without the knowledge or consent of the master of the vessel or any of its officers. It is stated, also, that the master of the vessel was not aware that the taking of the seals within the waters referred to was prohibited, having sailed from San Francisco in January last, and without having been advised of the instructions of this Department in the premises.

I would respectfully request that instructions be given by you to the United States attorney at said port to proceed promptly in said case and commence proceedings for condemnation of the vessel and for the statutory fine to be imposed upon the master under section 8 of the act of Congress of April 6, 1894.

Although the district attorney states certain facts which might properly be considered on a petition for remission of the penalty or forfeiture of the vessel by the Secretary of the Treasury, yet this Department believes that the necessary steps preliminary to imposing a fine and condemnation of the vessel should be at once taken. Article 4 of the regulations of the Paris award provides that every vessel killing seals within the award area must be provided with a special license. Section 3 of the act of Congress of April 6, 1894, prohibits all seal killing without said special license, and section 8 provides for a fine or imprisonment of the master, and also for forfeiture of the vessel offending. This statute is peremptory and would seem to admit of no discretion whatsoever. Under the provisions of subsection 2, clause 4, of the British legislation known as the Bering Sea award act of 1894, it is provided that if a master uses due diligence to enforce the act, and that the offense in question was committed by some other person without his connivance, he shall not be liable to any penalty or forfeiture. There is no such provision, however, in said act of Congress of April 6, 1894.

I would further say that the State Department has filed a formal protest with the secretary of state of foreign affairs of Great Britain because of the action of the British Government in discharging without condemnation proceedings the two British vessels *Wanderer* and *Favorite* seized by United States officers during last season. It therefore becomes of the utmost importance that we proceed vigorously against offenders of our own nationality. Should the court impose a fine upon the master or declare the vessel forfeited, a petition may then be filed with the Secretary of the Treasury to remit the fine or forfeiture. The matter can then be brought to the attention of the State Department and proper action may be taken.

I would further request that the district attorney be directed to give to the Secretary of the Treasury and to the Secretary of State due notice of the trial, in order that the British Government may be notified and be given an opportunity to appoint counsel to take part in the condemnation proceedings should it deem such action advisable. A request has already been made by our Government that counsel may be appointed to represent our interests in similar proceedings in British Columbia.

I have the honor, etc.,

J. G. CARLISLE, *Secretary.*

The ATTORNEY-GENERAL.

JULY 26, 1895.

SIR: Referring to my letter to you of the 18th instant in relation to the case of the sealing schooner *Sophia Sutherland*, charged with taking seals during the closed season within the limits of the area of the award of the Tribunal of Paris, and without the special license provided for in article 4 of said award and section 3 of the act of Congress dated April 6, 1894, wherein request was made that instructions be given by you to the United States attorney at San Francisco to proceed promptly in said case and commence proceedings for the condemnation of the vessel and for the statutory fine to be imposed upon the master, under section 8 of the act of Congress of April 6, 1894, I have the honor to invite your attention to the following statement concerning the schooners *Perkins* and *Puritan:*

On May 11 last the collector of customs at Port Townsend informed the Department, by wire, that the schooners named, which were licensed duly, had reported with ten or twelve seal skins, respectively, which were taken off Cape Flattery, a distance not exceeding 15 miles from the shore, and that both schooners were owned and manned entirely by Indians. The collector stated further that no log books were kept on said vessels, and he recommended that the entry of the skins be allowed. On the 15th of the same month the collector was instructed by wire as follows:

If cases of schooners *Perkins* and *Puritan* are not within article 8 of Paris award and section 6, act of Congress April 6, 1894, report matter to district attorney for proper proceedings under sections 8 and 9 of said act.

I have to request that instructions similar to those given to the United States attorney at San Francisco in the case of the sealing schooner *Sophia Sutherland* be given to the United States attorney for the State of Washington in the cases of the schooners *Perkins* and *Puritan*, and that he be impressed with the necessity for speedy action in the premises.

Respectfully, yours,

S. WIKE, *Acting Secretary.*

The ATTORNEY-GENERAL.

SEPTEMBER 7, 1895.

SIR: I have the honor to inform you that sealing vessels returning from the seal fisheries are beginning to arrive at United States and British Columbian ports.

In view of this fact, I would ask whether the British Government has as yet replied to the request of our Government to be permitted to send experts to British Columbian ports to inspect the official logs and to examine all seal skins landed as to sex. Inasmuch, also, as several American vessels and at least one British vessel have been seized this season for violation of the award of the Paris Tribunal, I desire to be informed whether or not the British Government has consented to the

appointment by our Government of counsel to represent in condemnation proceedings for the forfeiture of said offending vessels.

In this connection I would suggest that the question whether joint
regulations for the coming sealing season can be agreed upon by the
respective Governments be determined at once, as vessels will begin to
leave for the sealing grounds early in November, and there will consequently remain very little time in which to agree upon joint resolutions
should such a course be determined upon.

I have the honor, etc.,

C. S. HAMLIN, *Acting Secretary.*

The SECRETARY OF STATE.

DEPARTMENT OF STATE,
Washington, June 17, 1895.

MY DEAR Mr. HAMLIN: I beg to transmit to you herewith for your
immediate use a copy of a note of the 13th instant, from the British
chargé d'affaires ad interim inclosing a copy of a minute of the Canadian privy council, which is said to contain the information requested
in a letter of the Treasury Department of the 19th of January last,
relative to Canadian pelagic sealing in 1893 and 1894.

To complete the record a formal letter of transmittal, signed by the
Secretary of State, purporting to cover these inclosures will be sent to
your Department to-morrow.

I am, my dear sir, very truly, yours,

ALVEY A. ADEE.

Hon. CHARLES S. HAMLIN.

[Inclosure.]

NEWPORT, *June 13, 1895.*

SIR: With reference to the State Department note of the 23d January last, marked
No. 17, requesting certain information with regard to Canadian pelagic sealing in
1893 and 1894, I have the honor, acting under the instructions of the Earl of Kimberly,
to forward herewith copy of a minute of the Canadian privy council containing the
information asked for.

I have the honor, etc.,

HUGH GOUGH.

Hon. E. F. UHL.

---

*Extract from a report of the committee of the honorable the privy council, approved by
his excellency on the 20th April, 1895.*

The committee of the privy council have had under consideration the annexed
report from the minister of marine and fisheries, dated 20th April, 1895, in connection
with certain information touching Canadian pelagic sealing in 1893 and 1894, which
had been requested by the United States Government.

The honorable committee advise that your excellency be moved to forward a certified copy of this report, together with its appendices, to the right honorable the
principal secretary of state for the colonies.

All which is respectfully submitted for your excellency's approval.

JOHN J. MCGEE,
Clerk of the Privy Council.

[Annex A to O. C. 883 J, April 26, 1895.]

MARINE AND FISHERIES, CANADA,
Ottawa, April 20, 1895.

To HIS EXCELLENCY THE GOVERNOR-GENERAL, *in Council:*
The undersigned has the honor to revert to an approved minute of council dated 2d April, 1895 (750 J.).

In referring to a dispatch from his excellency Her Majesty's ambassador at Washington, dated 19th February, conveying the request of the United States Government for certain information touching Canadian pelagic sealing in 1893 and 1894, this minute of council incidentally announced that much of the information was already in the hands of Her Majesty's Government.

It was also stated that the undersigned had caused steps to be taken to procure from Victoria, British Columbia, such supplementary information in the direction indicated as might be obtainable.

The undersigned has now the honor to report to your excellency that he has received the information asked for, which he appends to this report, together with the correspondence with the collector of customs at Victoria on the subject, as follows:

1. Letter to Mr. A. R. Milne, March 8, 1895.
2. Letter from Mr. A. R. Milne, March 30, 1895, inclosing—
(a) Letter from Mr. J. C. Nixon.
(b) Summary of catch by British Columbia sealing fleet, 1893 and 1894.
(c) Detailed statement of catch in Bering Sea in 1894, showing latitude and longitude where taken and sexes of seals.
(d) Detailed statement, 1893, showing vessels, tonnage, crews, hunters (whether white or Indian).
(e) Detailed statement, 1894, showing vessels, tonnage, crews, hunters (whether white or Indian).

The undersigned would observe that the United States Secretary of State, in his request for information, desired to be informed whether the skins taken by British pelagic sealers were examined as to sex, by expert inspectors, as was done in the case of skins entered at United States ports.

Your excellency will observe, from the appendices to this report, that the undersigned in seeking the information asked for gave considerable prominence to this point, with the object of elucidating whether any practical benefit was likely to accrue from such a course, whether or not it had been hitherto practiced.

Information was sought as to the practicability and value of such a means, and its effectiveness toward establishing the sex of the animals from which the skins were taken. Also whether it was considered to be reliable in establishing the sexes of the seals killed, whether it could be adopted, and whether, in view of the log records on this particular point, demanded by the terms of the award, such a course, if practicable and effective, would be necessary or useful, even in insuring by the check it might afford more careful attention to the examination by the masters of the vessels of the seals killed at sea and the consequent greater accuracy in their log entries.

From the information elicited on this point, it appears that the skins taken by the Canadian pelagic sealers were not so examined by expert inspectors at the time of landing at Victoria and Vancouver.

There also appears to be some ground, on the Canadian Pacific Coast, for doubting that the skins landing in San Francisco and Puget Sound ports were examined as to sex by expert inspectors.

The collector of customs gathers that little credence is given to the statement that an expert examination of the skins was made, inasmuch as it would be unreliable and uncertain.

The separation of the female from the male skins at the time of landing must, it is stated, be mainly determined by the teats, which it is well known occur with male as well as female seals, while a further complication arises from the fact that quite a number of the females are barren, and the teats on the skins taken from such animals would not be more prominent than on those taken from males.

Your excellency's attention is also invited to the statement that both in San Francisco and at Victoria a young, inexperienced lad was engaged by the firm of Liebes Bros., the largest furriers on the Pacific Coast, to examine some seal skins which they were about to purchase (presumably as to sex) and the reported opinions of reliable sealers and furriers as to the nature of such an examination, in view of the lack of either experience or intelligence by the examiner, requisite to determine the point.

The undersigned would further refer to the statement in the collector's letter, that formerly the matter had been the subject of much consideration among those interested in the sealing business, with the result that the opinion prevailed that very few experts are able to determine the sex from an examination of the skins after they have been salted and mixed.

It seems that all the rules as to the color of the hair, and the whiskers of the animals, as well as to the condition of the fur, have proven unreliable. The positive assertion is ventured that the only time when the sex can be correctly determined is during the operations of skinning the animals, when each pelt could be ticketed.

The collector instances the opinion of the late Mr. Davis, representative in Victoria of the firm of Ullman & Sons, fur dealers, of New York and St. Paul. This gentleman is referred to as possessing expert knowledge in the purchase of furs, particularly seal skins.

In 1891 Mr. Davis met the sealing fleet at its rendezvous off Olitak Bay to transfer the skins to the steamer *Danube* previous to the departure of the sealing schooners for the Asiatic waters.

He is represented as having emphatically stated that it was virtually impossible to distinguish the sexes of the animals from which the skins were taken by the teats or otherwise, the only sure way being observations during the process of flaying, although in exceptional cases of very choice skins, the sex might be determined. In the case of barren females, however, the distinction was practically impossible.

A further instance is cited of an examination in 1892 of about 200 skins at Victoria, Mr. Macoun of Ottawa, Mr. Munsie of Victoria, and several others failing to determine the sexes of the animals from which these skins were taken, notwithstanding they were selected and pronounced by Mr. Koutzouer, an expert for Messrs. Boscowitz & Co., to be the product from male seals. The teats proved to be as prominent as those to be found upon the skins of any female seal.

It is admitted, however, that the pelt of a female seal killed while very heavy with young may be determined from its peculiar shape.

The collector expresses the opinion that expert examination of seal skins would be almost impracticable, while its effectiveness would be uncertain in establishing the sex of seals from which they were taken. Hence it would not be necessary or useful, but he believes that if the sealers, in addition to keeping accurate log entries as to their fishing operations, were compelled to label or tag each skin as to the sex of the animal at the time of flaying the most reliable evidence obtainable would be insured.

The undersigned would further invite your excellency's attention to the statement that 1,037 skins were landed in Victoria and sold from the State of Washington, United States of America, during 1894 which were not examined as to sex by experts.

Also to the letter from Mr. J. C. Nixon, of Seattle, stating that the skins landed at the Puget Sound ports were not examined by experts.

In the light of the evidence contained in the appendices to this report, the undersigned is of opinion that such an examination of the salted seal skins when landed at the home ports would prove of little utility in establishing the sexes of the seals killed.

The undersigned recommends that a copy of this report, if approved, together with its appendices, be forwarded to the right honorable Her Majesty's principal secretary of state for the colonies.

Respectfully submitted.

JOHN COSTIGAN.

[Annex B to O. C. No. 883 J, April 26, 1895.]

OTTAWA, *March 8, 1895.*

SIR: I have the honor to inform you that a request of the United States Secretary of State for certain information in respect of the Canadian sealing fleet and their operations during the seasons of 1893 and 1894 has been communicated to the Government by his excellency Her Majesty's ambassador at Washington.

(1) The total number of seals taken by British vessels in the North Pacific Ocean and Bering Sea, both on the Asiatic and American sides.

(2) The total number of skins landed at British ports by said vessels.

(3) The total number transshipped in Japanese or Russian ports, including any that may have been ultimately entered at Victoria.

(4) The number of skins landed as entered at Victoria by American vessels.

(5) A report as to the sex of all skins taken in Bering Sea and the North Pacific Ocean.

(6) Location of the place of catch by latitude and longitude.

(7) The names of all vessels employed, tonnage, number of crew, and number of seal hunters, indicating whether whites or Indians.

(8) The further request is made that information may be given as to whether the skins taken by the Canadian pelagic sealers were examined as to sex by expert inspectors, as was done in the case of skins entered in United States ports.

The honorable the minister of marine and fisheries would be pleased if you will obtain the information as above intimated and classified, or such of it as is procurable, and forward it to this department at your earliest convenience.

With regard to section 8, touching the expert inspection of skins when landed, with a view to determining the sex of the animals from which they were taken, it is presumed that no measures of this nature have hitherto been adopted at Victoria or Vancouver.

The department would, however, like to have your opinion as to the practicability and value of such a means and its effectiveness toward the end in view.

Also, if it is considered to be reliable in establishing the sexes of the seals killed, whether, in your opinion, it could be adopted, and whether, in view of the log records on this point required by the terms of the award, such a course, if practicable and effective, would be necessary or useful.

It may be that the adoption of an inspection of this character would, by the check it afforded, insure more careful attention to the examination by the masters of the vessels of the seals killed at sea and greater accuracy in their consequent log entries.

I have the honor, etc.,

JOHN HARDIE,
Acting Deputy Minister Marine and Fisheries.

A. R. MILNE, Esq.,
Collector of Customs, Victoria, British Columbia.

———

[Annex C to O. C. No. 883 J, April 26, 1895.]

CUSTOMS CANADA,
Victoria, British Columbia, March 30, 1895.

SIR: I have the honor to acknowledge the receipt of your letter of the 8th instant, conveying the information that a request had been made by the United States Secretary of State for certain information in respect of the Canadian sealing fleet and their operations during the seasons of 1893 and 1894, and that such request had been communicated to the Government by his excellency Her Majesty's ambassador at Washington.

In compliance therewith, I beg to transmit herewith the information asked for touching the operations during those two years and are arranged under the following headings:

(1) The total number of seals taken by British vessels in the North Pacific Ocean and Bering Sea, both on the Asiatic and American sides.

(2) The total number of skins landed at British ports by said vessels.

(3) The total number transshipped in Japanese or Russian ports, including any that may have been ultimately entered at Victoria.

(4) The number of skins landed as entered at Victoria by American vessels.

(5) A report as to the sex of all skins taken in Bering Sea and the North Pacific Ocean.

(6) Location of the place of catch by latitude and longitude.

(7) The names of all vessels employed, tonnage, number of crew, and number of seal hunters, indicating whether whites or Indians.

(8) The further request is made that information may be given as to whether the skins taken by the Canadian pelagic sealers were examined as to sex by expert inspectors, as was done in the case of skins entered in United States ports.

The skins taken by Canadian pelagic sealers were not examined as to sex by expert inspectors at the time of landing from the vessels at this port or at Vancouver.

A general denial is made that the seal skins were, on being landed at United States ports—namely, at San Francisco and Puget Sound—examined as to sex by expert inspectors.

No confidence is entertained here in the American statement made that an expert examination of the skins was held, as such would be unreliable and uncertain when separating the females from the males at time of landing, which could mainly be determined by the teats of the animals, it being a well-known fact that males have such as well as females, and quite a number of females have been barren—that is, have been barren during the season or longer—and the males have teats as prominent on the skin as the barren females.

Parties interested in sealing state that in San Francisco and here a young, inexperienced lad was sent by Liebes Bros., the largest furriers of that city, to examine some seal skins landed which they were about to purchase, but the reliable sealers and furriers say that such was a perfect farce, he not having the experience and intelligence requisite to determine such.

In the past this matter has been the subject of much discussion here among those interested in sealing, and the consensus of opinion is that very few experts are able to determine the sex from the skins when they have been salted and mixed, and all rules as to the color of the hair and the whiskers of the animal or the condition of the fur have been found to be unreliable and unsatisfactory, a seal skin being split from the lower jaw to the tail, even the tail itself, which is very short, being about an inch in length, is also split, and the sex can not be determined that way.

It is positively asserted that the only time to determine the sex is while skinning the animal while fresh, and that masters could then put a ticket on each skin.

Mr. Davis, the representative here of Messrs. Joseph Ullman & Sons, fur dealers, of New York and St. Paul, who resided here for a few years, since deceased, and who came here with expert knowledge in the purchase of skins, particularly seal skins, went north on the steamer *Danube* in 1891 to meet the sealing fleet which had a rendezvous off Alitak Bay, to transfer their skins, previous to their departure to the Russian side, from the schooners to the steamer, to be brought here, emphatically stated that it was practically impossible to tell the male from the female skin by the teats or otherwise, and the only sure way, in his opinion, was to see the animal skinned. It might, however, in exceptional cases of very choice skins be determined by the fur or whiskers, and to tell a barren female from a male was almost impossible.

Mr. Macoun, of Ottawa, with Mr. Munsie, a shipowner, and others, in 1892 examined in warehouse here about 200 skins, and they could not determine in that number the males from the females; and they selected skins said to be males by Mr. Kautzauer, an expert for Messrs. Boscowitz & Co., and they were found to have teats as prominent upon them as those found on any female.

It is known, however, that a skin off a female seal that has been killed while very heavy with young is broader in proportion to its length than the male skin, which is more oblong.

It is my opinion that an expert examination would be almost impracticable and its effectiveness uncertain in establishing the sex of the seals killed, and I consider would not be necessary or useful.

The adoption of an inspection of this character would, no doubt, if reliable and conducted with certainty, afford a check as to the accuracy of log entries; but this, I am certain, would not be done at the time of landing without much irritation and disputation regarding the sex of seals.

I can only therefore say that it appears to me that if the sealers in addition to keeping their logs accurately as to each day's fishing were compelled to label or tag each skin as to sex at the time of skinning and splitting the animal, would insure the most reliable evidence which could be obtained.

I might state that 1,037 seal skins were landed here and sold from the State of Washington during the year 1894, which certainly were not examined as to sex by expert inspectors, and those that were landed at other Puget Sound ports you will see by the inclosed letter I received from Mr. J. C. Nixon, who is largely interested in the seal industry, that no such examination took place at any Puget Sound ports during last year or any previous year.

I have the honor, etc.

A. R. MILNE, *Collector.*

JOHN HARDIE, Esq.,
Acting Deputy Minister of Marine and Fisheries, Ottawa.

SEPTEMBER 10, 1895.

SIR: I have carefully considered the report of Hon. John Costigan to his excellency the Governor-General of Canada, containing minute in council dated April 2, 1895, inclosed in your letter to me of June 17. Although the inclosures purport only to give certain information asked for by our Government by letter of January 23, 1895, yet certain statements and denials are contained therein which merit careful attention and reply by this Department.

In answering in the negative the questions whether the pelagic seal skins taken by British sealers were examined at the British ports of entry as to sex (as was stated in our communication to have been done to all American seal skins entered at United States ports), the collector of customs at Victoria, British Columbia, in his letter to Mr. Costigan inclosed in said report, enters a general denial that seal skins were examined as to sex by expert inspectors at San Francisco and Puget Sound ports. He further states that it is impossible to distinguish the sex of seals by inspection after the skins have been salted, and that any examination at port of entry therefore would be impracticable and use-less. While this denial of an examination at United States ports was not called for by any question propounded in your letter to the British Government, yet as it is couched in such a brusque manner and repeated by the Hon. Mr. Costigan, it would seem to merit a prompt reply by this Department.

The statement that all skins landed at United States ports during the season of 1894 were examined as to sex by expert inspectors was supposed by this Department to be true when it was made in our letter sent to you. Inspectors were duly appointed for this specific purpose and instructed to make such examination, and their returns were on file in the Treasury. It was discovered later, however, that at Port Town-send the inspector although examining and verifying the skins landed, yet made no examination as to sex. On the receipt of this discovery, to wit, on February 21, I immediately notified you, with the request that this fact be communicated to the British Government. I inclose a copy of this letter. It can only be assumed that the British Govern-ment in forwarding to you with its tacit approval the report of Mr. Costigan containing this brusque denial must have overlooked the fact that it had in its possession at the time the letter of our Government fully explaining the matter.

I would further state that the Department does not understand the reference contained in said report to the young and inexperienced lad employed by Liebes Bros. in the examination of seal skins at San Fran-cisco. Such an action, if it took place, was wholly unofficial, and had no connection whatsoever with that made by the United States Govern-ment. The official expert employed by this Department at San Fran-cisco is a practical furrier by occupation. He has been engaged constantly and exclusively for the past eleven years in the fur business, and has handled large quantities of salted seal skins and all kinds of raw furs. He was highly recommended by persons well qualified to judge of his ability, and is a man of great experience. Every seal skin landed at San Francisco was carefully examined by him as to sex.

As to the further statement contained in said communication, that the sex of seals can not be determined after the skins have been removed, I would say that the Treasury Department has consulted experts of rep-utation upon this question and is informed that any ordinarily intelligent person at all acquainted with the subject could in at least seven cases out of ten accurately determine the sex before the skins go to the dresser. That such is an admitted fact will be seen by reference to the inclosed extracts from the testimony before the Tribunal of Arbitra-tion at Paris, which I will thank you to return after reading.

I also inclose for your information a copy of a Treasury circular giv-ing instructions to customs officers as to ascertainment of sex.

I would further state that in a previous communication I requested you to obtain the consent of the British Government to the appoint-ment of expert inspectors to examine all skins landed at British Colum-bian ports. The reason for this request was that there is a great discrepancy in the British and American returns made by sealers as to the proportion of females killed, the American sealers reporting a very much greater proportion than the British. Although in many instances the British sealers were close to the Americans, yet the Americans reported from two to five times as many females as males, a result entirely at conflict with the British returns. Certain masters, more-over, of British vessels in Bering Sea explained to the agent of the

United States Fish Commission that the seals were skinned in canoes by Indians and the pelts thrown on board, and that under the circumstances they had no time to bother with inspecting skins minutely as to sex. The Department is of opinion that under such circumstances reports of British sealers are unreliable, and that the proportion of female skins taken by the Canadian fleet is much greater than that returned.

This would seem to be corroborated by sworn statements, now in the possession of the Department, of experts who personally inspected in London some of the largest consignments of seal skins taken in 1894, and found that from 85 to 90 per cent of them were females.

The Department is therefore of the opinion that examination by inspectors of all skins landed at British Columbian ports would greatly assist in arriving at a more thorough knowledge of seal conditions, and I would therefore ask that the British Government be urged to reply as soon as possible to the request as above already transmitted.

I have the honor, etc.,

C. S. HAMLIN, *Acting Secretary.*

The SECRETARY OF STATE.

DEPARTMENT OF STATE, *August 26, 1895.*

SIR: I have the honor to inclose for your information translations of two notes from the Russian chargé d'affaires ad interim at this capital, presenting the views of his Government on the subject of Department's note of January 23 last to the British ambassador at Washington concerning the regulation of fur-seal hunting in Bering Sea.

I have the honor, etc.,

ALVEY A. ADEE,
Acting Secretary.

The SECRETARY OF THE TREASURY.

[Inclosure.]

IMPERIAL LEGATION OF RUSSIA,
Washington, August 7–July 26, 1895.

Mr. SECRETARY OF STATE: The contents of the note of Mr. Gresham, late Secretary of State, to the ambassador of Great Britain at Washington, bearing date of the 23d of January last, concerning the regulation of fur-seal hunting, has been submitted to the examination of a special commission, which recognized the necessity of a uniform régime for fur-seal hunting on the high seas for all the northern portion of the Pacific Ocean, from the coasts of America to those of Asia. To this end the Federal Government proposes the appointment of a mixed commission, to be composed of the representatives of the United States of America, of Russia, of Great Britain, and of Japan, whose duty it shall be to examine this question. The Federal Government further proposes to secure the enforcement of the decisions of the Tribunal of Arbitration which sat at Paris relative to fur-seal hunting as far as 35° north latitude in the Pacific Ocean, and to prohibit hunting in Bering Sea until the commission shall have finished its labors.

While accepting, in principle, the suggestion concerning the appointment of the aforesaid commission, the Imperial Government attaches much greater importance to the modus vivendi, whereby the decisions of the Tribunal of Arbitration at Paris are to be enforced in all waters of the Pacific Ocean situated north of the thirty-fifth parallel of north latitude, including the Sea of Okhotsk.

The spirit of equity which actuates the Federal Government does not permit me to doubt that Your Excellency will be pleased to agree that the present state of things, in which the decision of the Tribunal of Arbitration at Paris is enforced only in the eastern part of the Bering Sea, the fur seals in the western part of the same sea being thus deprived of this protection, should no longer exist. In reality, all the good measures that have been taken in Bering Sea are paralyzed and productive of no results from the very fact that the western part is not within the protected zone.

Consequently, the Imperial Government, adhering to its view with regard to the necessity of establishing a uniform régime for all waters of the Pacific Ocean situated north of the thirty-fifth parallel of north latitude, is of opinion that it would be more practical to make no exception in the case of Bering Sea by prohibiting seal hunting until the termination of the labors of the aforesaid mixed commission, but to enforce the same uniform régime in that sea to which all waters of the Pacific Ocean north of the thirty-fifth parallel of north latitude are to be subjected.

In ordering me to convey its thanks to the Federal Government for the kind communication which it has been pleased to make to it, the Imperial Government instructs me to assure Your Excellency of its earnest desire to cooperate in the success of the aforesaid suggestions, as of any other suggestion tending to establish a uniform régime for the regulation of fur-seal hunting on the high seas in all parts of the Pacific Ocean north of the thirty-fifth parallel of north latitude.

Be pleased to accept, Mr. Secretary of State, the assurance of my very high consideration.

A. SOMOW.

His Excellency, RICHARD OLNEY,
  Secretary of State.

[Inclosure.]

IMPERIAL LEGATION OF RUSSIA,
Washington, August 7–26, 1895.

Mr. SECRETARY: In herewith transmitting to you a note bearing date of August 7–26, 1895, concerning the regulation of fur-seal hunting, I have the honor to send you a memorandum containing a statement of the result of the labors of the special commission instructed to report concerning the contents of the note addressed by Mr. Gresham, late Secretary of State, to the ambassador of Great Britain at Washington, under date of January 23, 1895, relative to fur seals.

For my own part, I desire to beg you, Mr. Secretary, to keep this legation informed, as you have hitherto done, with regard to the development of this question, thus facilitating a mutual understanding between our two Governments.

Be pleased to accept, Mr. Secretary, the assurance of my very high consideration.

A. SOMOW.

Hon. ALVEY A. ADEE,
  Acting Secretary.

[Memorandum.]

IMPERIAL LEGATION OF RUSSIA,
Washington, August 7–26, 1895.

The commission is of the opinion that our principal object, viz, to put a stop to the extermination of fur seals, could be more successfully attained not by the appointment of a technical mixed commission, but by the convocation of a regular diplomatic conference whose decisions should be obligatory upon all nations. Otherwise there is reason to fear that illicit fur-seal hunting will continue under other foreign flags, such as that of Sweden, of Italy, etc.

The representative of the ministry of finance has made the following remarks concerning the enforcement of the decisions of the arbitrators at Paris:

(1) It is to be desired that article 1 of these regulations, which prohibits fur-seal hunting at all times and by all means within a zone of 60 nautical miles around the Pribilof Islands, should be modified in such a way as to extend the zone in which hunting is prohibited to 200 miles around the Commander Islands and Fulenien (Robben Islands).

(2) As the Commander Islands are on the boundary line between Bering Sea and the Pacific Ocean, fur-seal hunting with firearms should be prohibited in all the waters of the Pacific Ocean as far as 35° north latitude along the Konsilsky and Commander islands and the coasts of Japan.

(3) As the fur seals that winter on the Russian islands leave their winter lairs earlier than those on the Pribilof Islands, the time during which sealing on the high seas is prohibited should be made to last for the former not until the 1st of May, but until the 1st of March, or, strictly speaking, until the 1st of April.

All these measures relate exclusively to fur-seal hunting on the high seas, since hunting on land in Russian territory, viz, on the Commander Islands and Fulenien, is under strict inspection, and the number of seals that are allowed to be killed is strictly limited.

SEPTEMBER 18, 1895.

SIR: I have the honor to acknowledge the communication of the Acting Secretary of State, dated August 26, 1895, inclosing translations of two notes from the Russian chargé d'affaires ad interim at this capital, presenting the views of his Government on the subject of the note of your Department of January 23 last, to the British embassy at Washington, concerning the regulations governing seal hunting in Bering Sea.

The first note inclosed in your communication states that the Russian Government accepts in principle the suggestion of an international commission, and agrees to the extension of the Paris award regulations along the line of the thirty-fifth degree of north latitude to the shores of Asia, but that it disagrees with our proposition to close Bering Sea to sealers pending a report of said commission.

The second note incloses a report of a Russian commission, to which the whole matter of regulations of fur-seal hunting was submitted. The commission in its report states that the diplomatic conference of nations, obligatory upon all, for the purpose of regulating the fur-seal fisheries, would be preferable to an international commission; that a zone of 200 miles should be established around the Commander Islands within which no sealing should be permitted, and that the use of firearms should be prohibited north of the thirty-fifth degree of north latitude. The said commission further reports that "as the fur seals that winter on the Russian islands leave their winter lairs earlier than those on the Pribilof Islands, the time during which sealing on the high seas is prohibited should be made to last for the former not until the 1st of May, but until the 31st of March, or, strictly speaking, until the 1st of April."

I have carefully considered said notes, and have the honor to reply that I do not quite understand the proposition contained in the first, to the effect that to Bering Sea should be applied the same law as to the North Pacific Ocean north of the thirty-fifth degree of north latitude. I would respectfully ask that you request of the Russian Government an expression of opinion whether it desires the prohibition of all sealing north of the thirty-fifth degree of north latitude pending the report of a commission, or simply that the regulations of the Paris award and the closed season thereof be extended to Bering Sea on the western as well as the eastern side. Nor do I understand the last suggestion in the second note quoted in full above that the closed season should last until the 1st of April. At first sight it would seem that the translator employed by your Department is in error, and that the word "after" should be subsituted for the word "until," as under the Paris regulations the open season lasts from August through April.

I would further state that it appears from the note of Lord Gough to you, dated August 19, of which a copy was sent to me by you on August 28 last, that the British Government refuses to recognize that Russia and Japan have any interest in the seal fisheries regulated by the Paris award, and that it can not take part in any inquiry on the Pribilof Islands in which these powers are associated. I would respectfully suggest that the Russian Government be informed of this refusal. I can not see any objection to a diplomatic conference such as is suggested by the Russian Government as to the fur-seal hunting in the western part of the North Pacific Ocean and Bering Sea, and feel that much good would result from such a conference.

I have the honor, etc.,

C. S. HAMLIN, *Acting Secretary*.

The SECRETARY OF STATE.

OCTOBER 3, 1895.

SIR: I have the honor to acknowledge the receipt of your communication of the 1st instant[1] in relation to the cases of the three schooners against which proceedings for violation of the act of April 6, 1894, giving effect to the award rendered by the Tribunal of Arbitration at Paris, have been commenced at San Francisco, and in reply to your inquiry state that this Department did instruct the collector at San Francisco on the 16th of April last as to the licensing of vessels clearing from his port in the language quoted by you.

The instructions sent to the collector were based upon articles 2 and 4 of the award of the Paris Tribunal and section 3 of the act approved April 6, 1894, giving effect to the provisions of said award. If, in your opinion, the construction placed upon the articles and section referred to is incorrect, I would thank you to so advise me, in order that suitable instructions may be sent to the collectors of customs on the Pacific Coast for their future information and guidance.

Respectfully, yours,

W. E. CURTIS, *Acting Secretary*.

The ATTORNEY GENERAL.

NOVEMBER 27, 1895.

SIR: On January 3 last the schooner *Kate and Anne* cleared from San Francisco for "hunting and fishing outside prohibited waters" without the special sealing license provided by the Paris award and the act of April 6, 1894. On May 30 this schooner arrived at Astoria, Oreg., where she was allowed to land her catch after an examination of the skins by the collector of the port. The vessel, it appears, then proceeded to San Francisco, where it is believed she is at present. A copy of the log book of the schooner was forwarded to this Department by the collector at Astoria, upon examination of which it appeared that the vessel had taken seals within the award area, and, not having been provided with a special sealing license, should have been seized, together with her catch, upon her arrival at Astoria. A letter has been addressed to-day to the collector at San Francisco, directing him to take the necessary steps to seize the vessel for violation of the act above cited, and to report the facts to the United States attorney for his action. I have the honor to request that the attorney be instructed to proceed promptly and vigorously in the action against the vessel.

Respectfully, yours,

S. WIKE, *Acting Secretary*.

The ATTORNEY GENERAL.

---

[1] Not furnished.

DEPARTMENT OF STATE,
Washington, November 21, 1895.

SIR: The Russian chargé d'affaires has formulated inquiries of which you spoke to me the other day and presents them in the paper herewith inclosed.

I should be glad of any suggestions from you as to the answers to be made.

With the answers please return queries and oblige,

Respectfully, yours,

RICHARD OLNEY.

Hon. CHARLES S. HAMLIN,
Assistant Secretary of the Treasury.

[Inclosure.]

RUSSIAN IMPERIAL LEGATION, *Washington.*

Queries:

(1) Was the United States Government aware that in the act of Parliament issued by the Government of Great Britain in pursuance of the Bering Sea award act in 1894 was omitted the clause 6 of the following regulations on seal fisheries:

"If during the period and within the seas specified by the order a British ship is found having on board thereof fishing or shooting implements or seal skins or bodies of seals, it shall lie on the owner or master of such ship to prove that the ship was not used or employed in contravention of this act."

(2) In considering the best way to protect the seal fisheries the United States Government thought desirable, pending the investigations of the seal fisheries by a special commission, to extend the rules of the Paris award to the seal fisheries on the seas lying to the north of the thirty-fifth degree of north latitude and·prohibit entirely the seal fisheries in the Bering Sea proper.

Is that prohibition of the seal fisheries in the Bering Sea intended for the purpose of protecting the breeding area of the seal, or has it any other purpose?

NOVEMBER 29, 1895.

SIR: I have the honor to acknowledge the receipt of your communication of the 21st instant, with which was inclosed certain inquiries from the Russian chargé d'affaires, concerning the seal fisheries (North Pacific) act, 1893 and 1895, and certain suggestions on the part of this Government as to extending the regulations of the Paris award to the shores of Asia, and in reply to state in answer to said inquiries, and in the order of their submission:

(1) The United States Government is aware that in the act of Parliament known as seal fishery (North Pacific) act 1895, clause 6 of the earlier act known as seal fishery (North Pacific) act 1893, was omitted.

The Treasury Department on June 25, 1895, received a copy of the proposed seal fishery (North Pacific) act 1895 from the Secretary of State, and on July 3, 1895, called attention to this significant omission, and received from you a copy of your letter of July 8, addressed to James R. Roosevelt, chargé d'affaires at London, in which you expressed deep regret that the clause referred to should not have been reenacted, and the earnest hope that some means may be provided yet whereby this omission may be remedied.

(2) The purpose of this Department in recommending verbally to the Secretary of State, through Assistant Secretary Hamlin, that the existing Paris award regulations be extended along the line of the thirty-fifth degree of north latitude to the shores of Asia, and that in addition

thereto Bering Sea should be closed to seal fishing pending the report of the international commission, was to secure protection for female seals in the breeding area, as it was demonstrated by indisputable evidence that the female leave their young on the Pribilof Islands, and frequently are found in search of food and rest hundreds of miles at sea.

Such a prohibition, it is suggested, would be of equal value in protecting the Russian herd frequenting the Commander Islands, as well as the herd frequenting the Pribilof Islands.

The communication from the Russian chargé d'affaires is returned herewith, in compliance with your request.

Respectfully, yours,

S. WIKE, *Acting Secretary.*

The SECRETARY OF STATE.

DEPARTMENT OF STATE,
Washington, January 22, 1896.

SIR: Inclosed please find memorandum left with me by the Russian minister at this capital.

I desire to ask whether there is any other objection to the proposed remodeling of the convention with Russia of 1894 except the obvious one of the inducements and advantages it would afford to Canadian sealers.

Could Great Britain's concurrence be secured! Would not an arrangement like that proposed by the Russian minister be an advantageous one for the United States?

Respectfully, yours,

RICHARD OLNEY.

Hon. JOHN G. CARLISLE,
Secretary of the Treasury.

[Inclosure.—Translation.]

The fur-seal herds frequenting the Russian islands in Bering Sea and the Sea of Okhotsk are threatened with complete extermination within a very short period. Their situation has been doubly bad since the Paris arbitration recognized the right of America to prohibit maritime hunting within 60 miles of the islands belonging to her, whereas the Russian herds are protected only within 30 miles of their islands, without taking into consideration the fact that the entire prohibition of hunting during several months and the regulation of the weapons employed, also established by the Paris arbitration, secure to America privileges of such importance that the poachers prefer to direct their efforts against the Russian seals, which are massacred in such numbers that the extinction of the race must infallibly result, and the more so because the females, which regularly seek food for their young beyond the 30-mile belt, are the most certain victims of the hunters who lie in wait for them on the high seas.

In view of the great injury resulting therefrom to the Imperial treasury; in view of the necessity of preserving the fur-seal race, which constitutes the only resource of the inhabitants of the above-mentioned islands, the Imperial Government, desirous of remedying this state of affairs, thinks it its duty to represent to the United States Government that it is urgent to extend the regulations established by the Paris arbitration to all the waters of Bering Sea and the Sea of Okhotsk, as well as to those of the Pacific from one Continent to the other down to the thirty-fifth degree of north latitude.

Such a provision would be equitable as regards Russia and advantageous to the two powers, as their interests are absolutely identical in this question.

The Imperial Government would be especially gratified if the time of the close season could be made to begin one month sooner and to end one month later.

In return for a remodeling of the convention of 1891 in this sense, the Imperial Government would cooperate earnestly with the United States Government in its efforts having for their object the enlargement of the regulations prepared by the Paris arbitration.

JANUARY 29, 1896.

SIR: I have the honor to acknowledge the receipt of your communication of January 22, inclosing a memorandum left with you by the Russian minister at this capital, and asking my opinion as to whether there is any objection to the proposed remodeling of the convention with Russia of 1894, except the obvious one of the inducements and advantages it would afford Canadian sealers; also, as to whether the arrangement proposed by the Russian minister would be advantageous for the United States, provided Great Britain's concurrence could be secured.

I have the honor to reply that I see no objection to the proposed remodeling of the convention with Russia of 1894 other than the one mentioned to you. It is my further belief that it would be for the distinct advantage of the United States if the existing regulations provided by the Paris Tribunal were extended so as to embrace all the waters from the Asiatic to the American shores in the North Pacific Ocean, the Sea of Okhotsk, and Bering Sea north of the thirty-fifth degree of north latitude.

Under the terms of the Paris award the killing of fur seals is prohibited during the three months of May, June, and July. In practice the sealers start out early in January and hunt seals along the coast of the United States and British Columbia until the closed season begins on May 1. Between January and May they are thus able to take seals, but on account of the inclemency of the weather the catch is necessarily restricted. Sealing can not be lawfully carried on after May 1 until the 1st of August. At this time the fur-seal herd has left the North Pacific Ocean and is in Bering Sea. The masters of sealing schooners have, however, learned by experience that they can spend the months of May, June, and July sealing on the Russian and Japan coasts and procure good catches, after which they can reach Bering Sea in time to hunt during the month of August, when sealing is again permitted under the Paris award regulations. If, however, the Paris award regulations should be extended so as to embrace the Asiatic shores, there could be no sealing whatsoever in the North Pacific and Bering Sea between May 1 and August 1. This would undoubtedly operate to make it more expensive to maintain a sealing fleet, and would probably result in a large falling off of the number of pelagic sealers. In this connection I would respectfully call to your attention the suggestions made by Hon. James C. Carter, to the effect that pelagic sealing be absolutely forbidden to American citizens, for the reason that such sealing (as was contended before the Paris Tribunal) is cruel and inhuman, a large portion of the seals killed being females heavy with young. His suggestions will be found in a letter of the late Secretary Gresham to Ambassador Bayard of October 6, 1893, which is No. 14 in Senate Executive Document No. 67, Fifty-third Congress, third session.

As regards the further suggestion of the Russian minister, that "the Imperial Government would be specially gratified if the time of the closed season could be made to begin one month sooner and to end one month later," I have the honor to state that I approve such a suggestion as being a step in the right direction.

The minister further makes reference in his communication to the 30-mile zone failing properly to protect the Russian seals, from the fact that the females seek food for their young beyond the 30-mile belt. This statement is undoubtedly true. In fact, the 60-mile zone established around the Pribilof Islands does not satisfactorily protect the seal herd.

I have the honor to inclose a memorandum, showing in detail the location of the catches of 7,879 seals killed during the past season in Bering Sea, from which it appears that female seals were killed as far as 600 miles from the 60-mile zone.

The Department will shortly be in possession of a report from Dr. Charles Townsend, of the United States Fish Commission, as the result of special study of this subject last summer in Bering Sea. Should you desire further information upon this subject this report will be sent to you.

Very respectfully,   J. G. CARLISLE,
Secretary.

The SECRETARY OF STATE.

FEBRUARY 10, 1896.

SIR: I have the honor to forward herewith for transmission to Her Majesty's Government, in compliance with the provisions of article 5 of the award rendered by the Tribunal of Arbitration at Paris, two copies of the entries made in their respective log books by the masters of American pelagic sealing vessels during the season of 1895. I would respectfully request that you obtain from the British Government as soon as possible the returns of the British sealing vessels for the same period.

Respectfully, yours,

S. WIKE, *Acting Secretary.*

The SECRETARY OF STATE.

DEPARTMENT OF STATE,
Washington, February 8, 1896.

SIR: I beg to inclose draft of preliminary arrangements between Russia and the United States submitted by the Russian minister at this capital, the same to take effect only when adherence of Great Britain is procured.

Do you see any objection to the United States acceding to such an arrangement?

Awaiting reply at your earliest convenience, I am, respectfully, yours,

RICHARD OLNEY.

Hon. CHARLES S. HAMLIN,
Assistant Secretary of the Treasury.

FEBRUARY 13, 1896.

SIR: I beg to acknowledge the receipt of your letter of February 8 inclosing a draft of preliminary arrangement between Russia and the United States, submitted by the Russian minister, in which you ask whether I see any objection to the United States acceding to such an arrangement.

I have carefully examined the inclosed draft, and while I find certain portions not perfectly clear I understand the proposals to be:

First. That the prohibited zone around the Commander and Robben islands shall be in the future 60 miles instead of 30, as now fixed by the agreement of 1894.

Second. That pelagic sealing shall be wholly prohibited from April 1 to October 15 in Bering Sea, the sea of Okhotsk, and the Pacific Ocean north of the thirty fifth degree of north latitude, from the Asiatic to the American shores.

Third. That all existing articles of the arrangement of 1894 not hereby abrogated shall remain in force.

I perceive no objections to the first and second propositions. The American herd is now protected under the regulations of the Paris award during May, June, and July. On July 31, when the season ends, the herd is in Bering Sea. The proposed arrangement providing for closing Bering Sea to fur-seal fishing between April 1 and October 15 would satisfactorily protect our herd, as after October 15 no sealing vessels could remain in Bering Sea on account of the inclemency of the weather.

Under the third suggestion, the prohibition of sealing to American vessels within 10 miles of the Russian Coast would still continue, as would also the provision relating to the seizure of American vessels found hunting fur seals within said prohibited area, outside of Territorial waters, by Russian naval officers, and for condemnation of such vessels by the courts of the United States. I would suggest that in order to bring the proposed arrangement into harmony with the terms of the Paris award the provision as to the 10-mile limit be omitted, and that there should be inserted in the draft a reciprocal provision that Russian sealing vessels, if any, may be seized by American naval officers and condemned by the Russian courts for breach of the convention. I would further suggest that the Russian Government should agree to accept and be bound by all the provisions of the Paris award and the legislation in respect thereto not abrogated by said proposed draft, which should be stated to be not in derogation of the award of the Paris tribunal, except where necessarily inconsistent therewith, but in addition thereto.

I notice that no suggestion is made as to obtaining the concurrence of the Japanese Government as to said proposed arrangement. During the past season five Japanese sealers were reported to have engaged in fur-seal fishing, making a catch of 2,960 seals. I have the honor to suggest that it might be advisable to obtain the consent of Japan. The principal injury suffered by the Russian herd at the hands of pelagic sealers is while the herd is off the shores of Japan, going north to the Commander and Robben islands. For example, in 1894 only a little over 7,000 seals were killed at sea in Russian waters, as compared with over 71,000 taken off the coast of Japan. In 1895 about the same number were taken in Russian waters, as compared with over 29,000 taken off the coast of Japan; and in the same year only six American vessels report making catches in Russian waters, to the total amount of 766 skins. In 1894 only 201 skins were reported taken by American pelagic sealers.

I take pleasure in inclosing a memorandum giving, on page 107, information as to the catch of seals and other kindred subjects, which

H. Doc. 92, pt. 2——23

is shortly to appear in the Secretary's report. I also return the draft submitted with your letter of the 8th instant.

Respectfully, yours,

C. S. HAMLIN, *Acting Secretary.*

The SECRETARY OF STATE.

DEPARTMENT OF STATE,
Washington, February 24, 1896.

SIR: The inclosed note (in translation) of the Russian minister at this capital speaks for itself. It seems to me the 10-mile-along-the-coast arrangement is not material. Do you think it is?

If not, the agreement as approved by the Treasury Department would, as I understand it, be sufficiently expressed by an instrument containing the provisions shown by a draft herein inclosed.

Please give me your views upon the matter, returning the draft with such alterations as you deem to be needed.

The Russian minister is exceedingly anxious that no time should be lost in initiating the proposed negotiations in London. I therefore hope to hear from you on the subject at once.

Very truly, yours,

RICHARD OLNEY.

Hon. CHARLES S. HAMLIN,
Assistant Secretary of the Treasury.

FEBRUARY 26, 1896.

SIR: I beg to acknowledge the receipt of your letter dated February 24, inclosing the draft of an agreement between the Russian Government and the United States regarding new regulations of the fur-seal fisheries from the Asiatic to the American shores. You state that in your opinion the suggestion of the Department regarding the abrogation of the 10-mile zone is not material, and you ask further expression of opinion from me on this question. In reply, I have the honor to state that this prohibition, if continued, will prevent American sealing vessels from participating in the fur-seal fisheries within 10 miles of the Russian and Japan coasts. Such a prohibition, if enforced also against the British Canadian sealers as well as those of Russia and Japan, would put the citizens and subjects of all of these nations on equal grounds. I see, therefore, no objection to the continuation of such prohibition.

In the draft you inclose there is evidently an error as to the closed season, it being made to read from April 1 to April 15. In the original suggestion contained in your letter of February 8, the agreement read from April 1 to October 15, and I assume that those are still the dates during which total prohibition of fur sealing is desired.

I will also call your attention to the statement in the preamble of the last draft of agreement, "that the object is to provide that the same law shall govern maritime hunting of fur seals in the waters frequented by both the Russian and American herds." This statement should be omitted, as the suggestion made by the Russian Government modifies, at least to the extent of the 10-mile prohibited zone, the regulations of the Paris award.

I take pleasure in inclosing a draft of an agreement which I think covers the suggestions made by the Russian Government.

Very respectfully, yours,

J. G. CARLISLE, *Secretary.*

The SECRETARY OF STATE.

[Inclosure.]

Agreement between the Government of the United States and the Imperial Government of Russia, supplementary to an agreement of May 4, 1894, relative to the fur-seal fisheries in Bering Sea and the North Pacific Ocean.

Whereas experience has shown that the provisions of said agreement are inadequate to the accomplishment of its purposes, the contracting parties do hereby agree to add to and supplement the same by the provisions following:

1. The extent of the prohibited zone around the Commander Islands and Robben Island shall be 60 nautical miles.

2. Maritime hunting of fur seals shall be wholly prohibited from April 1 to October 15 in Bering Sea, the Sea of Okhotsk, and the Pacific Ocean north of the thirty-fifth degree of north latitude, from one continent to the other.

3. The said agreement of May 4, 1894, as hereby added to and supplemented, is hereby declared (except so far as necessarily in consistence therewith) to be in addition to the regulations of the Paris award and not in derogation thereof; all of which said regulations, save as herein modified, and the acts of Congress dated April 6 and April 24, 1895, passed to carry out the regulations of said award, the said Imperial Government of Russia hereby accepts and agrees to carry into full effect as regards its subjects.

4. All Russian vessels engaged in hunting fur seals in violation of the agreement of May 4, 1894, as hereby added to and supplemented, or in violation of the regulations of the Paris award and the acts of Congress dated April 6 and April 24, 1894, in relation thereto, may be seized and detained by the naval or other commissioned officers of the United States as provided in said legislation, and all the provisions of said statutes shall be applicable to Russian vessels and shall be carried out by the courts of Russia.

5. All the provisions of said agreement of May 4, 1894, not inconsistent with the provisions of this supplemental agreement shall remain in full force and effect.

6. The provisions of this supplemental agreement shall be operative and effective as soon as the British Government and the Japanese Government shall have both agreed to be bound thereby.

MARCH 6, 1896.

SIR: I have the honor herewith to transmit, for your information and consideration, copy of a letter dated the 27th ultimo, and of the affidavit therein referred to from Mr. H. Liebes, of San Francisco, Cal., in relation to the unprofitable condition of the seal-catching business as conducted by sealing schooners, and protesting against the present practice of the owners of such vessels in exterminating the seal herds without resulting pecuniary gain.

Respectfully, yours,

S. WIKE, *Acting Secretary.*

The SECRETARY OF STATE.

[Inclosure.]

[H. Liebes & Co., importers of skins and manufacturers of fancy furs, Nos. 133, 135, 137 Post street.]

SAN FRANCISCO, CAL., *February 27, 1896.*

DEAR SIR: Referring to the conversation had with you while in Washington lately pertaining to the unprofitable condition of the seal-catching business, and upon your suggestion I now beg to inclose two affidavits covering the same, which, if not in the form you desire, can be changed as you may indicate. Should you wish, several others similar and from other parties can no doubt be obtained.

In fact, there will be no difficulty whatever in being able to prove by experts that the sealing schooners are pursuing the business at a loss, and it seems outrageous that they should persist in exterminating the herd without any pecuniary gain. The only reason they continue in the business is that the vessels so engaged are not fit for any other purpose, and further in the hope of an advance in the price of seals in the London market, which can only take place if the seal catch on the Pribilof Islands be reduced, but if on the contrary the same were increased it would so demoralize the trade by reducing present prices to such an' extent that financially they would not be able to fit out another season, and this to some extent would solve the problem and save the seal herd from entire destruction.

It is now an acknowledged fact that sealing in the North Pacific is much more unprofitable than in the Bering Sea, and would be abandoned entirely if the hunting could not be successfully continued into the Bering Sea, as the seals taken therein have a much larger commercial value than those taken in the North Pacific, which is accounted for by the same containing quite a number of young male seals, such as are not taken on the Pribilof Islands, but allowed to go back into the water. The skins of these young males, as stated above, are valued at very much more than the skin of the cow seal, and if they (the young males) were taken on the islands and not allowed to go back into the water to fall into the hands of the poachers, their catches would be worth considerably less commercially, besides greatly reducing the quantity they could take.

These facts are borne out by the sworn statements made of the percentage of male skins contained in the Bering Sea catch of last year, which, I understand, numbered about 15,000, and had the lessees been permitted to take these seals the Government would have received a revenue on same and the sealers would have been so demoralized that very few, if any, would have fitted out the present season.

I trust I have not encroached upon your valuable time, and if I can be of any assistance to you please command me.

Yours, truly,

H. Liebes.

C. S. Hamlin, Esq.,
    Assistant Secretary, United States Treasury,
        Washington, D. C.

Alexander McLean, being duly sworn, deposes and says: I have been engaged as captain of sealing schooners for thirteen years last past, and for seven years commanded a British sealing schooner sailing from the port of Victoria, British Columbia, and have, while so engaged, hunted for fur seals in the waters of the North Pacific Ocean and Bering Sea; am thoroughly familiar with all the details as to the cost of such sealing vessels, their outfits, and every expense attached thereto. I am also fully informed as to the prices fur-seal skins sell for in the London market each year, and that the price so realized the last two years, taking the average catch per schooner for the same period of time, has left a loss to the vessel so engaged in the sealing business, both American and British, and in consequence of the slaughtering of the cow seals in the past years, I found during my last cruise to the North Pacific and Bering Sea that the quantity of seals has been so greatly reduced that the average catch per schooner will necessarily be considerable less the coming season than in the past, so that it will not pay to continue the business, and for the first time in thirteen years I shall not engage in the same the present season.

ALEXANDER McLEAN.

Subscribed and sworn to before me this 21st day of February, A. D. 1896.

[SEAL.]   MARK LANE, *Notary Public.*

A. P. Lorentzen, being duly sworn, deposes and says: That he is secretary and general manager of the Pacific Trading Company, a corporation existing under the laws of the State of California; that said company is the owner of the American schooners *Bonanza, Edward E. Webster, Herman,* and *Alton;* that said schooners were engaged in the sealing business during the past two years, hunting fur seals in the waters of the North Pacific Ocean and Bering Sea, and in the same locations visited by both the American and British Columbia sealing fleets; that said schooners left this port at the usual time and returned here with their cargoes at the end of each season; that they were engaged in the taking of fur seals during these respective seasons, consisting of from eight to nine months each year; that the average catches of said schooners for the past two years were more than the average catches per schooner of the entire sealing fleet, both American and British, during the same period; that the catches of seals above referred to were shipped to London and sold

there at public auction in the usual manner, and the amount realized for the said seal skins was much below the cost attached to the catching of the same, and in each case left a loss to the company; that he personally superintended the fitting out of the above-named schooners for both years above stated; that the same was done as economically as possible, and at as low a cost as anyone else could have fitted out said or similar vessels for, either in this port or in any port in British Columbia; that the parties engaged as hunters, etc., on above vessels were fully competent and equal to any engaged on any other vessel out of this port or any port in British Columbia; that during the present year (1896) the schooners *Bonanza*, *Webster*, and *Herman* have been withdrawn from the sealing business for the reason that the same has proven decidedly unprofitable, and for the further reason that there is a prospect that their catches would be decreased in consequence of the diminution of the seal herd, which would undoubtedly leave a larger loss than in former years; that from statistics gathered through his own experience in this business in the past years it is shown that the average cost of each seal taken in the waters of the North Pacific Ocean and Bering Sea in the usual manner, computed on an average catch of the entire sealing fleet, is from $10 to $11.50 landed in London; that such cost varies according to the size of the vessel's catch; that as the per capita catch of these vessels must necessarily be largely reduced from now on the seals so taken will naturally cost comparatively more, making the business of catching seals still more undesirable and unprofitable.

In witness whereof the said A. P. Lorentzen has hereunto set his hand and seal this 20th day of February, 1896.

A. P. LORENTZEN.

Subscribed and sworn to before me this 20th day of February, 1896.

HENRY B. MADISON, *Notary Public.*

MARCH 6, 1896.

SIR: I have the honor to state for your information that in the last annual report of J. B. Crowley, special agent in charge of the seal islands, it appears that by actual count 28,000 seal pups died in the Pribilof islands during the last season from starvation, their mothers having been killed at sea. A careful estimate, based upon a partial count, placed the number of pups which died from starvation during the season of 1894 at 20,000. The count for 1895 was carefully verified by an agent of the North American Commercial Company upon the Pribilof islands. The report of Agent Crowley, above referred to, with other papers, was recently transmitted to the Senate, in compliance with a resolution of that body, and is now in the hands of the Government Printer, its publication having been ordered. I desire to call your attention also to the unprecedentedly large catch of seals in Bering Sea during the past season. The total was 44,169, as compared with 31,585 during the season of 1894. This is by far the largest catch ever made in Bering Sea, and it is believed that another catch of similar size for the coming season will almost completely exterminate the fur-seal herd. The greater portion of seals killed at sea were females. The total catch during the last season in the North Pacific and Bering Sea from the American herd was 56,991, as compared with the total for 1894 of 61,838; the small falling off being due to the fact of the inclemency of the weather between January and May along the northwestern coast, and also to the diminution of the seal herd. On the other hand, the catch in the Bering Sea increased very largely, as the inclosed figures show.

I would respectfully call these figures to your attention, in the hope that the British Government may consent for the coming season to some further regulation of the fur-seal fishery in order to save the herd from extermination.

I have the honor, etc.,

S. WIKE, *Acting Secretary.*

The SECRETARY OF STATE.

MARCH 13, 1896.

SIR: The act approved February 21, 1893 (27 Stat., 472), extends the provisions of section 1956 of the Revised Statutes, relating to the killing of fur-bearing animals in Alaska, to that portion of the North Pacific Ocean covered by the international agreement reached as a result of the Tribunal of Arbitration at Paris. The act provides also that when such agreement is concluded it shall be the duty of the President to declare that fact by proclamation and to designate the portion of the Pacific Ocean to which it is applicable.

I have the honor to inquire whether the proclamation of the President, dated April 9, 1894, promulgating the provisions of the act of April 6, 1894, containing the award of the Paris Tribunal, is the proclamation issued in compliance with the act of February 21, 1893, above cited, or whether it is deemed necessary that another should be issued to meet the requirements of that act.

In order that the necessary instructions may be issued to officers of this Department, I have the honor also to request that I be advised whether or not it is proposed to issue for the current year a proclamation similar to that of February 18, 1895 (28 Stat., 1258), prohibiting all persons from entering the waters of Bering Sea within the dominion of the United States for the purpose of violating the provisions of section 1956 of the Revised Statutes.

Respectfully, yours,

S. WIKE, Acting Secretary.

The SECRETARY OF STATE.

DEPARTMENT OF STATE,
Washington, April 20, 1896.

SIR: In a private and confidential telegram from the British Foreign Office to the British Ambassador at this capital it is stated, among other things, that Her Majesty's Government is fully in accord with the United States Government in desiring that all necessary and practicable measures should be taken to prevent the destruction of the seal herd. To that end they propose to employ an additional cruiser the forthcoming season on patrol duty, while notice has been issued by the Canadian Government that in future seasons nursing females should be distinguished from any others.

It is added that, with a view to the consideration of the question of further restrictions to be imposed in future years, Her Majesty's Government desires, with the concurrence of the United States, to commission a naturalist from England to reside during the forthcoming season on the Pribilof Islands; that the Canadian Government also desires to again send Mr. Macoun to continue his investigations; that these gentlemen are expected to reach the islands in the early part of June; and that Her Majesty's Government hopes that they will enjoy the cooperation of the United States authorities and will receive all necessary facilities for the accomplishment of the purposes of their mission.

It is suggested that the company leasing the fur-seal catch might be willing to make arrangements to permit the gentlemen named to proceed in the company's steamer.

Will you kindly give me your views upon the foregoing propositions, and especially upon the point whether there is any objection to a residence upon the Pribilof Islands of British or Canadian commissioners, as suggested.

Very respectfully, yours,

RICHARD OLNEY.

The SECRETARY OF THE TREASURY.

DEPARTMENT OF STATE,
Washington, April 22, 1896.

MY DEAR SIR: I should like an answer at your earliest convenience to mine of a day or two since, asking your views respecting the request of the British Government that a naturalist, appointed by that Government, be permitted to reside on the Pribilof Islands the coming season, and respecting the other matters stated in my letter.

Very truly, yours,

RICHARD OLNEY.

Hon. CHARLES S. HAMLIN,
Assistant Secretary of the Treasury.

APRIL 23, 1896.

SIR: I have the honor to acknowledge your note of April 20, informing the Department of the desire of Her Majesty's Government to commission a naturalist from England to reside during the forthcoming season on the Pribilof Islands, and of the further desire of the Canadian Government to send again Mr. Macoun to continue his investigations. You suggest, also, that the company leasing the fur-seal islands might be willing to make arrangements to permit the gentlemen named to proceed in the company's steamer.

I have the honor to reply that I see no objection to a suitably qualified zoologist residing on the islands this season, but I beg to impress upon you the necessity, if such permission be given, of also sending a zoologist to represent the United States Government. I would further suggest that it be understood that this permission shall in no way be construed as a waiver of the position of the United States that an immediate change in the regulations of the Paris award is necessary to preserve the fur-seal herd from destruction during the coming season, or as an abandonment of its efforts in this direction. Inasmuch as the closed season begins May 1, and continues until the 1st of August, when the fur-seal herd is in Bering Sea, there will be ample time to complete negotiations for such a change in the regulations, should the British Government consent thereto. As to the desire of the Canadian Government to send Mr. Macoun to the islands, I would state that I have made careful inquiries as to Mr. Macoun's standing and abilities as a zoologist. I beg, in this connection, to inclose a copy of a note from Mr. J. Stanley-Brown, agent of the North American Commercial Company, the present lessee of the seal islands, from which it would seem to appear that the interests of both the United States and of Great Britain might be subserved better by the designation of some other person than Mr. Macoun. I do not, however, see the necessity of having any zoologist to represent Canada, other than the one representing the British Government.

I would further respectfully suggest that before any permission be given, the name of the zoologist be submitted to me, in order that the Department may be satisfied as to his capacity. Mr. Brown requests that his letter be considered confidential.

I have the honor, etc.,

S. WIKE, *Acting Secretary.*

The SECRETARY OF STATE.

———

1318 MASSACHUSETTS AVENUE,
Washington, D. C., April 22, 1896.

SIR: In reply to your note of this date, I have to say that Mr. J. M. Macoun is a botanist officially connected with the Canadian geological survey. He is an agreeable and amiable gentleman to whom personally I feel most kindly, but in the sense of having knowledge of animals he is not a naturalist at all. In 1892, while studying seals and their habits on the Pribilof Islands, he confirmed a report of Dr. Everman's to the effect that they had in company seen a number of dead female seals at Northeast Point, St. Paul Island. The investigation which was immediately made showed that the creatures they had seen were the pups of the sea lion and at least three out of five were males.

Upon his return to Canada, Mr. Macoun prepared a report which was printed as pages 134 to 156 of the appendix of the Counter Case of Great Britain, vol. 1. This report is so filled with his own version of conversations and statements of other individuals, insufficient observations, conjecture, and statements made with the evident intention of misleading those unfamiliar with the subject, and is so lacking in the frank and honest spirit of scientific investigation that he should not be again received by the United States as a representative of Great Britain in any capacity. The whole report was made under the supervision of Dr. Dawson, one of the British Bering Sea commissioners, and was designed to form an adjunct to and bolster up that extraordinary and thoroughly discreditable document, the report of the British commissioners. Mr. Macoun is a subordinate of Dr. Dawson's, and any future investigations by him will be carried out in the same spirit and within the same purpose as were those of 1892.

The natural history facts relating to the fur seal, so far as they can be ascertained from residence on the Pribilofs, are well known, and it is not a naturalist who is needed so much as the services of an honest man of high standing who will faithfully count the dead pups upon each rookery after September 20. All sealing in Bering Sea should be suspended, and it would be a simple matter to determine from the count made in September by representatives of the United States and Great Britain in concert (this is important) whether the enormous numbers of dead pups counted last year is a normal, annual occurrence unaffected by the frightful slaughter of mothers in Bering Sea. There is still ample time to save the terrible waste of seal life which has occurred during the past two summers—that waste which has destroyed alike the Alaskan and Russian seal herd.

Very respectfully,

J. STANLEY-BROWN.

Hon. C. S. HAMLIN,
Assistant Secretary of the Treasury.

———

APRIL 25, 1896.

SIR: I have made inquiries of Mr. J. Stanley-Brown, superintendent of the North American Commercial Company, who informs me that the company's steamer *Homer* will sail from San Francisco between the 15th and 20th of May for the islands. The steamship is a commodious one, having forty staterooms, and there will be no trouble in accommodating the representatives of the British Government. The steamer will reach Unalaska probably about June 3, and continuing its course, will reach the islands about July 7. The whole seal herd will not be on the islands before July 10, and it would not be practicable, therefore, to undertake to count the number of seals before that date, and this counting would take probably ten days. It would be, therefore, nearly

August 1 before any report of said representatives could be sent back and reach the United States or Great Britain.

The only effective mode would be to agree in advance, as I have suggested in another letter of this date, upon a modus vivendi closing Bering Sea, and prohibiting all killing on land and sea during the remainder of this season, except food skins for the natives. If, however, the British representatives were authorized to declare such a modus, if the result of their investigation warranted such action, it would be possible to notify the sealing vessels at Unalaska, where they rendezvous, prior to going into Bering Sea. It would be impossible, however, to send a report back to the United States and Great Britain in time to receive an answer and to notify said vessels of the conclusion of the respective Governments.

I have the honor, etc.,

S. WIKE, *Acting Secretary.*

The SECRETARY OF STATE.

DEPARTMENT OF STATE,
Washington, April 25, 1896.

SIR: I have the honor to transmit to you herewith two copies of a report from the Canadian minister of marine and fisheries respecting the catch of the Canadian sealing fleet during the season of 1895, and of the statistics annexed to the report, which supply the information required by article 5 of the Bering Sea arbitration award.

The documents in question were received with a note of the 23d instant from the British ambassador at this capital.

I have the honor, etc.,

RICHARD OLNEY.

The SECRETARY OF THE TREASURY.

[Inclosures.]

Report from the minister of marine and fisheries of Canada, with reference to the catch of the British sealers who operated on the North American coast during the year 1895.

MARINE AND FISHERIES, CANADA, OTTAWA,
January 4, 1896.

To His Excellency the GOVERNOR-GENERAL IN COUNCIL:

Reverting to the approved minute of council dated the 25th July last, communicating the catch of the British sealers which operated on the North American coast during the spring of 1895, the undersigned has now the honor to append, for the information of Your Excellency, a communication from the collector of customs at Victoria, dated the 7th ultimo, covering:

(*a*) List of the names and masters of all vessels licensed at Victoria for 1895.
(*b*) Statement of the catch of British sealing fleet for 1895.
(*c*) The catch of American schooners landed at Victoria.
(*d*) The catch of American schooners landed at Puget Sound.
(*e*) The catch of American schooners landed at Astoria, Oreg.
(*f*) The catch of American schooners landed at San Francisco.
(*g*) The catch of Pribilof Islands (American).
(*h*) The catch of Copper Islands (Russian).
(*i*) The boardings of British vessels in Bering Sea.
(*j*) Copies of official logs of all British vessels sailing from Victoria, giving location of each day's fur seal fishing operations.

The undersigned would observe that the catch of seals during the past seven years was—

	Number.		Number.
1889	35,310	1893	70,592
1890	43,325	1894	95,048
1891	52,365	1895	73,614
1892	49,743		

representing practically an average of 60,000 skins per annum.

It is thus demonstrated that the yield of the present year, notwithstanding the explanations of unpropitious weather and unfavorable circumstances, is largely in excess of the average take of the past seven years.

Prior to the extraordinary and abnormal take of 1894, under the most favorable conditions of weather and other circumstances, that of 1893 greatly exceeded any in previous years in the history of the sealing industry, yet the take of the present year is considerably larger than that was.

The undersigned is of the opinion that the significance of the decrease in the catch as compared with 1894 can not be so marked as might at first sight appear, if the results of the two years are considered alone.

He further believes that the contention that the seal herds are being rapidly exterminated, and that only a vestige of their former greatness remains, does not appear to gather much strength from an impartial examination of the figures representing the annual catch. Indeed, considered in the light of the explanations offered by the sealers, the result of the present year's operations may be regarded as affording reasonable grounds for an exactly opposite conclusion.

Perhaps one of the most noteworthy incidents in the industry this year is the catch by the schooner *Director* in the South Atlantic Ocean, off Falkland Islands, of 602 seals.

Inquiries were instituted for the purpose of collecting any information connected with the incident which might be of interest to the question of the sealing industry generally.

It was ascertained that Capt. Frederick W. Gilbert, of the schooner *Director*, 87 tons register, with a crew of 25 men, sailed from Halifax, Nova Scotia, on the 20th of December, 1894, bound for the Asiatic side of the North Pacific Ocean.

On reaching the tenth degree of south latitude the master was obliged to change his course by reason of his supply of provisions and water being insufficient to enable him to complete his voyage.

The run from Halifax to the Falkland Islands was made in forty-eight days. While off the southern end of the islands he encountered several groups of seals. He consequently devoted thirty-six days to sealing in that neighborhood, as well as off the east and west ends of Staten Island, resulting in the capture of 620 seals, which he took to the port of Victoria.

The captain reports that he was compelled to suspend his sealing operations owing to a change in the weather, which became quite stormy, and as it was getting late in the season he proceeded on his voyage to Victoria, reaching there on the 21st of May, 1895.

Captain Gilbert reported that all the seals were secured at sea, far distant from any of the sealing preserves, and were shot in the same manner as are those taken in the North Pacific Ocean by the Victoria sealing fleet.

He met with no interference. In reply to the inquiries made it was ascertained that no record existed of the landing in the past of any seal skins at a British Columbia port which did not form part of the catch of the sealers operating in the North Pacific Ocean, either on the American or Asiatic side thereof.

The skins are reported to have been in good condition, and to be of the same kind as those usually sold by Messrs. Lampson & Co., London, and are classed and known with the Lobos Islands seal skins from the mouth of the river Platte, and bringing about the same prices as those taken in the North Pacific Ocean.

The character of the skins is represented as being very similar to that of those usually secured by the British Columbian fleet.

The *Director*, under the command of Captain Gilbert, fitted at Victoria for the August and September season in Bering Sea, where a catch of 688 seal skins was secured.

The undersigned recommends that a copy of this report, if approved, together with its appendices, be forwarded to the right honorable the principal secretary of state for the colonies.

Respectfully submitted.

JOHN COSTIGAN.

CUSTOMS, CANADA, VICTORIA, BRITISH COLUMBIA,
December 7, 1895.

SIR: I have the honor to forward herewith a statement in complete form showing the following:

(1) The names and masters of all vessels licensed at Victoria for 1895.
(2) The catch of British sealing fleet for 1895.
(3) The catch of American schooners landed at Victoria.
(4) The catch of American schooners landed at Puget Sound.

(5) The catch of American schooners landed at Astoria, Oreg.
(6) The catch of American schooners landed at San Francisco.
(7) The catch of Pribilof Islands (American).
(8) The catch of Copper Islands (Russian).
(9) The boarding of British vessels in Bering Sea.
(10) The copies of official logs of all British vessels sailing from Victoria, giving location of each day's fur seal fishing operations.

In submitting the above report of the fur-sealing operations for the past year, I would beg to say that the figures have been collected with considerable care and trouble, and I desire here, for your better understanding of the year's work, to give a résumé of the operations of the fleet sailing from this port.

Results, 1894.—The catch of 1894 being the largest in the history of this port, amounting to 94,474 skins, the results, however, were generally unsatisfactory and unprofitable to the owners of vessels.

This loss was brought about partly by their own act in giving exorbitant prices to the seal hunters, paying them as high as $4.50 and $4.75 per skin, when the sales in London only brought about $7 to $8 per skin, thus bringing quite a number of owners into debt, and some to disaster.

I endeavored at the time of the engagement of the crews in 1894 to bring the owners to realize that there was great danger in paying this high figure to hunters, but there was such competition that all argument was unavailing.

The present year, however, the effect of the low prices brought the owners to realize the situation, and the vessels paid this year to all white hunters only one-fifth lay and to the Indians the same, with a small bonus.

Licenses.—There have been licensed during the past year 64 British vessels sailing from this port. Of this number 22 sailed during December, 1894, and January, 1895, to Japan; 33 were engaged in the British Columbia Coast catch, and 9 Indian schooners, which likewise confined their operations to the British Columbia Coast up to the 1st of May.

Crews.—There were engaged in this industry 705 white seamen and 854 Indians, showing this year a decrease in the white seamen and an increase in the number of Indians. The fact of the increase of Indians was owing to the demand for spearmen in Bering Sea, where firearms could not be used.

Boats and canoes.—The record shows that there were 210 boats and 421 canoes in use this year, exhibiting a decrease of 56 boats and an increase of 162 canoes. This, as with the crews, was owing to the number of Indian spearmen going to Bering Sea.

British Columbia Coast catch.—The figures show the total British Columbia Coast catch to be 9,853, exhibiting a decrease of 1,850 skins compared with 1894, although a larger number of vessels were engaged. The cause assigned for this decrease was chiefly owing to the boisterous weather which prevailed along the British Columbian Coast, and when the weather moderated the seals had passed northward, so that the larger herds could not be reached before the 1st of May, the commencement of the close season.

Japan catch.—The total result of the operations on the Japanese Coast show that there were only taken 18,687 skins, as against 49,483 in 1894, being a marked decrease of 30,796. This decrease caused me to make diligent inquiry from the masters and crews, and the conclusion arrived at was that stormy weather usually prevailed all along the Japanese Coast, preventing the schooners from lowering their boats sometimes for days together. Also, it seems to be the consensus of opinion among them that the schooners this year were in advance of the seal herd, which had apparently gone farther to the south than usual.

Copper Island catch.—Twenty schooners were sealing in the vicinity of Copper Island, which obtained a catch of 6,281, as against 24 vessels last year with a catch of 7,437.

Bering Sea.—In the month of June last 33 vessels cleared from this port for Bering Sea, proceeding to the west coast, where they obtained Indian hunters, and proceeded direct to Unalaska, where they were all instructed, if they reported themselves to the custom-house there, that no difficulty was anticipated. On the clearance papers from this port it was plainly stated that they had no firearms, nets, or explosives, and that their hunting outfit consisted only of spears. On the 1st of August they all left the port of Unalaska, United States of America, and at once sailed to the sealing grounds.

There were also 8 vessels which entered Bering Sea from the westward which had been engaged in sealing on the Japanese side, making a total of 41 British vessels fishing in Bering Sea, the result of their fishing being 35,918, as against 27 vessels in 1894, with a catch of 26,425—an increase of 9,493 seals.

The weather was good, and seals reported to be fairly plentiful; but in this regard some conflicting accounts were given, no doubt measured by their individual success.

Boardings.—The whole force of the American Revenue Service, consisting of 5 or 6 vessels, was congregated in Bering Sea, and from the evidence given and the entries made in the official logs, they were continuously on the patrol outside the 60-mile zone, as you will see by the inclosed statement, they having made 106 boardings from the 3d of August to the 20th of September.

The records of the boardings show that several of the vessels were boarded from 3 to 5 times, and nearly all were boarded more than once, and no exemption was made to those vessels which had cleared from Victoria with only spears, as it was understood that in searching the vessel and overhauling the skins the main contention was whether the skins had been shot.

This our sealers, who had only spears and no firearms, considered a great hardship—that their skins should be so often overhauled and the skins and salt scattered over their provisions and coal, leaving them all over the hold without in one case offering to repack them, thereby causing danger to their preservation.

It was likewise reported that these frequent boardings took place when storm was threatening and danger was apprehended to the canoes and boats, sometimes at a distance of 5 to 10 miles away, pursuing seals.

It appears to me, however, from the excellent condition all the skins were in when landed here, that no damage had come to the skins from the boardings, and it may possibly be that the frequent saltings may have added to their preservation.

Apart from the interruption to the voyage while being boarded, and the inconvenience and trouble in repacking the skins, no other damage was sustained, so far as I could learn.

Seizures.—During the past year there have been three seizures, which were all duly reported to you.

(1) The schooner *Shelby*, which was seized by the United States ship *Corwin* for having seals and unsealed sealing implements on board in prohibited waters. This vessel was seized in latitude 32° 59' north, longitude 134° 10' west, while on her homeward voyage from the sealing ground, having proceeded about 200 miles and reached as far south as Queen Charlotte Island, British Columbia, when she was arrested and ordered to proceed to Sitka, with a prize crew on board, and from that place she was directed to proceed to Victoria and report to me. On the return of the vessel to this port, where she has lain in the harbor awaiting the action of the vice-admiralty court, which sat here in November, and the vessel being adjudged guilty, was ordered to pay a fine of £100 and all costs.

(2) The *Beatrice*, which was seized in Bering Sea on the 20th of August for violation of the fifth article of the regulations of the Paris award (as set forth in the Bering Sea award act of 1894)—that is, that she had not entered up her log book from the 12th to the 20th of August. This vessel was tried and acquitted; the judgment carried all costs. This vessel has since been restored to her owners.

(3) The *E. B. Marvin*, which was seized in Bering Sea by the United States revenue cutter *Rush* for violation of article 6 of the regulations of the Paris award (Bering Sea award act, 1894)—that is, that she had used firearms in killing seals. This schooner had on board at the time of seizure 386 skins, one of which was selected which appeared to have a hole resembling a shot hole, and which was the grounds on which this schooner was seized and sent back to Victoria for trial. This vessel was tried and the seizure discharged without costs.

While the case of the *Marvin* was in court I was very kindly requested by the defendant to view the skin which was the cause of seizure, the defendant having obtained permission to view the same, which was then in possession of the court, and it was only from curiosity, and not that I was required to give any evidence, that I, with a few others, viewed the skin, and I came to the conclusion, seeing that it was a very difficult matter for the experts to determine precisely where the shot hole was, that the charge against the vessel was a very uncertain one.

The owners of the schooners *Beatrice* and *E. B. Marvin* have suffered severe loss by the seizure and interruption of the season's voyage. The actual loss of the *E. B. Marvin* alone, presuming that she would have taken about the same number of skins as other vessels of similar size and equipment, will be not less than $10,000.

Disasters.—There were only three serious disasters to the sealing fleet during the past year, viz, the schooners *Rosie Olsen*, *Brenda*, and *Walter A. Earle*.

The *Rosie Olsen* was wrecked on the 18th of June while entering the port of Hakodate for water, and became a total loss. The crew and seal skins were all saved.

The *Brenda* was wrecked on the 1st of July last. When going into the Little Kurile Straits for water she struck on a rock and became a total loss. The schooner *Geneva* hove in sight and rescued the crew, skins, and some portions of the rigging.

The *Walter A. Earle* was capsized on the 14th of April in latitude 58° north and longitude 139° west. This vessel had a crew of 28 persons, and was sealing on the British Columbia coast when the storm overtook her. All hands were lost. She was subsequently found bottom up in the neighborhood of Cape Tonki, and was towed into Kodiak and 15 bodies were found in the ship's hold.

Falkland Island catch.—The schooner *Director* arrived here from Halifax, Nova Scotia, on the 21st of May, and while in the vicinity of the Falkland Islands, or between that and Staten Island, obtained 620 seal skins, which were brought to this port. The circumstances of this were conveyed to you in my letter of the 10th of August. I have seen Capt. F. W. Gilbert, and he has nothing further to add to the information conveyed to you in my letter, only that he affirmed that the hunting was done on the sea, and that no attempt was made to take skins on the land; that he was outside the jurisdiction of any foreign Government, and that if the weather had been more moderate he could have obtained many more, but he desired not to delay there, as he wished to proceed to Bering Sea.

Sealing of arms.—After careful inquiry in regard to this matter, I find that the majority of our sealers would desire to have their firearms sealed up before entering Bering Sea, to avoid interruption and seizure for trivial and doubtful causes, but this, I do not think, they would mean to apply to the Pacific Ocean proper.

Official logs.—All vessels leaving this port have been carefully notified to keep their official log books in conformity with the instructions of article 5 of the Paris regulations, and they have been fairly kept, recording each day the number of males and females taken; but I might suggest that an additional sealing log or an addition to the present official log be made whereby the sealers might record daily some other items of interest that might be of use to the Government.

Agreement with Russia.—I had the honor, on the 16th of January last, to receive your letter, and, in compliance with the wishes of the honorable the minister, I promulgated in every possible manner the renewal of the provisional agreement with Russia, providing a protective zone of 30 miles around the Komandorski Islands, Tulenew Islands, or Robben Reef, in the Okhotsk Sea, and a protective zone of 10 miles along the shores of the Russian mainland.

Results of 1895.—The catch for the British Columbia fleet of British vessels being 71,359, and there was likewise landed from American vessels 2,255, making a total of 73,614 skins which have passed through this port.

The ruling price during the month of August and the early part of September was $9, but during the latter part of September and October the price went up to $10.50. I thought the price was very good, being a considerable advance on last year, and those who were fortunate enough to sell at the latter price will do very well, as information has been received here from London that the skins will not reach that figure after deducting the expenses.

It may be that some of the skins may be kept over for higher figures, but I think it safe to say that the value of the present year's sealing will amount to about $700,000, or, say, $730,000.

I have, etc.,

                                        A. R. MILNE, *Collector.*

WM. SMITH, Esq.,
    Deputy Minister of Marine and Fisheries, Ottawa.

APRIL 29, 1896.

SIR: I have the honor to acknowledge the receipt of your note, inclosing a report from the minister of marine and fisheries of Canada, with reference to the catch of British sealers, and copies of British log books for the season of 1895.

In this connection I desire to call your attention to two significant statements in the report of Mr. A. R. Milne, collector at Victoria, to the deputy minister of marine and fisheries, as bearing on the implied claim of the British Government, contained in a letter and inclosures from the British ambassador to yourself, dated April 23, 1896, of which you inclosed me a copy, that the fur-seal skins were injured by the frequent searches of our revenue cruisers.

On page 4 of said report the collector states that, from the excellent condition of the skins when landed at Victoria, no damage, in his judgment, had come to them from the searches by revenue cruisers. On page 5 the further statement is made that, after careful inquiry, the collector found that a majority of the British sealers desired to have their firearms sealed up before entering Bering Sea, to avoid interruption and seizure for trivial and doubtful causes.

I have the honor, etc.,

S. WIKE, *Acting Secretary.*

The SECRETARY OF STATE.

DEPARTMENT OF STATE,
Washington, April 29, 1896.

SIR: I beg to inclose herewith copy of note just received from the British ambassador at this capital; also copy of note I have sent to him in reply.

In reply to the inquiry addressed to me by the Treasury Department in a communication of a day or two since, viz, whether in my judgment there was any prospect of the British Government agreeing to any modification of the Bering Sea regulations for the coming season, I desire to say that, in my judgment, it is quite improbable that any such agreement will be made.

On the other hand, I have good reason for believing that if the report of the British and Canadian agents proposed to be sent to the Pribilof Islands, as above stated, should confirm the position of this Government as to the threatened immediate extermination of the fur-seal herd, a modus vivendi might be entered into which would insure the protection of the fur-seal herd during the coming season.

Respectfully, yours,

RICHARD OLNEY.

Hon. CHARLES S. HAMLIN,
Assistant Secretary of the Treasury.

[Inclosures.]

BRITISH EMBASSY,
Washington, April 27, 1896.

SIR: With reference to your note No. 344 of the 11th ultimo, in which you urge the adoption of some further restrictions on pelagic sealing in Bering Sea for the coming season in view of the alleged imminent extermination of the fur-seal herd, I have the honor to inform you that the contents of your note have received the careful consideration of Her Majesty's Government.

I am instructed by Her Majesty's principal secretary of state for foreign affairs to state that the apprehensions of the United States Government on this head appear to be founded mainly on the fact that by actual count 28,000 dead pups were found in the island last year, and on the assumption that the deaths of these pups were the direct result of their mothers having been killed at sea.

But, from the exhaustive discussion of the question in the report and supplementary report of the British Bering Sea commissioners, it has not been satisfactorily established that the mortality of the pups is caused by the killing of seals at sea. The date, moreover, which the arbitrators fixed for the opening of Bering Sea pelagic sealing and the radius within which sealing was prohibited round the Pribilof Islands were determined, after full consideration, to be sufficient to protect nursing females whose pups were not able to provide for themselves.

It should also be borne in mind that in the Bering Sea catch of 1895 the proportion of males to females taken by Canadian sealers was about 45 per cent of males against 55 per cent of females, although the returns of the American sealers in that sea give an average of 3 females to 1 male.

In the meantime the admitted fact that the seals at sea show no apparent diminution of numbers, and that the sealers in Bering Sea were able to make practically as large catches last year as in the previous year, does not point to the imminent extermination of the seals.

The returns show that the Canadian sealing vessels all kept well outside the 60-mile radius, and as there seems little doubt that during the period when sealing is allowed in Bering Sea the great bulk of the seals are inside that limit, the natural deduction is that less than half the herd is at any time exposed to capture, and that the danger of extermination by pelagic sealing must therefore be comparatively remote.

It is observed that on the islands 15,000 seals were killed last season, as compared with 16,000 in the season of 1894, but in the reports which have been received on this point it is not stated whether any difficulty was experienced in obtaining that number of skins, nor from what class of seals the skins were taken.

Taking into account the catch on the islands, the whole catch from the Alaskan herd was 71,300 in 1895, as compared with 71,716 in 1894, being only about half the total catch taken in 1889 and previous years, and though it may be the case that a slaughter of some 70,000 a year is more than the herd can properly bear for a series of years, Her Majesty's Government see no reason to believe that it is so large as to threaten early extermination.

The necessity for the immediate imposition of increased restrictions, to take effect during the coming season, does not, therefore, appear to be established, and it must be borne in mind that at this late period it is no longer possible to give effective warning of any change in the regulations to the large number of vessels which have already cleared for the Japan Coast fishery, and which will, after that is concluded, proceed to Bering Sea for the opening of the fishery in August. The imposition of restrictions without due warning would cause great confusion and hardship, and would undoubtedly give rise to large claims for compensation on ground which could not, with justice or reason, be disputed.

But Her Majesty's Government fully share the desire so strongly expressed by your Government that all necessary and practicable measures should be taken to prevent the possible extermination of the seals.

As a precaution for the strict observance of the regulations prescribed by the Tribunal of Arbitration and now in force, they will give directions for the employment of an additional cruiser this season in policing the fisheries, although, as far as they have been able to judge, the force employed up to the present time has been sufficient.

In accordance with the desire expressed by you in your note No. 317 of the 6th February, Her Majesty's Government have requested the Dominion government to issue a notice to the effect that the returns which the sealing vessels are required to furnish shall in future specify which of the females killed are barren and which are in milk, and a reply has been received from His Excellency the Governor-General of Canada that this will be done.

In order to investigate more completely the question of the necessity of further restrictions in future years, Her Majesty's Government are desirous at once to take the necessary steps for conducting an independent inquiry on the Pribilof Islands into the state of the herd, by an agent sent from Great Britain. This gentleman would be a naturalist possessed of the necessary scientific qualifications, and care will be taken to select a person who will be entirely free from bias in carrying out the mission intrusted to him.

The Canadian government are also desirous of sending Mr. Macoun again to the islands this season in order to continue his investigations. The British agent and Mr. Macoun would arrive at the islands early in June and remain until toward the end of September, and Her Majesty's Government would be glad if the United States authorities would grant them all necessary facilities and cooperate with them as far as possible.

It has been suggested that arrangements might perhaps be made with the company which leases the seal catch on the Pribilof Islands to allow the British agent and Mr. Macoun to proceed in their steamer as passengers.

I have the honor, etc.,

JULIAN PAUNCEFOTE.

APRIL 29, 1896.

SIR: I have the honor to acknowledge your favor of the 27th instant, being an answer to my note No. 344, of the 11th ultimo, wherein is urged the adoption for the coming season of further restrictions on pelagic sealing in Bering Sea in view of what this Government believes to be the demonstrated imminent extermination of the fur-seal herd.

Without at this time adducing any additional considerations in support of the position taken by this Government, I hasten to say that it welcomes an independent inquiry by the British Government into the present state of the fur-seal herd through the British and Canadian agents referred to in your note. They will be given all needful facilities for their investigations by this Government, which will request the North American Commercial Company to give them all convenient transportation facilities on its steamers.

I venture to also suggest that if the naturalist selected by the British Government could come to Washington on his way to Alaska and have a free and full conference with Assistant Secretary Hamlin the objects of his mission would probably be greatly promoted.

I avail myself, etc.,   RICHARD OLNEY.

His Excellency the Right Honorable Sir JULIAN PAUNCEFOTE.

MAY 5, 1896.

SIR: I have the honor to acknowledge your letter of April 29 inclosing copy of a note received by you from the British ambassador at this capital, and also a copy of your answer thereto.

In the note of the British ambassador it is stated that the whole catch taken from the Alaskan herd, including the land catch on the Pribilof Islands for the years 1894 and 1895, was, 71,716 and 71,300, respectively. While this statement is substantially correct for the year 1895, it would appear that in the year 1894 a large number was taken, namely, 76,871—61,838 at sea, and 15,033 on the Islands.

The further statement is made in said letter that the fur seals show no apparent diminution in numbers, and attention is called to the fact that the sealing vessels in Bering Sea made practically as large catches during the season of 1895 as in that of 1894, which fact the ambassador contends does not point to the immediate extermination of the fur-seal herd. The fact, however, that the seals on the islands have decreased at least one-half since 1890 would seem to answer this claim. A further answer will also be found in the report of the Secretary of the Treasury for 1895, on page cc, wherein it appears that the average catch per vessel on the Northwest Coast fell off 5 per cent in 1895 as compared with 1894, while the average catch in Bering Sea fell off 12 per cent as compared with 1894. At the same time, while the percentages of females killed in Bering Sea were the same for British vessels in 1894 and 1895, there was an increase from 69 to 73 per cent for American vessels in 1895. That the seal catch is maintained at the figures cited is because of the fact that Bering Sea is a nursery for the herd while it is on the islands, and of the further fact that the seals can be killed easier while in Bering Sea than when traveling off the Pacific Coast toward the islands.

The statement of the ambassador that the total land and sea catch from the Alaskan herd in 1895 was only about one-half of what the same was in 1889 would seem to be a further convincing argument as to the decrease in the seal herd. In this connection I would state that in 1889 the catch on land and sea was about 132,000, of which 102,000 were taken on the Pribilof Islands and 30,000 at sea, the pelagic catch being about 22 per cent of the total. In 1895, on the other hand, the pelagic catch, 56,291, had increased to 78 per cent of the total, 71,291. From 1880 to 1895 the pelagic catch increased from about 8,000 to 56,000, or 600 per cent, while the Pribilof Islands catch decreased from 105,000 to 15,000, or 86 per cent.

It is stated also in said letter that it would now be too late to give effective warning of any change in the regulations, and that vessels which have cleared already for the Japanese Coast would be seriously injured by any change at this late date. I have the honor, however, to call your attention to the fact that the modus vivendi of 1891 was agreed upon as late as June 15.

Respectfully, yours,

S. WIKE, *Acting Secretary.*

The SECRETARY OF STATE.

MAY 5, 1896.

SIR: I have the honor to state that the Department is informed by the Secretary of State that the British ambassador has requested permission in behalf of the British Government to send two naturalists to the Pribilof Islands to investigate the present condition of the fur-seal herd during the coming season. It will, therefore, in the opinion of this Department, be necessary to secure the services of some eminent zoologist in this country to conduct a similar investigation on behalf of the United States in cooperation with the officers of the United States Government who may be detailed for this purpose by the President. I have the honor to suggest that an appropriation of $5,500 be made for this purpose, and herewith inclose a form of appropriation for your consideration.

Very respectfully, yours,   C. S. HAMLIN,
Acting Secretary.

Hon. WILLIAM B. ALLISON,
Chairman Committee on Appropriations, United States Senate.

The Secretary of the Treasury is hereby authorized, out of any moneys not otherwise appropriated, to expend not exceeding $5,500 for investigation of the condition of the fur-seal herd in the North Pacific Ocean and Bering Sea during the fiscal years 1896 and 1897. He is further authorized, in addition to said $5,500, to pay the necessary expenses of any officials of the Government who may be detailed for said investigation by the President.

UNITED STATES COMMISSION OF FISH AND FISHERIES,
Washington, D. C., May 9, 1896.

SIR: I have the honor to inform you that the Fish Commission steamer *Albatross* is now being fitted out at the Mare Island navyyard, California, for fishery investigations on the Alaskan coast, and that preparations are also being made in connection therewith to conform to the provisions contained in the act of Congress approved March 3, 1893, as follows, namely:

And the Commissioner of Fisheries is authorized and required to investigate, under the direction of the Secretary of the Treasury, and when so requested to report annually to him regarding the conditions of seal life upon the rookeries of the Pribilof Islands.

My reason for bringing this matter to your attention is due to the fact that the investigations called for are subject to your authorization, and I beg to be informed if you desire to have the work taken up again this season. In this event, I would respectfully request that the chief Treasury agent at the Pribilof Islands be notified of the proposed inquiries, and that he be directed to facilitate the work of the Fish Commission party to such an extent as he may be able to. I would also ask that the party be allowed the same accommodations at the Government building as have been accorded it in the past, and that it be permitted to participate in the mess furnished by the sealing company, the same as the Treasury agents.

Very respectfully,
J. J. BRICE, *Commissioner.*

The SECRETARY OF THE TREASURY,
Washington, D. C.

H. Doc. 92, pt. 2——24

MAY 14, 1896.

SIR: Acknowledging the receipt of your letter of the 9th instant, wherein you request to be advised whether, under the act of March 3, 1893, this Department desires an investigation and report to be made by the Fish Commission as to the condition of seal life on the rookeries of the Pribilof Islands during the coming summer, I have the honor to inform you that the Department considers such investigation and report most desirable, particularly in view of information received from the honorable the Secretary of State, that an English and a Canadian naturalist will be permitted to reside on the islands during the coming summer. I feel, therefore, that the assistance desired can be rendered by our Commission in examining into the condition of the rookeries, taking a careful census of the seals, examining the bodies of the dead pups to ascertain the cause of death and making other similar investigations.

I would thank you to inform me of the earliest date on which the *Albatross* will be ready to sail, so that I may have a personal interview with you before that date.

Respectfully, yours,
C. S. HAMLIN,
Acting Secretary.

The COMMISSIONER OF FISH AND FISHERIES,
Washington, D. C.

MAY 14, 1896.

SIR: The report of the minister of marine and fisheries, sent by the British Government, inclosing a copy of the log-book entries of British vessels for the season of 1895, as required by article 5 of the regulations of the Paris tribunal, contains copies of such entries only as relate to the catch of the British vessels in Bering Sea, and omits all entries showing the killing of the 9,853 seals taken by the British vessels on the Northwest Coast.

I have the honor to request, therefore, that you ask the British Government to furnish the log-book entries containing the record of the Northwest Coast catch.

Respectfully, yours,
C. S. HAMLIN,
Acting Secretary.

The SECRETARY OF STATE.

MAY 26, 1896.

SIR: The Department has been informed by the honorable the Secretary of State that two English naturalists will shortly go to the Pribilof Islands as representatives of the British Government to enter upon an investigation of the present condition of the fur-seal herd.

In my judgment it would be necessary to send some eminent zoologist to accompany these British representatives to make an independent investigation in behalf of the United States Government.

It will also probably be expedient to detail one or more officers or employees of the United States to take part in said investigation.

I inclose a form of joint resolution authorizing such investigation, and I would ask that you expedite its passage through the House.
Respectfully, yours,

O. S. HAMLIN,
Acting Secretary.

Hon. NELSON DINGLEY, Jr.,
House of Representatives.

JOINT RESOLUTION to authorize a scientific investigation of the fur-seal fisheries.

Be it resolved by the Senate and House of Representatives of the United States of America in Congress assembled, That the Secretary of the Treasury be, and is hereby, authorized to expend from any moneys in the Treasury not otherwise appropriated a sum sufficient to provide for the employment of persons to conduct a scientific investigation during the fiscal years 1896 and 1897 of the present condition of the fur-seal herds on the Pribilof, Commander, and Kooril islands, in the North Pacific Ocean and Bering Sea, said amount not to exceed for both said years the sum of $5,000.

The President is also authorized to detail for the purposes of assisting in this investigation any officer or officers or employees of the United States Government, their actual expenses and the expenses of the person or persons employed under the preceding paragraph to be paid by the Secretary of the Treasury out of any moneys in the Treasury not otherwise appropriated.

The President may detail a vessel of the United States for the purpose of carrying out this investigation.

DEPARTMENT OF STATE,
Washington, June 9, 1896.

SIR: Referring to the Department's letter to you of the 27th of May, 1895, transmitting to you a copy of a communication from the British Foreign Office, dispatch No. 93, of May 17, 1895, to Sir Julian Pauncefote, on the subject of the necessity of further provisions to preserve the fur-seal herd of the Northern Pacific Ocean and Bering Sea; and referring also to the Department's note on the 24th of June, 1895, to Lord Gough in regard to the same subject (see Foreign Relations, part 1, 1895, pp. 649–653), I have the honor to inclose for your information and consideration a copy of a further note of the 4th instant, from the British ambassador at this capital in regard to the matter in question.

I have the honor, etc.,

RICHARD OLNEY.

The SECRETARY OF THE TREASURY.

[Inclosure.]

WASHINGTON, *June 4, 1896.*

SIR: With reference to the question raised in your note to Her Majesty's chargé d'affaires, No. 133, of June 24, 1895, whether the computation made by the British Bering Sea commissioners of the seal catch of 1891 within the awarded area included the number of seals caught on the Asiatic side of Bering Sea, I have the honor to transmit you herewith, by direction of Her Majesty's secretary of state for foreign affairs, a report of the collector of customs at Victoria, giving full particulars of the catch for the year in question and showing the respective localities from which the yield was secured.

The total Asiatic catch was 6,595 seals. The deduction of this figure leaves the total for the award area at 43,361, including the catch of the United States schooner *City of San Diego,* landed at Victoria and taken on the American side, which amounted to 641 skins.

The figures given for 1894, however, include the Indian catch on the British Columbia Coast, viz, 3,989 skins. This figure was very properly added, since these skins were secured from animals belonging to the so-called American herd. For purposes of comparison it is necessary to include those taken by Indians in 1891, amounting to 1,953, and this would raise the figure for 1891 to 45,614, showing a balance in favor of that year, as against the total of 38,044 for 1894, of 7,570 skins.

The collector of customs also points out that the number of British vessels engaged during 1891 in sealing within the award area was 50, while, in 1894, 59 vessels were so employed.

According to the statement taken from the books of the United States custom-house, no less than 41 United States vessels were engaged in the seal fishery during 1891, making a total for that year of 91, instead of 115, as stated in your note of Viscount Gough, above referred to.

The total fleet mentioned in that note, for the year 1894, was 95 vessels. The deduction of 59 British vessels would leave the number of United States vessels at 36.

It is, moreover, apparent that the catch of 1891 would have been still larger but for the interference with and expulsion of sealing vessels from Bering Sea under the modus vivendi.

I have the honor, etc.,

JULIAN PAUNCEFOTE.

Hon. R. OLNEY.

Report by the collector of customs, Victoria, British Columbia, to the deputy minister of marine and fisheries, Ottawa.

CUSTOMS, CANADA, VICTORIA, BRITISH COLUMBIA,
February 20, 1896.

SIR: I have the honor to revert to your letters of the 19th of September and 10th of December last, inclosing for my information an extract from a communication from the United States Government touching the catch of seals as taken from the statistics supplied through me.

I beg leave to observe that the contention appears to be hinged on the relative catches of the years 1891 and 1894, and that the British statement in the main is that, notwithstanding the large take of 1894, that of 1891 was as large, so far as the so-termed American herd was concerned.

Before dealing with the subject, I beg to premise that I regret that I had to tabulate the figures of the seal catch for the year 1891 from the reports of the masters of the several vessels as declared by them, to which no objection could be taken, as at that time there was no information given in the vessel's log book as to the locality or date of each fur seal fishing operation, the catch then being divided into three periods, viz, the lower coast, the upper coast to Sand Point, and the Bering Sea, and this division was made to agree with the landing periods when the skins were taken or shipped home to his port. Thus, after the vessels had left their skins at Sand Point, Alaska, to be shipped to Victoria, they included all seals taken thereafter as being from Bering Sea and adjacent waters; the imaginary water line obviously arranged between the United States and Russia for other reasons than to give dominion over any part of Bering Sea, other than territorial waters, was then of little concern to our sealers, and thus, when they returned to their home port, the declaration on their inward and special sealing report designated their entire catch as being made in Bering Sea, without any regard to location, and the figures were so regarded in our statistical books.

The manifests and special reports of all British sealing vessels arriving back at this port during the season of 1891, when compiled, gave the total catch as follows:

Lower Coast catch 3,565
Sand Point, or Upper Coast catch 17,162
Bering Sea catch.. 28,489
Kurile Islands (Asiatic) catch 399

Total.. 49,615
Caught by Indians on British Coast.. 1,953

Total skins for 1891... 51,568

These figures as given in the tabulated statements then supplied your department are beyond conjecture, having been compiled with the greatest care, and the number of skins landed also having been verified at the time by actual count by the local customs officers.

The promulgation of the modus vivendi of 1891 was an important feature in the history of the sealing industry, when a number of our British vessels were warned out of the American side of Bering Sea, between the dates of 30th June and 16th August of that year.

It is also well known, in connection with the history of pelagic sealing, that 11 vessels of the British Columbia fleet, who were notified of the operation of the modus vivendi, left Bering Sea and adjacent waters, going over to the Russian side of Bering Sea, and sealed a short season in the vicinity of Copper Island, returning from them direct to this port.

It must be borne in mind that most of the 11 vessels which went to Copper Island had, on leaving the east side, a considerable number of skins, which they kept on board on their voyage to the Russian side, having no opportunity of landing them, so that when they did return to this port, the locality of catch then being deemed of little importance, the whole of the skins were entered on their reports as being from Bering Sea and were classified in our statistical books as such, and which is perfectly accurate, so far as the actual number of skins landed was concerned, and which agree with the tabulated returns. (Table A, page 205, of the report of the Bering Sea Commission, 21st June, 1892.)

Since the receipt of your letter of the 19th of September last, I have examined the log books and papers of the various sealers as to the exact number of skins taken by those 11 vessels in the vicinity of Copper Island, and after many interviews with masters and owners, much delay, and patient inquiry, I am now able to present to you an accurate report of the result of their sealing operations while in the neighborhood of the Commanderosky Islands in 1891, which was:

Name of schooner.	Number of skins.	Remarks.
C. H. Tupper	874	Copper Island.
Viva	731	Do.
Beatrice	300	Do.
Ocean Belle	1,170	Do.
Oscar and Hattie	1,062	Do.
Maud S	605	Do.
Katherine	624	Do.
Penelope	541	Do.
Teresa	387	Do.
Geneva	148	Do.
Umbrina	254	Do.
Total	6,196	
Borealis (catch near Kurile Islands)	399	
Total Asiatic catch	6,595	

The schooner *Borealis*, after her return to this port on the 31st August, was chartered by a man named Hansen and went from here to the Kurile Islands, and returned late in the fall with 399 skins. These figures are given in the above statement as belonging properly to the Asiatic side. Therefore the figures for the season of 1891 are as follows:

Lower coast catch [1]	3,565
Sand Point or Upper coast catch [1]	17,162
Bering Sea [1]	22,293
Copper Islands (Asiatic) [2]	6,196
Kurile Islands [2]	399
Total	49,651
Taken by Indians on British Columbia coast	1,953

It will be seen by the figures above given that the actual number of skins taken in 1891 within the area of the Paris award was 43,020, and 6,595 were taken on the Asiatic side, outside of the award area.

By the returns of 1894 you will observe that the total catch of seals within the area now covered by the Paris award was as follows: Eastern side Pacific Ocean, including Bering Sea, 38,044.

The foregoing figures relate solely to British vessels, but in the year 1891 the American schooner *City of San Diego* landed at this port 641 skins taken on the eastern side of Bering Sea, which should be added to the catch taken by British vessels, because they were landed here, making the total catch of seals landed at Victoria, taken in that year within the area of the Paris award, 43,361, and in comparing the year 1891 with 1894 the result is as follows: Total skins obtained within the area of the Paris award in 1891, 43,661; in 1894, 38,044, making 5,617 more skins in 1891 than in 1894.

It must be taken into consideration that the vessels in 1891 were very early in the

---

[1] Paris award area.   [2] Asiatic.

sealing season warned out, and had to leave Bering Sea, the larger number being compelled to relinquish their sealing operations in July, for had the vessels remained until August, with the good weather such as prevailed, the catch of 1891 would have been much larger.

The number of British vessels engaged in sealing within the area of the Paris award in the year 1891 was 50. The number of vessels engaged in sealing within the area of the Paris award in 1894 was 59.

The following statement shows the number of American vessels which cleared from American ports on sealing voyages (vide p. 206, United States No. 2, 1893, Report of the Bering Sea Commission, 21st June, 1892: San Francisco, 23[1]; Port Townsend, 9; Astoria, 2; San Diego, 2; other ports, 5.

Therefore, taking the American statement as taken from their custom-house books, exhibits the fact that 41 American vessels were engaged in sealing in the year 1891, and this number, added to 50 British vessels similarly employed (see returns, 1891), plainly shows that the entire fleet, British and American, consisted of 91 vessels in the year 1891, and it is incomprehensible how the United States authorities place the number at 115 for 1891.

The sealing returns from this port, which I think are beyond question, show that in the year 1894 there were engaged in sealing 59 British vessels, sailing from British Columbia ports (see sealing returns, 1894), and according to the American statement the entire sealing fleet, British and American, numbered 95 vessels for 1894; deducting therefrom the 59 British vessels would leave 36 United States vessels so employed.

You will find on examination of the sealing returns for the year 1894 that the crews and equipment of the vessels were considerably increased in comparison with the same in 1891, and you will likewise observe that in 1894 the greater number of seals were taken on the Japan coast.

It was estimated, with uncertainty, that the American catch in 1891 approximated 10,000, and this added to our British catch, 49,615, would make 67,615, or, in round numbers, 68,000 skins.

The returns for the year 1894 exhibit the fact that the following-mentioned number of skins were taken by British vessels on the Asiatic side (outside award area), viz:

Japan coast... 48,993
Copper Island.. 7,437

Total, 1894 (Asiatic)... 56,430
Within Paris award area, 1894................................... 38,044

Total, 1894.. 94,474

In following the argument advanced by the United States, on page 3 in the extract you send me, it is apparently admitted that our figures of the catch in 1894, within the Paris award area, is correct.

But it appears by United States Treasury Department tables, the details of which were mentioned in Mr. Gresham's note of 23d January, that there were taken 6,836 skins by American vessels, the locality of catch being undetermined.

I can not understand why it should be assumed by the United States Government that 6,152 skins, taken from those whose location of catch was undetermined, and added to those taken within the award area, when it is well known that the greater number of United States vessels went to Japan waters to engage in sealing in 1894.

It appears to be well known that there were few American vessels sealing in the North Pacific within the award area in 1894, as stated, the greater number having gone to Japan waters, for in the past a number of these usually visited this port for supplies, and to dispose of their skins, and I became aware of their movements; therefore it appears to me unjustifiable to assign 90 per cent of the undetermined catch in 1894 to the award area.

In regard to the statement made relative to the evidence taken before the Tribunal at Paris that the number of seals killed but not recovered was from two to five times as many as those secured. This is to me an extraordinary statement to introduce into the matter under consideration, and one which I can not concede in any way, for I am assured that as the seasons go by the seal hunter improves in skill and dexterity in pursuing the seal, and owners and masters are now so careful in selecting those competent to hunt that they will employ only those skillful as seamen and hunters.

There is no doubt that the lack of success of many American vessels is a good deal due to obtaining unskilled hunters and seamen, principally in San Francisco, while our sealers are very intelligent and competent men, mostly domiciled here, and to tell them that they lose from two to five times as many seals as they secure would amuse them.

Regretting that this has been so long delayed, I have, etc.

A. R. MILNE, Collector.

---

[1] The steamer Thistle, being a British vessel, not included in above.

Return of the number of skins taken by British vessels on the Asiatic side (outside award area) for the year 1894.

Name of schooners.	Tons.	British Columbia coast.	Japan coast.	Copper Island.	Bering Sea.	Total.
Victoria vessels.						
Agnes McDonald	107	....	1,707	471	....	2,178
Ainoko	75	467	....	....	1,657	2,124
Annie C. Moore	113	....	1,497	531	....	2,028
Annie E. Paint	82	....	1,197	....	91	1,288
Arietis	86	....	693	21	217	931
Aurora	41	358	....	....	1,160	1,518
Beatrice	66	303	....	....	1,149	1,452
Borealis	37	....	2,383	343	....	2,726
Brenda	100	....	1,947	....	....	1,947
Carlotta G. Cox	76	....	1,926	....	....	1,926
Casco	83	....	1,304	250	....	1,554
City of San Diego	46	....	1,961	433	....	2,394
Diana	150	....	2,584	....	....	2,584
Dora Siewerd	94	....	2,118	....	....	2,118
E. B. Marvin	96	....	1,254	314	....	1,568
Enterprise	69	606	....	....	1,240	1,846
Favourite	180	....	911	....	646	1,557
Fawn	159	92	....	....	....	92
Fisher Maid	21	....	96	81	....	177
Florence M. Smith	99	....	1,092	558	....	1,650
Geneva	92	315	....	....	767	1,082
Henrietta	31	79	....	....	867	946
Kate Katharine	58	269	....	....	1,059	1,328
Kilmeny	82	....	....	....	634	634
Kilmeny	19	308	....	....	560	868
Labrador	25	....	1,010	200	....	1,210
Libbie	93	....	1,909	86	457	2,452
Mary Ellen	63	....	874	250	....	1,124
Mary Taylor	43	....	558	....	545	1,103
Mascot	40	....	1,343	86	....	1,429
Maud S.	97	....	925	197	....	1,122
May Belle	58	....	1,603	505	....	2,108
Mermaid	73	488	....	....	1,665	2,153
Minnie	46	175	....	....	....	175
Mountain Chief	23	....	530	274	....	804
Ocean Belle	83	....	1,733	176	....	1,909
Oscar and Hattie	81	....	1,014	623	....	1,637
Otto	86	....	1,306	296	....	1,602
Penelope	70	418	....	1,263	....	1,681
Pionce	66	....	1,043	....	856	1,889
Rosie Olsen	39	....	1,783	171	....	1,954
Sadie Turpel	56	20	....	....	849	869
San José	31	535	....	....	2,105	2,640
Sapphire	109	170	....	....	668	838
Saucy Lass	38	34	....	....	377	411
Shelby	16	....	1,102	120	....	1,222
Teresa	63	1,320	....	....	3,240	4,560
Triumph	98	....	2,588	153	60	2,801
Umbrina	99	....	....	....	909	909
Venture	48	....	1,075	....	195	2,270
Vera	60	....	1,437	....	....	1,437
Viva	92	....	710	....	....	710
W. P. Hall	99	....	606	35	....	641
W. P. Sayward	60	....	1,471	....	672	2,143
Walter A. Earle	68	691	....	....	1,749	2,440
Walter L. Rich	76	400	....	....	....	400
Wanderer	25	3,989	....	....	....	3,989
Vancouver vessels.						
Beatrice	49	.!..........	1,703			1,703
C. D. Rand	51	357	....	....	....	357
Total	....	11,703	48,993	7,437	26,341	94,474

Catch of British Columbia schooners in the vicinity of Copper Island after they were warned out of Bering Sea, 1891, and included in tabulated statement with Bering Sea catch.

Name of schooner.	Number of skins.	Remarks.
C. H. Tupper	374	Copper Island.
Viva	731	Do.
Beatrice	300	Do.
Ocean Belle	1,170	Do.
Oscar and Hattie	1,062	Do.
Maud S	605	Do.
Katherine	624	Do.
Penelope	541	Do.
Teresa	387	Do.
Geneva	148	Do.
Umbrina	254	Do.
Total	6,196	
Borealis catch near Kurile Islands	399	
Total Asiatic catch	6,595	

JUNE 13, 1896.

SIR: I have the honor to acknowledge receipt of your note of June 9, inclosing a note dated June 4 from the British ambassador to yourself. Sir Julian's note is in reply to your letter of June 6, 1895. I have carefully considered the same and have the honor to give the following résumé of the correspondence leading up to said letter.

On January 23, 1895, the late Secretary Gresham in a communication to the British Government, stated that the slaughter of seals at sea in 1894, both American and Asiatic, was unprecedented in the history of pelagic sealing.

On May 17, 1895, the British Foreign Office by letter denied this statement, making the further assertion that in the season of 1891, 12,000 more seals were killed from the American herd than in 1894.

On June 24, 1895, you replied to the foreign office, calling to its attention a serious error in the returns cited by it to justify the above denial. This error consisted in the fact that in the figures cited by the British foreign office for 1891 in said letter (British vessels, 49,615; American vessels, 18,000; total, approximately, 68,000) there were included 8,432 seals killed from the Asiatic herd on the western side of Bering Sea by British and American vessels, warned from the eastern side by American cruisers under the modus vivendi. Of these seals 6,616 were estimated by you to have been killed by British vessels, and 1,816 by American vessels. You further pointed out that by deducting these Russian seals, there was left a total of 59,568, as the corrected pelagic catch for 1891; that, on the other hand, adding to the United States official figures of the catch for 1894 (55,686), the estimated number of skins taken from the American herd contained in the 6,836 skins landed at American ports and classed as "undetermined" in the American returns, there was left a total of 61,838 as the catch of 1894, fully sustaining the contention of your predecessor that the catch from the American herd of 1894 exceeding that of 1891 or any previous year.

The British ambassador in his letter of June 4, 1896, incloses a report from the collector of customs at Victoria. This report discloses the fact that the original official returns of the British Government for 1891, upon which the said Government based the above mentioned denial, were in error, as claimed by you, in that they included, as a part of the catch, from the American herd some 6,595 seals killed by 11 British vessels in Russian waters from the Asiatic herd.

Notwithstanding this admitted error, the original denial of the British Government is still maintained in said letter of June 4. Sir Julian calls attention therein to the facts that in the return of the pelagic catch for 1894, the Indian catch off the British Columbian coast (3,989) was included; he claims, therefore, that the Indian catch of 1891 should also be added to the returns of that year. This catch he states to be 1,953 skins. It should be noted, however, that although the report of the British Commissioners (Appendix F) contained in the proceedings of the Paris tribunal, does give 1,953 skins as purchased from the Indians at Victoria in 1891, in addition to those killed at sea by British vessels, yet the official report for 1891 of the Canadian department of marine and fisheries, page 171, states the total catch of Indians in canoes to be only 404. Assuming, however, the figures 1,953 to be correct, and adding these figures to the corrected British returns as shown by your letter of June 24, the total pelagic catch would be 61,521 for 1891, while that of 1894 was 61,838; thus even on the British contention the catch of 1894 was larger than in 1891.

Sir Julian in his letter deduces the conclusion from the report of the collector at Victoria that in 1891 7,570 more skins were taken by British vessels than in 1894. Your original statement, however, was not confined to British vessels, but to the total pelagic catch, both of British and American vessels.

The estimate of the number of American herd skins in the 6,836 skins entered at American ports as "undetermined," contained in your letter of June 24, is also disputed in said letter of Sir Julian. It is respectfully submitted, however, that said estimate is substantially correct. It was reached by dividing the said 6,836 skins in the same proportions between the American and Asiatic herds as the other skins landed at American ports where the location was definitely determined.

It would thus appear that your proposition that the slaughter of seals from the American herd in 1894 was greater than in 1891 is maintained.

As regards the number of vessels employed in the seal fisheries in 1891, the number stated, 115, was obtained from the appendix to the United States Government's case before the Paris tribunal, volume 1, page 591, and is believed to be as accurate a statement as can be made.

Respectfully, yours,

CHARLES S. HAMLIN,
Acting Secretary.

The SECRETARY OF STATE.

JUNE, 13, 1896.

SIR: Further instructing you as to the scientific investigation to be made by you of the present condition of the fur-seal herd on the Pribilof, Commander, and Kurile islands, I have the honor to state that Prof. D'Arcy W. Thompson and Mr. James M. Macoun have been designated by the British Government and Canadian government, respectively, to make an independent investigation relative to the same subject. Having found it impracticable to rely upon the ordinary means of reaching the fur-seal islands, they have been offered and have accepted transportation and accommodations on board the steamer *Albatross*, and will be granted the same facilities as yourself and party for conducting their independent investigations. As regards the investigation on behalf of the United States Government, you are charged with the arrangements of the details both of the field work and of the work to be performed by the gentlemen designated to assist you, reliance being placed upon your judgment to utilize to the best advantage the means supplied for accomplishing the objects of the expedition. You are authorized to direct the members of your party to act conjointly with you on all matters, or you may assign them severally to the study of separate subjects, or to different localities, as you consider most expedient. The advisability is suggested for your consideration of sending one of your party upon the *Albatross* to the Kurile and Robben islands. Should you need transportation during such absence of the *Albatross*, the commander of the Bering Sea patrol fleet, Captain Hooper, will be instructed to render you every facility.

Your final report will be expected to relate more specifically to the group of seals which resort to the Pribilof Islands, but the Asiatic herd may be investigated to such extent as seems advisable in order to afford the opportunity for instituting comparisons from which important deductions may be reached.

The principal object of this investigation is to determine by precise and detailed observations, first, the present condition of the American fur-seal herd; second, the nature and imminence of the causes, if any, which appear to threaten its extermination; third, what, if any, benefits have been secured to the herd through the operation of the act of Congress and act of Parliament based upon the award by the Paris Tribunal of Arbitration; fourth, what, if any, additional protective measures on land or at sea, or changes in the present system of regulations as to closed season, prohibited zone, prohibition of firearms, etc., are required to insure the preservation of the fur-seal herd.

Your inquiries should furthermore be extended, in so far as the time and circumstances permit, to embrace the consideration of all important questions relating to the natural history of the seals, both at sea and on the islands, with special reference to their bearing upon the sealing industry.

Your attention is specially directed to the following questions which should be treated in your report.

(1) The effect of pelagic sealing in the North Pacific Ocean and Bering Sea upon the fur-seal herd, due account being taken of the classes of seals killed.

(2) What effect, if any, has the annual removal of bachelor seals, which has taken place on the Pribilof Islands, had upon the fur-seal herd?

The solution of these two questions involves a study of the entire subject of the relations of the two sexes and the proportion of the male seals required to be preserved in order to maintain the stability of the herd.

(3) Whether killing on land or sea has interfered with the regular habits and occupation of the islands by the herd, or has operated to reduce the strength of the seal race as a whole by a natural selection?

(4) The propriety of existing methods of driving seals from the hauling grounds to the killing grounds, culling, and other practices connected therewith?

(5) The cause of the destruction of nursing pups upon the islands? During the seasons of 1894 and 1895 about 20,000 and 30,000 dead pups, respectively, were found upon the islands. You should specially consider the causes of their death, whether from starvation or other cause, preserving specimens whenever practicable.

(6) The extent, date, and causes or mortality on the islands of seals of all classes?

(7) The breeding habits of the seals, with special reference to the age at which the females begin and cease to breed, and the frequency of breeding, whether annually or at longer intervals?

(8) The condition of female seals taken at sea as to nursing and pregnancy?

(9) The distance which the several classes of seals go from the islands, and the direction which they take in search of food or rest at different times during the season?

(10) The actual decrease, if any, in the number of seals in each class on the Pribilof Islands which has occured during the past year, and also since the year 1890, and since the year 1870? A careful census of the rookeries should be taken this season for comparison with the enumeration made in 1895 and previous years.

(11) An examination of the question as to the character of the food of fur seals.

(12) Whether the Pribilof Islands herd of fur seals intermingle with the Asiatic herds of the Commander or Kurile islands.

(13) Whether nursery seals nurse other than their own pups on the islands?

These latter questions are merely suggestions to guide you in your examination and report.

I have the honor, etc.,

CHARLES S. HAMLIN,
Acting Secretary.

Dr. DAVID S. JORDAN,
Palo Alto, Cal.

JUNE 13, 1896.

SIR: I have the honor to inform you that Prof. David S. Jordan, president of the Leland Stanford Junior University, has been appointed to conduct a scientific investigation of the fur-seal herds in accordance with the joint resolution of Congress, approved June 8. There have been detailed to assist in this investigation the following gentlemen: Lieut. Commander Jefferson F. Moser; Leonhard Stejneger, esq.; F. A. Lucas, esq., both of the National Museum, and Charles H. Townsend, of the United States Fish Commission. The United States Fish Commission steamer Albatross has also been assigned for this purpose.

I have the honor to request that you send to Professor Jordan, care United States Fish Commission steamer Albatross, Seattle, Wash., a copy of the Foreign Relations of the United States for 1895, part 1; also copy of the proceedings, 14 volumes, before the Paris Tribunal of Arbitration, published under the auspices of the State Department.

I shall be able to advise you to-morrow when the Albatross will sail in order that you may communicate with the British Government as to the gentlemen who are to take passage thereon, in behalf of said Government.

Respectfully, yours,

C. S. HAMLIN, Acting Secretary.

The SECRETARY OF STATE.

DEPARTMENT OF STATE,
Washington, June 12, 1896.

SIR: Referring to your letter of the 27th of September last, relative to a report received by you from Capt. C. L. Hooper, of the Revenue-Cutter Service, commanding the Bering Sea fleet, concerning the circumstances attending the seizure of the British sealing schooner *Beatrice*, and with reference also to the subsequent correspondence on the subject as noted below, I have the honor to inclose for your information and consideration a copy of a note of the 9th instant, from the British ambassador at this capital in regard to the matter of prosecuting an appeal in the case.

You will observe that the ambassador states that Her Majesty's Government does not consider that it would be justified in proceeding with an appeal unless this Government is prepared to bear the cost of pursuing it and to satisfy any damages which the court of appeal may award.

Awaiting an expression of your views in regard to the subject, I have the honor to be, sir, your obedient servant,

RICHARD OLNEY.

The SECRETARY OF THE TREASURY.

[Inclosures.]

No. 208.] DEPARTMENT OF STATE, *Washington, October 1, 1895.*

EXCELLENCY: I have the honor to inform you that from a report dated the 21st ultimo, received at the Treasury Department from Capt. C. L. Hooper, R.C.S., commanding the Bering Sea fleet, it appears that on the morning of August 20 last, in latitude 54° 51' 03'' north, longitude 168° 31' 21'' west, the British sealing schooner *Beatrice*, of Vancouver, was boarded by two officers from the revenue steamer *Rush* and found to have 147 seal skins on board, while her official log recorded but 64, and that 4 of the skins showed evidence that the seals had been shot, and that he seized the *Beatrice*, her tackle, cargo, etc., for violations of the fifth article of the regulations of the Paris award, set forth in the British act of Parliament known as the Bering Sea award act, 1894.

In view of the report made by Captain Hooper as to the shooting of seals, the Treasury Department has instructed that officer to prepare and file an amended declaration with the commander of H. M. S. *Pheasant*, specifying the killing of seals with firearms by the crew of the *Beatrice* in Bering Sea in violation of the sixth article of the regulations referred to and of the Bering Sea award act.

I have the honor, etc.,

RICHARD OLNEY.

His Excellency Sir JULIAN PAUNCEFOTE, G. C. B., G. C. M. G.

No. 361.] DEPARTMENT OF STATE, *Washington, April 3, 1896.*

EXCELLENCY: Adverting to my note of October 1 last, I have the honor to request that Her Majesty's Government will direct that an appeal be taken to the proper court from the decision of the British Columbian court in the case of the British sealing schooner *Beatrice*, of Vancouver, seized by the United States revenue cutter *Rush*, on August 20, 1895, for violation of the regulations of the Paris award and the Bering Sea award act of 1894.

I have the honor, etc., RICHARD OLNEY.

His Excellency Sir JULIAN PAUNCEFOTE, G. C. B., G. C. M. G.

WASHINGTON, *April 7, 1896.*

SIR: I have the honor to acknowledge the receipt of your note, No. 361, of the 3d instant, requesting that Her Majesty's Government will direct that an appeal be taken to the proper court from the decision of the British Columbian court in the case of the British sealing schooner *Beatrice*, of Vancouver, seized by the United States revenue cutter *Rush* on August 20, 1895, for violation of the regulations of the Paris award and the Bering Sea award act of 1894.

I have not failed to bring this matter to the notice of Her Majesty's principal secretary of state for foreign affairs.

I have the honor, etc.,

JULIAN PAUNCEFOTE.

Hon. RICHARD OLNEY.

WASHINGTON, *June 9, 1896.*

SIR: With reference to my note of the 7th April last and to previous correspondence in regard to the case of the British sealing schooner *Beatrice*, I have the honor to inform you that I am in receipt of a dispatch from Her Majesty's secretary of state for foreign affairs, stating that he has considered, in communication with the secretary of state for the colonies, the request contained in your note to me, No. 361, of 3d April, that an appeal should be taken from the decision of the British Columbian court.

The Marquess of Salisbury observes that it will be seen, on referring to the text of the judgment, that the court distinctly stated that the delay in posting up the log was not unreasonable in the circumstances, and further implied that even if the proceedings had been taken against the master for a personal penalty under the merchant shipping act, a conviction would not have been obtained.

The legal point raised in the judgment is, however, a novel one, and it may be desirable to obtain a definite decision from a higher court as to whether the penalty for infringing the regulation requiring the entry in the official log book of particulars of every seal-fishing operation is determined by section 1 (2) of the Bering Sea award act, 1894, or by the provisions of the merchant shipping act as to the keeping of logs.

The intention of section 1 (3) of the Bering Sea award act would seem to have been to compel the keeping of logs by small seal-fishing vessels which are not required by the merchant shipping act to do so, rather than to define the penalty for breach of the award regulation, which prescribes special log entries; and it would seem to have been contemplated that the vessel should be liable for any breach of these regulations.

But the decision as regards the case of the *Beatrice* appears to Her Majesty's Government to have been substantially in accordance with justice, and if an appeal is to be taken in order to settle the above point it would not, in their opinion, be fair to throw upon the owners of the vessel the trouble and cost of defending the appeal.

I am instructed by the Marquess of Salisbury to state to you that for the reasons briefly indicated above, Her Majesty's Government do not consider that they would be justified in proceeding with an appeal unless the United States Government are prepared to bear the cost of pursuing it and to satisfy any damages which the court of appeal may award.

I have, etc.,

JULIAN PAUNCEFOTE.

Hon. RICHARD OLNEY.

DEPARTMENT OF STATE, *Washington, June 12, 1896.*

EXCELLENCY: With reference to my note to you of the 1st of October last, in relation to the seizure of the British sealing schooner *Beatrice*, and to the subsequent correspondence concerning the subject, I have the honor to acknowledge the receipt of your note of the 9th instant, in regard to the question as to an appeal of the case, and to inform you that the matter is receiving consideration.

I have the honor, etc.,

RICHARD OLNEY.

His Excellency Sir JULIAN PAUNCEFOTE, G. C. B., G. C. M. G.

JUNE 15, 1896.

SIR: I beg to acknowledge receipt of your letter of June 12, inclosing a copy of a note of the 9th instant from the British ambassador with relation to your request that the decision of the court in the case of the British sealing schooner *Beatrice* be appealed from. In said letter the British ambassador states that Her Majesty's Government does not consider that it will be justified in proceeding with an appeal unless this Government is prepared to bear the costs of the same and to satisfy any damages which the court of appeals may award.

Under these circumstances, I am of the opinion that our request for an appeal should not be pressed. My object in asking that this appeal be taken was because of the action of the court in referring the case to arbitrators to assess the damages to which the sealing schooner was entitled on account of the seizure thereof. I assume that the United States Government will in no event be liable for the amount found due by such a tribunal, it not having been a party to the case.

On this assumption I base my opinion that it would be unwise to press for an appeal, thereby becoming a party to the case.

Respectfully yours,

C. S. HAMLIN, *Acting Secretary.*

The SECRETARY OF STATE.

SALMON FISHERIES OF ALASKA.

REPORTS

OF

SPECIAL AGENTS PRACHT, LUTTRELL, AND MURRAY

FOR

THE YEARS 1892, 1893, 1894, 1895.

REPORT OF SPECIAL AGENT PRACHT.

WASHINGTON, D. C., *January 19, 1893.*

SIR: In accordance with the terms of my instructions, dated August 10, 1892, I have the honor to submit herewith a statement of my work for the partial season of 1892.[1]

A more extended report was made impracticable by the lateness of the action of Congress in making the necessary appropriation for the protection of the salmon fisheries of Alaska. The delay thus enforced upon the agent made it impossible to leave for the scene of my labors until the sailing of the September steamer for Alaska.

Immediately upon my arrival at Sitka, I caused to be published a "notice to packers of salmon within the district of Alaska," securing 300 extra copies of the issue of the newspaper containing it, and the same has been placed in the hands of every owner, agent, or manager having connection with the salmon fisheries of Alaska. The same is herewith attached, marked Appendix A.

By rapid traveling, made possible by my thorough knowledge of the numerous waterways, I succeeded in visiting all but two of the canneries operated in southeast Alaska during the past season, having prearranged meetings with the managers of those that I was not able to reach before the cessation of active operations.

BARRICADES AND OBSTRUCTIONS.

Within the letter and the spirit of the law, barricades or other obstructions, such as are described in the act of Congress approved March 2, 1889, copy of which is hereto attached (see Department Circular No. 131, Appendix A), were reported to me to have existed in a number of streams, and evidences of such having been removed previous to my visit were found by me at a number of points. Without exception, all the responsible managers cited to me that if the law was impartially enforced, the corporations having the larger interests would hail the result with satisfaction, and the assurance from the agent that all would be brought within the strict pale of the law led to general acquiescence.

In the larger streams, such as the Stikine, Unuk, Taku, Chilkoot, and Chilkat, effective barricades are a practical impossibility. Where tried, the forces of nature, such as drift and freshets, have carried them out. In several of these streams, traps connected with the shore have been used, but, as such have not extended into or beyond the channel, no obstruction can be said to have been maintained. In arriving at this conclusion, I have the advice of the United States attorney, Hon. Charles S. Johnson, Sitka, to whom was referred the case of the alleged obstruction of the Chilkat River.

In the smaller streams, however, the partial or complete obstruction, by means of fences, dams, fish wheels, or traps, has been more success-

---

[1] This report has been printed in Senate Document No. 31, Fifty-second Congress, second session.

fully maintained, and, while all such operated by white men had been removed previous to my expected arrival, the evidences to me were conclusive as to their previous existence. In most instances, the fact being known that an officer to enforce the law was appointed was sufficient to cause a hasty removal of all such obstructions as were unlawful, and I am led to believe that for the latter part of the season the salmon had unrestricted passage to the lakes which constitute their breeding grounds and which find their outlet to the ocean through the smaller, clear-water streams.

The most successful obstructionists of these small salmon rivers are the natives themselves. Those having proprietary rights to a salmon "chuck" find no difficulty in disposing of all the salmon they may deliver at the nearest canning establishment, and with them the first move is to barricade the stream a short distance above its mouth or just above the confluence of the tides with the current, so that seining for the mass of fish struggling to ascend the river is a matter of little labor and productive, to the native fishermen, of desirable results. To reach these violators of the law is practically impossible. The natives are mostly impecunious, and the collection of a fine is impracticable to a degree.

In this connection, and also to enable the officer having charge of this work to reach the more numerous and irresponsible violators of a law which they do not entirely comprehend, it is, in my opinion, necessary to revise and amend the law, so that an alternative punishment, by means of imprisonment, can be inflicted by the court. I am led to believe, as the result of formal interviews with several representatives of more pretentious establishments than those controlled by the natives, that the payment of the fine of $250 imposed by the present law would not be considered an insuperable hardship, and that in the height of the "run" it would be to their advantage to pay the fine, "if convicted," rather than lose the fish.

As to the possibilities of conviction under the present jury system of Alaska, I will leave the prosecuting attorney to speak for himself. My own previous experience has led me into the belief that in cases where the United States is the plaintiff the average Alaska jury is for the defense.

FISH HATCHERIES.

Several of the more experienced fishermen have attempted the propagation of salmon in the streams entirely controlled by them, and others are said to be anxious to provide for their future wants by a system of spawn hatching if the Government will sanction their means and methods. Application to the special agent in charge for permission could only result in a reference to the law, which, if interpreted literally and enforced impartially, would prevent the maintenance of the necessary dams in the hatching streams. A typical hatchery of this class has been in operation at the works of Calbreath & Co., at Point Ellis, on Kuiu Island, Chatham Straits, which can best be described as follows:

A dam has been constructed at a point just above extreme high tide, with a second dam a short distance above it, with access thereto by a suitable passageway, so that a person standing upon the lower dam, armed with a scoop net, can dip up the desired salmon from below and readily transfer them into the stream above the upper or second dam, beyond which there are no further obstructions, and the fish are left undisturbed to finish their journey to their breeding waters, never far removed.

It is estimated that out of 500 female salmon, to which must be added the requisite number of milters, there will be fry enough to furnish all the adult salmon required for such a cannery as the one operated by them (since burned), estimated at 15,000 cases of 48 tins each, holding one pound each, and yet make provision for the loss of young and adult fish from natural causes before it is time for them to return to propagate their kind in turn. There is much force in the assertion on the part of experienced fishermen that the unrestricted passage upstream of the hordes of ravenous trout, which always follow the run of salmon, is productive of more damage to the issue of the breeding salmon than all other causes combined, and that by a system of hatcheries, such as devised and operated at Point Ellis, the trout will not be able to ascend the streams, and a much larger percentage of the ova will hatch out.

THE SALMON AND HIS ENEMIES

Beginning with the trout, which follows the breeding salmon into the mountain streams and lakes in which he delights, the course of life for the salmon is uncertain and erratic. His enemies are numerous, and each in its way more than his match. If the male salmon succeeds in fighting off the trout and protects his mate while she deposits the ova in some apparently secure crevice in the rocks or in a hole scooped out of the bottom gravel with his battered nose, and again covered from sight, it is not yet safe from the marauding instinct of the sea gull and the pernicious search of the "saw-bill" duck. If escaping both of these, and in the early days of spring, the bunches of young fry, playing upon the surface of the water while drifting out to sea, escape the frequent dives of the kingfisher, they are in danger of being gulped by the schools of herring which come up into some of the estuaries to meet them; or, if by maneuvering along the shore in the reeds and grasses, a portion manages to escape these heretofore unheard-of enemies, more of them are destined to help make a dainty meal for the sea bass, whose upward rush scatters the terrified little shiners, whose number is reduced at each successive running of the gauntlet.

Once out to sea, lurking in the protecting fastnesses afforded by the rocks, the young salmon is not yet out of danger, furnishing food for the "big fish," not excepting the members of his own immediate family. Having escaped the teeth of his own kind and grown to a size affording protection as against them, we might follow him to the feeding grounds or banks, where shark and dogfish feast upon him and the members of the seal family are in unremitting pursuit. These enemies of the salmon follow the schools, when at the age of 4 years they are impelled by instinct and the promptings of nature to seek a fresh-water stream for purposes of reproduction, and when caught in the gill net of the fishermen, unable to flee, gorge themselves upon his delicate flesh. In addition to these, the birds of the air and beasts of the forests lay in wait for him as he appears in the shallows of the streams; the eagle, raven, crow, and hawk swoop down upon him from above; and the otter steals upon him from his hole in the rocks; the bear wades out among them and with a flip of his forefoot throws them out on the shore, there to be devoured at leisure. It would certainly seem that when this valuable fish has to contend with so many natural enemies the superior skill of man should be held in restraint and wise legislation for his protection be enforced.

THE SALMON PACK OF 1892.

The entire pack of salmon for the District of Alaska for the season just closed is, as is shown by the tables in Appendix B, 457,969 cases,

15,252 barrels, and 4,245 half-barrels of salted whole salmon; and of salted bellies, 35 barrels and 36 half-barrels. In ordinary parlance, it requires 3 barrels of salmon to produce 1 barrel of bellies.

As it will be seen, the total value of the aggregate pack at the market prices ruling in San Francisco, the principal port of distribution, is $2,064,340.05.

Based upon an average of cost of $2.73 per case and $6 per barrel (200 pounds to the barrel) and $11 for bellies, the profits of the industry, while not excessive, are for the season just past fair and satisfactory, and several establishments heretofore conducted at a loss are presumptively able to realize dividends this season. Among other causes leading to this result are to be mentioned the reduction in operating expenses, brought about by the more economical management incident to cooperation through a board of trustees known as the Alaska Packers' Association, and a very considerable falling off in the pack of the British Columbia canneries, which has resulted in a better market and larger demand abroad. Of these markets England and Australia are the principal ones.

TIN PLATE.

Tin plate, which enters so largely into the cost account of tinned salmon, was sold and delivered cheaper to the consumer for the season of 1892 than for the two seasons previous, and contracts for deliveries for the season of 1893 have been made upon a still lower basis, as will be seen by a reference.

Lowest reported price for season (duty paid, delivered at San Francisco and
Astoria):

1891	$6.80
1892	5.82½
1893	5.72

ALIEN LABOR.

A large percentage of the labor employed in the principal establishments is noncitizen. British Columbia and Europe furnish some, but the larger proportion are Chinese. The latter are employed principally in work requiring great manual dexterity, such as making cans, filling cans, labeling, and packing. In some few cases Chinese contractors employ native or Indian labor, and in a few minor instances natives do all the work usually done by the Chinese, but on the whole the system of contracting with a responsible Chinese firm for a certain number of "hands" or to put up a pack of a specified minimum number of cases for the season meets with the most approval. The introduction of improved machinery, which has taken the place of much hand work, such as can soldering and can filling, has brought the business within such limits as to have a restrictive operation upon the tendency to "strike." The native fisherman has not been slow to avail himself of the strike method as taught him by the more irresponsible European laborers, but the cooperative management has apparently had a depressing effect, and during the season just passed no strikes were reported.

SALMON STREAMS HELD BY ALIENS.

During the past season some difficulties arose among the native fishermen and a party of fishermen from British Columbia headed by an educated half-breed from Victoria. Complaint being made to me, I referred the matter to the United States attorney for his action and his decision has had the tendency to discourage any further irruption of like character. Upon this matter I also corresponded with the Treasury Department, and in Appendix C will be found a copy of the correspondence.

FISHING INDUSTRIES OTHER THAN SALMON.

While salmon heads the list as among the productive food fishes of Alaska, it is by no means the only profitable industry. A further reference to the tables, Appendix D, shows the catch of cod as reported by the two concerns engaged in this industry. Explorations by the United States Fish Commission steamer *Albatross* have definitely located a number of "banks" greater in area than those of Newfoundland, upon which feed innumerable codfish of good size and superior flavor, and it needs but a market within reach of the distributing point of San Francisco to insure a permanent and ample supply.

As compared with the fisheries of the Atlantic banks, those of the Gulf of Alaska and Bering Sea are to all intents and purposes to be preferred. The dangers to the fishermen are but few. The loss of a dory or a man is so infrequent, and the climatic conditions so favorable, as to reduce the risk to a minimum. Much of the fishing is conducted from shore stations, located in snug harbors. The men are comfortably provided for and well paid. A reduction in freights such as would ensue from the completion of an interoceanic canal would, in my estimation, place at the disposal of the millions of the people of the United States east of the Rocky Mountains the delicious quality and inestimable quantity of the Alaskan codfish.

OTHER FISHING INDUSTRIES.

Many of the inlets and lagoons to which access is had from the straits, sounds, and gulf of the Alexander Archipelago are, in their season, full of herring, smelt, and capelin, and they all have their share in the furnishing of the larders of the Alaskan housekeeper. There is but one concern engaged in the business of preparing marketable product from such sources—the Alaska Oil and Guano Company, located at Killisnoo, Admiralty Island—and the productions of this concern, such as herring oil and fish guano, find a ready sale in the United States, Hawaiian Islands, and quite recently shipments by means of sailing vessels have been made direct to England. As a fertilizer the guano is said to be superior to all others in the production of sugar cane, while the oil can be used by special preparation for all the ordinary purposes to which linseed oil is put in the preparation of paints. The output of this concern for the season is given in Appendix E.

HALIBUT FISHERIES.

Nearly all the inshore banks and sheltered bays contain halibut in large quantities, and sporadic efforts have been made to find a market for them. Canning has not met with success, but such would come into demand were the salmon output much reduced below the present limit. The fish when so prepared is delicate and toothsome. An occasional schooner has secured a load of fresh halibut, packing same in ice obtained from the near-by glaciers, meeting with more or less success by shipping same in refrigerator cars from ports on Puget-Sound direct to New York and Boston.

Sun-dried or smoked halibut is a staple article of food for the native Alaskan during the winter months, and salted napes and fins are esteemed a delicacy by the Caucasian epicure. Some day in the future the halibut of Alaska may supply the place of the Greenland article now sold in the Atlantic States. The difference in the rate of freights, when same are more nicely adjusted to competing circumstances, may safely be met by the fact that halibut fishing in Alaskan waters may be pursued with safety and comparative comfort during the entire season. Some recently discovered grounds in and about Cordova Bay and Dixons Entrance, along the southern extremity of Prince of Wales Island, have excited attention, and at this time of writing a steam schooner (*Francis Cutting*) is taking a fare, and the visit may result in the establishment of a station at or near Cape Muzon.

EULACHON, OR CANDLE-FISH.

This peculiar fish, a member of the smelt family, has achieved a world-wide reputation as the candle-fish of the Northwest Indians, and derived its peculiar cognomen from the fact that when sun dried or smoked it is so rich and oily that the application of flame to one end will cause it to burn as would a piece of pitch pine or "lightwood."

A few years ago a considerable quantity was smoked and shipped by a firm then located at old Fort Tongass, but the difficulty in obtaining same in the waters of Naas River, the same being in British Columbia, made the venture unprofitable. A small quantity is put up in salt pickle upon orders, and the natives of the southern end of the Alaskan panhandle secure by purchase and barter from the Tsimpsian Indians of British Columbia a considerable quantity of eulachon "grease," which takes the place of lard in their domestic economy. At infrequent periods this fish has been known to ascend the Stikine, Unuk, and Chilkat rivers, and may frequently be met with among the natives at Fort Wrangel, Juneau, and vicinity. Its peculiarity in selecting only glacial rivers in its spawning migrations and the fact that it can be caught only by the insertion of small-mesh gill nets through holes in the ice during the month of February tend to make it exclusive and expensive. As the eulachon "grease" is extracted by a system of putrefaction, its presence in a native house is indicated to the European nostril while he be yet afar off.

In addition to the foregoing, not less than 10,000 gallons of dogfish oil, so called, produced from the liver of the dogfish and shark, was extracted by the natives and sent to market through the medium of the trading store. This oil, because of its heavy body and freedom from grit, is a most desirable lubricant, and finds among the logging camps of the Puget Sound region a profitable market as "skid grease." With improved facilities, such as may be assumed will be at the service of the special agent during the season of 1893, an exact report of this industry may be expected, and for the season of 1892 the sum of $3,000 can safely be added to the amount heretofore reported, making a grand total of $2,257,939.55 received from the various fishing industries of Alaska, as follows:

Salmon	$2,064,340.05
Codfish	104,062.00
Herring oil and guano, salted herring	86,537.50
Dogfish oil	3,000.00
Total	2,257,939.55

Respectfully submitted.

MAX PRACHT,
Special Agent for the Protection of Alaskan Salmon Fisheries.
The SECRETARY OF THE TREASURY.

APPENDIX A.

NOTICE TO PACKERS OF SALMON WITHIN THE DISTRICT OF ALASKA.

In conformity with instructions from the Treasury Department, I desire to refer all firms or persons engaged in the taking of salmon within the limits of the District of Alaska to the following:

[Circular.—1892. Department No. 131, division of special agents.]

PROTECTION OF THE SALMON FISHERIES OF ALASKA.

TREASURY DEPARTMENT, OFFICE OF THE SECRETARY,
Washington, D. C., August 10, 1892.

To the officers of the customs in the Territory of Alaska and all other persons concerned:

The attention of the collector of customs and all officers of the United States in the Territory of Alaska, as well as all other persons concerned, is called to the terms of an act of Congress approved March 2, 1892, wherein it is provided: "That the erection of dams, barricades, and other obstructions in any of the rivers of Alaska, with the purpose or result of preventing or impeding the ascent of salmon or other anadromous species to their spawning grounds, is hereby declared to be unlawful, and the Secretary of the Treasury is hereby authorized and directed to establish such regulations and surveillance as may be necessary to insure that this prohibition is strictly enforced and to otherwise protect the salmon fisheries of Alaska; and every person who shall be found guilty of a violation of the provisions of this section shall be fined not less than $250 for each day of the continuance of such obstruction;" and also to an act approved March 3, 1891, entitled "An act to repeal timber-culture laws and for other purposes," which provides that the United States reserves the right to regulate the taking of salmon and to do all other things necessary to protect and prevent the destruction of salmon in all the waters of the lands granted under said act and frequented by salmon.

Officers of the customs service and officers appointed to enforce the provisions of the law referred to are hereby directed to report all cases of infraction of said laws to the United States attorney for the District of Alaska, with a view to the prosecution of offenders.

A. B. NETTLETON,
Acting Secretary.

And also to the following extracts from the letter of instructions accompanying my appointment:

TREASURY DEPARTMENT, OFFICE OF THE SECRETARY,
Washington, D. C., August 10, 1892.

SIR: Having been appointed as a special agent for the preservation of the salmon fisheries in Alaska, you are informed that it will be your duty to ascertain and report the location of every salmon cannery or saltery in Alaska; the capacity of the same in cases, barrels, half barrels, and kits; the pack in full for each season; the number of boxes of tin consumed and the cost of same; the number of employees in each cannery or saltery, and the total thereof, segregating whites, natives, Chinese, etc., male and female, adults and minors, and whether citizens or aliens. You should also include in said report the codfish, herring, herring oil and guano, and other such industries.

I inclose herewith for your information a copy of a circular, this date, relating to the provisions of the first section of the act approved March 2, 1889, entitled "An act to provide for the protection of the salmon fisheries of Alaska." It will be your duty to enforce the provisions of said act, and to warn all persons who have erected dams, or barricades, or other obstructions to remove the same forthwith, and in default thereof you will report all the facts with the proper proofs to the United States attorney for prosecution.

At the close of the season you will submit a full report of your labors and the result of your observations under these instructions, with such recommendations as you may deem advisable. Your official station will be Sitka.

Respectfully yours,

A. B. NETTLETON,
Acting Secretary.

. Mr. MAX PRACHT, *Washington, D. C.*

SITKA, ALASKA, *September 20, 1892.*

A copy of the above "circular" having been properly addressed to the person in charge of every cannery and saltery within the limits of the district, and the same consigned to the custody of the United States mails to be forwarded, all such are hereby informed that such is considered sufficient for purposes of "warning," and that proceedings in prosecution will be instituted against all persons found to be violating the law.

MAX PRACHT,
Special Agent in Charge.

APPENDIX B.

Alaska salmon pack—season of 1892.

[Collated by Max Pracht, special agent.]

Name of company	Location of works	Total pack — Cases of 48 pounds each, or 4 dozen cans	Barrels of 200 pounds each: Whole fish	Barrels: Bellies	Half barrels of 100 pounds each: Whole fish	Half barrels: Bellies	White: Male adults	White: Days employed	White: Compensation per month	White: With board	White: Without board	White: Transportation to and from work	Natives: Male adults	Natives: Days employed	Natives: Rate per day of 10 hours	Natives: Female adults	Natives: Days employed	Natives: Rate per day of 10 hours	Natives: Children	Natives: Days employed	Natives: Rate per day of 10 hours	Chinese: Compensation per case	Chinese: Number employed	Boxes of tin plate consumed: Number	Boxes: Cost at San Francisco
Alaska Packing Co.	Nushagak River, Bristol Bay	31,500					60	135	$65	Yes		Yes	45	90	$1.50	5	85	$1	5		.1	$0.40	95	2,387	$5.90
Alaska Salmon Packing and Fur Co.	Loring, Naha Bay	22,000					20	149	65	Yes		Yes	10	100	1.50							.40	48	2,300	5.90
Alaska Coast Packing Co.*	Kadiak Island																								
Alaska Oil and Guano Co.a	Killisnoo, Admiralty Island				625	4																			
Arctic Packing Co.	Alitak and Karluk (2 stations)	21,000	400				33	141	65	Yes		Yes	15	90	1.50							.40	60	2,250	5.90
Do.	Nakuek		1,650				12	135	45	Yes		Yes													
Do.	Cook Inlet																								
Aleutian Islands Fishing and Manufacturing Co.b	Kadiak Island																								
Bering Sea Packing Co.	Bristol Bay	31,500					90	135	65	Yes		Yes	40	90	1.60	10	60		1		.1	.40	90	3,395	5.90
Bristol Bay Packing Co.	Redfish Bay, Baranoff Island	10,400					7	90	75	Yes		Yes	19	90	1.50	6	60		1			.45	21	1,100	6.15
Baranoff Packing Co.	Glacier Bay																								
Bartlett Bay Packing Co.*	Yes Bay, Behm Channel																								
Boston Fishing and Trading Co.	Burro...a Bay	13,744	15				17	120	45	Yes		Yes	10	120	1.50	7	90					.44	31	1,845	6.00
Cape Lees Packing Co.*	Chilka...ver Inlet																								
Chilkat Packing Co.*	do	31,500					90	150	35	Yes		Yes			1.50							.41¼	60	3,100	5.90
Chilkat Canning Co.	Chignik Bay	22,500					50	120	35	Yes		Yes	20	100	1.50							.44	47	2,430	6.10
Chignik Bay Packing Co.	do	50,000					60	153	65	Yes	Yes	Yes										.40	140	5,358	5.95
Chignik Bay Co.*	Cook Inlet																								
Central Alaska Co.*																									

Glacier Packing Co.*	Stickine River															
Hume Packing Co.	Karluk, Kadiak Island	76,000		75	201	65	Yes.	Yes.31	150	1.50				.40	175	8,135	5.90
Hume G. W. (Flag)	Cook Inlet	20,500		40	185	65	Yes.	Yes.20	100	1.50				.40	70	2,088	5.90
Karluck Packing Co. g	Karluk, Kadiak Island		200	15												
Kadiak Packing Co.*	Kadiak, Kadiak Island			(1	365	75	Yes.	Yes.							1,500	6.00
Metlakahtla Industrial Co.h	Metlakahtla, Annette Island.	11,125	15	(4	95	50	No.	No.								
Moira Packing Co	Moira Sound, Cape Fox, Tongass Narrows, southeast Alaska.		1,120	10 15	90	75	Yes.	Yes.13	90	1.50	10	60	,1	5.60	.50	
Mexico Bay Co	George Inlet, southeast Alaska.		230						8	45	3	45	1			
Millar & Sons i	Hunters Bay, Prince of Wales Island.		1,720	6	4	365			5	95	2	90	1			
Do	Nichols Bay, Prince of Wales Island.		350						5	120						
Nakat Fishery (Tuxekan) i.	Nakat Inlet, Tongass		92	1	365	50	Yes.	Yes. 3		1.75	3					
North Pacific Trading and Packing Co. j.	Klawack	10,200		9	95										1,000	5.86
Northern Packing Co.*	Kenai															
Nushagak Canning Co.*	Nushagak River															
Pacific Packing Co.*	Prince William Sound															
Pacific Steam Whaling Co.*	Copper River															
Peninsular Trading and Fur Co.*	...do															
Prince of Wales Packing Co.	Lake Bay, Tolstoi Bay, and Karta Bay, Prince of Wales Island.	250	3,500	6	150	50	Yes.	Yes. 30	150	1.50	20	90	1	8.60	.75	
Pyramid Harbor Packing Co.	Pyramid Harbor, Chilkat Inlet.	30,000		50	149	65	Yes.	Yes.50	100	1.50				.40	65	3,215	5.90
Point Barry Packing Co.*	Point Barry		8	1	10			2	10	1.50						
Redfish Bay Co	Redfish Bay, Baranoff Island.	340	12	9	105	50	Yes.	Yes. 6	105	1.50	2	105	1			
Royal Packing Co.*	Afognak															
Russian American Packing Co.*	...do															
Shumagin Packing Co.*	Chignik Bay	1,500		10	90		Yes.	Yes.	90	1.50						
Thin Point Packing Co.	Thin Point	150		1	365	65	Yes.	Yes. 2								
Wrangel Fishery	Fort Wrangel	8,340														
All other fisheries k	At Bering Sea, Cook Inlet, Kadiak Island, and southeast Alaska not reached personally, but through agents at San Francisco and Portland.															

* Not operated in 1892.
a See special report on "Herring fishery."
b No report obtainable. Total pack included under head of "All other fisheries."
c Cost at the works.

d And 1¼ cents per case.
e Cost at Astoria.
f And 2 cents per fish.
g Operated conjointly with the Hume Packing Co.
h Cooperative work and by natives exclusively.

i Cooperative.
j Piecework and by natives exclusively.
k Detailed information not obtainable.

*Alaska salmon pack—season of 1892—*Continued.

RECAPITULATION.

Cases packed, 457,960, at $4.20 .... $1, 923, 400. 80
Barrels salted, 15,252, at $8.. 122, 016. 00
Half barrels salted, 4,245, at $4.25.. 18, 041. 25
Barrels of bellies salted, 35, at $15... 525. 00
Half barrels of bellies salted, 36, at $8... 288. 00

Total... 2, 064, 340. 05

Tin plate consumed (49,239 boxes, 108 pounds each)...........................pounds.. 5, 317. 812
Value of tin plate, duty paid, $291,660.60; duty at 2.2 cents per pound 116, 991. 86
Average cost per box, duty paid, for the season of 1892 .... 5. 92. 33
Average cost of canned salmon, per case of 4 dozen, in 1-pound tins, delivered at San Francisco.. 2. 73

NOTE.- Size of sheet of tin plate, 14 by 20 inches; 112 sheets (108 pounds) to a box. A box of tin makes 448 cans. The cost, as reported, is duty paid at San Francisco, Astoria, or at the works. The rate of duty is 2.2 cents per pound, or $2.376 per box. The amount of drawback allowed upon exportation equals about 25 cents per case, or $2 per box of tin.

APPENDIX C.

TREASURY DEPARTMENT, OFFICE OF THE SECRETARY,
Washington, D. C., January 16, 1893.

SIR: Referring to your report of the 2d ultimo, in relation to disputed claims upon Alaska fishery locations and particularly to your inquiry whether or not aliens may claim and hold salmon fisheries or control streams that carry salmon in Alaska, I inclose herewith for your information copy of an opinion, dated the 28th ultimo, of the Acting Solicitor of the Treasury, to whom the subject was referred.

Respectfully, yours,

O. L. SPALDING, *Acting Secretary.*

Mr. MAX PRACHT,
Special Agent, Sitka, Alaska.

DEPARTMENT OF JUSTICE,
OFFICE OF THE SOLICITOR OF TREASURY,
Washington, D. C., November 28, 1892.

SIR: Inquiry is made by Special Agent Max Pracht "whether aliens can claim and hold salmon fisheries, or control streams that carry salmon, in Alaska?"

In reply to your reference of said inquiry, I have to advise you that aliens have no such right. Besides, Congress has reserved to the United States the exclusive right to regulate the taking of salmon, and to prevent the destruction of salmon, in Alaska. See section 14, act of March 3, 1891 (26 Stat., p. 1095), and act of March 2, 1889 (25 Stat., p. 1005).

The letter referred to is herewith returned.

Very respectfully,

F. A. REEVE, *Acting Solicitor.*

The SECRETARY OF THE TREASURY.

APPENDIX D.

Codfish industry of Alaska.

[Collated by Max Pracht, special agent.]

Name of company and stations and trading posts.	Name of vessels employed.	Fish caught.	Weight.			Value per ton.	Total value.	Other products (tongues and sounds).	Value per barrel.	Cod-liver oil.*
			Green.	Dry.						
The McCollam Fishing and Trading Co.:			*Tons.*	*Tons.*				*Bbls.*		
Shumagin Islands, five stations; principal store, Pirate Cove.	Schooner Czarina, three trips during season.	210,000 240,000 83,000	344 365 150	310 330 135	$50 50 50	$94,350	53 60 20	$20 .... 20 .... 20	
Okhotsk Sea, station at Petropavlovski.	Schooner Hera	125,000	256	225	50	(†)	58	20	
Lynde and Hough Co.: Sand Point, Popoff Island.. Unga, Junga Nelson Island, Sanak Group	Schooner Venture.. BarkentineFremont Barkentine J. A. Falkenburg.								
Company Harbor, Sanak Island.	Schooner John Hancock.	655,000	985	887	50	(†)	‡166	‡12	
Ikatuk Station, Ikatuk Peninsula.	Schooner Arago ...								
New Station, Henderson Island.									
Squaw Harbor, Red Cove; salmon stations.								135	20	

* Report not ready. † Included in the above figures. ‡ Pickled fish.

NOTE.—Pack of 500 barrels included in the salmon statistics.

Number of men employed and compensation.

Total fishermen 142
Lay per 1,000 fish (including transportation and subsistence):
  Shumagin Group $27.50
  Other stations 25.00
Dress gang, per month 25.00
Splitters, per month 60.00
Salters, per month 50.00

RECAPITULATION.

Total value of dried codfish $94,350
Total value of pickled codfish 1,992
Total value of tongues and sounds 6,520
Total value of oil 1,200

Total 104,062

APPENDIX E.

Alaska herring fisheries.

[Statistics collated by Max Pracht, special agent.]

Corporation ..... Alaska Oil and Guano Co.
Location ..... Killisnoo, Admiralty Island.
Herring caught ..... 89,220 barrels.
Product:
  242,050 gallons oil, at 25 cents $60,512.50
  810 tons guano, at $27.50 ..... 22,275.00
  1,000 half barrels salted herring 3,750.00

  Total 86,537.50
Employees:
  White (including mechanics and crews of steamers) 49
  Natives (including fishermen and refinery operatives) 45
  Chinese (mess-house cooks) 5
Duration of season, five months (August to November, inclusive).

REPORT OF SPECIAL AGENT LUTTRELL.

CONDENSATION AND REARRANGEMENT OF DATA EMBODIED IN ANNUAL REPORT OF PAUL S. LUTTRELL, SPECIAL AGENT FOR THE SALMON FISHERIES IN ALASKA, YEAR 1893.

Mr. J. K. Luttrell, the former special agent, had made an extended tour through Alaska, visiting the various salmon canneries, and had laid the foundation of a complete and exhaustive report upon their condition, etc. Owing to his death, however, before the latter object had been accomplished, his son, Paul S. Luttrell, was delegated to collect data from the papers of the late J. K. Luttrell, and to construct therefrom as complete a report upon the condition of the salmon fisheries as was possible. Such report has been submitted, but was somewhat prolix, and in form but a verbatim copy of what letters he could find bearing upon the salmon packing industry, and written by various persons in Alaska upon the solicitation of J. K. Luttrell. It is the purpose, therefore, to present here, in a compact form, the substance of the report.

SALMON HATCHERIES.

It is the opinion of the packers on the Karluk River—the greatest breeding grounds of the salmon—that the supply of red salmon is rapidly decreasing, owing to the increased catch, and that some measures should be taken to artificially propagate this species, in addition to prohibitory measures. These Karluk River canneries erected a hatchery and achieved considerable success in the propagation of the salmon spawn, several millions of the young salmon having been hatched out. That hatchery, however, is now closed.

Mr. J. C. Callbreath has also erected a hatchery on Ethalene Island, and as the method employed at his hatchery is somewhat different from that usually pursued, a brief statement of his practice is here made: His hatchery is located on a small creek, practically useless, because few fish visit it. This creek is the outlet of a small lake. The creek is dammed completely across by two dams, one above the other, the first 5 feet high, just above tide water, while the other is 15 feet high, and 150 yards farther upstream. The hatchery is located between these two dams. No fish can pass either of these dams unaided. When the salmon try to ascend this stream, they are picked up with a dip net and passed over the first dam, leaving behind the trout, bullhead, and other fish that prey upon the salmon spawn. Owing to the fact, however, that the dams were erected so near salt water, many of the salmon, when they reached the barrier, did not attempt to pass, but lay in salt water until they were ready to spawn, and when taken up their eggs would not hatch out owing to the fact that they "ripened" in salt water. The hatchery will be moved to a point immediately on the lake, where there is an abundance of fresh water, and where it is expected much better results will be attained. As it was, Mr. Callbreath turned out over 700,000 young fish, where they had absolutely

397

no enemies until they go back into salt water. The lower dam will still be kept in position to exclude the pirate fish.

The gentleman recommends the passage of a law giving property rights to persons producing fish under these conditions, where it can be established that the fish are the product of private enterprise.

Bearing in mind the law which forbids obstructions placed over the entire width of a stream frequented by salmon, Mr. Callbreath makes a distinction between barricades and dams used as a means of wholesale slaughter of salmon and those placed for the purpose of detaining fish until they are ripe, and for preventing interference from pirate fishes. The latter class of obstructions, while a violation of the letter of existing law, is deemed not a violation in spirit. "Fencing" can be done only on small streams, and if but 10 per cent of the usual number of salmon were allowed to pass the barricade and spawn undisturbed in the waters above, the number of small fry hatched out would be greater than if no obstruction were offered, and the sea trout and other pirate fish allowed to work havoc among the salmon spawn and small fry.

HABITS OF THE NATIVES.

The natives, as a class, are intelligent, industrious, and peaceable, finding their entire means of support in hunting and fishing. A great portion of the lowlands of Alaska have abundant forests of spruce and pine, some trees 4 and 5 feet at the butt and running up for a hundred feet without a limb. There are excellent facilities for farming and herding, the climate being mild and moist. The thermometer seldom gets below zero in winter, and 60° F. is the average for summer.

While the native male population is engaged in hunting and fishing—there being a separate time for each—the women gather various kinds of roots, berries, and barks, which are preserved in seal grease, and eaten during the winter. The natives around Yukatat Bay catch about 1,600 hair seal every year, a portion of the flesh of which is dried, while the fat is boiled down into grease. Of this great quantities are used, everything they eat being cooked with it. They compare it to the "Boston man's butter."

The natives practice both polygamy and polyandry, although but few instances of the latter relation exist at present. This is due to a peculiar custom in vogue among them, namely, that when one of a married couple dies the relatives of the deceased take all the worldly goods the pair might have accumulated and divide them among themselves, leaving the survivor nothing but a heavy heart and the clothes on his back. To a woman left a widow with a half dozen children this practice works great hardship. To guard against this hardship a man generally becomes possessed of two wives or more, and when one of them dies the surviving wives still remain joint owners of his goods and chattels, and the involuntary division of his property among the relatives of the deceased is thereby avoided. Polygamy, it would seem, is a blessing to an industrious native, for on the death of one wife her relations can not step in and take away that which he has been a lifetime in accumulating.

The Swedish Missionary Society, whose headquarters are at Chicago, has been doing good work among these natives—establishing a school, and taking among them a number of native children to educate and raise. Unfortunately, their main building was destroyed by fire recently, and the mission practically closed. But the mission owns a small sawmill, and they expect soon to get out the lumber and build up again.

One of the most curious of the many remarkable customs of the natives is the "potlatch," and a description of it at some length would seem not out of place. The meaning of the potlatch is very broad, and signifies that which is given by one Indian to another, or by one tribe to another tribe, whether as a pure present, in payment for an assault committed, for an imaginary cause of death, for accidental homicide, or for murder.

In the first case, should a man desire to make himself popular and rise in the estimation of his fellows to the dignity of a big chief, he collects, by great efforts, a considerable quantity of blankets, camphorwood trunks, biscuit, molasses, Cabot W. blankets, etc., and distributes everything he has among the other members of the tribe, crowning the distribution with a feast, and, if hoocherioo can be obtained, a "drunk." When all has been consumed, his object has been attained and his prominence in the tribe is assured. He is a big man according to the size of his potlatch, which is not so barbarous after all.

In case of an assault, which is very rare, the friends of the man who is worsted will demand a potlatch from the conqueror to salve the wounded feelings or disfigured face, and they are always paid without regard to the merits in the case.

To illustrate a potlatch for an imaginary cause of death, there is instanced the case of a boy who owned a small skiff and who invited other boys to go with him in the skiff after berries. While away they all ate of some poisonous root, from the effects of which one of the boys died. His relations demanded payment from the father of the boy who owned the skiff, their argument being that if the boy had no boat he could not have taken the other boys with him, and, of course, none of them would have eaten the poisonous root and died. They got the potlatch, but it almost resulted in a fight.

Should an Indian accidentally kill another, his relations are made to pay heavily for the death, and if the two principals are members of different tribes the demand is much greater, amounting in some actual instances to more than $2,000. The whole tribe of the one who did the killing assist in the payment. For murder of a male Indian a similar payment is required, but if a female is killed a few blankets will suffice.

A potlatch is given upon the demolition of an old house, and also after the death of any member of the tribe. But it seems the most prolific source of potlatching is the erection of new houses. The location for the new building is selected at a "smoking council" of the tribe, after which the erection is commenced, the owner being assisted by such members of his tribe as are experts. As it nears completion another council is held, at which is decided the date of the potlatch. The whole tribe is notified, and each member is expected to contribute something toward the potlatch and the subsequent feast. On the eventful morning all assemble at the new house, each in his best, with the exposed portions of their bodies covered with paint and further embellished with wads of cotton pasted at irregular intervals on the face and in the hair. The festivities commence with a dance, the women executing a species of side shuffle, while the men augment the enthusiasm by stamping their feet. Everybody sings. When the song and dancing are finished, some one hands up a bolt of calico, or some blankets, handkerchiefs, soap, or what not, at the same time mentioning the name of the person or persons to whom the donor desires the present to be given. (It is well to mention, parenthetically, that the potlatch presents and feast are given to members of opposite tribes.) The present, whatever it may be, is divided or torn into as many portions as donees, and then presented, after which more singing and more presents until everything is given away. This may last twenty-four or forty-eight hours, the women during this time never leaving the house, and eating nothing save an occasional cracker which may have been presented to them, moistening their throats, as they become dry, with the juice of tobacco, made moist in a can of water.

After the potlatch comes the feast. Rice has been cooked and seasoned with molasses and seal oil; boxes of sugar and biscuit opened, and an abundance of the omnipresent seal grease provided. Every available receptacle, from a washtub to an old tin can, is used for passing around the food, and everybody eats until their stomachs rebel, go outside, relieve themselves by vomiting, and return to the attack until all has been consumed. They know no such thing as stopping at an intermediate point. The potlatch and subsequent feast must exceed the cost of the simple structure in honor of which it is given many times.

THE ISLAND OF AFOGNAC.

Mr. A. Lasey, United States deputy surveyor for Alaska, accompanied the late J. K. Luttrell during a greater portion of his travels among the various canneries, and especially among those on the Karluk River and on the island of Afognac. Mr. Luttrell had thoroughly discussed with him in regard to these canneries, and had communicated to him his ideas and the recommendations he would make in his report with reference thereto. He therefore presents, upon request, Mr. Luttrell's ideas and conclusions on the subject of the better protection of the salmon fisheries and the proposed establishment of a Government hatchery on the island of Afognac.

This island has been recently condemned and set apart as a Government reservation, the object being to use the same for the purpose of a hatchery. It is the second largest island in northwest Alaska, containing an area of over 600 square miles. It is mountainous, and the lower parts are covered with a thick growth of valuable pine, from which most of the small schooners and boats employed in hunting and trading have been built. On it fur-bearing animals, such as brown and black bear, silver-gray fox, and other small game, are found. There is on its coast one settlement of about 200 inhabitants, natives and creoles, and has a church, two stores, and a schoolhouse. The population depend for their living on hunting, fishing, and cutting wood for export to the more southerly points of Alaska, the peninsula and adjacent islands being entirely devoid of timber, even for domestic purposes. On it there are a half dozen canneries, the value of two of which exceed $100,000. The Afognac River, on which it is intended to erect the hatchery, is filled with rapids and natural obstructions, so that the number of salmon endeavoring to ascend this river to spawn is comparatively small. The greater portion of the fish caught in Afognac Bay are passing schools.

In marked contrast to this is the Karluk River and Lake on Kodiak Island—the great natural breeding ground of the salmon. Immense schools of the fish gather every summer at the mouth of the river, and in former years ascended unhindered to the lake. On Karluk Spit, a narrow tongue of land at the mouth of the river Karluk, are established 5 first-class canneries and fishing stations, and in close proximity on the south side of the river 2 more, making 7 canneries in all.

These canneries pack away every year from 200,000 to 250,000 cases of red salmon, each case requiring on an average 14 fish, making a grand total of from 2,000,000 to 2,500,000 salmon caught there every season.

The owners of these canneries, foreseeing or fearing that a few years would bring about a total destruction of the red salmon if no protective measures were adopted, agreed among themselves not to fish on one day in each week and on that day to leave open the mouth of the river to afford the fish an opportunity of ascending to the lake. In addition, they established a first-class hatchery, capable of turning out several millions of young fish every season. This hatchery was located on the bank of the Karluk River at the head of tide water, about 2 miles above the canneries. The hatchery, however, has not been in operation within the last two years. A very careful examination of it by Messrs. Luttrell and Lasey showed it to be in very good order, requiring only trifling repairs. A competent person had been employed by the canneries to superintend the hatchery, and had for his use a comfortable dwelling house. The experiment proved successful in so far as several millions of young salmon were hatched, but later on it was found that the water used in the hatchery and obtained from a ravine had become surcharged with impurities, covering the young fish with a species of parasite, eventually causing death. This difficulty can very easily be obviated by leading the waters from the Karluk River to the hatchery, a distance of about 300 yards, in an iron pipe or wooden flume, at a cost not to exceed $500.

This hatchery the canneries propose to turn over to the United States Government, providing the Government is willing to operate it instead of establishing a hatchery on the island of Afognac.

The object of Mr. Lasey was to make plain that Mr. Luttrell believed Afognac Island to be ill chosen as a place for establishing a hatchery. Its reservation would entail the destruction of several canneries located thereon, the value of two of which is estimated at over $100,000. These canneries have already suffered great pecuniary loss by reason of the compulsory closing of their establishments, and it would seem that a claim for damages against the Government would properly lie. The United States would be obliged to purchase these canneries and other improvements, and the amount necessary therefor would greatly exceed $100,000. On the other hand, the owners of the Karluk River canneries have erected and equipped a hatchery, and this these owners have agreed to transfer gratis to the Government, and stand ready so to do whenever the Government chooses to accept.

In view of these facts, Surveyor Lasey states that Mr. Luttrell, after repeated conferences with the owners of the several canneries, had decided to make the following recommendations bearing upon the subject:

First. To abandon Afognac Island as a Government reservation, for the following reasons:

(a) It does not require for the purposes of a hatchery an island containing an area of over 600 square miles.

(b) The natural resources of the island, particularly the timber, are needed not only by the inhabitants of the island, but by the whole of the peninsula and adjacent islands lying southwest down to Unalaska, and the closing of the island would seriously affect the whole country and its industries.

(c) The Government would be obliged to purchase the canneries and fishing stations and all improvements existing on the islands, the claimants of which have already made application for patents and deposited the amount of purchase money in the United States subtreasury at San Francisco.

(d) The value of these improvements will exceed $100,000.

(e) The shutting down of these canneries during the present summer has already caused the owners pecuniary losses, for which losses the Government will most likely be held responsible in addition.

(f) The erection of a hatchery will cost several thousand dollars more.

(g) The improvements of the Afognac River, to enable the fish to ascend, will cost a considerable sum.

(h) That the Karluk River and Lake, being the great breeding grounds of the salmon, ought to be the place for a hatchery, the more so when the immense catch there, season after season, is considered.

The second recommendation Mr. Luttrell intended to make was the following:

Second. To accept the offer of the Karluk Packing Company, to turn over to the United States Government this hatchery on the Karluk River under the conditions specified above, for the following reasons:

(a) It will save the expense of building a hatchery on Afognac Island.

(b) The estimated cost of repairing the Karluk hatchery will be trifling.

(c) Unless a hatchery is established at once at Karluk, and artificial means are resorted to for the propagation of the salmon, in addition to other preventive measures, the red salmon will soon be exterminated, the yearly catch diminishing perceptibly, although a greater number of fishermen are employed, and a great variety of seines are used.

(d) It will not be necessary to purchase any improvements or vested rights, saving thereby a great sum of money.

(e) No improvements, in the way of removing obstructions, are needed on the Karluk River.

Mr. Luttrell also intended recommending additional measures for the protection of the salmon fisheries:

First. To prohibit entirely fishing in the river except by natives for their own use.

Second. To suspend all fishing operations during two days of each week, or limit the season's catch to a specified number of cases during a certain number of years, to give the fish an opportunity to recuperate.

Third. To leave a space of 100 yards wide, from the mouth of the river to deep water, open at all times for the fish to enter the river.

Fourth. Not to tolerate any obstructions in the shape of dams or wire fences in the river.

Fifth. Violation of any adopted protective measures to be punished by a fine and imprisonment.

Sixth. To appoint a proper officer to reside during the fishing season at Karluk, whose duty it shall be to see that all protective measures are strictly observed.

No attempt has been made to afford data from which to base an estimate of the total number of cases of salmon packed by the various canneries, or the total value of all the canning plants. Hence it is utterly impossible to compile such statement from this report.

The suggestion is made that the attention of the Government should be directed to the wanton destruction of deer by the natives. Mr. Wadleigh, of the North Pacific Trading and Packing Company, of San Francisco, who makes the suggestion, states that it is no uncommon thing for a party of natives to go out and return in a few days with 25 or 50, and sometimes more, deerskins. No use whatever is made of the carcasses, and they are allowed to rot where the animal has been skinned. At the present rate of destruction it will be but a few years before deer will become extinct. He recommends the enactment of a law similar to that in force in British Columbia, prohibiting the exportation of deerskins.

The report contains the draft of a bill the object of which is to better protect the salmon fisheries in Alaska. A copy of such draft is hereto annexed.

AN ACT for the better protection of the salmon fisheries of Alaska.

Be it enacted by the Senate and House of Representatives of the United States of America in Congress assembled, That the erection of dams, weirs, barricades, or other obstructions in any of the rivers of Alaska with the purpose or result of preventing or impeding the ascent of salmon or other anadromous species to their spawning ground, or taking, catching, or fishing for salmon by any device, save and except by an Indian or Aluet spear up unnavigable streams more than one thousand yards from its confluence with the ocean, is hereby declared to be unlawful; and the Secretary of the Treasury is hereby authorized and directed to establish such surveillance and regulations as may be necessary to insure that this prohibition is strictly enforced, and every person or persons, corporation, or association or company who shall be found guilty of a violation of the provision of this section shall be fined not more than three thousand dollars nor less than one thousand dollars and two hundred and fifty dollars for each day of the continuance of such obstruction, and the half of such fines is hereby directed to be paid to the person or persons who may give the information leading to the conviction of the guilty party or parties.

SEC. 2. That for the purposes of this act all streams in Alaska shall be deemed unnavigable when vessels of six-foot draft can not ascend the same with safety at ordinary high water.

SEC. 3. That the Commissioner of Fish and Fisheries is hereby empowered and directed to investigate all charges of illegal fishing brought to his notice by responsible parties and shall, if he finds them well founded, instruct the United States attorney to proceed at once against the offender.

SEC. 4. Dams, weirs, barricades, or other obstructions shall be defined as being an obstruction when a distance of one hundred yards is not left open at all times for fish to ascend said rivers.

SEC. 5. That during each week one day of twenty-four hours shall be set apart, and fishing in any manner, shape, or form on said day will be illegal and unlawful and punishable by the fines as set forth in section 1.

SEC. 6. That for the purpose of preventing the further impairment or exhaustion of the valuable fisheries this act shall take effect from and after its passage, and all acts or parts of acts in conflict with this act are hereby repealed.

REPORT OF SPECIAL AGENT MURRAY ON THE SALMON FISHERIES IN ALASKA.

OFFICE OF SPECIAL AGENT, TREASURY DEPARTMENT,
Washington, D. C., February 1, 1895.

SIR: I have the honor to report that, pursuant to Department instructions dated June 12, 1894, I sailed to Alaska, and visited and inspected the salmon-canning establishments on many of the bays, rivers, and streams of that Territory, an account of which is herewith respectfully submitted for the information of the Department.

July 10, 1894, I sailed from San Francisco on board the United States revenue steamer *Rush*, Capt. C. L. Hooper commanding, and proceeded to Port Townsend, Wash., where we arrived on the 15th, and where we were afterwards joined by Hon. C. S. Hamlin, Assistant Secretary of the Treasury,

July 23 we sailed from Port Townsend and steered for the seal islands in Bering Sea, where we landed August 3, and on which we spent five days going over the rookeries, noting their condition and the condition and numbers of the fur seals, and making inquiry into matters of importance connected with the seal question.

August 8 we left the seal islands, reaching Unalaska on the 9th, where we remained one day to coal ship, and then, on the 10th, we sailed along the Aleutian chain and the Alaskan Peninsula, calling on the way at Akutan, Akun, Belkofsky, Sand Point, Coal Harbor, Karluk, Kadiak, Yakutat, Sitka, Taku Inlet, Juneau, Douglas City, Fort Wrangell, Kassan, Loring, Port Chester, or New Metlakahtla, St. Marys, Port Simpson, Nanaimo, and Vancouver City, where Mr. Hamlin left the ship and proceeded by rail to Washington. Continuing the voyage, I proceeded to San Francisco, calling in at Port Townsend, New Whatcom, and Astoria on the way down.

At Karluk, on Kadiak Island, we found what I consider the finest of all the salmon streams in Alaska, if not the finest on the whole Pacific Coast; most certainly the finest from which salmon are at present taken for canning purposes, quantity and quality being considered, for I find that nearly one-third of the entire Alaska pack for 1894 was put up at the mouth of the Karluk River.

Assuming, then, that it is the principal salmon stream in Alaska, I shall take it as a model for all of the others for the purposes of illustrating what I have to say about the salmon industry of Alaska and of the dangers by which it is beset.

Excepting the great Yukon, which is navigable for thousands of miles, the Kuskokwim and a few others of minor importance, the rivers of Alaska are small streams of from 20 to 200 miles in length, and many a stream that is rich in the finest of fish and of the utmost importance to the fisherman is only a few miles in length—a mere drain for a very limited watershed of high, rugged, and snow-clad hills, behind which small lakes of the clearest, purest, and coldest water are to be found, and in which the salmon deposit their eggs in season, and from which hundreds of millions of young salmon descend annually to the sea.

401

where they remain until maturity, after which they return to their native stream and deposit their eggs.

Karluk River, on Kadiak Island, is about 20 miles long, and flows between high hills and over many falls or rapids from the time it leaves the lakes above, at times widening out to a breadth of several hundred feet and again narrowing down to less than 100 feet at its mouth.

Notwithstanding its diminutive size, however, there are six canning plants erected there, which in the aggregate represent an outlay of $500,000, and in four of which was canned during the season of 1894 230.000 cases of 48 pounds per case, or 11,040,000 pounds of fish, or in round numbers about 3,220,000 salmon.

Appended will be found Exhibit A, in which is given the number of cases of salmon packed in Alaska from 1889 to 1893, both inclusive, each case containing 48 cans of 1 pound each.

Exhibits B and C give an itemized statement of the work done at the canneries in 1893 and 1894, the name and location of each cannery, the number of men employed by each (white, Indian, and Chinese), the apparatus used in fishing, the number of salmon taken and canned, the number salted and barreled, number of steamers, lighters, and boats used, the necessary sea-going vessels and their tonnage, and the value of each plant.

Exhibits B and C contain the names of the canneries which were running in 1893-94 only. Exhibit D gives the name of every known canning plant and saltery in Alaska—27 canneries, 14 salteries, and 1 herring fishery.

Exhibit E shows the distances a vessel would have to sail from Cape Fox, in southeastern Alaska, to Nushigak, Bering Sea, if she called at all the canneries en route, a total distance of nearly 5,000 miles.

Exhibit F shows the amount of tin consumed in the canneries in 1894, and also its price and the amount of import duty paid to the Government.

It is not claimed at all that any of the exhibits are absolutely full and complete; on the contrary, I found it very difficult to find the offices or headquarters of many of the canneries, and, when found, it was impossible to get my questions answered by many of them.

For most of the information received I am indebted to the Alaska Packers' Association, the R. D. Hume Canning Company, and to Mr. Barling, of the Alaska Improvement Company, all of San Francisco.

A comparison of the annual output from 1889 to 1894 shows that in 1891 the salmon-packing industry of Alaska reached its highest point, with an output of 807,999 cases of 48 pounds each, or in round numbers about 20,000 tons of fish.

The output fell off about one-half in 1892, since which time a gradual increase is perceptible, until in 1894 we have an output of 669,041 cases of 48 pounds each.

From the best information obtained in Alaska—and an earnest effort was made to gather it impartially—the salmon-packing industry within the section embraced between Cape Fox and the Nushigak River has attained the limit beyond which it is dangerous to pass; and that, if we would perpetuate the salmon industry and keep it up to its present grand proportions, measures of protection must be taken by which the streams and spawning grounds shall be kept open and undisturbed at all times, so that the fish may freely ascend and deposit their eggs in season.

With good care and a due regard for the future of the salmon industry, millions of fish may be taken from the Karluk River annually for all time without injury; but it should never be forgotten that there is a limit beyond which it is not safe to go, and that if we would reap an annual golden harvest we must also guard the source of supply, and see that nothing is done to either fish or stream that will change the natural order under which the fish have grown to such numbers and by which they may be perpetuated without abatement forever. Unfortunately, the conditions existing at Karluk are not for the best interests of the salmon industry, its growth or perpetuation; and unless the United States Government asserts its full rights in the premises by enacting and rigidly enforcing laws for the adequate protection of the salmon of Alaska, they, like the sea otter and fur seal, will soon be things of the past.

Paradoxical though it may appear, it is nevertheless true, that none are more anxious to save and perpetuate the salmon than the canners themselves, and yet their methods are such as, if continued, will very soon destroy them.

Let it be borne in mind that all the canning factories in Alaska are owned by three or four corporations in San Francisco, who have millions invested in the salmon-canning industry, but who have no interest in the development of Alaska, and who, as a matter of fact, do not add one dollar to the wealth of the young Territory from which they take millions of dollars annually.

These corporations are rivals in the salmon-canning business, and their rivalry is carried to such extremes betimes that bloodshed at any moment will not surprise those who know the real conditions existing there.

Now, this bitter rivalry of great and rich corporations, if allowed to continue, will eventually destroy the salmon, for, rather than allow A to make a good haul of fish, B will dam the stream and prevent the ascent of the salmon, or C will destroy the fish already on the spawning grounds and thus destroy the crop which would otherwise appear off the mouth of the stream four years hence; or A and B will join forces against C and actually destroy his nets and by force prevent his fishing.

We had barely cast anchor at Karluk before we were approached by the superintendent of one of the great canneries with a long list of wrongs perpetrated on his company during the peaceful and legitimate pursuit of their business.[1]

Landing, afterwards, we were met by a crowd of native fishermen who had complaints to make to the Government about the way they are treated by the whites, who take up all the streams and forbid the natives to fish there any more.

After the Indians came the superintendent of another of the canning establishments with a complaint that his rival over the river had broken the agreement mutually made by them some time before, by which a "close time" of twenty-four hours per week should be observed for the purpose of allowing the salmon to enter the stream and ascend to the spawning grounds for the purpose of reproduction.

This agreement was observed for awhile until a scarcity of fish in the bay threatened a short output of canned goods, and then orders were issued to not only ignore the "close time" in future, but to go into the river and take out all the fish that had reached the spawning beds, which was done at once, and some 225,000 salmon were captured and canned, and not a fish of that run was left to reproduce the species.

When the representatives of these great corporations tell us of the

---

[1] See affidavit in Appendix.

millions of money they have invested in the Alaskan salmon business, and ask us if we can possibly believe that they would permit the doing of anything which could injure the salmon or reduce the annual supply, it looks so reasonable that they should be fully alive to their own financial interests that at first it is hard to realize that the salmon are being destroyed very rapidly, and those who have not been on the ground to see it with their own eyes are not to be blamed for doubting the assertion.

It is nevertheless only too true, and a few words of explanation will make it quite clear to the doubting ones.

It is true in a general way that the canners themselves do not fence or dam the streams, but they buy the salmon from the men who do.

At Loring, for instance, Captain Hooper and Mr. Hamlin undertook to enter the stream in a boat, but were prevented by a dam clear across from bank to bank. That fence had been there for years, and the salmon, running up against it in their efforts to enter the stream every year for purposes of reproduction, were caught and canned until the regular supply was exhausted, and other streams were laid under tribute to keep the canneries going.

No one had even thought it worth while to remove the old dam.

The gentleman who gave me the information has been a resident of Loring for the past eleven years, and knew of what he talked.

Among many other things, he said: "Because of the bringing of whites and Chinese here from San Francisco the natives are crowded out, and only about 6 per cent of those formerly employed can now find work at Loring."

What is true of Karluk and of Loring is also true of every place in Alaska where salmon are canned—wherever two rival canneries are located on the same stream there are neither dams nor fences allowed, but neither is there time given the fish to enter and ascend the stream, and the consequence in either case is to destroy the salmon.

Wherever a cannery is located far enough away from rivals a dam, fence, or some other mode of trapping salmon is resorted to and relied on for a steady supply until the river is fished out.

Speaking to one of the superintendents at Karluk, and asking him for reliable information, he said:

Wherever rivalry does not exist on any fishing river in Alaska there generally exists a dam, barricade, etc., to wit, Chignik Bay, on the north side of the Aleutian Peninsula, is fished by means of a fence. There is a fence at Loring, in southeastern Alaska, and there exists a fence in the small tributary stream at its confluence with the Nushigak River, Bristol Bay, Alaska. Up in Cooks Inlet, 3 miles below the mouth of the Copper River, there exists a weir extending out into the inlet 400 yards, to which is attached a pound net.

I asked him, "How can the Karluk River be made self-sustaining?" to which he replied:

(1) By establishing a weekly close season.
(2) By prohibiting fishing in the river (excepting Indians with hook or spear).
(3) When it is too rough to fish on the ocean beach fish will enter the river if let alone.

The fish naturally run to the river on the turn of the ocean tide, but when it is storming the fish are unable to enter the river by crossing the bar on account of the flying gravel, which scares them offshore.

At half tide, however, when the bar is covered by 6 or 8 feet of water, the fish make a break across the bar and enter the calmer water of the river.

When the salmon first enter the river they do not go directly up to the spawning grounds or lakes, but remain for weeks in the brackish waters until they are ripe and ready to spawn, and for this, if for no other reason, the Government ought to prohibit the taking of the fish that have once escaped the nets below and entered the brackish or tide waters of the river.

I have ever been consistent in my advocacy of no fishing in the Karluk River, and in 1888 I started out to stop it.

One of my rivals, Mr. ——, came to me lately and said, "You're a fool not to enter into some agreement with us; bring —— in along with you. Discard all these expenses, boats and steamers, and save coal. It's money we're after in this Territory. We do not come up here to this God-forsaken place for fun. Form one grand big fishing pool, even though you do not wish to can together, and work as I do for my company at Chignik." I said, "How do you do it at Chignik? What economical device have you got down there?" He replied, "I've got a fence in the river; I've got a pound net on this side, and a pound net on that side, and one day I use one, and one day I use the other. It fishes day and night, and it is the slickest thing you ever saw, and it's a dead open and shut game. All I say is, 'Jimmie, go up to the trap and bring me down 15,000 fish;' all they've got to do is to take a gang of men on the lighter and she comes down with the catch pro rata, and the next tide with 15,000. The next day I say, 'Well, boys, go up and bring me down 18,000 fish,' and they go and get them out of the other trap, for while they are working one side the other side is fishing." I said, "Why, Billy, that is against the law of 1889." "I know that," he replied, "but we are not up here for our health." I continued, "Then I am to infer from this conversation that you would have me enter a pool along with you and ——, decide on the number of fish to be taken, divide the catch pro rata, lessen our expenses, hire fewer men, use less seine, fewer boats, and double our profits by simply building a wire fence across the river?" "That's the whole thing," said he, "in a nutshell."

The story told by my friend from Karluk is only a repetition of several others, to the same effect, told by the leading citizens of every settlement where we called, and so well did all of them agree in the main, I have no doubt whatever of their absolute truth.

Were it necessary I could mention the names of all who furnished the information, but for obvious reasons I will only mention the names of men who are in the service of the Government.

While we were at Sitka Captain Burwell, commanding the U. S. gunboat *Pinta*, called on Mr. Hamlin, and during the conversation said:

It is a common occurrence when attempting to ascend a river or stream in a steam launch to find traps, dams, and wire screens obstructing them in many places to the utter destruction of the salmon.

I should have been happy to destroy them if I only had orders from anyone in authority to do so, but no one ever asked me to meddle with them in any manner.

Speaking of the illicit distilleries and liquor smuggling in Alaska, and of the impossibility to find a jury to convict for such crimes, a prominent official of Sitka said:

The same thing is true of the fisheries law, the land law, or any other law that would control white men; a jury of white men can not be found here who will convict a criminal of that class.

Traps are set, streams are dammed, salmon are prevented from ascending the rivers to the spawning grounds, and are destroyed by men who have no interest whatever in the development of the Territory, and yet is impossible to find a jury to convict the guilty ones, for *the salmon men will stand by the liquor men, and the liquor men will stand by the salmon men.*

I could fill a volume with testimony like this; testimony given voluntarily by disinterested men and reputable citizens; but enough has been said, I think, to show the necessity of the Government taking steps to control the streams and save the salmon from extinction.

During our stay at Karluk we landed and visited the establishments of the Alaska Packers' Association and also that of the Alaska Improvement Company. Owing to bad weather we were unable to reach the R. D. Hume Cannery.

We conversed with all sorts of men, from the superintendents down to the native Kadiak fishermen, and they were all agreed that salmon were decreasing in the Karluk River, and that unless the United States Government interfered to prevent it they would continue to decrease.

Some of the men went so far as to say that in order to keep up the regular supply of canned salmon some very inferior fish were being packed at some of the canneries that would not have been looked at or used for any purpose a few years ago.[1]

One of the most remarkable things I noticed at Karluk was the number of foreigners engaged as fishermen. Scandinavian, Dane, and German predominated on one side of the stream, and Italians on the other, while Chinese, exclusively, were employed within the canneries, cleaning and canning the fish and preparing the cases for market.

It seemed, too, that the bitter rivalries of the corporations are sometimes taken up in a more intensified form by the men and carried to the point of explosion.

However that may be, it is true that the foreigners are brought from San Francisco to fish the streams of Alaska, and that they actually look upon the streams and fish as their own individual property.

The unfortunate native Aleuts, whose fathers owned Alaska and all its riches of stream and forest long before Columbus was born, are hustled out of the way of these Mediterranean fishermen with scant ceremony, and forbidden to fish in their native streams.

They must obey. Appeal? To whom are they to appeal? There is no one within reach who would listen to them.

Dimly, in a sort of dazed way, they know something of a Great Father away, away off in a place called Washington; but how are they to reach him? Whenever the American flag appears they fly to the vessels carrying it to present a petition and recount the wrongs and the injustice which they suffer.

Who cares anything for poor, dirty, ignorant creatures like them? Who believes their story? No one.[2]

Landing at Karluk we met a committee of native men who, through an interpreter, told us of how they were denied the right to fish for themselves, and refused employment by the canners as well. It seems that owing to the fact that seines were stretched across the mouth of the river the salmon could not ascend the stream and consequently there were no fish for the natives to get whenever they did attempt to get any; and being refused employment as regular hands along with the foreigners, they could not make a living.

That the natives may possibly exaggerate the wrongs inflicted upon them; that they may magnify their suffering whenever they meet a person who will stop and listen to their tales of woe, is possibly true enough; but it is equally true that the conditions existing on the Alaskan streams, from which so many millions' worth of beautiful fish are taken, are not the sort of conditions that will benefit the native Alaskan either morally, physically, or financially.

Nor is it either just or right that his best interests should be left dependent upon the whim of foreigners who may come in and camp down beside his stream and monopolize its treasures, while refusing him either employment to earn or the right to fish to make a living.

The other side of the story is told, however, by the superintendent of one of the canneries:

KARLUK, *August 17, 1894.*

GENTLEMEN: In allowing the natives only to fish in the river I would say that at certain times of the tide we are compelled to lay our seines from the mouth of the

---

[1]See letter of Commissioner of Fisheries in Appendix.
[2]Incidentally, a letter from an Alaskan canner to Hon. Marshall McDonald has been referred to the Department and to the special agent for the protection of the salmon fisheries in Alaska; and as its story fully illustrates my meaning I have appended it to this report. It tells its own story.

river so that they will swing with the tide, or to avoid their doubling up or swinging too far. We have oven to cross the river at its mouth to work our seines to the best advantage.

This season our white fishermen have not caught any fish to speak of above what virtually might be called the mouth of the river. Our white fishermen are quite willing, and have been for the past two years, to give the natives the privilege of fishing the river above the mouth. The same is true of the Italians fishing for the Alaska Improvement Company. They have not done anything against the natives fishing.

The decrease in fish caught in the river is something too large to number.

Yours, truly,

S. B. MATTHEWS.

Messrs. CHAS. S. HAMLIN and JOS. MURRAY.

We asked Mr. Matthews whether the salmon were decreasing in the Karluk River, and his reply was, "The decrease in fish caught in the river is something too large to number."

His words have been corroborated by everyone to whom I mentioned the subject, and there were many who suggested the establishment of "hatcheries" for the propagation of salmon, so that the present supply might be continued indefinitely; and some of the canners offered to donate to the Government a "hatchery" already prepared on the Karluk River, on condition of its being worked at the expense of the Government; while others suggested a tax of 5 cents per case and 10 cents per barrel, on every case and barrel of salmon taken in Alaska, on condition that the Government would enact laws, and appoint agents to enforce them, for the full protection of the salmon streams and the perpetuation of the fish.

Before leaving Karluk Mr. Barling, of the Alaska Improvement Company, sent the following letter:

KARLUK, August 17, 1894.

DEAR SIR: Herewith appended you will find a few suggestions necessary to the protection of the salmon of Karluk River:

(1) Prohibiting all fishing in the river above the first rapids, save and except by Aleuts, and their catch of salmon should be limited.

(2) To prohibit fishing from Friday 6 p. m. until Saturday 6 p. m.

(3) Regulating the size of seine mesh used—not to be less than 3¼ inches stretched mesh.

(4) Punishing the anchoring of set nets at or near the mouth of the Karluk River. This will insure the immediate and uninterrupted laying out of all nets.

(5) The use of purse nets should be declared illegal. Upon the head of purse nets, I can assure you their use has often been deplored since their introduction upon the Atlantic Coast.

(6) That a tax of 5 cents per case be levied against the total pack of Alaska; same to be collected as the Treasury sees fit to decide. This tax would raise upward of $25,000.

(7) Operate the hatchery at Karluk out of the funds above raised.

Hoping the above will meet with your kindly consideration, I remain,

Respectfully, yours,

H. J. BARLING.

Hon. C. S. HAMLIN.

The idea of levying a tax on salmon packed in Alaska was first broached to me by Mr. Barling, and, with a single exception, I have found it favorably received by the canners.

After my return to Washington, and while collecting data for my report, I compared several bills which had been introduced into Congress, or prepared for that purpose, by the friends of the several rival establishments in Alaska, for the protection of salmon. I had been assured, too, by every salmon canner I had met that they were deeply interested in the matter of full and adequate protection, and that they hoped to see a bill passed and the law most rigidly enforced to that end. That I might succeed in framing a satisfactory bill I read many that had been drawn or proposed in the House by the attorneys and friends of the canners themselves, as well as the reports of the honorable Commissioner of Fisheries, from whose valuable Report for 1892 I have made lengthy quotations.[1] I made diligent inquiry into the salmon laws of Oregon and Washington, and that we might benefit by the experience of those who have been protecting salmon for the past thousand years I read Bund's Law of Salmon Fisheries in England and Wales and A Treatise on the Law of Scotland Relating to Rights of Fishing, by Stewart.[2]

The following bills, which I respectfully submit to the consideration of the Department, are the result of my investigations:

A BILL to amend an act entitled "An act to provide for the protection of the salmon fisheries of Alaska."

Be it enacted by the Senate and House of Representatives of the United States of America in Congress assembled, That the act approved March second, eighteen hundred and eighty-nine, and entitled "An act to provide for the protection of the salmon fisheries of Alaska," is hereby amended and reenacted as follows:

"SECTION 1. That the erection of dams, barricades, fish wheels, fences, traps, pound nets, or any fixed or stationary obstructions in any part of the rivers or streams of Alaska, or to fish for or catch salmon or salmon trout in any manner or by any means with the purpose or result of preventing or impeding the ascent of salmon or salmon trout to their spawning ground within one hundred yards of the mouths of such rivers or streams, is declared to be unlawful, and the Secretary of the Treasury is hereby authorized and directed to remove such obstructions and to establish and enforce such regulations and surveillance as may be necessary to insure that this prohibition and all other provisions of law relating to the salmon fisheries of Alaska are strictly complied with.

"SEC. 2. That it shall be unlawful to fish, catch, or kill any salmon or salmon trout of any variety, except with rod or spear, above the tide waters of any of the creeks or rivers or their tributaries in the Territory of Alaska, or to lay or set any drift net, set net, or seine for any purpose, across the tide waters of any river or stream, for a distance of more than two-thirds of the width of such river, stream, or channel, or lay or set any seine or net within one hundred yards of any other net or seine which is being laid or set in said stream or channel, or to take, kill, or fish for salmon in any manner or by any means in any of the waters of the Territory of Alaska, either in the streams or tide waters, from noon on Friday of each week until six o'clock postmeridian of the Saturday following, or to fish for or catch, or kill in any manner, or by any appliances, any salmon or salmon trout in any stream of less than one hundred yards in width in the said Territory of Alaska between the hours of six o'clock in the morning and six o'clock in the evening of the same day, of each and every day of the week.

"SEC. 3. That the Secretary of the Treasury may, at his discretion, set aside certain streams as spawning grounds, in which no fishing will be permitted; and when, in his judgment, the results of fishing operations on any stream indicate that the number of salmon taken is larger than the capacity of the stream to produce, he is authorized to establish weekly close seasons, to limit the duration of the fishing season, or to prohibit fishing entirely for one year or more, so as to permit the salmon to increase.

"SEC. 4. That to enforce the provisions of law herein, and such regulations as the Secretary of the Treasury may establish in pursuance thereof, he is authorized and directed to appoint one inspector of fisheries at a salary of three hundred dollars per month, and two assistant inspectors at a salary of two hundred and fifty dollars per month, and he will annually submit to Congress estimates to cover the salaries and actual traveling expenses of the officers hereby authorized, and for such other expenditures as may be necessary to carry out the provisions of the law herein.

"SEC. 5. That any person violating the provisions of this act, or the regulations established in pursuance thereof, shall, upon conviction thereof, be punished by a fine not exceeding one thousand dollars, or imprisonment at hard labor for a term of ninety days, or both such fine and imprisonment, at the discretion of the court; and further, in case of the violation of any of the provisions of section one of this act, and conviction thereof, a further fine of five hundred dollars per diem will be imposed for each day that the obstruction or obstructions therein are maintained."

---

[1] See extracts in Appendix.   [2] See Appendix.

A BILL entitled "An act to provide for the protection of the salmon fisheries of Alaska."

SECTION 1. That from and after the passage of this act every person or corporation engaged in the business of taking salmon in the waters of the Territory of Alaska for salting or canning purposes shall, on the first day of December of each year, file with the Secretary of the Treasury of the United States a sworn statement of the number of barrels, packages, or cases of salmon so packed, salted, or canned by him or them, and shall pay annually to the Treasury of the United States the sum of five cents per case of forty-eight pounds or less, and ten cents per barrel for each case or barrel of salmon so canned or salted by him or them.

SEC. 2. That the returns provided for in section one shall be made under regulations to be prescribed by the Secretary of the Treasury; and all provisions of existing law as to omitted or false returns of persons or corporations and as to penalties, civil or criminal, for such omission or false return under the provisions of the law providing for an income tax, act of August twenty-eighth, eighteen hundred and ninety-four, shall, so far as the same may be applicable, be in full force and virtue as to this act.

Bill No. 1 was drawn as nearly in conformity with the Oregon statutes as the difference in the size of the Columbia River and the salmon streams of Alaska would warrant; and a glance at the bill and at the Oregon statutes will show that a yearly close time of three months and a weekly close time of twenty-four hours during the season in Oregon is much more oppressive, in comparison, than a weekly close time of thirty hours in Alaska.

I have no desire to injure the men who pack salmon in Alaska; on the contrary, I wish to see the canneries flourish and multiply and their owners prosper so long as there is no danger of destroying the source from which their prosperity is derived. But I do know that the history of salmon in America is a history of hurried devastation and extinction of the species, and I am anxious that the Government shall step in in time to prevent its destruction in Alaska, and, like Scotland, enact laws, and enforce them, by which the salmon may increase and multiply and be perpetuated for all time.[1]

Bill No. 2 was drawn at the request of Mr. Barling and other large canners, and it has been warmly indorsed by R. D. Hume & Co., the Alaska Improvement Company, and others who are deeply interested in the Alaska salmon industry.

The question of its constitutionality has been raised in certain quarters and may possibly vitiate it for all practical purposes, for which I should be very sorry indeed, for, looking at the matter from the practical standpoint solely, I say there ought to be full and ample protection given to the food-fishes of Alaska; and if those who make millions out of them and have millions invested in the business are willing to pay for such protection, they ought to be allowed to do so.

Should those who are opposed to legislation looking to the protection of salmon in Alaska succeed in defeating the proposed bills, however, it will still be the duty of the Department to do everything within the existing law that can be done for the perpetuation of the salmon.

History teems with evidence of the fact that from the tenth century till now the Scotch have had to wage a continuous legal battle for the constant protection of their fisheries, and that their immense salmon interests of to-day owe their origin, growth, and world-renowned success to the tireless efforts of the men who labored for their protection.

---

[1] A comparison of the American and Scotch systems of salmon culture is given by a friend, who says: "From the time of the settlement of the State of Maine by the whites until there was not a salmon left in the streams, which, previously had always been full of them, was about two hundred years; and the population was not yet 1,000,000 souls. Scotland, on the contrary, with a population of 3,000,000 souls, has more salmon now than she had one thousand years ago, when she very wisely enacted laws, which have always been enforced, to protect them"

As my sole aim throughout this inquiry has been to elicit truth for the purpose of laying down a basis of action for the sure protection and perpetuation of the Alaska salmon, without injury to any legitimate enterprise, I sent copies of the prepared bills, with the following letter, to the principal Alaska canners, and their replies and criticisms are subjoined:

WASHINGTON, D. C., *February 1, 1895.*

GENTLEMEN: Please find inclosed copies of two bills about to be introduced in Congress for the protection of the salmon fisheries of Alaska.

They are essentially the production of all that has been suggested for that purpose from time to time by the representatives and friends of all of the corporations interested in the Alaskan salmon-canning business.

The bills introduced by Messrs. Stewart, Mitchell, Robbins, and others; the suggestions made by Messrs. Hirsch, Hume, and Barling, and the reports and recommendations of the Hon. Marshall McDonald, Commissioner of Fish and Fisheries, have been diligently examined by me, and it has been my honest intention and sole aim to frame a bill that would be as nearly just and equal to all interested in the perpetuation of the salmon fisheries as it is possible to be.

If I have not succeeded in doing all that should be done, or if I have suggested something that would be unfair or injurious, I beg of you to point it out immediately, and I promise to give your suggestions the most respectful attention.

The proposition to levy a tax of 5 cents per case and 10 cents per barrel comes to me directly from the canners themselves, and, from what I observed while at the canneries, I am in favor of the tax, and I think it will prove a blessing to you who have millions invested in the business.

The amount of the tax wisely expended by the Government in propagating and protecting salmon will be of lasting benefit to all concerned, but more especially to you who are deeply interested.

Very respectfully,

JOSEPH MURRAY,
Special Agent for the Protection of the Salmon Fisheries in Alaska.

ALASKA ——— ASSOCIATION,
San Francisco, Cal.

The following replies from the several firms addressed and from Mr. Barling of the Alaska Improvement Company were received by me, and are given in full for the information of the Department:

SAN FRANCISCO, *February 11, 1895.*

DEAR SIR: We have the pleasure to acknowledge receipt of yours of the 1st instant, covering a bill as proposed by yourself, for the protection of salmon fisheries of Alaska, and in response to your request if you have suggested anything that was unjust that we should point it out immediately, we wired you as follows: "Letter received, with thanks. Bill objectionable. Same explained by mail."

And in confirmation of same beg to say that in section 1 you specify that the erection of dams, barricades, fish wheels, fences, traps, pound nets, etc., in any of the waters of Alaska shall be prohibited. So far as dams, barricades, fish wheels, and fences, we agree with you most strongly, but, as you are aware, the fishing grounds of Alaska cover a large amount of territory, and what might be just and best for one section are not for another; for instance, in Prince William Sound, Cook Inlet, and Bristol Bay, where the waters at point of fishing are from 5 to 30 miles wide, the use of traps and pound nets are necessary to make the business remunerative. And as it might be necessary that the law be general, we would suggest that the use of traps and pound nets be permitted in the waters of Alaska, but not to extend over one-third the width of any stream—thus leaving two-thirds the width free for the uninterrupted passage of the fish.

To make a close season from noon on Friday of each week until 6 o'clock p. m. of the Saturday following would work a very great hardship in a district like Bristol Bay, where the pack of red salmon is made in fifteen to sixteen days at the outside.

Referring to a special tax, we hardly feel it just that the Government should impose same, when it is a fact that the Alaska salmon packers pay yearly upward of $100,000 from import duties. There may be locations that would warrant propagation, and that a tax should be levied for that purpose and for that direct location would certainly be proper.

We beg to ask that, in making laws for the protection of the salmon, you do not lose sight of the fact that the canners who have large sums invested in property,

which is immovable and of no value otherwise, can not afford to fish streams or the
waters of Alaska in a way to the depletion of the fish; and also that they be not
asked to catch fish in so expensive a manner that they who are not producers of over
one-third the quantity of salmon which is canned shall have to do so at an expense
which precludes their competition with other localities, especially British Columbia,
as the fish from that river are superior to a certain extent and will always demand a
somewhat increased price, and therefore would comparatively shut out the United
States product if an increased cost of catching is demanded.

Another point which the canners of Alaska come into competition are the waters
of both Oregon and Washington, in both of which the use of traps and pound nets
are allowed.

Hoping you will consider our suggestions in the tenor in which they are intended,
we remain,

Yours, truly,   W. B. BRADFORD, Secretary.

Col. JOS. MURRAY,
    Fish Commissioner of Alaska, 1321 N street NW., Washington, D. C.

SAN FRANCISCO, CAL., February 12, 1895.

DEAR SIR: Your letter (and inclosures) of February 1 came duly to hand and have
particularly noted its contents, and in reply would say:

The act of 1889, would, if carried out, result in what we have no doubt the Government has in view, viz: The preservation of the salmon in Alaskan waters, and give
to each and all of her citizens equal chance and right to take salmon.

The bill of two sections, to provide for the protection of salmon fisheries in
Alaska, is, in our opinion, an equitable and just measure, and should become a law,
inasmuch as the salmon industry pays no tax and needs protection for which it
ought to pay, and we have grave doubt in our mind if any can be found to oppose
such a measure who wish equal and just protection.

The draft of the other bill, with all due deference to you, we herewith return, with
erasures that we think ought to be made.

Section 4, we would suggest, instead of inspectors, that you provide for policing
the rivers by United States vessels, and information that may be laid by responsible
parties, and on proof of violation of the law give half of fine to the informer. By
such a method we think the law could be enforced.

Hoping that the Government will accept our suggestions in the spirit in which
they are written, we remain,

Yours, very respectfully,   ALASKA IMPROVEMENT CO.,
                                             By JAMES EVA, President.
                                             By JAMES MADISON, Secretary.

Hon. JOS. MURRAY,
    Special Agent of the Fisheries in Alaska.

WASHINGTON, D. C., February 14, 1895.

DEAR COLONEL: In compliance with your request for my views as to wherein the
proposed act, entitled "An act to amend an act entitled 'An act to provide for the
protection of the salmon fisheries of Alaska,'" is injurious to the salmon fishing
industry, I have the honor to submit the following:

Section 1 of the proposed act provides, among other things, that it is unlawful to
"erect dams, barricades, fish-wheels, fences, traps, pound nets, or any fixed or stationary obstructions in any part of the rivers or streams of Alaska, or to fish for or
catch salmon or salmon trout in any manner or by any means with the purpose or
result of preventing or impeding the ascent of salmon or salmon trout to their
spawning ground within 100 yards of the mouth of such rivers or streams," etc.

As manager for the Alaska Improvement Company my duties compel my attendance
at the fishing grounds during the entire fishing season, and by virtue of my long
experience as manager and director of the industry in the immediate neighborhood
of Karluk, I believe I have acquired more than a superficial knowledge of the whole
subject of salmon fishing as well as the peculiarities and characteristics of the fish
which inhabit those waters during the season for catching the same.

If the purpose of the Government is to construe the above-quoted provisions of
said section 1 so as to embrace within its scope the hauling of seines or nets, and
such seine or net hauling is to be considered as "preventing or impeding the ascent
of salmon or salmon trout to their spawning ground within 100 yards of the mouths

of such rivers or streams," such an inhibition or prohibition will tend to cause incalculable injury to the salmon industry, as well as to those now engaged in the same, who, by reason of their large interests, are endeavoring to promote the progress and prosperity of the Territory.

In this connection it will not be inappropriate to detail one of the peculiarities of the fish in their ascent of the streams or rivers for the purpose of spawning.

It is the nature of the anadromous species to enter a stream or river and follow the eddy formed by the juncture of the fresh and salt water. This habit is better illustrated by an examination of the accompanying diagram.

e e represent the outflowing fresh water from the Karluk River, which, by reason of the rapids near its mouth, gives it a tremendous impetus for the last 300 yards of its descent.

b represents the eddy of salt water formed by the outflow of the fresh water from the river, on the Alaska Packing Association's side, the current of said eddy having a trend toward the latter's beach.

a represents the eddy of salt water also formed by the outflow of the fresh water from the river, on the Alaska Improvement Company's side, the current of the eddy in this instance trending away from the latter's beach.

d c represent the directions in which the fish "stand" inshore on their way to the river and spawning grounds.

It will be readily seen that the entry of the fish in the eddy on the Alaska Packer's side is in the direction which necessarily brings them closer to the beach; while the opposite is the case on the Alaska Improvement Company's side, the fish in that instance, and on account of the offshore trend of the eddy, are compelled to "stand" inshore on a line parallel to the fresh-water current.

If we are prohibited or prevented from hauling our nets within the proposed proscribed distance of 100 yards of the mouths of such rivers and streams, we will practically be legislated out of existence, and the salmon-fishing industry will be ruined.

The gist of the whole controversy, developed within the past year, and consequent on the abuses of the salmon-fishing industry, is that the salmon which enter the rivers and streams and sport about in their natural playground should remain unmolested and free from any avaricious motives on the part of those unprincipled fishermen who, in order to further their own selfish interests, would conduct this industry in a way such as would very soon cripple if not completely destroy the species of fish which now frequent Alaskan waters to spawn.

Section 2 of the proposed act provides, among other things, for a weekly close season of thirty hours, or in the language of the section, "from noon on Friday of each week until 6 o'clock postmeridian of the Saturday following."

The number of fish entering any of the rivers or streams during the close period of thirty hours each week is augmented by the number of salmon that will "run" into said rivers and streams before the "laying out" or hauling of the nets, as well as during the interim of these acts.

About one-fifth of the entire fishing season is stormy, during which time it is impossible to "lay out" or haul a seine or net; but the storms do not prevent or obstruct the entrance of the fish in the rivers and streams after the "half tide," and consequently the number is thereby enormously augmented also.

Under these various conditions a sufficient number of salmon will have entered the rivers and streams to render extinction of the species impossible.

Special attention is called to the fact that it is above and not below the mouth of the stream or river that it is essential, important, and necessary to protect the salmon.

It can not surely be the desire or purpose of the Government to injure or destroy the fishing industry of Alaska—an industry which is indissolubly connected with the future progress and advancement of the welfare of the Territory.

I desire further to call your attention to the fact that the Alaska Improvement Company was the one that forced its competitors at Karluk to fish the ocean beach, as against the former practice of fishing in the river, and against the dam erected therein, and it is the purpose of that company to aid the Government in any way possible in the enforcement of all laws which have for their object the perpetuity of the chief industry of Alaska, viz, the salmon fisheries.

While as a matter of fact the Alaska Improvement Company has no objection to urge in opposition to the weekly close season of thirty hours, it is manifest to anyone who knows the conditions at places in Alaska other than Karluk that such a provision to a greater or less extent may be a hardship and an injustice to many other companies whose plants are not as favorably situated as is that of the company which I represent.

The operators at Karluk have the advantage of their competitors in having a much longer period in which to operate in the line of their business. Some of their competitors are restricted, naturally, by reason of the fact that the "run" of the salmon embraces a period of from thirty-six to forty-five days only, a period less than half of that enjoyed by the Alaska Improvement Company.

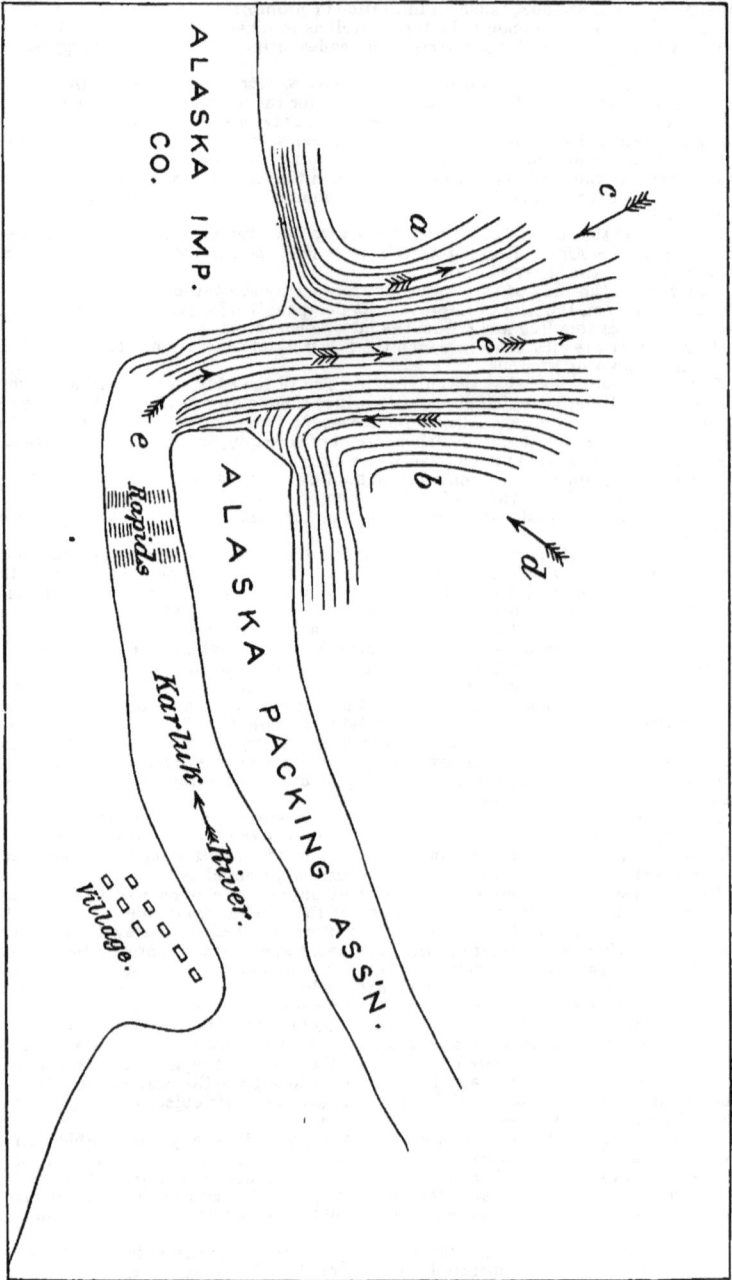

I desire again to impress upon you the fact that it is not within the proscribed 100 yards that the salmon need protection, but in those portions of the streams and rivers above their mouths and hereinbefore mentioned as the playgrounds of the fish, where their movements are such as to tempt the cupidity of those who, if left to their selfish devices, would soon annihilate the species.

In view of the foregoing I respectfully suggest, recommend, and urge that the 100-yard limit, as proposed to be enacted into law, be eliminated from said section 1 of the proposed act.

Respectfully submitted.

H. J. BARLING,
Manager of the Alaska Improvement Company.

JOSEPH MURRAY,
Special Agent for Protection of Salmon Fisheries in Alaska,
Washington, D. C.

BROOKLYN, N. Y., *February 18, 1895.*

DEAR COLONEL: Supplementary to my letter of the 14th instant, I beg to state that the bill meets with the approval of our president, Eva. He requests, however, in justice to all the various canning interests concerned, that the words in section 2, "or to fish for or catch, or kill in any manner or by any appliances, any salmon or salmon trout in any stream less than 100 yards in width in the said Territory of Alaska between the hours of 6 o'clock in the morning and 6 o'clock in the evening of the same day of each and every day of the week," be eliminated. He lays stress on the fact that those words would be detrimental to the interests of quite a few, and he claims that section 3 amended so as to read, "set aside certain parts of streams," etc., would be sufficient, and at the same time it would not limit the discretionary powers conferred upon the honorable Secretary of the Treasury, who could then designate any certain stream or streams which were in danger of exhaustion or impairment of their run of salmon.

Another reason he had in mind was the difficulty which would beset the Government in enforcing the law, inasmuch as some catch most of their fish at night and others during the day.

By leaving it to the discretion of the Secretary he could, as he saw fit, absolutely prohibit fishing in any stream or only partially so, such as is suggested by the specific language which Mr. Eva desires stricken out in section 2.

Yours, respectfully, H. J. BARLING.

Col. JOSEPH MURRAY,
United States Fish Commissioner for the District of Alaska.

SAN FRANCISCO, *February 7, 1895.*

DEAR SIR: Your valued favor of 1st instant just received, also the bills, for protection of salmon in Alaska, referred to, in regard to which will say that I can discover nothing in them but that which will be a benefit to all. You are to be congratulated for having framed such a bill, and if you are successful in having it become a law will deserve much credit. These bills will afford the necessary protection and mean the maintenance of a permanent industry in the Territory.

With my best wishes for your success, I remain,

Yours, truly, R. D. HUME.

Hon. JOSEPH MURRAY,
Washington, D. C.

It is not necessary, I think, to criticise all the objections raised by my friends to the bills proposed; for if once given a fair trial the faulty parts will easily be detected and quickly altered without injury to anyone.

The main point to be considered is that the Alaska fisheries are of great extent, immense value, and deserving of the greatest care; that

H. Doc. 92, pt. 2——27

the history of salmon in America is a history of wanton destruction and
· waste, and that unless we begin now to enact laws and rigidly enforce
them for the protection of the species it is only a matter of time until
the same destructive methods pursued from Maine to Oregon will produce the same results in Alaska.

An industry that has produced in five years 3,850,466 cases of canned
salmon—184,822,368 pounds of the very finest and richest of human
food—is undoubtedly one that is deserving of the very highest consideration from us all, and ought to have the best possible protection from
the General Government.

Particular attention is invited to the report for 1892 of the honorable
Commissioner of Fisheries, from which I have largely quoted; for there
he shows, beyond the possibility of contradiction, what destructive
methods have been followed in other localities, and which, if continued
in Alaska, will bring about the same direful and irreparable results.

Knowing the possibilities that await Alaska in the near future if her
natural resources are not allowed to be frittered away, and also knowing how easy it is to make a bad or a good beginning, I respectfully
recommend that all possible safeguards be thrown around those natural
resources, consistent with the best interests of all who have investments
made and business established in the Territory.

A revenue cutter, reenforced by half a dozen steam launches, ought
to be sent to and kept in Alaska for the purpose of enforcing the revenue laws—patrolling the inland waters, and carrying the officers of the
Government from place to place in the prosecution of their duties. The
appointment of an inspector of Alaskan fisheries and two assistants, to
visit and reside at the canneries during the fishing season, would be
productive of great good to all those who are in favor of law and order
and good government.

All of which is very respectfully submitted.

JOSEPH MURRAY,
Special Agent for the Protection of Salmon Fisheries in Alaska.

Hon. JOHN.G. CARLISLE,
Secretary of the Treasury.

EXHIBIT A.

Alaska salmon pack, 1889 to 1893.

Firms.	Location of canneries.	Number of cases.				
		1889.	1890.	1891.	1892.	1893.
Alaska Packers' Association Canneries.						
Aleutian Island and Mining Co...	Karluk	53,500	39,308	39,300		
Hume Packing Co	....do	28,000	36,000	36,000	76,000	59,959
Karluk Packing Co	....do	63,145	39,114	66,483	67,500	59,220
Kodiak Packing Co	....do	27,600	47,000	32,800		30,138
Arctic Packing Co	Eyak	37,000	44,000	33,100		
Royal Packing Co	Afognak	16,000	11,000			
Russian-American Packing Co	....do	25,000	26,434	25,300		
Arctic Packing Co	Alitak	13,000	17,400	20,100	21,000	25,777
Kodiak Packing Co	....do		16,250	22,000		
Alaska Packing Co	Nushagak River	20,000	25,000	31,071	31,500	37,188
Bristol Bay Canning Co	Bristol Bay	32,000	31,000	29,400	32,100	34,750
Nushagak Canning Co	....do	27,000	26,000	30,000		
Arctic Packing Co	....do			30,900		35,848
Chignik Bay Packing Co	Chignik Bay	22,000	44,000	75,000	50,000	57,553
G. W. Hume	Cooks Inlet		13,000	21,000	21,200	
Northern Packing Co	....do	18,500	16,000	17,000		
Arctic Packing Co	....do	31,000		20,000		31,665
Central Alaska Co	Thin Point	1,748	7,017	7,000		
Thin Point Packing Co	....do	27,000	2,400	4,000		
Pacific Packing Co	Copper River	5,000	13,000	22,000		28,999
Aberdeen Packing Co	Wrangell	13,800	15,800	16,000		22,728
Alaska Salmon Packing and Fur Co.	Loring	25,500	22,780	22,800	21,000	25,153
Chilkat Packing Co	Chilkat River	12,000	17,327	13,375		
Pyramid Trading and Fishing Co.	Pyramid Harbor	16,000	17,000	13,500	28,700	13,668
Total		514,793	576,830	629,220	349,000	462,646
Canneries not belonging to the Alaska Packers' Association.						
Alaska Improvement Co	Kadiak	26,000	27,000	52,000	52,000	38,795
Astoria Packing Co	Ruin Island		9,600	16,200		
Baranoff Packing Co	Baranoff Islands	3,700	10,475	7,949	10,200	9,609
Bering Sea Packing Co	Ugashik			5,000		
Boston Fishing and Trading Co	Yes Bay	7,000	9,327	17,335	13,741	15,102
Chilkat Cannery Co	Chilkat River	19,000	20,000	20,940	22,500	19,418
Metlakahtla Industrial Co	Metlakahtla			6,000	11,300	12,500
North Pacific Trading and Packing Co.	Klawak	11,370	10,108	9,281		
Pacific Steam Whaling Co	Copper River	15,000	16,000	24,000		35,000
Peninsula Trading and Fishing Co.	....do	2,531	12,119	20,074		15,270
R. D. Hume & Co	Karluk					15,492
Total		84,601	114,629	178,779	109,741	161,186
Grand total		599,394	691,459	807,999	458,741	623,832

Exhibit B.

Statistics of Alaska salmon pack, season 1893.

ALASKA PACKERS' ASSOCIATION

Name.	Location.	White.	Na-tive.	Chi-nese.	Appara-tus used.	King.	Red.	Silver.
		Men employed.				Salmon taken.		
Bristol Bay Canning Co.	Nushagak	60	41	87	Gill nets.	15,000	260,000	22,000
Alaska Packing Co...	...do	62	40	90	...do....	16,000	290,000	24,000
Arctic Packing Co...	...do	58	45	85	...do....	13,000	290,000	28,000
Do..............	Naknek	12			...do....		100,000	
Thin Point Packing Co	Thin Point	18	15		Seine....		60,000	
Karluk Packing Co...	Karluk	90	30	153	...do....		800,000	
Hume Packing Co....	...do	80	28	150	...do....		800,000	
Kodiak Packing Co...	...do	50	25	75	...do....		400,000	
Arctic Packing Co....	Alitak	33	15	60	...do....		300,000	
Arctic Fishing Co....	Kussiloff	42	30	68	Gill nets.	30,000	170,000	34,000
Chignik Bay Packing Co.	Chignik	60	20	140	Gill nets and seine.		600,000	64,000
Pacific Packing Co...	Prince Williams Sound.	65	23	58	...do....		220,000	72,000
Pyramid Harbor Pack-ing Co.	Pyramid Harbor..	70	50	64	Gill nets.		140,000	
Glacier Packing Co...	Fort Wrangell....	19	45	55	...do....	6,000	85,000	96,000
Alaska Salmon Pack-ing and Fur Co.	Loring...........	15	32	50	Seine....		42,000	160,000
Ugashik Fishing Sta-tion.	Selina River.....	20			...do....		200,000	
Total..........		754	439	1,135		80,000	4,757,000	500,000

Name.	Cases.	Barrels.	Steam-ers.	Light-ers and boats.	Value.	Num-ber.	Value.	Sail ton-nage em-ployed.	Value of plant.
						Nets.			
Bristol Bay Canning Co.	34,750		1	40	$12,000	80	$4,000	831	$91,000
Alaska Packing Co....	37,188		1	42	12,000	84	4,200	1,072	91,000
Arctic Packing Co....	35,848		1	41	10,000	82	4,100	611	91,000
Do		2,000	1	8	13,000	16	800	555	2,500
Thin Point Packing Co		1,232	1	10	6,000	4	800	175	20,000
Karluk Packing Co....	59,220	63	2	35	25,000	12	2,400	2,694	136,500
Hume Packing Co....	59,959		2	35	15,000	12	2,400	2,650	120,250
Kodiak Packing Co...	30,138		1	25	10,000	6	1,200	1,100	123,500
Arctic Packing Co....	25,777	73	1	20	10,000	4	800	771	52,000
Arctic Fishing Co....	31,665	200	2	40	16,000	80	4,000	1,376	32,500
Chignik Bay Packing Co.	57,553	32	2	24	19,000	90	4,500	1,625	71,500
Pacific Packing Co....	28,999		3	40	25,000	70	3,500	939	39,000
Pyramid Harbor Pack-ing Co.	13,668	8	1	25	16,000	50	2,500	1,187	30,000
Glacier Packing Co....	22,728		1	14	13,000	30	1,500	636	27,300
Alaska Salmon Pack-ing and Fur Co.	25,153	68		8	2,000	4	800	658	52,000
Ugashik Fishing Sta-tion.		1,970		6	1,200	3	600	233	1,500
Total............	462,646	6,496	20	413	205,200	627	38,100	17,113	990,550

Statistics of Alaska salmon pack, season 1893—Continued.

CORPORATIONS NOT IN THE ALASKA PACKERS' ASSOCIATION.

Name.	Location.	Cases.	Barrels.
C. F. Whitney & Co	Nushagak		1,400
L. A. Pederson	Naknek		2,600
Chas. Nelson	Selina River		2,700
Lynde & Hough	Shumagin Islands		205
Alaska Improvement Co	Karluk	38,795	
R. D. Hume & Co	....do	15,429	6
Oliver Smith	Kadiak Island		2,500
C. D. Ladd	Cooks Inlet		466
Pacific Steam Whaling Co	Prince William Sound	35,000	239
Peninsular Fishing and Trading Co	Copper River	15,270	
Chilkat Canning Co	Pyramid Harbor	19,418	
Foard & Stokes	Port Althorp		600
Baranoff Packing Co	Baranoff Island	9,609	1,006
North Pacific Fishing and Trading Co	Klawak	12,595	157
Cape Fox Packing Co	Cape Fox		2,000
Boston Fishing and Trading Co	Yes Bay	15,102	
Metlakahtla Industrial Co	Metlakahtla	12,500	
J. Macauley	Whale Bay		500
Various	Southeastern Alaska		1,400
Total		173,718	15,779

EXHIBIT C.

Statistics of Alaska salmon pack, season of 1894.

ALASKA PACKERS' ASSOCIATION.

Name.	Location.	Men employed.			Apparatus used.	Salmon taken.		
		White.	Native.	Chinese.		King.	Red.	Silver.
Bristol Bay Canning Co.	Nushagak	60	41	87	Gill nets.	3,000	283,000	18,000
Alaska Packing Co	....do	62	40	90	....do	4,500	270,000	15,000
Arctic Packing Co	.... do	58	45	85	....do	3,000	307,000	14,000
Do	Naknek	15			....do		128,550	
Thin Point Packing Co.	Thin Point	18	17		Seine		125,950	
Karluk Packing Co	Karluk	100	50	132	...do		1,066,000	
Hume Packing Co	....do	92	48	132	...do		1,066,000	
Arctic Packing Co	Alitak	35	17	52	....do		300,000	
Arctic Fishing Co	Kusiloff	56	40	75	Gill net..	15,500	283,000	19,000
Chignik Bay Packing Co.	Chignik	60	26	124	Gill net and seine.		600,000	
Pacific Packing Co	Prince William Sound.	65	33	60	....do	2,000	270,000	17,000
Pyramid Harbor Packing Co.	Pyramid Harbor..	74	61	70	Gill net.	7,000	340,000	11,000
Glacier Packing Co	Fort Wrangell....	31	51	55	....do	6,000	80,000	126,000
Alaska Salmon Packing and Fur Co.	Loring	25	36	65	Seine		37,000	205,000
Point Roberts Packing Co.	Koggiung	25			Gill net..		134,000	
Ugashik Fishing Station.	Selina River	34			Gill nets and seine.		112,850	
Total		810	505	1,027		41,000	5,403,350	425,000

Statistics of Alaska salmon pack, season of 1894—Continued. .

ALASKA PACKERS' ASSOCIATION.

Name.	Cases.	Barrels.	Steamers.	Lighters and boats.	Value.	Nets.		Sail tonnage employed.	Value of plant.
						Number.	Value.		
Bristol Bay Canning Co	30,999	....	1	40	$12,000	80	$4,000	940	$91,000
Alaska Packing Co ...	30,038	....	1	42	12,000	84	4,200	632	91,000
Arctic Packing Co....	30,413	420	1	41	10,000	82	4,100	1,072	91,000
Do	....	2,571	1	8	13,000	16	800	554	2,500
Thin Point Packing Co	....	2,519	1	10	6,000	4	800	263	20,000
Karluk Packing Co....	79,000	....	3	47	30,000	15	3,000	1,831	136,500
Hume Packing Co....	79,000	....	2	48	20,000	15	3,000	1,830	120,250
Arctic Packing Co	27,720	....	1	20	10,000	4	800	1,161	52,000
Arctic Fishing Co.....	34,033	354	2	40	16,000	80	4,000	1,129	32,500
Chignik Bay Packing Co.	55,352	....	2	24	19,000	90	4,500	1,536	71,500
Pacific Packing Co....	28,378	4	3	40	25,000	70	3,500	1,276	39,000
Pyramid Harbor Packing Co.	38,781	....	1	25	16,000	50	2,500	1,182	39,000
Glacier Packing Co....	25,250	20	1	14	13,000	30	1,500	636	27,300
Alaska Salmon Packing and Fur Co.	26,869	....	....	8	2,000	4	800	771	52,000
Point Roberts Packing Co.	....	2,680	....	6	1,200	3	.600	234	1,500
Ugashik Fishing Station.	....	2,257	....	6	1,200	3	600	310	1,500
Total............	485,833	10,825	20	419	206,400	630	38,700	15,357	868,550

CORPORATIONS NOT IN ALASKA PACKERS' ASSOCIATION.

Name.	Location.	Cases.	Barrels.
C. E. Whitney & Co	Nushagak	....	650
Prosper Fishing and Trading Co.............	Kvichak	....	2,000
L. A. Pederson	Naknek	....	2,640
Bering Sea Packing Co.....................	Ugashik	17,394	....
Chas. Nelson	....do	....	2,600
Norton, Teller & Co	....do	....	596
Lynde & Hough....................	Shumagin Islands	....	30
Oliver Smith......................	Kadiak Island.....................	....	2,000
Alaska Improvement Co	Karluk	44,300	45
R. D. Hume & Co.....................	Tanglefoot Bay	26,984	....
C. D. Ladd......................	Cooks Inlet.....................	....	2,064
Pacific Steam Whaling Co	Prince William Sound	29,000	....
Peninsular Fishing and Trading Co..........	Copper River	15,000	....
Baranoff Packing Co	Baranoff Island	10,910	....
North Pacific Fishing and Trading Co	Klawak.............................	13,620	61
Boston Fishing and Trading Co..............	Yes Bay	12,000	....
Metlakahtla Industrial Co....................	Metlakahtla	14,000	....
Miller & Co	Cardovia Bay	....	1,800
Cape Fox Packing Co	Cape Fox	....	2,000
Various	Southeastern Alaska	....	4,000
Tolstoi Salting Station......................	....	....	700
Total	....	183,208	21,186

EXHIBIT D.

Salmon packing stations in Alaska.

No.	Locality.	Name of company.	Cannery.	Saltery.	Herring.
1	Chilcat	Alaska Packers' Association	2		
2	Port Althorp	Ford & Stokes		1	
3	Killisnoo	Herring Fishery			1
4	Red Fish Bay	Baranoff Packing Co	1		
5	Fort Wrangell	Alaska Packers' Association	1		
6	Yes Bay	Boston Fishing and Trading Co	1		
7	Loring	Alaska Packers' Association	1		
8	Port Chester	Metlakahtla Industrial Co	1		
9	Klawak	North Pacific and Packing Co	1		
10	Cordovia Bay	Miller & Co		1	
11	Tolstoi Bay	....do		1	
12	Port Ellis	Kniu Island		1	
13	Cape Fox			1	
14	Copper River, Delta Peninsula	Fish and Trading Co	1		
15	Eyak Village	Pacific Steam Whaling Co	1		
		Alaska Packers' Association	1		
16	Cooks Inlet, Kussilo River	....do	1		
	West side of Cooks Inlet	C. D. Ladd & Co	1		
17	Afognak [1]		2		
18	Karluk River	Alaska Packers' Association	2		
	Alaska Improvement Co		1		
	R. D. Hume & Co		1		
19	Alitak Bay	Alaska Packers' Association (used up)	1		
20	Ugak Bay, Eagle Harbor	Oliver Smith	1		
21	Chignik Bay	Alaska Packers' Association	1		
22	Pirate Cove, Popoff	McCollum Trading Co		1	
23	Thin Point	Alaska Packers' Association		1	
24	Ugashik	Bering Sea Packing Co	1		
		Alaska Packers' Association		1	
		Sullivan River Packing Co		1	
		Johnson		1	
25	Naknek River	Alaska Packers' Association	1		
		Peterson		1	
26	Kvichak River	Alaska Packers' Association		1	
		Prosper Fish and Trading Co		1	
27	Nushagak	Alaska Packers' Association	3		
	Fort Alexander	Whiteney Company		1	
	Total		27	14	1

[1] Not in operation.

EXHIBIT E.

Sailing distances from Cape Fox to the different salmon canneries in Alaska.

[Figures in parentheses are map numbers.]

Localities.	Miles.	Localities.	Miles.
(13) Cape Fox to (10) Cordovia Bay	80	(15) Eyak Village to (17) Afognak	500
(13) Cape Fox to (8) Port Chester	50	(17) Afognak to (20) Ugak Bay, Eagle	
(10) Cordovia Bay to (9) Klawak	100	Harbor	75
(8) Port Chester to (11) Tolstoi Bay	60	(20) Ugak Bay to (19) Alitak Bay	100
(8) Port Chester to (7) Loring	60	(19) Alitak Bay to (18) Karluk River	100
(7) Loring to (6) Yes Bay	25	(18) Karluk River to (21) Chignik Bay	300
(11) Tolstoi Bay to (5) Fort Wrangell	100	(21) Chignik Bay to (22) Pirate Cove	200
(5) Fort Wrangell to (12) Port Ellis	100	(22) Pirate Cove to (23) Thin Point	150
(9) Klawak to (4) Red Fish Bay	150	(23) Thin Point to (24) Ugashik	500
(4) Red Fish Bay to (2) Port Althorp	150	(25) Naknek River to (26) Kvichak River	25
(2) Port Althorp to (3) Killisnoo	200	(26) Kvichak River to (27) Nushagak	100
(3) Killisnoo to (1) Chilcat Inlet	200		
(1) Chilcat Inlet to (14) Copper River Delta	1,000	Total	4,375
(14) Copper River Delta to (15) Eyak Village	50		

EXHIBIT F.

Amount of tin consumed in the salmon canneries of Alaska, cost, and import duty paid, for the year ending December 31, 1894.

Boxes 74,000
Cost $230,000
Duty paid $173,000

APPENDIX.

No. 1.—*Extracts from report on the salmon fisheries of Alaska, by Marshall McDonald.*

JULY 2, 1892.

ORIGIN AND DEVELOPMENT OF THE ALASKAN SALMON FISHERIES.

The marvelous abundance of several species of salmon in Alaskan waters has been long known, but in consequence of the remoteness of this region and its inaccessibility, the abundant supply in rivers nearer markets, and a disposition on the part of buyers to underrate Alaskan products, its fishery resources have not been laid under contribution for market supply within a few years, during which we have seen, as the result of reckless and improvident fishing, the practical destruction of the salmon fisheries of the Sacramento and the reduction of the take on the Columbia to less than one-half of what it was in the early history of the salmon-canning industry on that river. At present the streams of Alaska furnish the larger proportion of the canned salmon which find their way to the markets.

The pioneer in the early development of the salmon-canning industry in Alaskan waters was the Alaska Commercial Company, which in 1887 established a cannery on Karluk River, on the west side of Kadiak Island, and packed about 13,000 cases of salmon. The enterprise proved exceedingly profitable, and operations were rapidly extended so that the pack of this company on the Karluk River in 1888 aggregated 101,000 cases of 48 pounds each, representing a catch of over 1,200,000 blue backs or red salmon in the estuary of a small stream with a volume and drainage area not exceeding that of Rock Creek (the small stream flowing through the Zoological Park and discharging into the Potomac River within the city limits of Washington, D. C.). The enormous production of this year was secured by entirely obstructing the river by running a fence across so that no fish could pass up, and by continuing canning operations without intermission until late in October, when most of the fish were dark and unfit for food.

The immense pack made by the Alaska Commercial Company in 1887 and 1888, the fame of which quickly extended to San Francisco, had two important results. The attention of Congress was directed to the inevitable disaster that would overtake the salmon fisheries of Alaska unless prompt measures were taken to restrain the improvident and destructive methods employed for the capture of the salmon. Accordingly, upon the recommendation of the Commissioner of Fisheries, an act for the protection of the salmon fisheries was introduced into Congress and became a law on March 2, 1889, as follows:

AN ACT to provide for the protection of the salmon fisheries of Alaska.

"*Be it enacted by the Senate and House of Representatives of the United States of America in Congress assembled,* That the erection of dams, barricades, or other obstructions in any of the rivers of Alaska, with the purpose or result of preventing or impeding the ascent of salmon or other anadromous species to their spawning grounds, is hereby declared to be unlawful, and the Secretary of the Treasury is hereby authorized and directed to establish such regulations and surveillance as may be necessary to insure that this prohibition is strictly enforced and to otherwise protect the salmon fisheries of Alaska; and every person who shall be found guilty of a violation of the provisions of this section shall be fined not less than two hundred and fifty dollars for each day of the continuance of such obstruction.

"SEC. 2. That the Commissioner of Fish and Fisheries is hereby empowered and directed to institute an investigation into the habits, abundance, and distribution of the salmon of Alaska, as well as the present conditions and methods of the fisheries, with a view of recommending to Congress such additional legislation as may be necessary to prevent the impairment or exhaustion of these valuable fisheries, and placing them under regular and permanent conditions of production.

"SEC. 3. That section nineteen hundred and fifty-six of the Revised Statutes of the United States is hereby declared to include and apply to all the dominion of the United States in the waters of Bering Sea; and it shall be the duty of the President, at a timely season in each year, to issue his proclamation and cause the same to be published for one month in at least one newspaper, if any such there be, published at each United States port of entry on the Pacific Coast, warning all persons against entering said waters for the purpose of violating the provisions of said section; and he shall also cause one or more vessels of the United States to diligently cruise said waters and arrest all persons, and seize all vessels found to be, or to have been, engaged in any violation of the laws of the United States therein."

This act, though authorizing and directing the Secretary of the Treasury to establish such regulations and surveillance as should be necessary to insure that the prohibition would be enforced, neither prescribed the machinery nor appropriated the means to carry it into effect. Some restraint has doubtless been imposed upon attempts at violation of the law where they are likely to come under observation, but it is probably violated without hesitation or scruple where the chance of discovery is casual or remote.

STATISTICS OF THE FISHERIES.

The immense take of salmon in the estuary of the Karluk River in 1887 and 1888 had the additional result of attracting attention to a field promising such extravagant returns for the capital invested. More than 30 new canneries were established during the season of 1889. Five were located on the sand spit at the mouth of the Karluk River and 3 others so near as to draw their supplies from that source. Over 350,000 cases of red salmon, representing 4,000,000 fish, were taken from this insignificant rivulet in 1889 and sent into the markets of the world. During this season there were 36 canneries in operation in Alaska, and the value of the salmon pack amounted to $3,375,000.

The following table, showing the Alaskan salmon pack from 1883, when systematic canning operations were first instituted, to 1890, after they had probably reached their largest development, is very interesting as well as suggestive; interesting, as illustrating the wonderful wealth of the waters; suggestive because we know that it has been accomplished by irrational and destructive methods, and by improvident, willful, and contemptuous disregard of natural laws, whose aid and unobstructed operation are essential to the maintenance of a continuing and productive salmon fishery in Alaska:

The Alaska salmon pack from 1883 to 1890.

Year.	Number of cases.	Year.	Number of cases.
1883	36,000	1887	190,200
1884	45,000	1888	298,000
1885	74,850	1889	675,000
1886	120,700	1890	610,747

A review of the statistics of the salmon pack of Alaska from 1883 to 1890, compiled from data gathered by the division of fisheries of the United States Fish Commission, shows that the total yield of the salmon fisheries of this region from 1883 to 1890, both inclusive, was 2,050,497 cases of 48 pounds each, representing an aggregate production of 28,706,958 salmon within the period mentioned. During the first three years the pack was small, viz, 36,000 cases in 1883, 45,000 cases in 1884, and 74,850 cases in 1885. After this the increase in production was phenomenal, and in 1889 had reached the enormous amount of 675,000. Production in the subsequent years receded slightly, but the aggregate for 1890 and 1891 did not fall much short of the pack of 1889. Of the entire Alaskan yield, about one-half is taken from the estuary of the Karluk River. Adding the product of 1891 to the aggregate for previous years, we have a total yield of canned salmon since 1883, when regular canning began, amounting to nearly 2,750,000 cases, and a total value of $11,000,000.

Besides the canned salmon, the rivers of Alaska yield annually nearly 7,000 barrels of 200 pounds each of salt salmon. When we add to the above production the enormous quantities of salmon which are consumed by the natives in the fresh and dried condition, we shall be able to form some adequate idea of the immense value of the Alaskan salmon, and the importance of fostering and establishing conditions of permanence for this great resource.

In 1889 the salmon fishery gave employment to 66 vessels, including 13 steamers, 13 barks, 2 brigs, and 1 ship. Thirty-six canneries were in active operation, not counting a number of small establishments whose pack was light and incidental to general trading with the natives. The capital stock of these canning companies ranged from $75,000 to $300,000. The estimated capital was $4,000,000 and the value of the pack, $3,375,000.

PRESENT CONDITION OF THE FISHERIES—OBSTRUCTIONS IN THE RIVERS.

Early in April, 1890, information reached the Commissioner of Fisheries in regard to a salmon trap, the construction of which had been determined upon by four cannery firms located on the Nushagak River. About 25 miles from the mouth of this river is a tributary known as Wood River, into which most of the salmon entering the Nushagak make their way for the purpose of spawning in the two large lakes at its head. Believing that such action was a violation of the act of Congress approved March 2, 1889, providing for the protection of the salmon fisheries of Alaska, the Commissioner transmitted the information to the Secretary of the Treasury with the suggestion that the necessary steps be taken by some of the Treasury officials in that region. The matter was referred to the chief of the Revenue-Marine Division with the recommendation that, if possible, the captain of one of the revenue-marine steamers cruising in Alaskan waters be directed to make an investigation and, if necessary, have the obstructions removed and the guilty parties arrested and prosecuted. On April 12 the chief of the Revenue-Marine Division returned the correspondence to the Commissioner of Fisheries with the information that the commanding officers of the revenue-marine steamers cruising in Alaskan waters during the ensuing season would be instructed to enforce the law for the protection of the fisheries as far as circumstances would permit. He suggested, also, that the commanding officer of the Fish Commission steamer *Albatross* be instructed to investigate the complaint and enforce the law if found necessary. Inasmuch as the Commissioner of Fisheries did not have authority to give directions for the enforcement of the law, he wrote to the chief of the Revenue-Marine Division on April 17 that if the Secretary desired to confer the necessary authority upon the commanding officer of the *Albatross*, Lieut. Commander Z. L. Tanner, United States Navy, he would take pleasure in forwarding same. On the following day, therefore, the Acting Secretary of the Treasury, Hon. George S. Batcheller, forwarded to the Commissioner of Fisheries the following order, clothing the commander of the *Albatross* with the necessary authority to act in the matter, inclosing at the same time copies of Treasury circular of March 16, 1889, in relation to the matter:

TREASURY DEPARTMENT, OFFICE OF THE SECRETARY,
Washington, D. C., April 18, 1890.

SIR: You are hereby clothed with full power and authority to enforce the provisions of law contained in act of Congress approved March 2, 1889, providing for the protection of the salmon fisheries of Alaska, which prohibits the erection of dams, barricades, or other obstructions in any of the rivers of Alaska, with the purpose or result of preventing or impeding the ascent of salmon or other anadromous species to their spawning grounds.

Respectfully, yours,

GEO. S. BATCHELLER,
Acting Secretary.

Lieut. Commander Z. L. TANNER,
Commanding United States Fish Commission Steamer Albatross,
San Francisco, Cal.

*   *   *   *   *   *

This correspondence was referred to the ichthyologist of the Commission, who made the following report:

UNITED STATES COMMISSION OF FISH AND FISHERIES,
Washington, D. C., July 24, 1890.

SIR: After having considered the letters of Lieut. Commander Z. L. Tanner, United States Navy, dated June 15 and 18, 1890, referring to the construction of a trap in Wood River, Alaska, I respectfully offer my opinion that such a contrivance for the capture of salmon is of the nature of an obstruction which would impede and, in all probability, prevent the ascent of salmon to their spawning grounds. It is therefore clearly a violation of the act approved March 2, 1889, a portion of which is quoted herewith:

[Public No. 158.—An act to provide for the protection of the salmon fisheries of Alaska.]

"*Be it enacted by the Senate and House of Representatives of the United States of America in Congress assembled,* That the erection of dams, barricades, or other obstructions in any of the rivers of Alaska, with the purpose or result of preventing or impeding the ascent of salmon or other anadromous species to their spawning grounds, is hereby declared to be unlawful, and the Secretary of the Treasury is hereby authorized and directed to establish such regulations and surveillance as may be necessary to insure that this prohibition is strictly enforced and to otherwise protect the salmon fisheries of Alaska; and every person who shall be found guilty of a violation of the provisions of this section shall be fined not less than two hundred and fifty dollars for each day of the continuance of such obstruction."

It has been demonstrated that traps in salmon rivers will speedily exterminate the salmon. Newfoundland furnishes a satisfactory illustration of this fact. So well is this matter understood that British Columbia forbids altogether the capture of salmon in narrow reaches of streams, and the rivers are guarded to see that the close time and other regulations are observed; the length of nets and their size of mesh are fixed by law; even the offal from canneries is not allowed to lie in the way of ascending fish.

The Alaskan salmon firms are in the Territory to get fish. They prefer to get them without injury to the future of the business if possible, but get them they must or be overcome by financial disaster. In their efforts to win success they have often stretched nets across the mouths of small streams and prevented the salmon from going up until a sufficient number had collected to make a good seine haul possible. They have erected traps in rivers in such a way as to stop every salmon from ascending, and, in some cases, actually built impassable barricades to prevent the ascent of fish entirely until the demands of the canneries were satisfied. Even when fishing regulations were adopted by mutual agreement among the firms interested individual infractions of the rule were only too frequent.

The trap men on Wood River are building upon the well-known habit of the quinnat (or king salmon) of following along the shores in shallow water to escape from enemies. All the conditions, both natural and invented, will favor the entrance of salmon into the great inclosure at the end of the leader of netting. In all probability few salmon will swim in mid-channel and reach the upper waters and lake sources of the river, and it will always be possible to cut off this remnant in the manner suggested by Lieutenant-Commander Tanner, and actually practiced by fishermen on occasions, that of stretching a seine across the open water. If the Government should interpret its acts so as to allow the use of traps, in spite of the unfortunate outcome of such appliances in neighboring countries, it should then prescribe regulations for the conduct of the fishery and appoint agents to see that the laws are enforced. If these matters are left solely to the discretion of the individuals having a financial interest in this fishery there will soon be no salmon to protect.

Very respectfully,

T. H. BEAN,
Ichthyologist, United States Fish Commission.

Col. MARSHALL MCDONALD,
United States Commissioner of Fish and Fisheries.

Absolute prohibition of the capture of salmon by the use of any kind of nets or traps within 100 yards of the mouths of the rivers would assure that some proportion of each run of salmon would succeed in entering the streams and reaching the spawning grounds.

The prohibition of the use of more than one seine in the same berth would prevent that actual and effective obstruction of the approaches to the rivers which is now accomplished by the use of seines in pairs sweeping the same area and succeeding each other so continuously as to capture every fish coming within the seine berth.

The above requirements, reasonably and uniformly enforced, would probably be sufficient to maintain regular conditions of production and render permanent this great food source. Should they be supplemented by recourse to artificial propagation on an adequate scale, it will be possible not only to maintain the present supply, but probably greatly to increase the annual production. The enforcement of the regulations and requirements above indicated would, however, demand constant minute supervision and the employment of a large personnel and difficult administration.

It is believed that better results and more satisfactory administration could be accomplished by limiting the catch in each stream to its actual productive capacity under existing conditions, and by leasing the privileges of taking the salmon to the highest bidder. The lessees of any river would see that there was no trespassing upon privileges for which they paid. The limitation of the catch being kept safely within the natural productive capacity of the stream, greater care would be exercised by the canners, the quality of the products would be improved, and stability of prices assured by reason of the fact that the total production would be approximately known in advance of the season.

The number of cases packed would be a matter of easy and accurate ascertainment by the Government agent charged with that duty. Should the funds obtained from the lessees be applied first to the administration of the regulations of the fishery, and the balance devoted to systematic fish-culture, it is probable that the revenues from these fisheries will not only suffice for their rational management, but will permit and provide for such extensive fish-cultural operations as will not only maintain present conditions and production, but also greatly increase the annual output.

Very respectfully,

MARSHALL MCDONALD, *Commissioner.*

No. 2.—*Oregon Statutes, vol. 2, of fishing for salmon.*

SEC. 3489. It shall not be lawful to take or fish for salmon in the Columbia River or its tributaries, by any means whatever, in any year hereafter during the months of March, August, and September; nor at the weekly close times in the months of

April, June, and July; that is to say, between the hour of six o'clock in the afternoon of each and every Saturday until six o'clock in the afternoon of Sunday following; and any person or persons catching salmon in violation of the provisions of this section, or purchasing salmon so unlawfully caught, shall, upon conviction thereof, be fined in a sum of not less than five hundred dollars nor more than one thousand dollars for the first offense, and for each and every subsequent offense, upon conviction thereof, shall be fined not less than one thousand dollars, to which may be added, at the discretion of the court, imprisonment in the county jail for a term not exceeding one year.

SEC. 3490. It shall not be lawful to fish for salmon in the Columbia River or its tributaries during the said months of April, May, June, and July with gill nets the meshes of which are less than four and one-eighth inches square, nor with seines whose meshes are less than three inches square, nor with weir or fish traps whose slats are less than two and one-half inches apart.

Nothing herein contained shall prevent fishing in said river or its tributaries with dip nets during the fishing season as established and defined by section thirty-four hundred and eighty-nine.

Every trap or weir shall have in that part thereof where the fish are usually taken an opening at least one foot wide, extending upward from the bottom toward the top of the weir or trap five feet, and the netting, slats, and other material used to close such aperture while fishing shall be taken out, carried upon shore, and there remain during the said months of March, August, and September, and the weekly close time in the months of April, May, June, and July, as prescribed in section thirty-four hundred and eighty-nine, to the intent that during said close time the salmon may have free and unobstructed passage through such weir, trap, or other structure, and no contrivance shall be placed in any part of such structure which shall tend to hinder such fish.

In case the inclosure where the fish are taken is furnished with a board floor an opening extending from the floor five feet toward the top of the weir or trap shall be equivalent to extending the said opening from bottom to top.

Any person or persons violating the provisions of this section or encouraging its violation by knowingly purchasing salmon so unlawfully caught shall be deemed guilty of misdemeanor, and upon conviction thereof shall be fined for the first offense not less than five hundred dollars nor more than one thousand dollars, to which may be added imprisonment in the county jail for a term not exceeding one year.

SEC. 3491. The person or persons making complaint of any violations of the provisions of this act shall, upon conviction of the offender, be entitled to one-half the fine recovered; and any prosecuting attorney who shall, upon complaint being made to him of the violation of this act, fail to prosecute the party accused shall be deemed guilty of a misdemeanor in office, and upon conviction thereof shall be fined in the sum of five hundred dollars for each and every offense.

SEC. 3492. This act shall not be so construed as to interfere in any way with any establishment or enterprise for the propagation of salmon, whether by the United States Government or any regularly organized company or society for that purpose, located or operated upon said Columbia River or any of its tributaries.

SEC. 3493. It shall be unlawful for the proprietor of any sawmill on the Columbia River or any of its tributaries, or any employee therein, to cast the sawdust made by such sawmill, or suffer or permit such sawdust to be thrown or discharged in any manner, into said river or its tributaries below the Cascades of the Columbia River and falls of the Willamette River.

For each and every willful violation of this section the party guilty of such violation shall be liable to a fine of fifty dollars for each and every such offense, to be recovered before a justice of the peace of the proper county.

SEC. 3494. Any party convicted of any violation of the provisions of this law shall be sentenced to pay the fine and costs adjudged, and in default of paying or securing the payment thereof, he shall be committed to the county jail until such fine and costs shall be paid or secured, until he shall have been imprisoned one day for every two dollars of such fine and costs. But execution may at any time issue against the property of the defendant for whatever sum may be due of such fine or costs.

Upon payment of such fine or costs, or the balance after deducting the commutation by imprisonment or securing the same, the party shall be discharged.

All fines and penalties collected for violation of this act shall constitute a fund for the maintenance of hatching houses for the propagation of salmon, and be disbursed in accordance with the provisions of an act entitled "An act to encourage the establishment of hatching houses for the propagation of salmon in the waters of the Columbia River."

SEC. 3495. All fines and penalties hereby or herein imposed shall be enforced and collected as other fines and penalties, and jurisdiction to enforce such fines not herein given to the justices' courts shall be vested in the circuit court of the proper county.

SEC. 3496. It shall be unlawful to catch salmon fish with net, seine, or trap, in any stream of water, bay, or inlet of the sea, or river of this State, at any season of the year between sunset on Saturday and sunset on the Sunday following of each and every week.

SEC. 3497. Any person who shall violate this act, either by fishing with the means and appliances aforesaid or hiring others to do so, shall be guilty of misdemeanor, and be fined in any sum not less than fifty nor exceeding one hundred and fifty dollars, and by imprisonment in the county jail of the proper county not less than five days nor more than ten days.

SEC. 3498. Justices of the peace shall have concurrent jurisdiction in such cases.

No. 3.—*Extracts from a treatise on the law of Scotland relating to rights of fishing, by Stewart.*

SALMON FISHING—POACHING AND OTHER OFFENSES.

Fishing by means of a light, etc.—By the act 1868 (31 and 32 Vict., c. 123, 17) it is enacted "that every person that shall use any light or fire of any kind, or any spear, leister, gaff, or other like instrument, or otter, for catching salmon, or any instrument for dragging for salmon, or have in his possession a light or any of the aforesaid instruments under such circumstances as to satisfy the court before whom he is tried that he intended at the time to catch salmon by means thereof, shall be liable to a penalty not exceeding £5, and shall forfeit any of the aforesaid instruments and any salmon found in his possession; but this section shall not apply to any person using a gaff as auxiliary to angling with a rod and line."

Dynamite.—No person may kill fish in the United Kingdom by means of dynamite or any explosive (40 and 41 Vict., c. 65.)

Catching salmon leaping at a fall.—By the act 1868 (31 and 32 Vict., c. 123, 15, subsec. 5) it is enacted that "every person who 'sets or uses, or aids in setting or using, a net or any other engine for the capture of salmon when leaping at or trying to ascend any fall or other impediment, or when falling back after leaping,' shall be liable to a penalty not exceeding £5, and to a further penalty not exceeding £2 for every salmon taken, and shall forfeit the salmon so taken; he shall further be liable in the expenses of the prosecution."

With regard to this prohibition, it seems only necessary to remark that it extends only to machinery of a fixed nature, and imports no prohibition of dragging pools lying at the foot of falls.

Taking or destroying the young salmon or obstructing their passage, or disturbing spawning beds.—By the act 1868 (31 and 32 Vict., c. 123, 19) it is enacted that "every person who shall willfully take or destroy any smolt or salmon fry, or shall buy, sell, or expose for sale, or have in his possession the same, or shall place any device or engine for the purpose of obstructing the passage of the same, or shall willfully injure the same, or shall willfully injure or disturb any salmon spawn or disturb any spawning bed, or any bank or shallow in which the spawn of salmon may be, or during the annual close time shall obstruct or impede salmon in their passage to any such bed, bank, or shallow, shall be liable to a penalty not exceeding £5 for every such offense, and shall forfeit every engine used in committing such offense, together with any smolt or salmon fry found in his possession."

The clause goes on to declare that this provision shall not "apply to acts done for the purpose of the artificial propagation of salmon or for other scientific purposes, or in the course of cleaning and repairing any dam or mill lade, or in the course of the exercise of rights of property in the bed of any stream." It provides also that the district board may, with the consent of all the proprietors of salmon fisheries in any river or estuary, adopt such means as they think fit for preventing the ingress of salmon into narrow streams, in which the fish or the spawning beds are, from the nature of the channel, liable to be destroyed, but always so that no water rights used or enjoyed for the purpose of manufactures, of agriculture, or of drainage shall be interfered with thereby.

"No fixed engine of any description shall be placed or used for catching salmon in any inland or tidal waters; and any engine placed or used in contravention of this section may be taken possession of or destroyed; and any engine so placed or used and any salmon taken by such engine shall be forfeited; and in addition thereto, the owner of any engine placed or used in contravention of this section shall, for each day of so placing or using the same, incur a penalty not exceeding £10; and for the purposes of this section, a net that is secured by anchors or otherwise temporarily fixed to the soil shall be deemed to be a fixed engine; but this section shall not affect any ancient right or mode of fishing as lawfully exercised at the time of the passing of this act by any person by virtue of any grant or charter or immemorial usage: *Provided, always,* That nothing in this section contained shall be deemed to apply to fishing weirs or fishing milldams."

No. 4.—*Extracts from Bund's law of salmon fisheries in England and Wales.*

[Salmon-fishery act, 1861.]

(1) Causing or knowingly permitting liquid or solid matter to be placed into any waters containing salmon, or into the tributaries of such waters, that poisons or kills fish. Penalties: First offense, £5; second, not less than £2 10s.; not more than £10 and £2 a day; third, not less than £5, not more than £20 a day, from the date of third conviction; fourth, not less than £20 a day.

(2) Using or having in possession lights, otters laths, jacks, wires, snares, stroke halls, snatches, or other like instruments (except gaffs as auxiliary to a rod and line) for taking salmon, trout, or char. Penalty: Forfeiture of instruments; first offense, £5; second, not less than £2 10s., not more than £5; third, not less than £5, or imprisonment for not less than one or more than six months.

(3) Using any fish roe for fishing, or buying, selling, or having in possession any salmon, trout, or char roe. Penalty: Forfeiture of roe; first offense, £2; second, not less than £2; third, imprisonment for not less than one or more than six months.

(4) Using any nets with a less mesh than two inches, unless a smaller size is allowed by bye-law. Penalty: Forfeiture of nets; first offense, £5; second, not less than £2 10s., or more than £5; third, not less than £5.

(5) Placing or using any fixed engine not lawfully in use in 1857, 1858, 1859, 1860, and 1861, for catching, or facilitating the catching, or deterring or obstructing the free passage of salmon. Penalty: Forfeiture of engine; first offense, £10 a day; second, not less than £2 10s. in whole, not exceeding £10 a day; third, not less than £5 in whole, not exceeding £10 a day; fourth, not less than £10 a day.

(6) Using any dam, except legal fishing weirs and fishing milldams, for catching, or facilitating the catching, of salmon. Penalty: Forfeiture of all traps, nets, and contrivances, and all salmon caught; first offense, not exceeding £5 and £1 for each salmon caught; second, not less in the whole than £2 10s., and not exceeding £5 and £1 for each salmon; third, not less than £5, and not exceeding £5 and £1 for each salmon caught; fourth, not less than £5 and £1 for each salmon caught.

(7) Fishing for any salmon within 50 yards above or 100 yards below any dam, or in the head, tail, or race of any mill, unless the dam has a fish pass, approved by the home office, with such a flow of water as will enable salmon to pass up and down. Penalty: Forfeiture of all salmon caught and nets used; first offense, £2 and £1 for each salmon caught; second, not less in whole than £2 10s., and not exceeding £2 and £1 for each salmon caught; third, not less than £5, and not exceeding £5 and £1 for each salmon caught; fourth, not less than £5 and £1 for each salmon caught.

(8) Refusing to place a grating, approved by the inspectors, across any artificial channel for supplying towns with water, or any inland navigation. Penalty: Not exceeding £5 a day for the first offense; second, not less than £2 10s., not exceeding £5 a day; third, not less than £5, and not exceeding £5 a day; fourth, not less than £5 a day.

(9) Refusing to maintain such grating. Penalty: Not exceeding £1 a day for the first offense; second, not less than £2 10s., and not exceeding £1 a day; third, not less than £5, and not exceeding £1 a day; fourth, not less than £1 a day.

(10) Taking, killing, injuring, or attempting to take, buying, selling, exposing for sale, or having in possession for sale, unclean or unseasonable salmon, trout, or char. Penalty: Forfeiture of fish; first offense, £5 and £1 for each fish; second, not less than £2 10s., not exceeding £5 and £1 a fish; third, not less than £5 and not exceeding £5 and £1 a fish, or imprisonment for not less than one or more than six months; fourth, not less than £5 and £1 a fish, or imprisonment.

(11) Taking or destroying, buying, selling, or exposing for sale, placing any device for obstructing the passage of or willfully injuring the young of salmon, or disturbing any spawning bed on which the spawn of the salmon may be. Penalty: Forfeiture of all young salmon, rods, lines, nets, etc.; first offense, not exceeding £5; second, not less than £2 nor more than £5; third, not less than £5.

(12) Disturbing or attempting to catch any salmon spawning or near the spawning beds. Penalty: First offense, not exceeding £5; second, not less than £2 10s. or more than £5; not less than £5.

(13) Fishing for salmon during the annual close season. Penalty: Forfeiture of salmon and nets, or instruments used in fishing; first offense, not exceeding £5 and £2 for each fish caught; third, not less than £5 and not exceeding £5 and £2 for each fish caught, or imprisonment for not less than one nor more than six months; fourth, not less than £5 for each fish caught, or imprisonment.

(14) Not removing fixed engines and temporary fixtures from a fishery within thirty-six hours after close time begins. Penalty: Forfeiture of all engines and temporary fixtures, etc., first offense, not exceeding £10 a day; second, not less than £2 10s. and not exceeding £10 a day; third, not less than £5 and not exceeding £10 a day; fourth, not less than £10 a day.

(15) Fishing for salmon during weekly close time. Penalty: Forfeiture of all nets or movable instruments used; first offense, not exceeding £5 and £1 for each fish; second, not less than £2 10s. and not exceeding £5 and £1 for each fish; third, not less than £5 and not exceeding £5 and £1 for each fish; fourth, not less than £5 and £1 for each fish.

(16) Not maintaining an opening through cribs and traps during the weekly close time. Penalty: Forfeiture of fish; first offense, £5 and £1 a fish; second, not less than £2 10s. and not exceeding £5 and £1 a fish; third, not less than £5 and not exceeding £5 and £1 a fish; fourth, not less than £5 and £1 a fish.

(17) Obstructing any person authorized by the home office to make a fish pass. Penalty: First offense, £10; second, not less than £2 10s. and not exceeding £10; third, not less than £5 and not exceeding £10; fourth, not less than £10.

(18) Injuring any fish pass made under the authority of the home office. Penalty: The expense of making good the injury; first offense, not exceeding £5; second, not less than £2 and not exceeding £5; third, not less than £5.

(19) Doing any act whereby salmon are prevented passing through a fish pass, or taking salmon passing through a fish pass. Penalty: Forfeiture of salmon and instruments used in taking them; first offense, £5; second, not less than £2 and not exceeding £10; third, not less than £5 and not exceeding £10; fourth, not less than £10.

(20) Not affixing a fish pass to any new dam or to any old dam raised or altered so as to create increased obstruction to fish. Penalty: Expenses of making the fish pass, and not exceeding £5 for first offense; not less than £2 nor more than £5 for second; and not less than £5 for the third.

(21) Not keeping the sluices that draw off the water from a dam shut on Sundays and when the water is not wanted for milling purposes. Penalty: Frst offense, not exceeding 5s. an hour; second, not less than £2 10s. and not exceeding 5s. an hour; third, not less than £5 and not exceeding 5s. an hour; fourth, not less than 5s. an hour.

(22) Not making a legal gap in a fishing weir. Penalty: First offense, not exceeding £5 a day; second, not less than £2 10s. and not exceeding £5 a day; third, not less than £5 and not exceeding £5 a day; fourth, not less than £5 a day.

(23) Not maintaining a legal free gap or altering the bed of the river so as to reduce the flow of water through a legal free gap. Penalty: First offense, £1 a day; second, not less than £2 10s. nor more than £1 a day; third, not less than £5 nor more than £1 a day; fourth, not less than £1 a day.

(24) Placing any obstruction, using any contrivance, or doing any act whereby salmon are deterred in passing up and down a free gap. Penalty: First offense, not exceeding £5; second, not less than £2 10s. and not exceeding £10; third, not less than £5 and not exceeding £10; fourth, not less than £10.

(25) Using any box or crib in any fishing weir or fishing milldam, the upper surface of the sill of which is not level with the bed of the river, and the bars or inscales of which are nearer than 2 inches and not placed perpendicularly. Penalty: First offense, not exceeding £5 a day; second, not less than £2 10s. and not exceeding £5 a day; fourth, not less than £5 a day.

(26) Not maintaining a box or crib in such state. Penalty: First offense, not exceeding £1 a day; second, not less than $2 10s. and not exceeding £1 a day; third, not less than £5 and not exceeding £1 a day; fourth, not less than £1 a day.

(27) Using any box or crib in any fishing weir or fishing milldam, having any spur, tail wall, leader, or outrigger of a greater length than twenty feet from the upper or lower side of such box or crib. Penalty: First offense, not exceeding £1 a day; second not less than £2 10s. and not exceeding £1 a day; third, not less than £5 and not exceeding £1 a day; fourth, not less than £1 a day.

[Salmon fishery act, 1865.]

(28) Fishing for salmon with a rod and line without a license. Penalty: First offense, not less than double the amount of the license duty and not exceeding £5; second, not less than £2 10s. and not exceeding £5; third, not less than £5.

(29) Fishing for salmon within any fishing weir, fishing milldam, putt, putcher, net, or other instrument or device other than a rod and line. Penalty: First offense, not less than double the license duty payable and not exceeding £20; second, not less than £2 10s. and not exceeding £20; third, not less than £5 and not exceeding £20; fourth, not less than £20.

(30) Any person fishing refusing to produce his license on being asked by a conservator, water bailiff, or licensee. Penalty: First offense, not exceeding £1; second, not less than £2 10s.; third, not less than £2.

(31) Fishing for trout or char between the 2d October and the 1st February following, both inclusive. Penalty: Forfeiture of fish; first offense, not exceeding £2, second, not less than £2 10s.; third, not less than £5.

(32) Not entering salmon intended for exportation with the proper officer of customs before shipment between the 3d September and the 30th April. Penalty: First offense, not exceeding £2 a fish; second, not less than £2 and not exceeding £2 a fish; third, not less than £5 and not exceeding £2 a fish; fourth, not less than £2 a fish.

[Salmon fishery act, 1873.]

(33) Clerk of the peace omitting to send notice of the names and addresses of the conservators appointed by different counties where the district comprises more than one county to the clerk of the board within fourteen days of the appointment. Penalty: First offense, £2; second, not less than £2 10s.; third, not less than £5.

(34) Clerk of the justices not sending certificate of any conviction against the salmon fishery acts to the clerk of the board of conservators within one month. Penalty: First offense, not exceeding £2; second, not less than £2 10s.; third, not less than £5.

(35) Shooting any draft net for salmon across a river or across more than three-quarters of its width within 100 yards of any other draft net not drawn in and landed. Penalty: First offense, not exceeding £5; second, not less than £2 10s. and not exceeding £5; third, not less than £5.

———

No. 5.—*Letter of L. A. Pederson, showing condition existing on Naknek River, Alaska.*

SAN FRANCISCO, *January 28, 1895.*

DEAR SIR: Hope you will pardon my taking this liberty, sir, but Mr. Alexander, fish commissioner for this coast, speaking in reference to my cannery site in Alaska, recommended that I write full particulars to you personally.

Mr. Alexander stated that he was about to leave for Washington and will also bring the matter before you. He has been on the ground and is personally acquainted with the whole affair.

I have been to Alaska regularly for the last nine years, and for the last five years have been salting salmon for myself on the west side of Naknek River, Bristol Bay.

Having but little money, I was obliged to start alone on a small scale at first and only put up 250 barrels. I did this without any assistance from anyone.

The company who allowed me to take passage on their vessel charged $600 for the round trip. A moderate figure would have been $200.

The next year I packed 450 barrels with the assistance of one man and a little help from the natives.

For the third year I had a contract made to pay $700 for my passage, but at this time the Alaska Packers' Association was formed, which, as you no doubt know, is a combination of all the Alaska canneries, excepting two or three.

I went to them and endeavored to obtain a passage, but they refused to take me up and told me that if I could do anything alone to go ahead. This was rather discouraging to me, but, nothing daunted, I decided to charter a small schooner, *Golden Fleece* by name, and after many hardships succeeded in coming home with 1,200 barrels. These were packed with the assistance of 12 men and the natives.

The fourth year I chartered the schooner *Prosper*, and with the assistance of 25 men and the natives packed 2,600 barrels.

The fifth year I chartered the schooner *Sailor Boy*, and with the assistance of 29 men and the natives came home with 2,650 barrels. This was for the year 1894, but I have not succeeded in selling all the salmon as yet, owing to action of the Alaska Packers' Association.

Knowing that most salting expeditions finally result in a cannery being put up, it has been their policy right along to discourage salting as much as possible, and last year they made a master stroke by deciding to put up as much salt salmon as they could and then sell it for much less than cost. They reduced the price from $8 to $5 per barrel, which, of course, ruined the profit I had been making each year. Besides, I am unable to get rid of the salmon.

The only thing left for me to do is to start a small cannery and I am now making the necessary preparations. Before coming to this conclusion I appealed to the Alaska Packers' Association and endeavored to sell them my plant, failing which, I agreed to pay them for the use of their side of the river a good round rental each year. They also refused this, and in fact I have not been able to come to any understanding with them whatever.

They are also making preparations to put up a cannery across the river from me, and I learn from good authority that their idea is to put traps on my side of the river also, so that I will be entirely shut out. The situation is so that without traps the fish can not be caught.

I had my side of the river duly surveyed last summer, and what I particularly desire and pray for is that you restrict them from fishing on my side of the river and on the land that I have had surveyed. Of course, my survey only goes down on the beach as far as high-water mark, and it seems to me that they can be restricted from fishing on any land in front of my survey, and which is dry at low water.

I would be perfectly willing to stay on my side of the river if they would stay on theirs, and they have the better side. Of course, I shall be dependent entirely for my living on what I do in this river, while they are a large corporation with $5,000,000 of capital, and have cannery sites all over Alaska, so that any little inconvenience they might suffer by bothering me, or any small loss which they might incur by so doing, would really cut no figure in their business.

If they are allowed to block me in with traps on my side of the river I shall certainly be driven to the wall, and not only I, but many of the poor natives, who depend upon the work which they obtain from me for their living.

For the last three years I have given them $1 a day and board. They are also becoming more civilized each season and seem more willing to work. It has been my policy right along to encourage them in this and to teach them as much as possible.

Each year, as soon as our vessel is sighted out in Bristol Bay, a score or more of the natives start right out in canoes and board us many miles from our anchorage. They are always anxious for provisions, and I deal out to them chests of crackers, clothing, and provisions, all of which seem to delight them very much.

All I have made in and out of Alaska I have put in improvements at my cannery site, and it seems a strange law to me which will allow a huge corporation like the Alaska Packers' Association to down a poor man.

Since first starting in Naknek with a very small capital I have certainly had uphill work and a varied experience. It has been nothing but constant work and trouble. This will, of course, all count for naught if these people are allowed to crush me, and my whole prospect in life will be spoiled.

I appeal to you, sir, for protection, and hope you will do all in your power to assist me in seeing that they keep within the law, and that they do nothing toward their fellow-men (even if the law can be evaded in so doing) except what is just.

I know the Government likes to protect the natives as much as possible, and if I am ruined the natives will be injured beyond measure also. In addition to this, all the men that I have been employing each year, and have taken from here, will of course be out of employment, and if I am successful in constructing my cannery and protected in catching fish on the land which I have had surveyed, I will be able to employ many more than any year before.

I feel satisfied that with your assistance I can pull through, and this large corporation can easily be kept within the proper bounds. They have certainly no right to molest me and have no reason for being jealous of me, as my cannery is not in opposition to theirs, for I was in Alaska long before the Alaska Packers' Association was ever thought of.

Would it be convenient for you to send a steamer by the river before the fishing season commences, say about middle of June?

Am very anxious to hear from you, sir, and hope you will be kind enough to let me know what the prospect is as soon as convenient.

Thanking you in advance and anxiously awaiting your reply, I remain,

Your obedient servant,

L. A. PEDERSON,
722 Harrison street.

Col. MARSHALL McDONALD,
United States Commissioner of Fish and Fisheries, Washington, D. C.

KARLUK, ALASKA, *August 16, 1894.*

STATEMENT OF MR. ARTHUR L. DUNCAN.

Arthur L. Duncan, superintendent of the Hume Canning and Trading Company, Tangleioot Bay, near Karluk, Kadiak Island. My business is catching and canning salmon in shore seines drawn from the shore.

On July 9, 1894, we made our first lay out with the purse seine, under direction of Mr. James Williams, who was then our boss fisherman (purse seine), and who is now at San Francisco.

We first started to fish below Julia Ford Point, but the Alaska Packers' Association did not trouble us below that point. On July 9 we started to lay out our purse seine, and after we got our line run out this party came out from the shore and Mr. Williams picked up our line and then moved his whole gear, lighter and all, farther down toward the mouth of the river; he did this because he thought they were coming out to interfere with his net. Previous to this a notice was found signed "Fishermen of Karluk," outside our cannery, nailed to the flag pole outside of my house. This notice was to the effect that if we fished within the limits of Seven Mile Point and Julia Ford Point, with a purse seine, it would meet with the same fate that the traps did used by Barker in the Karluk River. (Destroyed.)

On the second occasion, when we went to lay out our purse seine, we were about three-quarters of a mile from the mouth of the river, and our net could not in any way have covered the mouth of the river, and they run a net out so that we could not close in without going over their net. On this occasion, Mr. Williams shortened up his circuit and run his net into the barge, just around their seine skiff, thus completing the one circuit. To do this, however, he could not get the full net out, but only a part of it. Then he made a haul, which, of course, was spoiled by not being able to get out his full net. The men pulled their net out, and cursed, and said the next time they would fix me. Our men went to the men, and shortly afterwards two steam launches and two seines came down to where the purse seine was lying, and every time that Mr. Williams would make a move these men would follow with the steam launches and seines.

The next day Mr. Williams said he would have to overhaul his gear and shorten up his net; if he was going to be molested he could not work such a long net. About 1.30 p. m. we were ready to start out again and get down to the point (close to the slide), and on passing the Alaska Improvement Company's steam launch noticed two men coming out from the shore in a seine boat, evidently to get up steam, and afterwards noticed they were getting up steam in one of the Alaska Packers' Association Company's steam launches which was lying at her mooring. About 2 p. m. we started to run out a purse seine, and in the meantime a seine skiff started off Karluk belonging to the Alaska Packers' Association and arrived just as we got our purse seine haul out about 2 fathoms to each wing. Capt. Harry Newman, of the Alaska Packers' Association, was in charge of this boat, and he ran over our net with his seine skiff and I warned him not to do it, and to keep away. He didn't notice this and was starting in to cross the line and held up his anchor evidently with the intention of dropping it inside our net and we called to him not to do it, but he took no notice of it and dropped his anchor over into the middle of our net. He then circled around the inside of our net and then crossed over the cork line and then passed the painter of his seine skiff to one of the Alaska Packers' Association steam launches which had come down in the meantime. The steam launch then towed his boat with his anchor still down in our net. I warned him not to do this, but he simply pointed to the seine skiff and to Newman.

Then another steam launch came down and passed over our net, and then Mr. Barling, of the Alaska Improvement Company, came down with a steam launch and had some dories, I think. He then passed the line of his seine skiff to the other Alaska Packers' Association's steam launch which had passed over the top of our net and I warned him to keep away, but he ran right through it and passed over the other side and came to the back of our lighter. Then he came around to the front of the net where the Alaska Packers' Association's steam launch was turning the seine skiff with the anchor and he took the line from the other steam launch and began to tow; that is, Barling's steam launch began to tow the skiff with the anchor in place of the other steam launch; then the other two steam launches hitched on to Barling's launch and all three towed. I was in our steam launch, and finally, after two attempts, succeeded in cutting the line. We then hauled in the balance of the purse seine to the lighter, and the anchor, which we found entangled in the web of the purse seine, we threw overboard, and after getting the balance of the web we started off and went down to the waterfall, which is about 2 miles from the mouth of the river, and two of the other steam launches followed us with a fishing gang and gear. We went to see if one of our fishing gangs had got any fish and then we started home. Within a few days after we laid our purse seine in front of our cannery, but they didn't trouble us any this time. We fished several times after this with our purse seine, but we were not troubled in front of our own cannery. They said we must not use our purse seine between Julia Ford Point and Seven Mile Point. Julia Ford Point lies just next to our cannery, between it and Barling's. Seven Mile Point is a point about 7 miles north of the Karluk River. They also limited us to a mile and a half offshore. This notice, however, simply applied to our purse seine.

Once before this our men had gone over to the mouth of the river and were about to start to lay a shore seine, but Barling informed them as often as they did this he would lay another seine within theirs and scoop the catch.

We have never used our purse seine within the limits laid down in the notice of the Karluk fishermen since the disturbance. We fish now exclusively with the shore seines directly in front of our establishment to the beach and sometimes down at the waterfall. We never go to the river at all.

The purse seine we found could not be worked with advantage off the shore in front of our cannery, and we have not used it regularly since the disturbance, but have made several trials. Ordinarily in using our shore seines we start with a line at Bridle Line, 15 or 20 fathoms; we then spread the seine in a semicircle, according to the way the fish have come in; we then haul it in on the other side by means of a donkey engine.

I desire to bring out especially the point if we were permitted to use the purse seine within the limits prescribed by the notice of the Karluk fishermen we could do so successfully; at least Mr. Williams claims this. Barling notified our boss fisherman on the day when he threatened to cork our lines above mentioned that any attempt on our part to land a shore net would be a failure.

The Indians are in the habit of drawing nets in the Karluk River and selling the fish to the Alaska Packers' Association.

To-day there are no obstructions in the Karluk River. The Indians merely drag the seines in the river, in my opinion.

There are more fish here this year than there were last, although they were late in coming.

We have 16,000 cases now and hope to get 28,000 during the season, and we have already cleared our expenses. We shut down last year September 16.

We fish at any time, regardless of tide.

There has been an understanding that there shall be no fishing by seine or otherwise from Friday at 6 p. m. to Saturday at 6 p. m., but all the canners at times have disregarded this.

The Alaska Packers' Association have four canneries. Two are now in operation. We have 65 Chinese in our employment and make our contract with one Chinaman at San Francisco. We guarantee him 25,000 cases and he is paid 44 cents per case and put up 800 a day, good and merchantable, and lacquer and label them.

All our Chinese are registered except one, and it is stated in our contract that if they are not registered the contractor is to pay the fine. We take them up and down. They return about September or October, after we have finished our season.

We have 31 white men, Swedes and Germans, and no natives, and have about 110 men in all.

The Chinese feed themselves, mainly on rice and fish. We merely give them quarters and fuel.

I think there ought to be some limitation at the mouth of the river.

I have worked in a hatchery and know of no reason why we should not succeed up here.

KARLUK, ALASKA, August 16, 1894.

Then personally appeared the within-mentioned Arthur L. Duncan and made oath that the statements herein contained were true, to the best of his knowledge and belief.

C. L. HOOPER,
Notary Public, District of Alaska.

REPORT OF JOSEPH MURRAY, SPECIAL TREASURY AGENT, FOR THE YEAR 1895.

DIVISION OF SPECIAL AGENTS,
TREASURY DEPARTMENT,
Washington, D. C., December 20, 1895.

SIR: I have the honor to report that pursuant to Department instruc tions dated April 4, 1895, I proceeded to the Pacific Coast and sailed from Seattle April 23, on board the regular mail steamer for Sitka, where I arrived May 1 and learned that court was about to be held at Juneau, to which city I immediately returned for the purpose of looking after the interest of the Government, as it might appear in the ex-Marshal Porter case, one of whose deputies, Mr. Adolph Myer, was about to be tried on charges of forgery, embezzlement, stealing public records, and several others of like nature.

My written instructions are as follows:

TREASURY DEPARTMENT, OFFICE OF THE SECRETARY,
Washington, D. C., April 4, 1895.

SIR: You are directed to perfect your arrangements with a view to your departure for Sitka, Alaska, with as little delay as practicable. It will be your duty to ascertain and report the location of every salmon cannery or saltery in Alaska; the capacity of the same in cases, barrels, half-barrels, and kits; the pack in full for each season; the number of boxes of tin consumed and the cost of the same per box at place of purchase; the approximate or actual selling price of the product of each fishery in the market to which the same may be consigned; the number of employees in each cannery and the totals thereof, segregating whites, natives, Chinese, etc., male and female, adults and minors, and whether citizens or aliens. You should include, also, in said reports the codfish, herring, herring-oil, guano, and other such industries. It is desired that you investigate the alleged taking and destruction of the eggs of game wild fowl in Alaska, as well, also, as to the alleged wanton destruction of game birds, deer, fox, and other animals, and also the advisability of adopting suitable regulations as to close seasons, in order to prevent such destruction in future.

You should visit, if possible, every cannery in Alaska, and, when practicable, the necessary journeys should be made on vessels of the United States. This instruction is not to be construed, however, as forbidding the use of other means of conveyance when necessary. You are expected to report to the nearest collector of customs any infraction of the revenue laws which may come to your notice. You should report, also, to the Department any violation of the laws relating to the introduction of firearms or of liquors into the Territory of Alaska.

For your information I inclose herewith copy of the circular dated August 10, 1892, pertaining to the erection of dams, barricades, or other obstructions in the rivers of Alaska for the purpose or result of preventing or impeding the ascent of salmon or other anadromous species to their spawning grounds. It will be your duty to enforce the provisions of said circular and to warn all persons who have erected dams, barricades, or other obstructions to remove the same forthwith, and in default thereof you should report the facts, with the proper proofs, to the United States attorney for prosecution.

You should submit reports to the Department from time to time showing the result of your work, and at the close of the fishing season you should forward a full report, covering said season and stating the result of your observations under these instructions. Any recommendations you deem advisable may be embodied in your reports. Any official communication which the Department may find necessary to address to you hereafter will be mailed to Sitka, Alaska. In this connection you are informed that in addition to your duties as an agent for the salmon fisheries you are to hold yourself in readiness to make such other investigations or render any service which the Department may require of you. If practicable, you should at some time during the ensuing season visit the seal islands of St. Paul and St. George for the purpose of inspecting the rookeries thereon and of comparing their condition with that of the season of 1894, with which you are familiar.

Respectfully, yours,

J. G. CARLISLE,
Secretary.

Mr. JOSEPH MURRAY,
Special Agent for the Protection of the Salmon Fisheries,
Fort Collins, Colo.

In addition to the foregoing, I was verbally instructed (time permitting) to attend court during the trial of the ex-Marshal Porter case and to take particular notice of how jury trial was conducted in Alaska, and to learn what I could from reliable sources about the manufacture and importation of spirituous liquors.

Finding it was as yet too early for salmon fishing and that I could not find transportation to the nearest cannery for several weeks, and as I was in the midst of the best part of Alaska and of its best and most energetic citizens, where I could procure most of the information asked for in my instructions, I resolved to attend court until the arrival of the Bering Sea patrol fleet off Sitka, and then continue my journey to the westward.

During our travels through Alaska in 1894, Hon. C. S. Hamlin, Assistant Secretary of the Treasury, and I were informed at every important point we touched and found white men that, " because of its nonenforcement, the law is looked upon as a farce," and that " it is impossible to find a jury to convict for smuggling or violating the revenue law," and I am sorry to have to report that it is only too true.

For three weeks I was present at every session of the court, and in that time I learned beyond a doubt that not only were juries to be had to return verdicts of " not guilty" in behalf of every violator of the revenue law, but also for any crime, if one only knew the particular attorney to employ.

Mr. Adolph Myer had been a deputy for Marshal Porter; had absolute control and personal charge of the marshal's office, books, and money, and for years served his superior faithfully and well. But under the evil influence of bad and wicked men he was led step by step from one crime to another until forgery and embezzlement were reached, and then the end.

When the case was about to come to trial, I was in daily, hourly communication with the district attorney, whom I advised to stand up for the right against all of the vile methods that might be used against him, and that in doing so he would be supported by the Government. He said he was afraid of bodily injury, of his personal safety; that unless he could secure the joint services of a certain attorney whom he named and whose strength and worth lay in his power to influence juries, it would be useless to try the case before a jury, for most of the jurymen would be personal friends of the prisoner and many of them participators in his crime; that although the prisoner was guilty of enough crime to keep him imprisoned twenty years, if he could not influence the jury he would be turned loose on a verdict of "not guilty."

Not knowing how to influence the jury for the purposes indicated, and being unable to control the district attorney, I was necessarily obliged to remain a silent spectator of a compromise between the parties interested, the terms of which were that on condition of the withdrawal of the plea of "not guilty" and the substitution of the plea of "guilty" the prisoner would be let off with a small fine and light sentence, which was done by the district attorney stating that a fine of $50 and twenty-eight months' imprisonment would be satisfactory.

As soon as he was sentenced he was taken from his cell to the grand jury room to testify against his former employer and superior officer, ex-Marshal Porter, and he actually did testify to Porter's having embezzled or stolen a sum of money from the Government, sent from the Department of Justice by check, amounting to some $1,120.32, which amount was part of the money drawn by Deputy Myers from the Department during the temporary absence of the marshal, and for which he had just been convicted.

And yet, on testimony of that sort and from such a source, ex-Marshal Porter was indicted by the grand jury of Alaska for embezzlement. He was approached in my presence by the district attorney as a friend, and asked to acknowledge that the Government owed the money to the marshal's office, or to be disgraced in his old age by an indictment by the grand jury.

Porter answered that he would die before he would consent to rob the Government, and the next day he was indicted.

LIQUOR AND SMUGGLING.

Liquor cases were called and disposed of with the regularity of clockwork, and always with the same result; the witnesses were Indians and half-breeds, the prisoner was a white man, and his friends and chums were in the jury box to acquit him.

"Can you render a verdict according to the law and testimony," said the judge to a man who was being sworn as a juror. "I can," said the fellow, "unless the testimony is that of an Indian."

The testimony of Indians is not valued in Juneau, although many of them are brought in here as witnesses, and supported at the expense of the Government.

Within sight of the court-house were 30 public saloons open and doing a public business, some of the more pretentious ones keeping open house all night, and there was not a Government officer in Juneau who could be found to interfere with them.

On one technicality or another it seems the laws are not sufficiently explicit to make it the plain duty of any particular officer to raid a saloon without the cooperation of other officers, who are, as a rule, not on hand when wanted.

Speaking to a customs officer at Juneau, I said, "How on earth do you account for the existence of so many saloons in Juneau, and many larger ones in course of erection, if you men do your duty?" To which he replied, "Mr. Murray, I know you are justified in asking such a question, but you do not know anything about the real situation here or you would not blame me personally. When I first came here I was zealous and watchful, and I raided a smuggler's den and captured some 10 barrels of liquor, but what was the result? The district attorney came into court and moved to have that smuggler discharged on paying a fine of $50."

Meeting the district attorney, I asked him for his side of the story, and he said, "Yes, I did let the fellow go on a small fine, for I found that because he was not in the inner circle of smugglers and vendors he had been selected as a victim and his whisky seized, taken to the custom-house, and sold at private sale to one of the inner ring for less than one-third its real value."

And so the story continued to the end of the chapter; one officer willing to lay all the blame on the other, while between them the interests of the Government are left to suffer, and the law, that was intended to do good, become a subject of derision and contempt.

At Juneau many influential professional and business men—whose names can be given if necessary—expressed themselves to me in substance as follows: "There are 30 saloons here doing an open, public business, and the governor is being very badly deceived by men high in public affairs who are all more or less financially interested in the liquor business. We favor the fearless enforcement of the law or its unconditional repeal. We think that the true solution of the liquor question in Alaska is high license—say $1,000 in Sitka and Juneau and in proportion in smaller places."

One of the most prominent attorneys at the Juneau bar said: "I have faith in the future of Alaska, and I think I can give some reliable information about the country and its needs. 1 believe the Government is to blame because, for ten or twelve years, no effort has been made to enforce the law, until now the average man has no idea of having any law enforced. Courts, juries, and lawyers are looked upon with contempt. Juries can not be found here, even among our best people, to convict for smuggling or violating the revenue laws. Perjury is common; and I should advise the taking away the jury system of trial in cases where the excise laws are in question. I would say that all petty cases should be tried without a jury. Everything here—cost, distance, and sparse settlement—is against it. The whole system needs an overhauling. Things are done in such a slipshod manner that Government interests are neglected and the weak attempts made to uphold the law are a complete farce. No serious attempt has been made to enforce the liquor law, and liquor is sold here publicly. Charge $1,000 for license and then enforce the law. Had I the power to do it, I would enforce the law at any cost; for, as now carried on, we are teaching the rising generation to utterly disregard all law, and they are growing up to be our dangerous classes."

The foregoing are sample conversations with the best people in Alaska, and I could quote scores of them were it necessary.

Complaint was made on all sides by men of that large class who are too poor to purchase liquor in large quantities and are not influential enough to get permits from the customs authorities to bring it in on the mail steamer. That only a few favored ones—mostly liquor dealers—were allowed this privilege seemed to be a source of much indignation.

Exhibit marked B, handed me by the district attorney, shows the quantity of liquor that entered by permit from January 1, 1894, to March 10, 1895—fourteen months; during which time permits were issued to 34 persons to bring in several hundred barrels of distilled and malt liquors.

It seemed that the necessity to obtain a permit had ceased to exist when I was in Juneau in May, for representative salesmen for wholesale liquor houses at San Francisco, Seattle, and Portland were offering to deliver the liquor into the saloons at Juneau before they would ask pay.

This, in brief, is a true outline of the liquor question in Alaska, nor can it be remedied unless the Government goes to work to enforce or repeal the present prohibitory law relating to the liquor traffic in the Territory.

So long as the Government does not own or control a boat of any sort in a stretch of country 1,500 miles long, where the only road is a waterway—so long as Government officers are compelled, because of lack of boat service, to stand helpless on shore while the smuggler plies his illegitimate trade beneath their very eyes, so long will the present state of affairs continue to curse Alaska and to be a disgrace to our whole country.

While at Juneau in May I was informed of an attempt that was about to be made to land a cargo of liquor destined for the Yukon Valley trade, and one of the most energetic inspectors in Alaska was on the watch to capture it if possible. He did not succeed, however, for by the time he secured a boat to transport him to the rendezvous of the smugglers he found he was twenty-four hours late. Speaking of the affair afterwards he said to me: "If I only had a boat that was always at my disposal I think I could break up a great deal of this smuggling; but, hampered as I am now, I am powerless, for no sooner do I hire a boat to go anywhere than the signal flies over the district. It will require the presence of a revenue cutter and half a dozen steam launches to kill off smuggling in Alaska." Every word of which I indorse.

The peculiar conditions surrounding the Alaskan liquor question have not been taken into account by many men of extreme views who have written or spoken on the subject; indeed, I question if they ever understood it. The truth is that if there is a climate under the sun where liquor is a necessity to man that climate is in Alaska, and consequently white men demand and must have it at any cost and in spite of all obstacles.

This is the reason we find 99 per cent. of the white population bitterly opposed to the present prohibitory law. This is why no officer can be found to attempt to enforce the law or a jury to uphold it. And where public sentiment and public opinion are so plainly against a law, no matter how well intentioned or good in itself, it is wise to heed the sign and amend or repeal it. During a conversation with the assistant district attorney, Mr. Hoggert, on this subject he said: "During the past four years $148,000 were spent in Alaska on cases of Indians and half-breeds who had gotten drunk or had peddled whisky without any lasting or definite results. Had we had a high license during that time we could have saved that expense to the Government and collected revenue enough to make the Territory self-supporting."

DESTRUCTION OF GAME-FOWL EGGS.

The stories told of the wanton destruction and the systematic stealing of wild game-fowl eggs have no foundation in fact.

I have traveled over thousands of miles of the coast line of Alaska, making diligent inquiry into this matter, without finding one person who knew anything about it. I have conversed with men who spent twenty to thirty years in the interior of Alaska, mining, hunting, and trading, men who had gone over every mile of habitable land in the Territory, without ever hearing of such a thing until I asked them. I have written to traders whose business takes them to the Upper Yukon country, far into the British possessions, men who travel from the source to the mouth of the great river; I have written to missionaries whose labors call them into all the native settlements on the Yukon, Kuskoquim, and other rivers, and the unvarying reply is, "We never heard anything about such things."

As a matter of fact, it is not yet known for certain where the wild fowl lay their eggs. They certainly find some island, marsh, morass, swamp, or tundra where man can not penetrate, or, at all events, where he has not as yet gone nor is likely to go until the inducement is something of far more value than wild-fowl eggs.

DESTRUCTION OF DEER.

The destruction of deer in southeastern Alaska and in all the timbered portion of the Territory from Cape Fox to Port Moller, a distance of, say, 1,200 to 1,500 miles, is carried to such excess that it would hardly be credited in a civilized community.

I saw bales of the dried deerskins at many of the trading posts awaiting shipment, and when I asked what use had been made of the carcasses, I was told the deer were shot for their hides only.

I was informed by many men—officers and citizens—that, as the weather became warmer in the early spring, the smell from decaying deer carcasses became horribly offensive around the towns and villages. White men go out and kill the animals for fun, just to see who can knock down most in a given time. The natives kill them, because they can get a drink of whisky, valued at 25 cents, for every skin secured.

That such things have been allowed to continue at any time is to be deeply regretted; but that it is still allowed to continue after the natives on the seal islands have become a burden on the Government, and other tribes to the northward soon will be because of the wanton waste of their natural food supply on land and water, passes the comprehension of every sensible citizen who understands the present situation.

To the northward we are endeavoring to procure and foster the reindeer for a future food supply for the natives of that barren region, and it is a very laudable enterprise; but at the same time we allow the continued wanton destruction of the deer that covers the whole timbered part of Alaska—an empire as large as Texas. In the winter, when the snow is deepest and the animals can not make a way through the dense undergrowth beneath the timber, the so-called sportsmen assemble, and with dogs drive them out on the seashore, whose beaches are kept clean by the tides, where riflemen are ready, stationed in boats offshore, to begin the manly sport of shooting down helpless creatures, who can neither resist nor escape.

The following letter from an eyewitness explains itself:

STEAMER ALBATROSS, *Unalaska, August 28, 1895.*

MY DEAR SIR: I have not been able to unearth the notes I had on deer killing in Alaska. Briefly, their slaughter has been very great. During the winter of 1894 deer were killed and wasted in southeastern Alaska. Snow was unusually deep and the deer were forced to the beaches, which were left clear by the tides. Shooting was done from boats and canoes by both whites and Indians. I know of three Indians killing 175 deer from canoes in two days. Many whites shot for hides alone, and at many places hides could be bought for 35 cents each. I do not think that Indians should be prevented from shooting all kinds of game for their own needs, but killing for hides alone is certainly reprehensible, and if the rate of slaughter that has been going on for the past few years is continued, there will be very few deer left.

As the hides are of comparatively little value, their exportation might be stopped without causing any serious hardship to anyone, and of course when the hides become unsalable, the Indians will not kill many more than they need.

I have never heard of any destruction of birds or birds' eggs and can not imagine how there could be any remarkable waste of that nature, although I am familiar with the natural history of a considerable portion of the Territory.

Very truly, yours

C. H. TOWNSEND.

Col. JOSEPH MURRAY.

FOXES.

After we pass the timber belt to the westward we find but very little game, the only valuable land animal on the Aleutian chain of islands being the fox, which until recently was a source of income to the natives, who spent the greater part of the winter hunting and trapping the animal.

All that has been said about the wanton destruction of deer can be said with equal truth about the wholesale poisoning by which whole islands are stripped of their foxes in one winter, and the native hunter and his children left to starve. So systematically is the work done and so desperate are the gang engaged in it that those who know them best are very careful to say least about them.

Members of the gang are to be found wherever there is money to be made suddenly by illegitimate means. In the fishing season they dam the streams, capture the salmon by the quantity, and sell them to the nearest cannery for what they will bring. They never take the trouble to tear down the dams. They are to be found in schooners in the early spring hunting the sea otter in forbidden waters. They go to Bering Sea after seals, and last season some of them made a successful raid on a trading post and robbed it of some 15 or 20 fine sea-otter skins, valued at $7,000 to $10,000.

Generally they wind up the year's plunder by selecting a group of islands, where they spend the winter poisoning foxes and securing the pelts. These are the men who are armed to the teeth with the best modern breech-loading arms; men who own swift-sailing schooners, in which they carry cargoes of whisky from British Columbia, and, following the Alaskan coast and Indian settlements, peddle it out to natives for whatever skins and trinkets they may have to spare, and having made them drunk, they slip in and rob them of everything.

No effort has ever been made to break up their nefarious business, and now they swagger into court as though the Government were an intruder, and listen awhile to the proceedings; just long enough to assure themselves that their tools at the bar and in the jury box are doing their duty—to the gang.

The perpetual presence of a revenue cutter that would patrol the inner waters of Alaska from Cape Fox to Chilcat and Sitka, aided by armed steam launches stationed at convenient points along the route, is the only practical method that I know of by which the present dangerous bands of outlaws can be suppressed.

With boats at his disposal whenever needed, the marshal could enforce the law, the collector could follow the smugglers to their rendezvous and break up the whole business at one blow. As it is now, all the officers in Alaska are utterly powerless to do anything, and the consequence is the laws are defied and derided and spat upon.

THE SEA OTTER.

The most valuable of all the fur-bearing animals in Alaskan waters and the most widely distributed is undoubtedly the sea otter, which, if properly protected by the Government, is capable of giving profitable employment to the native hunters for all time.

Beginning at Sitka they were to be found till very recently all around the coast and Aleutian Islands as far westward as Attou, a distance of nearly 5,000 miles; but now, after a few years of hunting by the modern methods of steamers and steam launches, they are seldom found outside a few favorably secluded spots. The steamer and the steam launch carry crews of white hunters into every nook and cranny on the coast and otter-hunting grounds where an animal is to be found, and every one of them is either killed or chased away from home—chased out to sea in many instances, where, if they happen to elude the hunter, they die of starvation, for they can not go down for food in deep water.

None but native hunters should be permitted to hunt sea otter, because it is almost the only support of all the native people from Cooks Inlet to Attou Island, and, if left to them exclusively, their simple methods of hunting on the water in skin boats, in which they dare not venture far from land, can not possibly drive the animal away from its customary haunts nor exterminate it.

I include in the term native hunters all whites who were married to Indian women prior to 1893, when the ruling was changed. The original ruling of the Department, made some twenty years ago, remained in force until 1893, and in the meantime many white hunters married native women, made homes, and raised families, and became natives of Alaska to all intents and purposes.

All their earthly possessions are invested in sea-otter hunting property, their families have been brought up to that business exclusively, the men themselves have made it their life work, and are now too old to change or to go away from home to attempt to make a living at any other business, and therefore it would be an act of gross injustice to disturb them at this late day. With the white man who married a native woman after the Department had given fair warning that he would not be given the rights of a native hunter the case is altogether different, and in his case the ruling of the Department ought to stand. The farther away from the native settlements the average Alaskan white hunter can be kept the better for the natives.

FUR SEALS.

Sailing from Sitka June 2, on board the U. S. revenue cutter *Rush*, Capt. C. L. Hooper commanding, I landed at St. George June 18, where I learned that the preceding winter had been one of unusual severity, that ice had lain around the island until June 15, and that, up to the date of my landing, very few female seals had appeared upon the rookeries.

The same story was repeated on St. Paul Island, where I spent the 19th and 20th of June visiting the principal rookeries and hauling grounds, after which I sailed away and visited many of the native settlements along the Aleutian chain, particulars of which will be given in my report on the condition of the native tribes.

I returned to the seal islands early in July and spent the 6th, 7th, and 8th on the rookeries observing their daily growth and expansion, as the cows were now arriving and the harems were well defined and the pups becoming numerous.

Being well aware of the fact, however, that it is not till about July 20 the rookeries are full for the season, I continued to follow the instructions which called me to other fields until July 18, when I returned to the seal islands, where, all being ready, I entered on the most careful and thorough inspection of the rookeries ever made by me. The result is shown in the inclosed table marked Exhibit A.

Beginning at St. Paul Island July 21, and completing the work at St. George August 14, I walked over the several rookeries and counted every individual breeding male or bull seal who had a harem, noting and counting very carefully, too, every idle bull, or, in other words, every bull whose youth, strength, and vigor fitted and qualified him for a harem had there been cows to be found in sufficient numbers to supply them, which, unfortunately, there were not. So carefully and so systematically was the counting done that I feel I can recommend the figures as being as nearly correct and reliable as it is possible to get them.

Under the head of bachelors, or young males, are included all the seals on the islands other than those on the breeding rookeries, many of them being young females, too young to go on to the breeding grounds.

The bachelors have been estimated by me in the usual manner of estimating a bunch of seals, and they may very possibly run a thousand or two more or less than the figures given.

The number of breeding females or cows is based on an arbitrary average of 40 to the harem, or 40 cows to every breeding bull, as was adopted in and followed since 1891, though I am of the opinion it was an overestimate and that the harems never did and do not now contain an average of 40 cows each.

Having adopted that number, however, and having used it so long in our estimates, it was necessary to use it in the present instance for the sake of making fair comparisons when considering the steady annual decrease of the seal herd and the shrinkage of the rookery area.

Admitting the average number of cows in a harem to be less than 40—and I believe all who know anything about seal life on the rookeries will admit it is—then the total number of seals in the herd, as estimated by me, will be that much less in proportion.

By way of explanation I will say that when we first attempted to count the bulls, in 1891, for the purpose of getting, approximately, at the number of seals on the islands, it was deemed best to run the risk of overestimating the herds, lest Great Britain should object to our figures and insist on a recounting and, possibly, discover an error upon which to base an argument against us for the purpose of showing our anxiety to prove the wicked wastefulness of pelagic sealing.

As the seals were at that time too numerous and the harems too compact to admit of our going through and among them as we can now, we simply aimed to count every bull we could see and multiply the number found by two, on the ground that it was not possible to penetrate the mass far enough to see more than one-half of them.

And, lest that was not enough, we allowed an average of 40 cows to each harem, although we were quite certain it was too high.

I have gone over the rookeries every year, in season, since 1891, and I have noted the steady decrease of the herd from 500,000 then to 237,800 in 1895, when, because of the decrease, I was able to go in among the herd at the height of the season and count every bull on the islands.

Whether we erred in our estimates in our first crude efforts to get at the facts is of no consequence now, for the fact remains that, no matter what the actual numbers were in 1891, more than one-half of the whole herd has been exterminated since then.

Taking it for granted that the estimates were wrong, the proportion is still correct for all practical purposes, so that if we take the 500,000 of 1891 against the 237,800 of 1895, we find an average annual decrease of 52,440 for the five years beginning with 1891 and ending with 1895.

That the average annual loss has been greater than this can be demonstrated from the statistics on file in the Department which show a pelagic catch of Alaskan seals to have been as follows:

1891	45,491
1892	46,642
1893	28,613
1894	55,668
1895 (estimated)	40,000
Total for five years	216,864

to which I add 60 per cent for the loss of pups that died on the rookeries because of the killing of their dams at sea during the nursing season.

I base the proportion of pups on what I witnessed this year in Bering Sea, where the logs kept by the sealers showed a killing of 60 per cent females for the season: 216,864 plus 60 per cent equals 346,982 seals taken or destroyed in five years by pelagic sealers who pay nothing whatever for the care of the animals.

I have estimated 40,000 as the catch for 1895. I left Bering Sea September 18, when 31,216 seals had been taken by pelagic sealers, of which number 18,868 or 60 per cent were females as per the logs of the several vessels. These females were nursing mothers in milk, whose young were left upon the rookeries while they went out to sea for food and rest, instead of which they met the pelagic sealer who, according to law, killed them and carried off their skins and left their helpless young to bleat themselves to death upon the rookeries.

In a former report I pointed out the absurdity of the regulations that would protect the female seals from the pelagic sealer during the months of May, June, and July, most of which time they are on the islands and beyond his reach, and that would give him a clear and free field in August, as soon as the mother seal takes to the water in search of much needed food and rest and when, above all other times, she needs protection.

The taking of 31,000 seals in the month of August, 1895, proves the correctness of my position, and renders it needless to dwell upon the absurdity of the position the nation has been placed in by the present sealing regulations.

I therefore most respectfully call the attention of the Department to the five suggestions made by me in my report of last year, the adoption of which I believe will forever settle the seal question.

SALMON.

Owing to a lack of traveling facilities to the several canneries during the fishing season, and to the fact that the whole revenue fleet of the Pacific Coast had to do duty in Bering Sea, I found it impossible to visit many of the canneries beyond Karluk, where I found that one of the rival establishments had sold out to the Alaska Packers' Association and quit the business, thus leaving only two principal competitors on the river—the Alaska Improvement Company and the Alaska Packers' Association.

Much crimination and recrimination were indulged in on both sides as each endeavored to show it was the other one who violated the law, and a string of complaints was presented by the Indians similar to those presented by the same party in 1894, and of which I treated in my report for that year.

I found the fishermen with their nets in the narrowest part of the Karluk River, and so systematically do they work the nets that I could not see how it was possible for a fish to ever pass them to the spawn-ing grounds.

Remonstrating with the foreman about such flagrant violation of the law and of his own promise, made in 1894, that such methods should not be continued, he replied: "I was sent here to take fish; my orders are to take them wherever I can find them, and I am going to obey my orders."

He afterwards explained to me how, during the storms when the water is too rough to allow the spreading of nets, enough salmon pass into and up the river to supply twice the quantity of spawn required for perpetuating the stock.

His rival across the river indorsed him in all this, but added: "As soon as the storm ceases the fishermen follow the salmon upstream to the playground and capture every one of them."[1]

Exhibit H, which accompanies this report, is a copy of a bill which I would like to see become law, for I believe it would, if enforced, put an end to the present wasteful methods of salmon slaughter in Alaska without doing injury to any honestly conducted enterprise in the Territory.

Exhibit C is a detailed statement of the salmon pack in Alaska for 1895, showing the number of fish taken, the number of cases put up, and the number of men—white, native, and Chinese—employed; also the cost of the tin consumed in the business, the amount invested in each plant owned by the Alaska Packers' Association, and other data as per instructions. The only item of prime interest I have been unable to secure is the selling price of the product of each cannery in the market to which it is consigned.

Through the kindness and courtesy of the Alaska Packers' Associ-ation I have learned that the average selling price in San Francisco, where the greater bulk of the whole pack is sold, is as follows: Silver salmon, 82 cents per dozen; red salmon, 92½ cents per dozen; king salmon, 92½ cents per dozen, and barrels of 200 pounds net, $4.75.

Considering that only very few silver salmon are taken and packed, it is safe to say that the whole number of cases put up in 1895 averaged $3.60 per case, or a sum equal to $2,229,764.40, which, added to the price of 16,857 barrels at $4.75, makes a grand total of $2,326,968 as the price realized on Alaskan salmon in 1895.

Exhibit F is a summary of the salmon pack of the Pacific Coast and Alaska for 1895, showing a grand total of 2,040,016 cases of 48 pounds each, the largest yearly catch on record. An examination of the figures shows that about one-third of this catch was taken from the streams of Alaska.

That adequate protection should be given to these streams by which the salmon may be perpetuated indefinitely goes without saying, and yet I find it the hardest part of all to make men believe there is any danger in the present methods of fishing.

That I might not be accused of setting my own individual opinion against men of practical experience, I addressed letters of inquiry to many gentlemen who are deeply interested in Alaska, whose homes are there, and who have everything at stake in the success or failure of the Territory.

To Mr. William Duncan (Father Duncan), of Metlakahtla, I sent a series of questions which I requested should be submitted to his peo-ple for consideration and the answers given to me when I called at the

---

[1] The playground is that part of the stream where the salt and fresh waters meet and mingle, in which the salmon prefer to live for several weeks before spawning.

settlement in the fall. As I did not have the opportunity to return by that route, I could not call at Metlakahtla as I intended, so Mr. Duncan very kindly sent me the following letter:

METLAKAHTLA, ALASKA, *October 15, 1895.*

My DEAR MR. MURRAY: Your letter dated Unalaska, September 11, only reached me the latter part of last week. It had been detained at Kitchecan over a week through the lack of courtesy of the postmaster there.

On my arrival home last May, after the pleasant trip in your company to Sitka, I called a meeting of our people and propounded to them the several knotty questions you had suggested for our consideration. Last night we held another meeting, to a late hour, on the same business, and I was much pleased with the sensible way the natives took part in the discussion.

Question 1. "How to secure possession of Annette Island to our people and to such other Indians as may join them from surrounding bands, whether by individual or community title."

Our answer to this question is a unanimous voice in favor of a community title, and the town council being empowered to grant allotments of land for legitimate purposes to individuals as circumstances may arise calling for such action.

By this plan the present unity and regulations in the community could be preserved, whereas if individual titles of 160 acres were granted by the Government, the holder of each allotment being thus independent of the community, conflicting interests might result in a rupture which would be very prejudicial to the character and progress of the settlement.

Question 2. "How can the rights of the natives to the salmon streams be best secured and maintained?"

Our answer is, that, pending the Indians arriving at full American citizenship and responsibility, the Government might proclaim all salmon streams Indian reservations or Government property, and only allow fishing in them to proper persons and under proper regulations.

Such a law would prevent canning companies from taking exclusive control of the salmon streams, and might be made an important factor for bettering the condition of the natives.

At present Alaska is in danger of losing one of its greatest food supplies, through cannery operations. The Indians are born fishermen, and being permanent residents of the country fishing should, to a great extent, be in their hands, not as employees only, but as vendors of the salmon to the canneries.

Question 3. "How best to preserve salmon life in Alaska?"

Our answer to question 2 partly applies as answer to this. I will, however, enumerate our views:

(1) Let the salmon streams be declared Government property, and the fishing in them be absolutely controlled under Government regulations and by Government agents.

(2) Only permit a certain number of salmon to be taken from each stream, the number being decided by the capacity of the stream.

(3) Allow no modern barricades to be used in the streams, and even the simple ones which have always been used by the natives ought to be removed on Saturdays in each week.

(4) A limit should be placed to the pack of each cannery. I think 20,000 cases should be the limit. If, however, canneries can keep on increasing their pack and extending their time each year, as at present, fewer salmon each season will be left for reproduction.

(5) No cannery should be allowed to work on Sunday, and if fishing was forbidden after noon on Saturday till midnight Sunday of each week Sunday labor would cease. We strictly keep to this rule at Metlakahtla.

Question 4. "How best to suppress liquor traffic?"

(1) Our answer is, give the present liquor law a fair trial, and to that end every liquor saloon in Alaska should be suppressed and every drop of liquor now in it should be destroyed.

(2) Any person found smuggling or selling liquor in Alaska should be fined and imprisoned.

(3) Do away with the juries at the trial of liquor cases. Let the judges or commissioners appointed by the Government decide, upon certain given evidence, on the guilt of the persons arrested for offenses against the liquor law and an appeal allowed only to the supreme district court.

(4) Let every person found intoxicated be imprisoned, and in the case of natives the sentence to be commuted if the prisoner will give information leading to the conviction of the person or persons who supplied him with the liquors that intoxicated him.

(5) Let every informer against offenders in liquor cases (if his evidence leads to a conviction) be rewarded by receiving a part of the fine imposed on the offender.

(6) Let every commissioner in Alaska be instructed to swear in a goodly number of special police, without salary, especially among the natives, and let these be encouraged to assist in carrying out this law.

Question 5. "Should absolute title to land in Alaska be given to cannery corporations?"

Our answer to this is, we think that title to the land on which they have placed canneries should be given, but not to lands used only as fishing stations.

Question 6. "As to the granting of titles to land to whites in general."

We think that just so long as the Government refuses to give titles to land in Alaska the country will be overrun with an irresponsible floating population. Owing to the characteristics of the country this will be true, to a large extent, in any event; but the ownership of property would have a tendency to locate some permanent residents.

Question 7. "Should the exportation of lumber from Alaska be allowed?"

We think that until the country is more settled up the law forbidding the exportation of lumber, which now exists, should remain in force.

Question 8. "Indian citizenship."

We think that question had better be delayed. No doubt there are some natives ripe for the position, but the mass are not so. Let the missionary and school teacher continue their work till the goal be reached.

Yours, very respectfully,

W. DUNCAN.

Hon. JOSEPH MURRAY,
    Fort Collins, Colo.

I respectfully ask particular attention of the Department to this letter of Mr. Duncan, for I believe that the adoption of many of his suggestions would be a full and satisfactory solution of the many knotty problems at present perplexing all who feel a worthy and laudable interest in the present good and future welfare of Alaska.

Another and an entirely different phase of the salmon question was brought to my attention by Mr. John C. Callbreath, of Fort Wrangell, who has been endeavoring, single-handed, to introduce and propagate salmon in streams where they did not exist, or from which they had been driven before. I promised him in the spring that I would visit his hatchery in the fall during my stay in the vicinity, but I did not get an opportunity to return that way.

The following letter was written afterwards by Mr. Callbreath and deserves careful consideration. Particular attention is called to that portion of it treating of special "property rights to the producer for all fish in excess of the natural product of the stream."

What Mr. Callbreath wants is assurance that after he has successfully stocked a stream with salmon, where none or but very few existed before, he will be given rights in the fish as against all other claimants who might desire to establish canneries on the stream. But here is his letter to speak for itself:

SEATTLE, WASH., *December 10, 1895.*

DEAR SIR: I regret my inability to have forwarded you an account of salmon hatchery at an earlier day. Business in the interior, from which point there was no means of communication, detained me until late in the fall. I have, however, a trusted superintendent trained under my own care, who has made a complete success up to November 1. I shall return soon and give the business my personal attention until the young fry are out and placed in their respective preserves.

My process of hatching is the same as that followed by the Government hatcheries at Clackamas, in the State of Oregon, and need not be described here.

My hatchery is situated on the western side of Etholine Island, on a lake discharging through a small stream, a mere brook, into McHenry Inlet—and producing from 3,000 to 5,000 sukkesh (*Oncorhynchus nerka*) salmon, an amount too insignificant to be fished by the canuers or salters—and known among the Indians and fishermen as a "cuttus chuck," or worthless stream. The lake on which my hatchery is located is about three-fourths of a mile from tide water and contains about 500 acres.

I have built a dam 8 feet high across the creek a few yards above tide water, over which no fish unaided can pass. When the sukkesh start to ascend the stream for spawning, they are impounded in a trap below the dam, picked up with a dip net, and carefully placed above the dam, from whence they quickly proceed up to the lake, where they lie in the still, deep water until ripe for spawning, a period of from two to six weeks. They then take to the small clear streams running into the lake, where they are again impounded by means of a weir and trap, and are stripped of their eggs. The eggs are then fertilized by stripping the male over them, placed in baskets, and set in troughs in the way usual in all hatcheries. A peculiarity about this class of salmon, the *Oncorhynchus nerka*, is that they will not frequent a stream unless it has a lake where they can lie and ripen before spawning, although they never spawn in the lake; all the other species of Alaskan salmon frequent the streams where they can obtain spawning ground indiscriminately whether they have lakes or not.

My object in damming my stream near tide water is to keep back all enemies of the young fry, such as sea trout, bull heads, sculpins, sticklebacks, etc.; by this means I have my lake and streams cleared of these scourges of the young salmon, as they are all salt-water fish, and only go up to the lakes for plunder, returning to salt water when their season is over. Of course there were many of them in the lake and streams the first year, but when they passed down over the dam they could never return. To protect the young fry from their enemies in the fresh water I believe to be the great secret of successful salmon propagation. There is no bar to the number of young that can be produced at the Government hatcheries, where the spawn in unlimited quantities can be obtained. But unless protected from their enemies while young (and everything large enough to swallow them are their enemies) a large proportion of them are destroyed in fresh water. In my own case, however, the supply of fish is limited, and all are utilized.

My lake now fairly swarms with young salmon where heretofore scarcely one could be seen. I find, however, that the sea trout and others named are not their only enemies. Their older brothers feed on the young fry. The young salmon remain in the fresh water where they were hatched fourteen to eighteen months, so they have from two to four months to prey on their young brothers. Then, after going to sea, they will return for a short time to their native streams for a cannibalistic feast, and here again in my case my dam acts as protector to the little ones, as when once they pass down they can not return. I have seen them in vast numbers about the size of sardines, and packed almost as close, below the dam, trying to get up, but they soon disappear and return to salt water. In connection with this matter of protecting the younger from their older brothers, I last year commenced an experimental process, which I feel encouraged to believe will prove successful; that is, by turning out a portion of my young fry in streams, on which there are lakes that fall into the sea by falls, over which no fish can pass. As a consequence, most of these streams and lakes are entirely barren of fish of any kind. There are three streams and lakes of this description contiguous to my hatchery. In the winter of 1894–95 I placed 1,000,000 young fry into one of these lakes, and the present season of 1895–96 will place 2,000,000 in another lake, and so keep on alternating until I prove whether they will return to these streams. At the same time I will keep on stocking my own hatchery lake with as many as I think it will sustain. If my experiment of stocking these heretofore barren lakes and streams proves successful, and I can see no reason why it should not, it will prove of great value to the salmon fisheries of this coast, as these lakes abound all along the Alaskan coast.

I commenced my hatchery in the fall of 1892, but owing to the impure water of the creek, which contains a large amount of impurities, had but indifferent success, turning out only about 600,000. I then moved my hatchery up to the lake, three-fourths of a mile, where I found streams of pure water and even temperature, 45° in summer and never below 38° in winter, and then the fish ripened more healthily, as they were in their natural water. In the fall of 1893, however, there was but a small run of salmon, but the eggs hatched much better and I turned out about 1,700,000 young fry. The season of 1894–95 we had a better run and turned out 1,500,000 in one hatchery lake and 1,000,000 in the barren lake before mentioned. The present winter of 1895–96 we will turn out 4,000,000 or over, having had a much larger run than usual, which we will distribute between our hatchery lake and two other barren lakes. These three barren lakes are situated, respectively, 3, 6, and 9 miles from our hatchery, and entails a good deal of labor and expense cutting trails and carrying the young fish in buckets to their nursery. There are a few cohoes (*Oncorhynchus kisutch*) that frequent our stream, but never more than 400. As they are a good fish, although not as valuable as the sukkesh, we also pass them over the dam and strip them. Their time of running is about six weeks later than the sukkesh.

Owing to the smallness of our hatchery stream, we have opportunities of observing the habits of the salmon with greater accuracy than on large streams. From close observation made for a number of years, I am of the opinion that no salmon return to the sea after ascending for propagating purposes, unless their natural habits of copulating are interfered with. I am, however, of the opinion that some of the males will return if they are kept from the spawning beds and from performing the functions of nature for which they ascend; that is to say, if left to their natural state they will all die. And the females will all die anyway. But the males, if they do not connect with a spawning bed, their milt in some cases does not liquify, but remains solid, and some of them will return to the sea. But had they not been barred from the spawning beds their milt would liquify and they would all die.

We have discovered what seems to us to be a new variety of sea trout, quite similar to the rainbow, with the addition of a gristly hook, or turning up of the lower jaw, and fitting into a recess in the end of the upper jaw, completely covering the end of the snout and fitting in the recess so neatly that it will not be observed unless the jaws are open. I can find no description of a similar trout in the treatise sent me in June, 1894, by the Fish Commissioner, Hon. Marshall McDonald, which gave a full description of all the different species of trout. We have no alcohol or would have preserved a specimen.

According to the rule generally accepted by scientific men, the salmon will return four years from the time their parents enter the stream for spawning purposes, which will bring my first salmon back the coming summer, when I will be able to give you a more definite account of my venture.

I think, in cases like my own, where hitherto worthless streams are built up and made to produce large quantities of valuable fish that will assemble in the bays or inlets at the mouths of the streams, where they have been bred, that hitherto produced none, so to speak, a law of Congress should be passed giving property rights to the producer for all fish in excess of the natural product of the stream. It is held by legal men that I have consulted on the subject that I will have a property right in such fish, but it would be far better if such rights were reenforced by an act of Congress.

I have kept a careful account of all the sukkesh and cohoes that we passed over the dam from day to day, so that I can tell exactly the number of fish that the stream normally produced.

Hoping I may have the pleasure of meeting you on your return,

I remain, yours, truly,

JOHN C. CALLBREATH.

Hon. JOSEPH MURRAY,
Special Agent for the Protection of Salmon Fisheries in Alaska.

P. S.—I will be at Fort Wrangell during February, and should be pleased to hear from you.

J. C. C.

Now, here is a man who, though not wealthy, has spent money and many years' valuable time making experiments in one of the most useful and honorable of the arts—the production of human food.

Such men deserve a patient hearing and every possible encouragement, and in the hope of his getting both I respectfully recommend his very timely and practical letter to the serious consideration of the Department.

Exhibit D gives the names of the canneries and packing stations, Exhibit E shows the sailing distances one must travel from cannery to cannery in order to see all of them, and shows conclusively, I think, that in order to see all of them in one season it will be necessary to detail a revenue cutter to carry the agent.

A revenue cutter could make the trip between June 1 and July 20, completing the journey at the canneries in Bering Sea in ample time to report for patrol duty in August.

Exhibit G gives a summary of the Alaskan and Pacific Coast salmon pack from 1866 to 1895, both inclusive, showing at a glance that the Alaskan streams were drawn upon to their utmost capacity in 1891, when 789,294 cases of 48 pounds each were packed, with the result of a falling off of 40 per cent the following year.

The wisdom of protecting an industry that has yielded in the thirteen years of its existence 5,505,002 cases of salmon, worth $22,000,000, should not be lightly questioned or set aside, and when it is remembered that, excepting the civilized Indians with Mr. Duncan at Metlakahtla, there is not a resident cannery owner in Alaska, and that not one dollar of all the millions taken from her streams is left or spent in the Territory, it will be conceded, I think, by all fair-minded men that the least

the General Government can do is to protect the fish against extermination, and the native, dependent solely upon a salmon diet, in his right
to an abundant supply of salmon for food. These two things are easy
of accomplishment if immediate steps are taken. but if neglected much
longer the task will be a most difficult one.

In order to protect the salmon streams the laws must be enforced,
and it is, unfortunately, only too true that up to date there has not
been a united attempt made to enforce them.

The United States commissioner at Fort Wrangell—one of the few
fearless ones who only know their duty—wrote me a full account of how
the law is disregarded, evaded, and not enforced by officers whose duty
it is to uphold the law at all hazards. The letter covers a wide field,
and is quite plain and outspoken, using men's real names, the printing
of which in my report is not now considered necessary; I will therefore
suppress names and quote only a few passages relating to salmon
matters.

Case after case has been compromised at the instance of attorneys for their clients
in criminal cases. Take one example: In July, 1893, Mr. J. G. Brady, who was then
acting United States attorney in the absence of Mr. Johnson, entertained a complaint of the natives against cannery men for obstructing salmon streams. Mr. Brady
prevailed upon Commander Burwell, of the U. S. S. *Pinta*, to convey him and other
necessary officers to Klawak and Loring. * * * They came via Fort Wrangell and
requested me to accompany them. The *Pinta* reached Loring in the night, and we
found the river "fenced" from shore to shore, and about 4 tons of salmon in the traps
and nets. We caused the arrest of Mr. Heckman, the superintendent of the cannery,
and placed him under bond of $3,000, requiring him to appear before the United
States district court at its next session.

The law, as you are aware, imposes a fine of $250 for every day a stream is
obstructed. * * * Court did not convene until after the retirement of the district attorney, Mr. ——, and the appointment of his successor, Mr. ——, who
agreed to accept the nominal fine of $100. * * * Superintendent Wadleigh, of
the Klawak cannery, was also placed under bonds, and although two terms of court
have since been held he has not been required to appear. The —— is lending a
hand in helping to whitewash his case. * * *

These two cases, including the trip of the *Pinta*, have cost the Government more
than $1,000.

The officers who placed the men under bonds have been humiliated, while the violators of the law, aided by officials disloyal to the Government, have won a great
victory. * * *

I could cite many cases in which official positions are used to shield crime and
defeat the ends of justice.

Your obedient servant, WM. A. KELLY, *Commissioner*.

Were it necessary to add to or confirm the commissioner's words, I
would say that while at Loring in 1894 Hon. C. S. Hamlin, Assistant
Secretary of the Treasury, accompanied by Capt. C. L. Hooper, commanding the revenue cutter *Rush*, attempted to go up the river in a
small boat, but soon found themselves barred out by the identical
"fence" mentioned in the commissioner's letter.

The Wadleigh case referred to was called in court at Juneau last
May, and in my presence his attorney arose and said: Mr. Wadleigh
had written and offered to pay as much as it would cost him to travel
back and forth on the steamer from Klawak to Juneau, some $40, on
condition that the Government dropped the complaint.

Now, here is a case where the man was taken red-handed in the act—
he does not attempt denial—and yet, although it happened in 1893, he
has not been brought to trial, but instead of answering the summons
of the court he impudently writes back his ultimatum, which was seriously considered by the district attorney, who would have accepted it
had I not been present and strongly protested against the whole farce.

In justice to the present district attorney for Alaska, I will say all
these things happened prior to his appointment.

CONCLUSION.

Enough has been said I think to show the necessity of some radical changes in Alaska, the first of which should be the enforcement of the law.

I therefore most respectfully recommend the following:

First. The repeal of the present prohibitory liquor law and the substitution of high license.

Second. That a revenue cutter and three armed steam launches be permanently located in Alaskan waters.

Third. That the custom-house on Mary Island be discontinued and removed to a more desirable, because more useful, location in the Tongas Narrows.

Fourth. That Alaska be divided into at least two judicial districts, with one judge at Sitka and one at Circle City, on the Yukon.

Fifth. That three additional commissioners be appointed, one at Unga, one at St. Michaels, and one at Circle City.

Sixth. That a deputy collector (if not a custom-house) be located at Unga.

Seventh. That a marine hospital be erected at Unalaska. (Either of the trading companies will erect and furnish a building if the Department will furnish medicines and a physician.)

Eighth. That Alaska be allowed a Delegate to Congress.

There are many important matters that I have not referred to in this report, such, for instance, as the condition of the native tribes on the Aleutian Islands and in southeastern Alaska; schools and post-offices on the Yukon River and in the great interior—all of which will be dealt with in a future report.

The proposed changes are really necessary to the present and future welfare of Alaska, and, because of the rapidly increasing white population flocking to the rich gold diggings, it is absolutely necessary that the law should be rigidly enforced. The wealth of Alaska in furs, fish, and gold, if properly protected by the Government, will be of immense value, which may be made to increase annually, but which, if neglected by the Government and left to the present system of no law at all, or what is far worse, lawlessness, will soon end in disgrace and disaster.

Very respectfully submitted. JOSEPH MURRAY,
Special Agent for the Protection of Salmon Fisheries in Alaska.

Hon. JOHN G. CARLISLE,
Secretary of the Treasury.

EXHIBIT A.

Number of seals on St. Paul and St. George islands, season of 1895.

ST. PAUL ISLAND.

Rookery.	Bulls with harems.	Cows.	Bachelors.	Idle bulls.	Total.
Northeast Point	1,725	69,000	9,000	1,000	80,725
Halfway Point	350	14,000	2,000	200	16,550
Lukannon	300	12,000	1,000	200	13,500
Katavie	200	8,000	300	50	8,550
Reef	1,000	40,000	5,000	500	46,500
Lagoon	50	2,000	50		2,100
Tolstoi	400	16,000		250	16,650
Middle Hill			1,500		1,500
English Bay	100	4,000	800	100	5,000
Zapadnie	500	20,000	3,500	300	24,300
Total	4,625	185,000	23,150	2,600	215,375

Number of seals on St. Paul and St. George islands, season of 1895—Continued.

ST. GEORGE ISLAND.

Rookery.	Bulls with harems.	Cows.	Bach-elors.	Idle bulls.	Total.
Starry Arteel	60	2,400	300	40	2,800
North	100	4,000	500	50	4,650
East	80	3,200	3,000	40	6,320
Little East	25	1,000	50	20	1,095
Zapadnie	110	4,400	3,000	50	7,560
Total	375	15,000	6,850	200	22,425
Total on both islands	5,000	200,000	30,000	2,800	237,800

EXHIBIT B.

Liquors cleared from Puget Sound for Alaska, January 1, 1894, to March 10, 1895.

Date.	Per-mit No.	Kinds and quantities.	Consignee.	Name of vessel.	Date cleared.
Jan. 6, 1894	51	1 barrel bottled beer....	Adolph Myer	City of Topeka.	Jan. 15, 1894
Dec. 22, 1893	46	1 barrel rye whisky	do	do	Do.
Do	40	1 barrel Bourbon whisky, 10 barrels beer.	C. F. Fueher	do	Do.
Do	44	1½ barrels California brandy, 1½ barrels California claret, 5 barrels beer, 3 barrels ale, 3 cases porter, 10 gallons sherry, 10 gallons Irish whisky, 10 gallons Scotch whisky, 10 gallons gin, 10 gallons rum, 1 case imported brandy, 1 case imported whisky, 2 cases champagne.	William Nelson	do	Do.
Dec. 21, 1893	41	1 gallon port wine, 1 gallon brandy.	C. J. Kostromehuoff	do	Do.
Jan. 5, 1894	50	6 bottles China liquor...	Hung Sing Gee	do	Do.
Dec. 6, 1893	39	1 case whisky, 1 case wine, 1 case porter.	Duncan McKimon	do	Do.
Jan. 2, 1894	49	1 barrel beer	W. C. Mills	do	Jan. 30, 1894
Dec. 22, 1893	45	1 barrel porter	Chas. Giffey	do	Do.
Do	47	1 barrel whisky, 1½ barrels rum, 1½ barrels brandy, 1 barrel porter, 1 barrel beer.	Wm. Mulcahy	do	Do.
Dec. 6, 1893	37	30 gallons claret wine ..	C. S. Johnson	do	Jan. 15, 1894
Feb. 7, 1894	57	5 gallons whisky	Smeby Bros	do	Feb. 17, 1894
Feb. 8, 1894	59	2 barrels beer	E. De Groff	do	Do.
Feb. 7, 1894	58	1 barrel whisky, 5 barrels beer, 2 barrels porter, 2 barrels ale, 6 cases Irish whisky, 6 cases Hennesy brandy.	W. Mulcahy	do	Do.
Do	55	1 barrel whisky, 1 barrel brandy, 8 barrels beer, 1 case champagne, 2 barrels whisky.	William Nelson	do	Do.
Jan. 20, 1894	53	1 barrel gin, 5 gallons Jamaica rum, 2 gallons Madeira wine, 12 gallons sherry wine, 30 gallons claret, 1 case brandy, 1 case champagne, 1 barrel ale, 1 barrel porter, 5 gallons port wine, 10 gallons whisky, 5 cases whisky.	E. De Groff	do	Do.
Jan. 16, 1894	52	1 case wine	W. P. Mills	Mexico	Feb. 27, 1894
Jan. 22, 1894	54	50 gallons whisky, 20 gallons brandy, 20 gallons port wine, 10 gallons alcohol, 50 gallons claret, 6 barrels beer, 2 cases gin.	J. C. Koosher	City of Topeka..	Mar. 14, 1894

Liquors cleared from Puget Sound for Alaska, etc.—Continued.

Date.	Permit No.	Kinds and quantities.	Consignee.	Name of vessel.	Date cleared.
Feb. 23, 1894	61	2 barrels bourbon whisky, 10 gallons rum, 20 gallons brandy, 10 gallons port wine.	William Nelson	City of Topeka	Mar. 14, 1894
Feb. 24, 1894	63	3 barrels bottled beer...	E. DeGroff	do	Mar. 29, 1894
Mar. 9, 1894	64	1 case whisky	Robert Reid	do	Do.
Feb. 20, 1894	60	3 cases wine	Mrs. R. C. Rogers	do	Do.
Mar. 9, 1894	67	1 case whisky	H. F. Swift	do	Do.
Do	66	1 gallon rum	W. M. Taylor	Chilkat	Apr. 7, 1894
Do	65	1 case ale, 1 case porter	J. M. Davis	do	Do.
Do	68	1 barrel whisky, 1½ barrels brandy, 3 cases Irish whisky, 6 barrels beer.	W. Mulcahy	do	Do.
Feb. 23, 1894	62	10 gallons whisky, 1 case beer, 1 gallon brandy, 2 cases wine.	Robert Duncan, jr	City of Topeka	Apr. 13, 1894
Mar. 9, 1894	70	15 gallons port wine	Father Donskey	do	Apr. 28, 1894
Apr. 4, 1894	72	3 gallons whisky	Archy Campbell	do	May 14, 1894
Do	71	10 gallons alcohol	C. F. Feuher	do	Do.
May 19, 1894	93	5 gallons alcohol	F. D. Nowell	Rosalie	May 29, 1894
Do	96	2 barrels whisky, 1 barrel brandy, 2 cases gin, 20 gallons rum, 5 cases whisky, 5 cases brandy, 10 barrels beer, 10 cases porter, 10 cases ale, 2 cases champagne, 1 barrel port wine, 1 barrel sherry, 1 barrel claret, 1 barrel alcohol.	Max Endleman	do	Do.
Do	91	1 barrel beer	Mrs. Hammond	do	Do.
Do	92	do	F. Bach	do	Do.
June 4, 1894	100	3 barrels beer	E. De Groff	City of Topeka	June 12, 1894
Do	99	1 barrel beer	R. E. Rogers	do	Do.
May 9, 1894	84	4 barrels beer	E. De Groff	Queen	June 22, 1894
June 13, 1894	105	1 dozen bottles brandy	do	City of Topeka	June 29, 1894
June 18, 1894	107	15 barrels beer, 1 barrel sherry wine, 1 barrel whisky, 5 cases porter, 1 barrel port wine, 5 cases ale, 5 cases gin, 5 cases claret wine.	Max Endleman	do	Do.
June 28, 1894	108	10 barrels beer	E. De Groff	Queen	July 8, 1894
May 19, 1894	94	2 cases whisky	C. Spuher	do	Do.
July 1, 1894	116	20 barrels beer, 1 barrel brandy, 1 barrel port wine, 1 barrel sherry, 1 barrel claret, 3 cases champagne, 5 cases porter, 5 cases ale, 5 cases brandy.	Max Endlemen	City of Topeka	July 27, 1894
June 28, 1894	109	10 gallons whisky	E. De Groff	Queen	Aug. 6, 1894
June 18, 1894	113	1 case whisky	Karl Koehler	do	Do.
Aug. 3, 1894	120	10 barrels beer	E. De Groff	City of Topeka	Aug. 13, 1894
July 18, 1894	114	1 case whisky, 1 case wine, 1 case porter.	D. McKinnon	do	Do.
Aug. 3, 1894	119	1 case cognac, 10 cases port wine, 10 cases sherry, 10 cases whisky.	E. De Groff	Queen	Aug. 21, 1894
Aug. 13, 1894	123	1 case ale, 1 case porter.	J. Mont. David	City of Topeka	Aug. 31, 1894
Aug. 13, 1894	124	10 barrels beer	E. De Groff	do	Sept. 19, 1894
Sept. 6, 1894	129	10 barrels beer	do	do	Sept. 25, 1894
Aug. 13, 1894	122	30 barrels wine	R. C. Rogers	Mexico	Do.
Sept. 6, 1894	128	10 barrels whisky	E. De Groff	do	Do.
Oct. 4, 1894	139	10 barrels alcohol	do	City of Topeka	Oct. 11, 1894
Sept. 6, 1894	130	10 barrels white wine, 1 barrel claret.	do	Mexico	Oct. 25, 1894
Oct. 3, 1894	138	1 case whisky	R. C. Rogers	do	Do.
Sept. 25, 1894	132	1 barrel claret	Karl Koehler	do	Do.
Oct. 17, 1894	142	1 case whisky	Robert Reid	do	Do.
Do	145	5 barrels beer	Ed. De Groff	do	Nov. 10, 1894
Sept. 25, 1894	135	1 case whisky	M. Healy	do	Do.
Oct. 17, 1894	144	30 gallons whisky	Ed. De Groff	Chilcat	Nov. 21, 1894
Do	144	10 gallons rum	do	Mexico	Nov. 26, 1894
Dec. 19, 1894	150	1 bottle brandy, 1½ gallons whisky, 1 gallon port wine.	Dr. C. Theving	do	Do.

Liquors cleared from Puget Sound for Alaska, etc.—Continued.

Date.	Permit No.	Kinds and quantities.	Consignee.	Name of vessel.	Date cleared.
Nov. 16, 1894	147	1½ dozen Kimmel, 1½ dozen Benedictine, 1½ dozen absinth, 1½ dozen kirschwasser, 1½ dozen anisette, 1½ dozen maraschino, 1½ dozen Chartreuse, 1 case gin, 1 case champagne, 1 barrel porter, 1 barrel ale.	E. De Groff	Mexico	Dec. 28, 1894
Dec. 19, 1894	151	1 barrel bottled porter	Wm. Mulcahy	Chilcat	Jan. 3, 1895
Jan. 2, 1895	157	5 barrels bottled beer	E. De Groff	Mexico	Jan. 12, 1895
Do	157	12 bottles China liquor.	Sing Lee	do	Do.
May 4, 1894	82	5 gallons alcohol	E. Valentine	Chilkat	Jan. 23, 1895
Jan. 2, 1895	155	12 bottles assorted liquors.	R. C. Rogers	City of Topeka.	Do.
Do	162	10 barrels beer	E. De Groff	do	Feb. 7, 1895
Jan. 19, 1895	159	6 dozen bottles porter.	C. E. Tibbits	do	Do.
Jan. 2, 1895	154	1 barrel claret	C. S. Johnson	do	Do.
Dec. 19, 1894	152	5 gallons port wine	Fred Hall	do	Do.
Feb. 15, 1895	170	10 gallons whisky	E. De Groff	do	Feb. 24, 1895
Do	169	5 gallons alcohol	do	do	Do.
Jan. 21, 1895	161	1 dozen brandy	do	do	Do.
Jan. 19. 1895	160	20 gallons claret	George Kyrage	do	Do.

The within permits are signed by Benjamin Moore, collector of customs, Sitka, Alaska.
PORT TOWNSEND, WASH., *March 1, 1895.*

EXHIBIT C.

Statistics of Alaska salmon pack, season of 1895.

Name.	Location.	Number of men employed.			Apparatus used.	Number of salmon taken.		
		White.	Native.	Chinese.		King.	Red.	Silver.
Bristol Bay Canning Co..	Nushagak	63	44	95	Gill nets..	4,544	252,776	9,250
Alaska Packing Co	do	65	46	98	do	8,823	356,622	10,100
Arctic Packing Co	do	61	48	95	do	5,106	329,548	8,700
Do	Naknek	47	21	70	do	1,047	269,851	
Thin Point Packing Co..	Thin Point	28	20		Seine		23,453	
Karluk Packing Co	Karluk	84	48	142	do		589,090	
Tanglefoot Bay Packing Co.	do	35	21	65	do		172,049	
Hume Packing Co	do	84	52	142	do		603,421	
Arctic Packing Co	Alitak	31	18	52	do		174,568	8,321
Arctic Fishing Co	Kualloff	62	48	77	do	25,199	324,277	
Chignik Bay Packing Co.	Chignik	89	31	110	Gill nets and seine.		683,319	
Pacific Packing Co	Prince Williams Sound.	95	39	63	do	4,319	143,100	142,937
Pyramid Harbor Packing Co.	Pyramid Harbor..	80	62	77	Gill nets..	9,453	310,750	7,028
Glacier Packing Co	Fort Wrangell....	41	71	55	do	8,294	133,509	154,183
Alaska Salmon Packing and Fur Co.	Loring	30	80	70	Seine		14,733	435,368
Point Roberts Packing Co.	Koggiung	34	10		Gill nets..	405	143,800	
Ugashik Fishing Station.	Selina River	49	22		Gill nets and seine.		65,219	
Egegak Fishing Station.	Egegak	12			Gill nets.		54,321	
Toglak Fishing Station..	Togiak	12			do		1,800	
Total		1,002	681	1,211		62,190	4,646,215	775,887

Statistics of Alaska salmon pack, season of 1895—Continued.

Name.	Cases.	Barrels.	Steamers employed.	Lighters and boats.		Nets.		Sail tonnage employed.	Value of tin plate.
				Number.	Value.	Number.	Value.		
Bristol Bay Canning Co..	33,434	......	2	42	$18,000	80	$4,000	1,355	$19,360
Alaska Packing Co.......	34,632	......	1	44	13,000	84	4,200	1,100	19,100
Arctic Packing Co	33,631	......	1	43	12,500	83	4,100	1,040	19,600
Do	22,731	1,045	1	23	15,000	42	2,100	900	13,200
Thin Point Packing Co ..	....	395	1	10	6,000	4	800	255	....
Karluk Packing Co	48,379	......	3	47	30,000	15	3,000	....	....
Tanglefoot Bay Packing Co.	15,277	......	2	28	15,200	10	2,000	4,500	66,200
Hume Packing Co	47,500	......	2	48	21,000	15	3,000	....	....
Arctic Packing Co	15,331	......	1	23	11,000	4	800	1,350	9,100
Arctic Fishing Co	30,188	366	2	40	18,200	80	4,000	1,320	21,600
Chignik Bay Packing Co.	70,050	......	2	24	19,000	90	4,500	1,520	42,000
Pacific Packing Co	21,453	65	4	43	45,000	75	3,800	1,340	13,200
Pyramid Harbor Packing Co.	35,373	......	1	31	18,000	50	2,700	1,132	21,000
Glacier Packing Co	27,416	......	1	17	14,200	30	1,500	776	16,200
Alaska Salmon Packing and Fur Co.	32,554	......	1	11	13,100	4	800	771	19,200
Point Roberts Packing Co.	....	3,142	1	12	6,200	7	1,400	234	....
Ugashik Fishing Station.	....	1,354	......	13	2,500	6	1,200	555	....
Egegak Fishing Station..	....	1,048	......	3	600	3	600	126	....
Togiak Fishing Station..	....	40	......	2	400	2	400	130	....
Total...............	473,949	7,455	26	504	278,900	684	44,900	18,404	279,760

Statistics of Alaska salmon pack, season of 1895—Continued.

Name.	Location.	Cases.	Barrels.
C. E. Whitney & Co	Nushagak	....	1,043
Prosper Fishing and Trading Co..................	Kvichak......................	....	2,300
L. A. Pederson	Naknek	14,253	300
Bering Sea Packing Co..................	Ugashik	12,007	....
Norton, Teller & Co	....do	....	220
Lynde & Hough.....................	Shumagin islands..........	....	75
Alaska Improvement Co	Karluk	26,000	....
C. D. Ladd.....................	Cooks Inlet......................	....	350
Pacific Steam Whaling Co..................	Prince William Sound......	25,037	....
Peninsular Fishing Co..................	Copper River	15,000	....
Baranoff Packing Co	Baranoff Island..............	14,805	....
North Pacific Fishing and Trading Co	Klawak......................	12,228	104
Boston Fishing and Trading Co	Yes Bay	14,100	....
Metlakahtla Industrial Co.	Metlakatta	12,000	....
Miller & Co	Cordova Bay..........	....	1,800
Cape Fox Packing Co	Cape Fox	....	1,200
Various	Southeastern Alaska........	....	2,000
Total.....................	....	145,430	9,392

EXHIBIT D.

Salmon packing stations in Alaska.

No.	Locality.	Name of company.	Cannery.	Saltery.	Herring.
1	Chilcat	Alaska Packing Association	2		
2	Port Althorp	Ford & Stokes		1	
3	Killisnoo	Herring Fishery			1
4	Red Fish Bay	Baranoff Packing Co	1		
5	Fort Wrangell	Alaska Packing Association	1		
6	Yes Bay	Boston Fishing and Trading Co	1		
7	Loring	Alaska Packing Association	1		
8	Port Chester	Metlakahtla Industrial Co	1		
9	Klawak	North Pacific Packing Co	1		
10	Cordovia Bay	Miller & Co		1	
11	Tolstoi Bay	do		1	
12	Port Ellis	Kniu Island		1	
13	Cape Fox			1	
14	Copper River, Delta Peninsula	Fish and Trading Co	1		
15	Eyak Village	Pacific Steam Whaling Co	1		
		Alaska Packing Association	1		
16	Cooks Inlet, Kussilo River	do	1		
	West side of Cooks Inlet	C. D. Ladd & Co	1		
17	Afognak (not in operation)		2		
18	Karluk River	Alaska Packing Association	2		
	Alaska Improvement Co		1		
	R. D. Hume & Co		1		
19	Alitak Bay	Alaska Packing Association (used up)	1		
20	Ugak Bay, Eagle Harbor	Oliver Smith	1		
21	Chignik Bay	Alaska Packing Association	1		
22	Pirate Cove, Popoff	McCollum Trading Co		1	
23	Thin Point	Alaska Packing Association		1	
24	Ugashil	Bering Sea Packing Co	1		
	do	Alaska Packing Association		1	
	do	Sullivan River Packing Co		1	
	do	Johnson		1	
25	Naknik River	Alaska Packing Association	1		
	do	Peterson		1	
26	Knichak River	Alaska Packing Association		1	
		Prosper Fish and Trading Co		1	
27	Nushagak	Alaska Packing Association	3		
	Fort Alexander	Whiteney Co		1	
	Total		27	14	1

EXHIBIT E.

Sailing distances from Cape Fox to the different salmon canneries in Alaska.

[Figures in parentheses are map numbers.]

	Miles.
(13) Cape Fox to (10) Cordovia Bay	80
(13) Cape Fox to (8) Port Chester	50
(10) Cordovia Bay to (9) Klawak	100
(8) Port Chester to (11) Tolstoi Bay	60
(8) Port Chester to (7) Loring	60
(7) Loring to (6) Yes Bay	25
(11) Tolstoi Bay to (5) Fort Wrangell	100
(5) Fort Wrangell to (12) Port Ellis	100
(9) Klawak to (4) Red Fish Bay	150
(4) Red Fish Bay to (2) Port Althorp	150
(2) Port Althorp to (3) Killisnoo	200
(3) Killisnoo to (1) Chilcat Inlet	200
(1) Chilcat Inlet to (14) Copper River Delta	1,000

	Miles.
(14) Copper River Delta to (15) Eyak village	50
(15) Eyak village to (17) Afognak	500
(17) Afognak to (20) Ugak Bay, Eagle Harbor	75
(20) Ugak Bay to (19) Alitak Bay	100
(19) Alitak Bay to (18) Karluk River	100
(18) Karluk River to (21) Chignik Bay	300
(21) Chignik Bay to (22) Pirate Cove	200
(22) Pirate Cove to (23) Thin Point	150
(23) Thin Point to (24) Ugashik	500
(25) Naknik River to (26) Kvichak River	25
(26) Kvichak River to (27) Nushagak	100
Total	4,375

EXHIBIT F.

Summary of salmon pack, 1895.

	Cases.
Columbia River	617, 460
Alaska	619, 379
British Columbia	512, 877
Outside rivers	290, 300
Total	2, 040, 016

		Cases.
Columbia River		617, 460
Alaska (16 locations)		619, 379
British Columbia:		
Fraser River	347, 674	
Skeena River	66, 983	
Lowe Inlet	8, 500	
Nass River	19, 000	
Rivers Inlet	61, 720	
Alert Bay	5, 500	
Clayoquot	3, 500	
		512, 877
Outside rivers and bays:		
Nehalem River	6, 300	
Sinslaw River	8, 552	
Coquille River	9, 468	
Umpque River	10, 300	
Tillamook River	5, 000	
Alsea Bay	5, 000	
Coos Bay	10, 380	
Puget Sound (4 locations)	157, 000	
Grays Harbor	18, 000	
Shoalwater Bay	16, 000	
Rogue River	14, 000	
Sacramento rivers	24, 000	
California rivers	6, 300	
		290, 300
		2, 040, 016

EXHIBIT G.

Alaskan and Pacific Coast salmon pack, from 1866 to 1895, both inclusive.

Year.	Columbia River.	Outside rivers and bays.	British Columbia.	Alaska.	Total.
1866	4, 000				4, 000
1867	18, 000				18, 000
1868	28, 000				28, 000
1869	100, 000				100, 000
1870	150, 000				150, 000
1871	200, 000				200, 000
1872	250, 000				250, 000
1873	250, 000				250, 000
1874	350, 000	2, 500			352, 500
1875	375, 000	3, 000			378, 000
1876	450, 000	33, 900	9, 847		493, 747
1877	460, 000	46, 300	67, 387		573, 687
1878	460, 000	66, 500	113, 601		640, 101
1879	480, 000	61, 000	57, 394		598, 394
1880	630, 000	86, 200	61, 300		779, 500
1881	551, 000	229, 700	175, 675		956, 375
1882	541, 300	249, 300	255, 061		1, 045, 661
1883	629, 400	198, 000	243, 000	36, 000	1, 106, 400
1884	656, 179	122, 800	138, 945	54, 000	971, 924
1885	524, 530	100, 250	106, 865	74, 850	806, 495
1886	454, 943	170, 400	163, 004	120, 700	909, 047
1887	373, 800	231, 900	201, 990	190, 200	997, 890
1888	367, 750	212, 000	135, 600	427, 372	1, 142, 722
1889	325, 500	265, 734	414, 400	709, 347	1, 714, 981
1890	433, 500	102, 123	400, 464	688, 332	1, 633, 419
1891	390, 183	82, 447	814, 813	789, 294	1, 576, 737
1892	481, 900	160, 800	221, 797	461, 482	1, 325, 979
1893	425, 200	209, 496	590, 220	645, 545	1, 870, 470
1894	511, 000	214, 896	494, 470	678, 501	1, 898, 867
1895	617, 460	290, 300	512, 877	619, 379	2, 040, 016

EXHIBIT H.

A BILL to amend an act entitled "An act to provide for the protection of the salmon fisheries of Alaska."

Be it enacted by the Senate and House of Representatives of the United States of America in Congress assembled, That the act approved March second, eighteen hundred and eighty-nine, entitled "An act to provide for the protection of the salmon fisheries of Alaska," is hereby amended and reenacted, as follows:

SECTION 1. That the erection of dams, barricades, fish wheels, fences, traps, pound nets, or any fixed or stationary obstructions in any part of the rivers or streams of Alaska, or to fish for or catch salmon or salmon trout, in any manner or by any means, with the purpose or result of preventing or impeding the ascent of salmon or salmon trout to their spawning ground, is declared to be unlawful, and the Secretary of the Treasury is hereby authorized and directed to remove such obstructions and to establish and enforce such regulations and surveillance as may be necessary to insure that this prohibition and all other provisions of law relating to the salmon fisheries of Alaska are strictly complied with.

SEC. 2. That it shall be unlawful to fish, catch, or kill any salmon or salmon trout of any variety, except with rod or spear, above the tide waters of any of the creeks or rivers of less than five hundred feet wide in the Territory of Alaska, or to lay or set any drift net, set net, or seine for any purpose, across the tide waters of any river or stream for a distance of more than one-third of the width of such river, stream, or channel, or lay or set any seine or net within one hundred yards of any other net or seine which is being laid or set in said stream or channel, or to take, kill, or fish for salmon in any manner or by any means in any of the waters of the Territory of Alaska, either in the streams or tide waters, from noon on Saturday of each week until six o'clock post meridian of the Sunday following, or to fish for, or catch, or kill in any manner, or by any appliances, except by rod or spear, any salmon or salmon trout in any stream of less than one hundred yards in width in the said Territory of Alaska between the hours of six o'clock in the morning and six o'clock in the evening of the same day of each and every day of the week.

SEC. 3. That the Secretary of the Treasury may, at his discretion, set aside certain streams as spawning grounds, in which no fishing will be permitted; and when, in his judgment, the results of fishing operations on any stream indicate that the number of salmon taken is larger than the capacity of the stream to produce, he is authorized to establish weekly close seasons, to limit the duration of the fishing season, or to prohibit fishing entirely for one year or more, so as to permit the salmon to increase.

SEC. 4. That to enforce the provisions of law herein, and such regulations as the Secretary of the Treasury may establish in pursuance thereof, he is authorized and directed to appoint one inspector of fisheries at a salary of ten dollars per day, and two assistant inspectors at a salary of eight dollars each per day, and he will annually submit to Congress estimates to cover the salaries and actual traveling expenses of the officers hereby authorized and for such other expenditures as may be necessary to carry out the provisions of the law herein.

SEC. 5. That any person violating the provisions of this act, or the regulations established in pursuance thereof, shall, upon conviction thereof, be punished by a fine not exceeding one thousand dollars, or imprisonment at hard labor for a term of ninety days, or both such fine and imprisonment, at the discretion of the court: *And provided further,* That in case of the violation of any of the provisions of section one of this act, and conviction thereof, a further fine of two hundred and fifty dollars per diem will be imposed for each day that the obstruction or obstructions therein are maintained after notice to remove the same. Said notice may be given by any Government officer or private citizen.

APPENDIX.

Murray, 1894: Page 11.

That no dead pups were found upon the rookeries in 1894 in the early part of August was due, not to their absence, but to the fact that no close inspection was made. It is impossible without actually going on the breeding grounds and driving off the living cows and pups to get an idea of the number of dead pups. Such an inspection was not made in 1894 nor in any year prior to 1896; consequently the facts regarding the phenomenon of dead pups were never known until that time. The dead pups seen on Tolstoi Rookery in 1891 and 1892 belonged, in the latter year wholly and in the former partly, to this early mortality, which occurs before pelagic sealing begins. What has heretofore been said regarding this estimate of starved pups in connection with the reports of Messrs. Hamlin and Crowley applies here also. In the quotation here ascribed to Mr. Crowley appears the statement that in the count of dead pups an effort was made to distinguish the recently dead from those long dead. If this is true, it would increase the value of the figures as a measure of starvation; but this statement does not occur in Mr. Crowley's report, and in any event, granting that the figures included only starved pups, they still fall short of the facts.

Murray, 1894: Page 15.

The several estimates by Mr. Elliott and others here quoted or mentioned will be discussed in connection with the reports from which they are taken, which appear in later volumes of this series.

Mr. Murray's estimate for 1891, here given in detail for St. Paul Island, represents a broad and general personal impression rather than an accurate enumeration, as undoubtedly does also that for 1894, the details of which are not given. The elements of weakness in these estimates lie in the assumed average size of harem and in the arbitrary doubling of the number of bulls seen in order to account for others supposed to exist but not seen. The size of harem assumed (40) is more than double that of the average number of animals ever seen at one time in a harem (17) and is one-fourth larger than the actual number (30) of cows, including absent ones, which the investigations of 1896-97 show to belong to the average harem. These figures therefore must be taken with a good deal of allowance and can be held only to represent in a very general way the relative condition of the herd. It may be noted that no higher accuracy was claimed by Mr. Murray for these and subsequent estimates made by him.

Murray, 1894: Page 23.

The discussion of dead pups on this and subsequent pages of this report has but little value, because built on the assumption that all had died of starvation. This was the common belief until the investigations of 1896 were made. The fact that a large natural mortality, due to

461

totally different causes, occurs prior to August 10, and has probably occurred for centuries, must be kept constantly in mind in reading all early discussions of dead pups.

Murray, 1894: Page 27.

The figures for the pelagic catch here given include also the seals taken on the Asiatic side, a fact which is not made clear.

Murray, 1895: Page 452.

This detailed estimate of seals for 1895 is doubtless the most elaborate and accurate which Mr. Murray has made. It, however, contains manifest inconsistencies, as for example, Lagoon Rookery is estimated at 50 harems and 2,000 cows. This rookery was counted in the same season both by Mr. True and by Mr. Townsend. The latter found 80 harems and 1,216 cows, the former 82 harems and 1,264 cows. Again, on Kitovi Rookery 200 harems and 8,000 cows are found, whereas Messrs. True and Townsend in the same season found 145 harems and 2,640 cows. Moreover, the figures themselves show that no account is taken of numbers less than 50 in the enumeration of harems. But the most serious defect in the enumeration arises from the date at which it was made. Mr. Murray assumed that the rookeries were at their height by the 20th of July and, beginning his enumeration at this time, completed it on August 14. Our investigations for the past two seasons show that the height of rookery development falls about the 15th of July; that by the 20th the harems are beginning to break up, and that the mating season for adult seals is practically over by August 1. Counts and observations made after the 20th of July give no true idea of conditions in the height of the season, and those made during the first half of August show wholly different conditions. Then the original harems are broken up. The regular bulls are gone, and their places are filled with young and idle bulls controlling transient harems of virgin cows. This enumeration of the seals, therefore, has only the value of a personal estimate made at an unfavorable time and under a misapprehension of the facts of rookery development.

We may here contrast the various estimates offered for the season of 1895, and express our regret that such variant and contradictory results should be reached and published by duly accredited agents of the Government:

Agent.	Harems.	Cows.
True	4,402	70,423
Crowley	5,552	99,936
Murray	5,000	200,000

It may, however, be observed that all this work was conscientiously and intelligently done. The trouble lay in the methods employed. It is a curious fact that the estimate most carefully and accurately worked out is farthest from the truth. This resulted chiefly from the vitiating assumption that practically all the cows were present on the rookeries at the height of the season.

Correspondence: Page 357.

The estimates of starved pups here given include also pups which died of natural causes prior to the beginning of pelagic sealing. Reference should be made to notes upon this subject appended to the reports of Messrs. Hamlin, Crowley, and Murray, where the subject is discussed at length.